PHYSICAL CHEMISTRY
An Advanced Treatise

Volume IXB/Electrochemistry

PHYSICAL CHEMISTRY
An Advanced Treatise

Edited by

HENRY EYRING

*Departments of Chemistry
and Metallurgy
University of Utah
Salt Lake City, Utah*

DOUGLAS HENDERSON

*IBM Research Laboratories
San Jose, California*

WILHELM JOST

*Institut für Physikalische
Chemie der Universität
Göttingen
Göttingen, Germany*

Volume I /Thermodynamics
    II/Statistical Mechanics
    III/Electronic Structure of Atoms and Molecules
    IV/Molecular Properties
    V/Valency
    VI/General Introduction to Kinetics: Gas Reactions
    VII/Reactions in Condensed Phases
VIII/Liquid State
    IX/Electrochemistry (In Two Parts)
    X/Solid State
    XI/Mathematical Methods

PHYSICAL CHEMISTRY
An Advanced Treatise

Volume IXB/Electrochemistry

Edited by

HENRY EYRING
Departments of Chemistry
and Metallurgy
University of Utah
Salt Lake City, Utah

 1970

ACADEMIC PRESS · NEW YORK / LONDON

ACADEMIC PRESS, INC.
111 Fifth Avenue, New York, New York 10003

United Kingdom Edition published by
ACADEMIC PRESS, INC. (LONDON) LTD.
Berkeley Square House, London W1X 6BA

LIBRARY OF CONGRESS CATALOG CARD NUMBER: 66-29951

PRINTED IN THE UNITED STATES OF AMERICA

List of Contributors

Numbers in parentheses indicate the pages on which the authors' contributions begin.

George E. Blomgren, Union Carbide Corporation, Consumer Products Division, Research Laboratory, Cleveland, Ohio (859)

John O'M. Bockris, Electro-Chemistry Laboratory, University of Pennsylvania, Philadelphia, Pennsylvania (611)

David B. Chang, Department of Physiology and Biophysics, University of Washington, School of Medicine, Seattle, Washington (903)

Aleksandar R. Despić, School of Technology, University of Beograd and the Institute for Chemistry Technology and Metallurgy, Beograd, Yugoslavia (611)

M. Eisenberg, Electrochemica Corporation, Menlo Park, California (773)

Edward M. Eyring, Department of Chemistry, University of Utah, Salt Lake City, Utah (731)

William L. Hardy, Department of Physiology and Biophysics, University of Washington, School of Medicine, Seattle, Washington (903)

J. Horiuti, Research Institute for Catalysis, Hokkaido University, Sapporo, Japan (543)

V. G. Levich, Institute of Electrochemistry, Moscow, USSR (985)

Michael C. Mackey, Department of Physiology and Biophysics, University of Washington, School of Medicine, Seattle, Washington (903)

Stephen H. White, Department of Physiology and Biophysics, University of Washington, School of Medicine, Seattle, Washington (903)

J. Walter Woodbury, Department of Physiology and Biophysics, University of Washington, School of Medicine, Seattle, Washington (903)

Foreword

In recent years there has been a tremendous expansion in the development of the techniques and principles of physical chemistry. As a result most physical chemists find it difficult to maintain an understanding of the entire field.

The purpose of this treatise is to present a comprehensive treatment of physical chemistry for advanced students and investigators in a reasonably small number of volumes. We have attempted to include all important topics in physical chemistry together with borderline subjects which are of particular interest and importance. The treatment is at an advanced level. However, elementary theory and facts have not been excluded but are presented in a concise form with emphasis on laws which have general importance. No attempt has been made to be encyclopedic. However, the reader should be able to find helpful references to uncommon facts or theories in the index and bibliographies.

Since no single physical chemist could write authoritatively in all the areas of physical chemistry, distinguished investigators have been invited to contribute chapters in the field of their special competence.

If these volumes are even partially successful in meeting these goals we will feel rewarded for our efforts.

We would like to thank the authors for their contributions and to thank the staff of Academic Press for their assistance.

Henry Eyring
Douglas Henderson
Wilhelm Jost

Preface

This second part of the volume on electrochemistry introduces the reader to several important fields. Many electrode processes involve gas evolution, others involve metallic deposition which are the subjects of the two initial chapters. The book concludes with a seventh chapter on bioelectrochemistry. Fused salts, fuel cells, fast ionic reactions, and electron transfer round out the bill of fare. Many additional areas might have been treated but this suffices to illustrate the methods of electrochemistry.

We have again been successful in obtaining distinguished specialists as authors of each of the chapters. We want to express our appreciation to them, to the staff of Academic Press, and to all who have helped to bring this treatment of electrochemistry to completion.

HENRY EYRING

April, 1970

Contents

LIST OF CONTRIBUTORS   v

FOREWORD   vii

PREFACE   ix

CONTENTS OF PREVIOUS AND FUTURE VOLUMES   xv

Chapter 6 / Gas Evolution Reactions
J. Horiuti

I.	Statistical Mechanical Method for Treating Electrode Reactions.	544
II.	Hydrogen Evolution Reaction	550
III.	Halogen Evolution Reaction	587
IV.	Oxygen Evolution Reaction	595
V.	Oxidation of Simple Organics with Evolution of CO_2	601
	References	607

Chapter 7 / The Mechanism of Deposition and Dissolution of Metals
John O'M. Bockris and Aleksandar R. Despić

I.	Introduction	611
II.	Basic Steps in Metal Deposition	613
III.	Crystallographic Aspects of Metal Deposition	660
IV.	Deposition of Alloys	697
V.	Kinetics of Metal Dissolution	708
	References	723

Chapter 8 / Fast Ionic Reactions
Edward M. Eyring

I.	Introduction	731
II.	Diffusion Controlled Reactions between Ions in Solutions	732
III.	Diffusion Controlled Ionic Reactions: Relaxation Experiments	741
IV.	Somewhat Slower Fast Reactions	749
	References	769

Chapter 9 / Electrochemical Energy Conversion

M. Eisenberg

I.	Introduction	774
II.	Thermodynamics of Galvanic Cells and Related Systems	775
III.	Electrode Kinetic Aspects of Energy Conversion	787
IV.	Theory of Porous Electrodes.	797
V.	Energy Conversion in Fuel Cells and Related Dynamic Systems	824
VI.	Regenerative Fuel-Cell Systems	842
VII.	Batteries	846
	List of Symbols	853
	References	854

Chapter 10 / Fused-Salt Electrochemistry

George E. Blomgren

I.	The Nature of Fused Salts	859
II.	EMF Measurements of Fused Salts	874
III.	Electrical Transport Processes	887
IV.	Electrochemical Kinetics	893
	References	899

Chapter 11 / Bioelectrochemistry

J. Walter Woodbury, Stephen H. White, Michael C. Mackey, William L. Hardy, and David B. Chang

I.	Introduction	904
II.	Electrolytes in Living Organisms	908
III.	Membranes: Steady-State Ion Transport	914
IV.	Membrane Structure	927
V.	Excitability: Voltage-Dependent Permeability	939
VI.	Ion Channel Characteristics	955
VII.	Membrane Properties: Some Theoretical Approaches	961
	References	977

Chapter 12 / Kinetics of Reactions with Charge Transport

V. G. Levich

I.	Reactions with Charge Transport Solutions	986
II.	Physical Pattern of Reaction with Electron Transfer in Polar Solvents without the Breaking of Chemical Bonds	989
III.	Electrode Reactions and Electrode Current	998
IV.	Model of a Polar Solution	1004
V.	Adiabatic Approximation and Quantum Transition Probability in a System Consisting of Heavy and Light Particles	1010

Contents

   VI.   Probability of Electron Transfer 1015

  VII.   Transition Probability in Reactions without Chemical Bond Breaking . . 1019

 VIII.   Flow of Current through the Metal–Solution Interface 1024

   IX.   Flow of Current through the Semiconductor–Solution Interface 1031

    X.   Physical Pattern of Charge Transfer Reaction in a Polar Medium, Which Is Accompanied by the Breaking or Forming of Chemical Bonds. 1039

   XI.   Quantitative Description of a Charge Transfer Reaction with Chemical Bond Breaking 1041

  XII.   Transition Probability 1046

 XIII.   Current Density 1049

 XIV.   Other Cases of Electrode Reactions 1060

  XV.   Critical Consideration of Theoretical Concepts about the Mechanism of the Elementary Act of the Electrode Process 1063

 XVI.   Conclusions 1070

       References 1072

AUTHOR INDEX 1075

SUBJECT INDEX 1087

Contents of Previous and Future Volumes

VOLUME II

Chapter 1 / CLASSICAL STATISTICAL THERMODYNAMICS
John E. Kilpatrick

Chapter 2 / QUANTUM STATISTICAL MECHANICS
D. ter Haar

Chapter 3 / CRYSTAL AND BLACKBODY RADIATION
Sheng Hsien Lin

Chapter 4 / DIELECTRIC, DIAMAGNETIC, AND PARAMAGNETIC PROPERTIES
William Fuller Brown, Jr.

Chapter 5 / ELECTRONS IN SOLIDS
Peter Gibbs

Chapter 6 / REAL GASES
C. F. Curtiss

Chapter 7 / EQUILIBRIUM THEORY OF LIQUIDS AND LIQUID MIXTURES
Douglas Henderson and Sydney G. Davison

Chapter 8 / ELECTROLYTIC SOLUTIONS
H. Ted Davis

Chapter 9 / SURFACES OF SOLIDS
L. J. Slutsky and G. D. Halsey, Jr.

AUTHOR INDEX—SUBJECT INDEX

VOLUME III

Chapter 1 / BASIC PRINCIPLES AND METHODS OF QUANTUM MECHANICS
D. ter Haar

Chapter 2 / ATOMIC STRUCTURE
Sydney G. Davison

Chapter 3 / VALENCE BOND AND MOLECULAR ORBITAL METHODS
Ernest R. Davidson

Chapter 4 / ELECTRON CORRELATION IN ATOMS AND MOLECULES
Ruben Pauncz

Chapter 5 / ATOMIC SPECTRA
W. R. Hindmarsh

Chapter 6 / ELECTRONIC SPECTRA OF DIATOMIC MOLECULES
R. W. Nicholls

Chapter 7 / ELECTRONIC SPECTRA OF POLYATOMIC MOLECULES
Lionel Goodman and J. M. Hollas

Chapter 8 / PI ELECTRON THEORY OF THE SPECTRA OF CONJUGATED
MOLECULES
G. G. Hall and A. T. Amos

Chapter 9 / IONIZATION POTENTIALS AND ELECTRON AFFINITIES
Charles A. McDowell

Chapter 10 / ELECTRON DONOR-ACCEPTOR COMPLEXES AND CHARGE
TRANSFER SPECTRA
Robert S. Mulliken and Willis B. Person

AUTHOR INDEX—SUBJECT INDEX

VOLUME IV

Chapter 1 / THE VARIETY OF STRUCTURES WHICH INTEREST CHEMISTS
S. H. Bauer

Chapter 2 / ROTATION OF MOLECULES
C. C. Costain

Chapter 3 / VIBRATION OF MOLECULES
Gerald W. King

Chapter 4 / VIBRATIONAL SPECTRA OF MOLECULES
J. R. Hall

Chapter 5 / SPECTRA OF RADICALS
Dolphus E. Milligan and Marilyn E. Jacox

Chapter 6 / THE MOLECULAR FORCE FIELD
Takehiko Shimanouchi

Chapter 7 / Interactions among Electronic, Vibrational, and
        Rotational Motions
    Jon T. Hougen

Chapter 8 / Electric Moments of Molecules
    A. D. Buckingham

Chapter 9 / Nuclear Magnetic Resonance Spectroscopy
    R. M. Golding

Chapter 10 / ESR Spectra
    Harry G. Hecht

Chapter 11 / Nuclear Quadrupole Resonance Spectroscopy
    Ellory Schempp and P. J. Bray

Chapter 12 / Mössbauer Spectroscopy
    N. N. Greenwood

Chapter 13 / Molecular Beam Spectroscopy
    C. R. Mueller

Chapter 14 / Diffraction of Electrons by Gases
    S. H. Bauer

Author Index—Subject Index

VOLUME V

Chapter 1 / General Remarks on Electronic Structure
    E. Teller and H. L. Sahlin

Chapter 2 / The Hydrogen Molecular Ion and the General Theory
        of Electron Structure
    E. Teller and H. L. Sahlin

Chapter 3 / The Two-Electron Chemical Bond
    Harrison Shull

Chapter 4 / Heteropolar Bonds
    Juergen Hinze

Chapter 5 / Coordination Compounds
    T. M. Dunn

Chapter 6 / σ Bonds
    C. A. Coulson

Chapter 7 / π BONDS
    C. A. Coulson

Chapter 8 / HYDROGEN BONDING
    Sheng Hsien Lin

Chapter 9 / MULTICENTERED BONDING
    Kenneth S. Pitzer

Chapter 10 / METALLIC BONDS
    Walter A. Harrison

Chapter 11 / RARE-GAS COMPOUNDS
    Herbert H. Hyman

Chapter 12 / INTERMOLECULAR FORCES
    Taro Kihara

AUTHOR INDEX—SUBJECT INDEX

VOLUME IXA

Chapter 1 / SOME ASPECTS OF THE THERMODYNAMIC AND TRANSPORT
        BEHAVIOR OF ELECTROLYTES
    B. E. Conway

Chapter 2 / THE ELECTRICAL DOUBLE LAYER
    C. A. Barlow, Jr.

Chapter 3 / PRINCIPLES OF ELECTRODE KINETICS
    Terrell N. Andersen and Henry Eyring

Chapter 4 / TECHNIQUES FOR THE STUDY OF ELECTRODE PROCESSES
    Ernest Yeager and Jaroslav Kuta

Chapter 5 / SEMICONDUCTOR ELECTROCHEMISTRY
    Heinz Gerischer

AUTHOR INDEX—SUBJECT INDEX

VOLUME X

Chapter 1/ DIFFRACTION OF X-RAYS, ELECTRONS, AND NEUTRONS ON THE
        REAL CRYSTAL
    Alarich Weiss and Helmutt Witte

Chapter 2 / DISLOCATIONS
P. Haasen

Chapter 3 / DEFECTS IN IONIC CRYSTALS
L. W. Barr and A. B. Lidiard

Chapter 4 / THE CHEMISTRY OF COMPOUND SEMICONDUCTORS
F. A. Kröger

Chapter 5 / CORRELATION EFFECTS IN DIFFUSION IN SOLIDS
A. D. Le Claire

Chapter 6 / SEMICONDUCTORS: FUNDAMENTAL PRINCIPLES
Otfried Madelung

Chapter 7 / SEMICONDUCTOR SURFACES
G. Ertl and H. Gerischer

Chapter 8 / ORGANIC SEMICONDUCTORS
J. H. Sharp and M. Smith

Chapter 9 / PHOTOCONDUCTIVITY OF SEMICONDUCTORS
Richard H. Bube

Chapter 10 / ORDER–DISORDER TRANSFORMATIONS
Hiroshi Sato

Chapter 11 / PRECIPITATION AND AGING
M. Kahlweit

AUTHOR INDEX—SUBJECT INDEX

Chapter 6

Gas Evolution Reactions

J. HORIUTI

I. Statistical Mechanical Method for Treating Electrode Reactions 544
II. Hydrogen Evolution Reaction 550
   A. Introduction 550
   B. Historical Survey 551
   C. Double Layer and Hydrogen Intermediates 559
   D. Dependence of i_+ on η 560
   E. Dependence of i_+ on $\mu(H^+)$ 568
   F. Separation Factor 571
   G. The Stoichiometric Number, ν_r, of the Rate-Determining Step, r . . . 577
   H. Transient Phenomena 579
   I. Dependence of i_0 on Electrode Materials 582
   J. Concluding Remarks 586
III. Halogen Evolution Reaction 587
   A. Introduction 587
   B. Characteristics of Halogen Evolution Reaction 588
   C. Chlorine Evolution Reaction 591
   D. Iodine and Bromine Evolution Reactions 594
IV. Oxygen Evolution Reaction 595
   A. Introduction 595
   B. Reversible Oxygen Electrode 595
   C. Oxygen Deposit 597
   D. Kinetics 598
   E. Concluding Remark 600
V. Oxidation of Simple Organics with Evolution of CO_2 601
   A. Introduction 601
   B. Oxidation of CO 603
   C. Oxidation of HCO_2H 604
   D. "Reduced CO_2" 605
   E. Oxidation of CH_3OH 605
   F. Oxidation of Hydrocarbons 606
   References 607

This chapter deals with electrolytic gas evolution reactions.

A brief, preparatory account of the statistical mechanical theory of heterogeneous reactions is given in Section I and applied to particular examples of gas evolution reactions in subsequent sections. An extensive treatment is given of the hydrogen evolution reaction in Section II as it exemplifies the great variety of electrolytic gas evolution reactions for which experimental data have accumulated. The halogen and oxygen evolution reactions are dealt with briefly in Sections III and IV, reliable experimental data being rather scanty in these cases, although they too are of practical importance. The evolution of carbon dioxide by anodic oxidation of simple organics is briefly discussed in Section V.

I. Statistical Mechanical Method for Treating Electrode Reactions

Electrolytic gas evolution reactions are heterogeneous reactions occurring on the surface of an electrode, which acts as a catalyst. For their treatment we require the theory of heterogeneous reactions. Most of the current theories of electrochemical processes are based on the "activated complex theory" or "transition state method," which is developed in the famous book by Glasstone *et al.* (1941), on the basis of the statistical mechanics of an assembly of independent molecules, namely, a perfect gas; the theory thus leads to the well-known rate formulas for gas phase reactions, which are extended to heterogeneous reactions, reactions in solutions and so on by analogy. The original accuracy is, in consequence, not claimed for this theory when applied to heterogeneous reactions (Horiuti, 1956). This situation obliged the author to develop a *generalized theory of absolute reaction rate* (Horiuti, 1948) which is founded more firmly on the general theory of statistical mechanics (Gibbs, 1902) and is applicable to the treatment of heterogeneous reactions without ambiguity, especially in those cases where the interactions among species on the surface have to be taken into account. This theory has first been applied to the hydrogen electrode reaction with success (Okamoto *et al.*, 1936). The analysis of the hydrogen evolution reaction summarized in Section II of this chapter is based on this theory, which will be referred to as the "generalized theory" in what follows.

Accounts are given here of only a few important formulas derived from the generalized theory, which are of common use in subsequent sections, leaving the details to previous papers (Horiuti, 1948; Horiuti and Nakamura, 1967). Assume an assembly (a macroscopic mass) kept at temperature T in which an elementary reaction, or step for short, is going on. The generalized theory leads to the following equation for the forward rate, v_+

$$v_+ = \kappa(kT/h) \exp(-\mu^{\ddagger}/RT)/\exp(-\mu^{\mathrm{I}}/RT), \tag{1.1}$$

where k, h, and R are the Boltzmann, Planck, and gas constants, κ is the transmission coefficient, and $\mu^{\ddagger}$ and μ^{I} are the chemical potentials of the critical system, $\ddagger$, and of the initial system I (see below), respectively, of this step. If the transmission coefficient κ is unity (Glasstone *et al.*, 1941; Horiuti and Nakamura, 1967), Eq. (1.1) appears as

$$v_+ = (kT/h)p^{\ddagger}/p^{\mathrm{I}} = (kT/h) \exp[-(\mu^{\ddagger} - \mu^{\mathrm{I}})/RT], \tag{1.2a}$$

where $p^{\ddagger}$ and p^{I} are the statistical mechanical functions related to $\mu^{\ddagger}$ and μ^{I} as *

$$p^{\ddagger} = \exp(-\mu^{\ddagger}/RT), \qquad p^{\mathrm{I}} = \exp(-\mu^{\mathrm{I}}/RT).$$

Similarly, the backward rate v_- of the same step is given as

$$v_- = (kT/h)p^{\ddagger}/p^{\mathrm{F}} = (kT/h) \exp[-(\mu^{\ddagger} - \mu^{\mathrm{F}})/RT], \tag{1.2b}$$

where the superscript F on p and μ refers to the final state of the system.

The quantity $\mu^{\ddagger} - \mu^{\mathrm{I}}$ or $\mu^{\ddagger} - \mu^{\mathrm{F}}$ in Eqs. (1.2) represents Avogadro's number, N_{A}, times the free energy increment owing to the conversion of the initial or final system, respectively, into the critical system. If this free energy increment is identified with the free energy increment $\varDelta F^{\ddagger}$ ("free energy of activation") in the book of Glasstone *et al.* (1941), Eq. (1.2a) becomes the well-known rate formula in the activated complex theory, i.e.,

$$v_+ = (kT/h) \exp(-\varDelta F^{\ddagger}/kT). \tag{1.3}$$

In the activated complex theory, the initial system of a step [e.g., the left-hand side of step (1.7a) or (1.7b) below] is supposed to be converted into the final system (e.g., the right-hand side of the above step) *via* the activated complex, which is situated at the saddle point of the potential energy on the reaction path; the potential energy is that of the set of particles involved in the step [e.g., $H^+ + e^-$ or $2H$ in case of step (1.7)]. The reaction rate is given by the velocity with which the activated complex travels over the saddle point (Glasstone *et al.*, 1941). In the generalized theory, the concept of a critical system or critical state plays a role similar to that of the activated complex in the activated complex theory.

* The $p^{\ddagger}$ and p^{I} are by definition the reciprocals of the absolute activities of the critical and initial systems, respectively.

It is of essential importance, however, that the critical state is defined by the condition that the "activation free energy," $\mu^{\ddagger} - \mu^{\mathrm{I}}$, is a maximum (Horiuti, 1961) which is not identical with the activated complex defined as above except in special cases.

The concept underlying this definition of the critical state is similar to that involved in the variational theory of reaction rates developed by Wigner (1937) and Keck (1960). Secondly, aside from the distinction between the critical system and the activated complex, it is to be remembered that Eqs. (1.1) and (1.2) have been derived from the general theory of statistical mechanics, while Eq. (1.4) is based, as described in the current textbooks (Glasstone *et al.*, 1941), on elementary statistical mechanics sketched above, and is of limited applicability.

Suppose now that the heterogeneous step on the surface of an electrode, which provides $N^{\ddagger}$ physically identical sites of the critical system, $\sigma^{\ddagger}$. Denoting the probability of the $\sigma^{\ddagger}$ being unoccupied (i.e., ready for accommodating a critical system) by $\theta(\sigma^{\ddagger}, 0)$, the generalized theory gives the relation,*

$$p^{\ddagger} = \exp(-\mu^{\ddagger}/RT) = N^{\ddagger}\theta(\sigma^{\ddagger}, 0) \exp(-\epsilon^{\ddagger}/RT), \qquad (1.4)$$

where $\epsilon^{\ddagger}$ is N_{A} times the free energy increment caused by the addition of a critical system ‡ from outside the assembly to a definite, unoccupied site $\sigma^{\ddagger}$. Equations (1.2) are now written as

$$v_{+} = (kT/h)N^{\ddagger}\theta(\sigma^{\ddagger}, 0) \exp[-(\epsilon^{\ddagger} - \mu^{\mathrm{I}})/RT] \qquad (1.5a)$$

and

$$v_{-} = (kT/h)N^{\ddagger}\theta(\sigma^{\ddagger}, 0) \exp[-(\epsilon^{\ddagger} - \mu^{\mathrm{F}})/RT]. \qquad (1.5b)$$

A chemical reaction consists not always in a single step but in general is composed of a set of steps. The theory has been developed for the steady state as a composite overall reaction (Horiuti and Nakamura, 1967); comments are given on a few results of the theory which are useful in the following sections. Suppose, for example, that the electrolytic hydrogen evolution reaction,

$$2\,H^{+} + 2e^{-} = H_{2} \qquad (1.6)$$

consists of the two steps

$$H^{+} + e^{-} \rightarrow H(a) \qquad (1.7a)$$

and

$$2\,H(a) \rightarrow H_{2}, \qquad (1.7b)$$

* The proof of Eq. (1.4) is found on p. 18 of Horiuti and Nakamura (1967), where the quantities $\theta(\sigma^{\ddagger}, 0)$ and $\exp(-\epsilon^{\ddagger}/RT)$ are denoted by $\theta^{\ddagger}(0)$ and $q^{\ddagger}$, respectively.

where (a) denotes an adsorbed state on the electrode surface. Multiplying steps (1.7a) and (1.7b) by 2 and 1 respectively and summing up, we have the conversion of reaction (1.6). The multiplier, 2 or 1, is called the *stoichiometric number* of step (1.7a) or (1.7b), respectively, which is the number of relevant steps to occur for every act of reaction (1.6).

The forward and backward rates, v_{+s} and v_{-s}, are given by Eqs. (1.5) as

$$v_{+s} = \frac{kT}{h} N_s{}^{\ddagger} \theta(\sigma_s{}^{\ddagger}, 0) \exp\left[-\frac{\epsilon(\ddagger_s) - \mu(I_s)}{RT}\right] \qquad (1.8a)$$

and

$$v_{-s} = \frac{kT}{h} N_s{}^{\ddagger} \theta(\sigma_s{}^{\ddagger}, 0) \exp\left[-\frac{\epsilon(\ddagger_s) - \mu(F_s)}{RT}\right], \qquad (1.8b)$$

signifying the relevance to step s with suffix s; e.g., I_s denotes the initial system of step s, $\mu(I_s)$ its chemical potential, and $\sigma_s{}^{\ddagger}$ the site of the critical system of step s. One of important conclusions drawn immediately from Eqs. (1.8) is that

$$v_{+s}/v_{-s} = \exp(-\Delta G_s/RT), \qquad (1.9)$$

where $-\Delta G_s$ is the free energy decrease accompanying step s, i.e., the affinity of step s,

$$-\Delta G_s \equiv \mu(I_s) - \mu(F_s). \qquad (1.10)$$

The affinities of steps (1.7a) and (1.7b) are by definition

$$-\Delta G_1 = \mu(H^+) + \mu(e^-) - \mu[H(a)]$$

and

$$-\Delta G_2 = 2\mu[H(a)] - \mu(H_2),$$

while the affinity of reaction (1.6) is

$$-\Delta G = 2\mu(H^+) + 2\mu(e^-) - \mu(H_2).$$

We find immediately that $-\Delta G$ is related to $-\Delta G_1$ and $-\Delta G_2$ as

$$-\Delta G = \nu_1(-\Delta G_1) + \nu_2(-\Delta G_2), \qquad (1.11)$$

where ν_1 and ν_2 are the stoichiometric numbers of steps (1.7a) and (1.7b), i.e., 2 and 1 respectively (see above); step (1.7a) has thus to occur twice as frequently as step (1.7b), if reaction (1.6) is to be kept steady. This is formulated as

$$v_1/\nu_1 = v_2/\nu_2 = V, \qquad (1.12)$$

where v_1 or v_2 is the net rate, or the excess of the forward rate over the backward one, for steps 1 and 2, respectively, and V is the steady rate of reaction (1.6).

Equations (1.11) and (1.12) are generalized as follows by denoting the stoichiometric number of step s by ν_s. The affinity $-\varDelta G$ of an overall reaction composed of S steps is given (Horiuti and Nakamura, 1967) as

$$-\varDelta G = \sum_{s=1}^{s} (-\varDelta G_s)\nu_s, \qquad (1.13)$$

while the steady rate, V, of the overall reaction is

$$v_s/\nu_s = V, \qquad s = 1, 2, \ldots, S, \qquad (1.14)$$

where v_s is the net rate of step s or the excess of v_{+s} over v_{-s}, i.e.,

$$v_s = v_{+s} - v_{-s}. \qquad (1.15)$$

Let us now suppose step r among the above S steps to be rate-determining. The remaining $S - 1$ steps may then be regarded as in partial equilibrium; hence according to Eq. (1.9)

$$-\varDelta G_{s(\neq r)} = 0. \qquad (1.16)$$

It follows from Eqs. (1.13) and (1.16) that

$$-\varDelta G/\nu_r = -\varDelta G_r. \qquad (1.17)$$

The forward and backward steady rates, V_+ and V_-, of the overall reaction are reasonably defined in this case as

$$V_+ = v_{+r}/\nu_r, \qquad V_- = v_{-r}/\nu_r \qquad (1.18a), (1.18b)$$

(Horiuti and Nakamura, 1967); hence according to Eqs. (1.14) and (1.15)

$$V_+ - V_- = (v_{+r} - v_{-r})/\nu_r = v_r/\nu_r = V. \qquad (1.19)$$

Combining Eqs. (1.18) and (1.9), we also find

$$V_+/V_- = v_{+r}/v_{-r} = \exp(-\varDelta G_r/RT). \qquad (1.20)$$

What we directly observe in electrochemical experiments is the electric current through an electrode as well as its potential. Let i be the current per unit area of the electrode surface, i.e., current density. Denoting the number of elementary charges transferred by one act of the overall reaction by z, we have

$$i = zeV = z(F/N_A)V, \qquad (1.21)$$

provided that V is reckoned per unit area, where e and F are the elementary and the Faraday charge, respectively; z equals 2 in the case of reaction (1.6). It is convenient to define the forward and backward current densities, i_+ and i_-, as

$$i_+ = zeV_+ = z(F/N_A)V_+, \quad i_- = zeV_- = z(F/N_A)V_-. \quad \text{(1.22a), (1.22b)}$$

From (1.21), (1.22), (1.19), (1.20), (1.18), and (1.8), we have now

$$i = i_+ - i_-, \tag{1.23}$$

$$i_+/i_- = V_+/V_- = \exp(-\Delta G_r/RT), \tag{1.24}$$

$$i_+ = \frac{2F}{\nu_r N_A} v_+ = \frac{2FkT}{\nu_r N_A h} N_r^{\ddagger}\theta(\sigma_r^{\ddagger}, 0) \exp\left[-\frac{\epsilon(\ddagger_r) - \mu(I_r)}{RT}\right], \tag{1.25a}$$

and

$$i_- = \frac{2F}{\nu_r N_A} v_- = \frac{2FkT}{\nu_r N_A h} N_r^{\ddagger}\theta(\sigma_r^{\ddagger}, 0) \exp\left[-\frac{\epsilon(\ddagger_r) - \mu(F_r)}{RT}\right], \tag{1.25b}$$

where the number of sites, $N_r^{\ddagger}$, is understood to be referred to the unit area of the electrode surface (see above).

We define overvoltage η of the gas evolution reaction in terms of its affinity as

$$-\Delta G = zF\eta; \tag{1.26}$$

hence according to Eqs. (1.17) and (1.24)

$$\nu_r = zF\eta/RT \ln(i_+/i_-). \tag{1.27a}$$

At $\eta = 0$, we have $i_+ = i_- = i_0$ or $\nu_r = 0/0$. The limiting value of ν_r at $\eta = 0$ is obtained by the usual method with reference to Eq. (1.15) as

$$\nu_r = i_0 zF/RT(\partial i/\partial \eta)_{\eta=0}. \tag{1.27b}$$

The η thus defined is a directly observable quantity as shown in subsequent sections.

We have alternatively from Eq. (1.27a) by differentiating with respect to η

$$\tau_+ + \tau_- = z/\nu_r, \tag{1.28\nu_r}$$

where

$$\tau_+ = (RT/F) \partial \ln i_+/\partial \eta, \quad \tau_- = -(RT/F) \partial \ln i_-/\partial \eta. \quad \text{(1.28}\tau_+\text{), (1.28}\tau_-\text{)}$$

We are now ready to go into the detailed discussion of the electrolytic gas evolution reactions in subsequent sections.

II. Hydrogen Evolution Reaction

A. INTRODUCTION

Hydrogen gas is evolved from the surface of an electrode kept at sufficiently low electric potential against an aqueous solution of acid or base, associated with inflow of electricity from the solution. The same happens if the solute is replaced by some neutral salt, e.g., sodium sulfate, or if the solvent is replaced by alcohol or liquid ammonia. These phenomena are called the *hydrogen evolution reaction* (HER) and the electrode concerned is called a *hydrogen electrode*.

The HER is formulated, describing the inflow of electricity as an outflow of conduction electrons, as

$$2\,H^+ + 2e^- = H_2, \tag{2.1}$$

where H^+ represents constituent protons of the solution feasible to the reaction.

The HER stops when the electrode is electrically isolated. Platinum and nickel electrodes are known to assume a definite potential against the solution under the latter condition, provided the solution concerned is saturated with hydrogen gas of definite activity. The electrode under the latter condition is called a *reversible hydrogen electrode* and its potential is the *reversible potential* as regards reaction (2.1).

The significance of the overvoltage η is illustrated below with special reference to reaction (2.1). We have thermodynamically, for a reversible hydrogen electrode,

$$2\mu(H^+) + 2\mu(e^-)_{rev} = \mu(H_2), \tag{2.2}$$

where $\mu(H^+)$, etc., are chemical potentials of H^+, and the subscript rev indicates the relevance of the subscripted quantity to the reversible hydrogen electrode. The affinity of the HER, $-\varDelta G$, is by definition

$$-\varDelta G \equiv 2\mu(H^+) + 2\mu(e^-) - \mu(H_2), \tag{2.3a}$$

where $\mu(e^-)$ is the chemical potential of the conduction electron in the hydrogen electrode in question. We have from the above two equations

$$-\varDelta G = 2[\mu(e^-) - \mu(e^-)_{rev}] = 2F\eta, \tag{2.3b}$$

in accordance with Eq. (1.26). We have further from Eqs. (2.3a) and (2.3b)

$$2\mu(H^+) + 2\mu(e^-) - \mu(H_2) = 2F\eta. \tag{2.3c}$$

The overvoltage and the relevant rate are electrically measurable and the rate is thus followed with precision over a range of 10^{10}. Our main problem is to understand how the rate depends on the experimental conditions inclusive of electrode materials. The elucidation of the problem promises a useful pattern of investigating heterogeneous catalyses, reaction (2.1) being itself a heterogeneously catalyzed combination of constituents of the hydrogen molecule in the presence of the electrode material as catalyst. Section II describes the status of our progress in this direction.

B. Historical Survey

1. Tafel's Law

What was early called the overvoltage of hydrogen was the overvoltage, η, defined by Eqs. (2.3) at the first appearance of hydrogen bubbles in the course of a gradual lowering of the electrode potential. This was regarded as a characteristic of the electrode material, although the observations were by no means reproducible. Tafel (1905) found that the current density, i, was a fairly reproducible function of the electrode potential, for a steady evolution of hydrogen bubbles, but not otherwise. He showed that under the latter conditions, η varied reproducibly with current density i according to the equation

$$\eta = a + b \log i, \tag{2.4}$$

where a and b are constants, thus establishing the modern concept of overvoltage. The b was found to have a value around 0.12 V for many electrodes; its up to date values are shown in terms of $\tau_+ \equiv 2.30RT/Fb$ in Fig. 1. The linear relation of Eq. (2.4) is called Tafel's law.

Tafel tried to explain his experimental results, admitting that reaction (2.1) followed the sequence of steps,

$$H^+ + e^- \rightarrow H(a), \qquad 2\,H(a) \rightarrow H_2; \tag{2.5a, 2.5b}$$

H(a) represents a chemisorbed hydrogen atom which fits in adequately with that Tafel called the "hydrogen modification," although the concept of chemisorption was not established in his time. He admitted that step (2.5a) was fast enough as it was an ionic reaction and consequently that the other constituent step, (2.5b), was rate-determining. This mechanism of HER is called the catalytic mechanism. His reasoning is reproduced below in modern terminology. The forward unidirectional rate, i_+, was

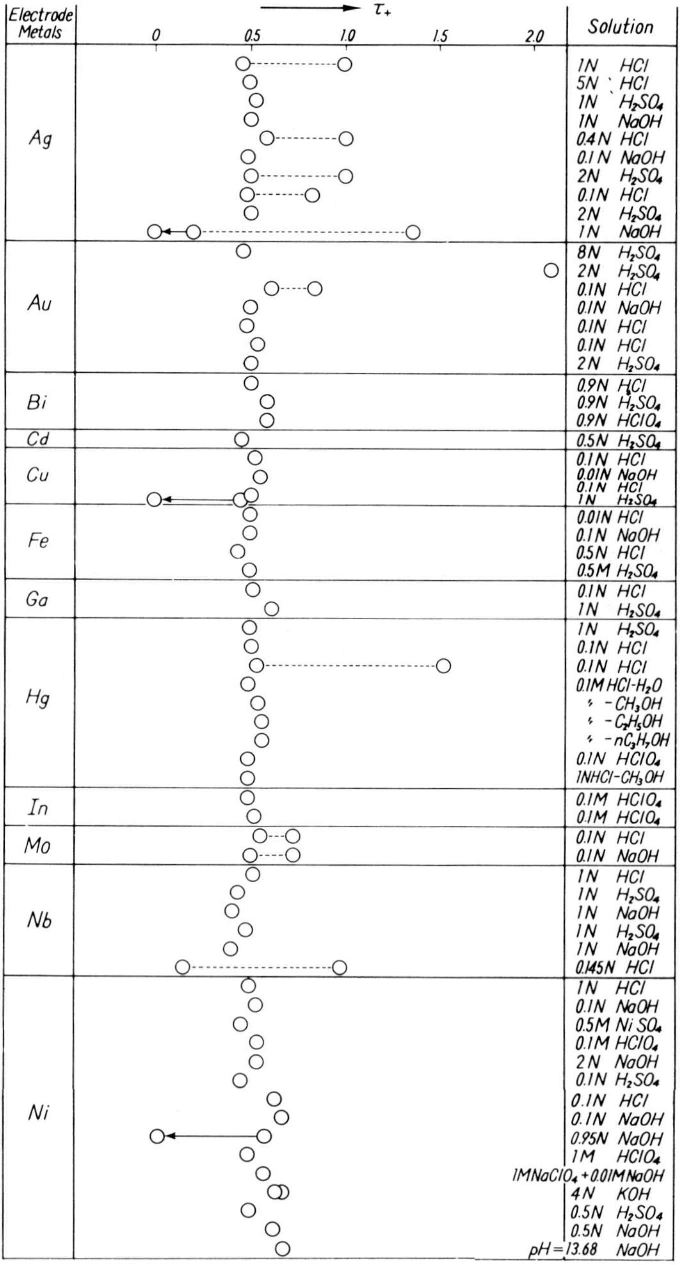

Electrode Metals
τ₊
0
0.5
1.0
1.5
2.0
Solution
Ag
1N HCl
5N HCl
1N H₂SO₄
1N NaOH
0.4N HCl
0.1N NaOH
2N H₂SO₄
0.1N HCl
2N H₂SO₄
1N NaOH
Au
8N H₂SO₄
2N H₂SO₄
0.1N HCl
0.1N NaOH
0.1N HCl
0.1N HCl
2N H₂SO₄
Bi
0.9N HCl
0.9N H₂SO₄
0.9N HClO₄
Cd
0.5N H₂SO₄
Cu
0.1N HCl
0.01N NaOH
0.1N HCl
1N H₂SO₄
Fe
0.01N HCl
0.1N NaOH
0.5N HCl
0.5M H₂SO₄
Ga
0.1N HCl
1N H₂SO₄
Hg
1N H₂SO₄
0.1N HCl
0.1N HCl
0.1M HCl-H₂O
-CH₃OH
-C₂H₅OH
-nC₃H₇OH
0.1N HClO₄
1NHCl-CH₃OH
In
0.1M HClO₄
0.1M HClO₄
Mo
0.1N HCl
0.1N NaOH
Nb
1N HCl
1N H₂SO₄
1N NaOH
1N H₂SO₄
1N NaOH
0.145N HCl
Ni
1N HCl
0.1N NaOH
0.5M Ni SO₄
0.1M HClO₄
2N NaOH
0.1N H₂SO₄
0.1N HCl
0.1N NaOH
0.95N NaOH
1M HClO₄
1MNaClO₄+0.01MNaOH
4N KOH
0.5N H₂SO₄
0.5N NaOH
pH=13.68 NaOH

Fig. 1.

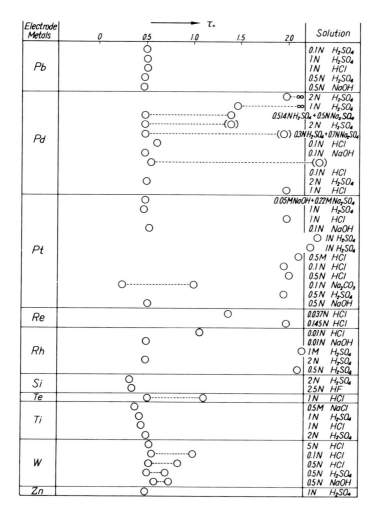

FIG. 1. τ_+-values of hydrogen evolution reaction. $\bigcirc$: uncertain values; ----: η-log$i$ plot consists of two linear parts respectively, having slopes as marked at both ends; $\leftarrow$: τ_+ approaches zero at higher values of η.

assumed, as the rate-determining step (2.5b) was "bimolecular" with respect to H(a), as

$$i_+ \propto \theta^2 \qquad (2.6a)$$

and the chemical potential of H(a), $\mu(H)$, was given by

$$\mu(H) = \mu(H)_1 + RT \ln \theta, \qquad (2.6b)$$

where θ is the population of H(a) and $\mu(H)_1$ is a constant at constant temperature; Eq. (2.6b) was assumed by analogy with the chemical potential of a dilute component in a homogeneous phase.

We have, on the other hand, by virtue of the partial equilibrium of step (2.5a),

$$\mu(H^+) + \mu(e^-) = \mu(H). \qquad (2.7)$$

Eliminating θ, $\mu(H)$, and $\mu(e^-)$ from Eqs. (2.6), (2.7), and (2.3b), we have

$$\eta = (RT/2F) \ln i_+ + \text{const} \qquad (2.8a)$$

or, identifying i_+ with i, as is valid under certain conditions referred to in Section II,B,4,

$$\eta = (2.30RT/2F) \log i + \text{const} \qquad (2.8b)$$

in terms of the common logarithm. Equation (2.8b) reproduces the linear relation of Eq. (2.4) except for the coefficient of $\log i$, which is calculated to be 0.029 V at room temperature or ca. one-fourth of the experimental b-value, 0.12 V.

Haber and Russ (1904) attributed this deviation to the inapplicability of the law for dilute components, represented by Eqs. (2.6), to H(a), which might be heavily compressed electrically through step (2.5a), and this was agreed to as a possible case by Tafel (1905).

The coefficient b has thus been a touchstone of the mechanism.

2. Catalyzed Recombination of Hydrogen Atoms

Bonhoeffer (1924) has compared the activity of different metals in catalyzing the recombination of atomic hydrogen. He wrapped the bulb of a thermometer with the foil of the metal being studied, kept it in a flow of hydrogen gas which had first been infused with atomic hydrogen by the Wood method (cf. Farkas (1939)), and measured the rise of the thermometer reading caused by the heat of recombination which was catalyzed by the metal foil. He thus found the catalytic activity as measured by the rise of this reading approximately in the reverse order of η for a common value of i. The latter reverse order is now of the order of magnitude of i at a common value of η, insofar as i increases with η according to Eq. (2.4) with a common value of b. The catalytic activity of electrode metals is in consequence of the same order as i for a common value of η or, according to Eq. (2.3b), for a common value of affinity. This result is quite in harmony with the catalytic mechanism whose rate is determined by step (2.5b), i.e., the recombination of hydrogen atoms.

3. *Catalyzed Exchange of Hydrogen Isotopes*

Horiuti and Polanyi (1933) found that hydrogen containing deuterium in excess of normal content lost the excess, when shaken with water or an aqueous solution of normal deuterium content in the presence of platinum black or reduced nickel. What happens ought to be the transfer of the excess deuterium in hydrogen gas to solution through the backward act of reaction (2.1), electrically balanced by the counter transfer of protium, insofar as these acts are the only processes allowed to proceed under the conditions. It then follows that protium as well as deuterium is transferred from solution to hydrogen gas and conversely in equilibrium in confirmation of the dynamical theory of equilibrium at the HER.

This conclusion verifies Eq. (1.23), theoretically derived, which expresses i as the difference of i_+ and i_-.

4. *Ratio of i_+ to i_-*

Equation (1.27) is written for reaction (2.1), where $z = 2$, as

$$i_+/i_- = \exp(2F\eta/\nu_r RT). \tag{2.9}$$

It follows from Eq. (2.9) that i_+ predominates over i_- for sufficiently large η. The lowest overvoltage as observed by Tafel was 0.05 or 0.44 V in the case of a platinum or a nickel electrode, respectively; the overvoltage was exceptionally low on the platinum electrode, while experiments on other electrodes were conducted at overvoltage much higher than 0.44 V as seen from Fig. 2. The ratio, i_+/i_-, at $\eta = 0.05$ or 0.44 V is calculated by Eq. (2.9) to be 50.3 or 10^{16}, respectively, for $\nu_r = 1$ and 7.3 or 10^8, respectively, for $\nu_r = 2$ at the temperature, 20°C, of Tafel's experiment. It has been concluded, as seen in Section II,G, that $\nu_r = 1$; hence $i_+/i_- \gg 1$ or according to Eq. (1.23)

$$i = i_+ \tag{2.10}$$

practically, which has been assumed as regards Eq. (2.8b).

5. *Exchange Current Density*

Equation (2.4) states, according to Eq. (2.10), that a linear relation exists between η and $\log i_+$, i.e.,

$$\eta = a + b \log i_+, \tag{2.11}$$

by which $\log i_+$ is extrapolated at $\eta = 0$, as

$$\log i_0 = -a/b. \tag{2.12}$$

We have in general $i_+ = i_-$ at $\eta = 0$ according to Eq. (2.9); the balanced value of the current density in both directions is called the exchange current density and denoted by i_0. Equation (2.12) gives its value, extrapolated by Eq. (2.11) assumed as valid down to $\eta = 0$. An exact determination of i_0 is referred to in Section II,G,3.

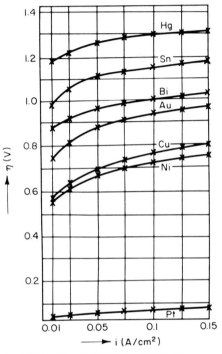

FIG. 2. Tafel lines in 2 N H$_2$SO$_4$, 20°C (Tafel, 1905). The overvoltage η was calculated by subtracting 0.674 V, according to Tafel (1905), from the electrode potential as measured against a saturated mercurous sulfate electrode.

6. Slow Discharge Mechanism

Erdey-Grúz and Volmer (1930) and Erdey-Grúz and Kromrey (1931) observed the HER on silver in 1 N H$_2$SO$_4$ by the galvanostatic transient method. They thus found a linear rise or fall of the electrode potential with time, depending on whether the constant current flowed from solution to electrode or in the reverse direction. The current was assumed to have been spent in changing the electrode potential almost exclusively, on the ground that the steady current density of reaction (2.1) or its reverse was negligible at the observed overvoltage as compared with the galvanostatic current density applied.

They concluded as discussed below that the experimental results disproved the catalytic mechanism, assuming tacitly that H(a) behaved according to Eqs. (2.6), hence as a dilute component. According to the catalytic mechanism, H(a) is accumulated or consumed through step (2.5a) in an amount proportional to the lapse of time, on application of the galvanostatic current in either direction. The population of H(a), θ, is given by eliminating $\mu(H)$ and $\mu(e^-)$ from Eqs. (2.6b), (2.7), and (2.3b) as

$$F\eta = RT \ln \theta + \mu(H)_1 - \mu(H)_0, \tag{2.13a}$$

where

$$\mu(H)_0 \equiv \mu(H^+) + \mu(e^-)_{rev} \tag{2.13b}$$

is the chemical potential of H(a) on the reversible hydrogen electrode; hence it equals $\mu(H_2)/2$. Equation (2.13a) states that η, and hence the electrode potential, should vary logarithmically with time rather than linearly as observed.

The deviation of the coefficient of log i in Eq. (2.8b) from the observed b-value was the additional basis for their argument against the catalytic mechanism. It may be noted that both their arguments against the catalytic mechanism are based on an assumption which was early questioned by Haber and Russ (1904) and by Tafel (1905), as mentioned ir Section II,B,1. Erdey-Grúz and Volmer (1930) thus proposed the slow discharge mechanism with step (2.5a) determining the rate and suggested, as a possible alternative to step (2.5b), step $H_2{}^+ + e^- \rightarrow H_2$ to conclude reaction (2.1).

They assumed that the depression of the electrode potential, E, by $-\Delta E$ reduced the activation energy by $\beta F \Delta E$, so that $\ln i_+ = -\beta FE/RT + C_+$, with an increase in the activation energy of the reverse reaction by the same amount, so that $\ln i_- = \beta FE/RT + C_-$, and they determined the constant β to be $\frac{1}{2}$, fitting the equilibrium relation derived, i.e., $i_+ = i_-$ or $2\beta FE/RT = C_+ - C_-$, to the thermodynamic requirement that $2\beta = 1$, where C_+ and C_- are quantities independent of E. The log i_+ is in consequence given as $\log i_+ = \beta F\eta/2.30RT +$ const; hence $\eta = (2.30RT/\beta F) \log i_+ +$ const, so that the coefficient of log i_+ is four times as large as that in Eq. (2.8b), i.e., $0.029 \times 4 = 0.12$ V, which agrees with experiment.

The derivation of the b-value from the slow discharge mechanism has been quantum mechanically elaborated by Gurney (1931) and Horiuti and Polanyi (1935).

7. *Dual Mechanism*

The electrolytic separation factor of deuterium, $S_+(D)$, is the ratio of the specific rate of the unidirectional transfer of protium from solution to that of deuterium in the course of the HER. Horiuti and Okamoto (1936) observed $S_+(D)$ on various metal electrodes in N H_2SO_4 aq. at room temperature and found that electrodes fell into two groups, i.e., that of Ni, Au, Ag, Cu, and Pt electrodes with $S_+(D)$ around 6–7 and that of Sn, Hg, and Pb electrodes with $S_+(D)$ around 3. A Pb electrode in N KOH aq. fell in the first group. These experiments were traced and confirmed by Walton and Wolfenden (1936, 1938). $S_+(D)$ greater than 10 had often been reported earlier, but the above values have been accepted in the course of time.

Horiuti and Okamoto (1936) inferred qualitatively that the separation factor was an indication of the rate-determining step, hence of the mechanism, as has since been established,* suggesting the appropriate mechanism of the respective group as follows. The maximum possible value, $S_+(D) = 19$, was worked out by Topley and Eyring (1934); this would arise in a critical system consisting of a quasifree hydrogen atom. A similar conclusion is to be expected if the critical step is (2.5a); Horiuti and Okamoto rejected the slow discharge mechanism, as was confirmed later by the detailed calculation of Keii and Kodera (1957). They attributed the catalytic mechanism to the group of $S_+(D) = 6–7$, which included such metals as Ni or Pt which show outstanding catalytic activity for the recombination of hydrogen atoms, as was established by later investigations and is illustrated in Section II,F. Step

$$H(a) + H^+ + e^- \rightarrow H_2 \qquad (2.14)$$

was first assigned as rate-determining to the other group of $S_+(D) \doteqdot 3$, which complies with its alternative operation in acid solution on a lead electrode. It was found later by detailed quantum-mechanical calculation (Horiuti et al., 1951) that hydrogen nuclei of H(a) and H^+ involved in the step were situated between a metal atom, M, and a water molecule, OH_2, lying symmetrically around the line through the centers of M and OH_2 resonating between the two electronic states of equal energy

$$
\begin{array}{ccc}
\text{Me}\diagup^{\displaystyle H}\diagdown_{\displaystyle B} & & \text{Me}\diagdown_{\displaystyle B,} \\
H^+ & \text{and} & H
\end{array}
\qquad (2.15)
$$

* Cf. Section II,F.

similar to the hydrogen molecule-ion H_2^+ in gas. The theory has thus led to the rate-determining step,

$$H_2^+(a) + e^- \rightarrow H_2, \tag{2.16a}$$

i.e., the neutralization of chemisorbed hydrogen molecule-ion, $H_2^+(a)$, which happened to be in line with the alternative to the step occurring subsequent to the rate-determining one as suggested by Erdey-Grúz and Volmer (1930); the preliminary formation of $H_2^+(a)$,

$$2\,H^+ + e^- \rightarrow H_2^+(a), \tag{2.16b}$$

was in consequence assumed in partial equilibrium, and hence

$$2\mu(H^+) + \mu(e^-) = \mu(H_2^+), \tag{2.17}$$

where $\mu(H_2^+)$ is the chemical potential of $H_2^+(a)$. The latter revised mechanism has since been called the electrochemical mechanism. The catalytic and the electrochemical mechanisms were thus concluded to be alternatively operative, depending upon the experimental condition of the HER inclusive of the electrode material. This system of alternative mechanism was called the *dual mechanism* by Walton and Wolfenden (1936, 1938).

C. DOUBLE LAYER AND HYDROGEN INTERMEDIATES

The hydrogen intermediate involved in either of the alternative mechanisms must be stable enough to keep it operating.

Horiuti *et al.* (1951) have theoretically investigated the specifically adsorbed or chemisorbed states, $H^+(a)$ and $H_2^+(a)$, of H^+ and H_2^+ on the electrode surface, concluding that they were hydrated even in the latter states, if they existed stably at all. It hence followed that it was not practical to make a distinction between the specifically adsorbed and the hydrated states of these intermediates.

Horiuti (1954) developed on this ground a theory of the boundary layer of the hydrogen electrode using a model plane, called the P-plane, parallel to the electrode surface; this furnished sites for specific adsorption of intermediates in the hydrated state as well as the boundary of nearest approach of hydrated ions to the electrode surface as shown in Fig. 3. The electrostatic potential on the P-plane referred to the potential in the bulk of solution is denoted by E_p.

Adsorption isotherms of $H(a)$, $H^+(a)$, and $H_2^+(a)$ were formulated, admitting the equilibria of steps, $H^+ \rightarrow H^+(a)$, (2.5a) and (2.16b). It was thus concluded that $H_2^+(a)$ is exceedingly stable as compared to its dismembered state, $H(a) + H^+(a)$, so that the coexistence of $H(a)$ and

$H^+(a)$ was practically excluded. Possible coexistent states are thus $H(a)$ with $H_2^+(a)$ or of $H_2^+(a)$ with $H^+(a)$. It was shown that either type of coexistence could occur only in a limited range of η on account of the different dependence of the chemical potentials of the intermediates on η. We will deal mainly with the case where only one class of intermediates predominates, giving rise to the relevant mechanism except in special cases.

On the basis of the above formulation, the experimental aspects of the HER are dealt with in what follows.

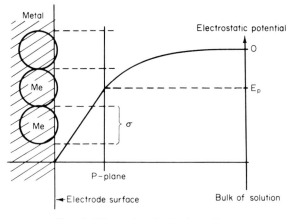

FIG. 3. Electrode-solution boundary.

D. DEPENDENCE OF i_+ ON η

The dependence is given by Eq. $(1.28\tau_+)$. We have over the region of η, where Tafel's law applies, $\tau_+ = 2.30RT/bF$ according to Eq. (2.11) or $\tau_+ = 0.49$ for $b = 0.12$ V at $20°C$. Experimental results on τ_+ as summarized in Fig. 1 are theoretically interpreted in subsequent sections on the basis of the dual mechanism.

1. Physical Meaning of τ_+

The forward and backward unidirectional rate is given by Eqs. $(1.25a)$ and $(1.25b)$ in terms of the respective current densities. The $\mu(I_r)$ in Eq. $(1.25a)$ is expressed as

$$\mu(I_r) = \mu(H^+) + \mu(e^-)$$

for the slow discharge mechanism, step (2.5a) being rate-determining in accord with Section II,B,6, whereas we have

$$\mu(I_r) = 2\mu(H^+) + 2\mu(e^-)$$

for either alternative of the dual mechanism on account of the partial equilibrium of step (2.5a) or (2.16b). Since according to Eqs. (2.7), (2.3b), and (2.13b)

$$\mu(H) = \mu(H^+) + \mu(e^-) = F\eta + \mu(H)_0, \qquad (2.18H)$$

we have

$$\mu(I_r) = F\eta + \mu(H)_0 \quad \text{or} \quad \mu(I_r) = 2F\eta + 2\mu(H)_0 \qquad (2.18s), (2.18d)$$

for the slow discharge or the dual mechanisms, respectively. It follows that $\mu(I_r)$ increases with increase of η.

We have from Eqs. $(1.28\tau_+)$ and (1.25a) with reference to Eqs. (2.18),

$$\tau_+ = (RT/F) \, \partial \ln \theta(\sigma_r^{\ddagger}, 0)/\partial\eta - \partial\epsilon(\ddagger_r)/F \, \partial\eta + n, \qquad (2.19)$$

where $n = 1$ or 2 for the slow discharge or the dual mechanism, respectively.

Erdey-Grúz and Volmer (1930) evaluated the second term at $-1/2$ as seen from Section II,B,6, noting that $\partial/\partial\eta = -\partial/\partial E$ by definition, tacitly ignoring the first term, which is reasonable, since the coverage of $H^+(a)$ increases quite slowly with increase of η, proportionally to the potential of the Helmholtz double layer.

2. τ_+ of the Catalytic Mechanism with Interactions Absent

The value of θ increases with η according to Eq. (2.13a), whereas $\theta(\sigma^{\ddagger}, 0)$ in Eq. (2.19) is hardly affected for sparse H(a). The second term remains zero throughout the change in θ on account of the postulated absence of interaction. The last term, n, is thus in full swing, so that $\tau_+ = 2$ as deduced by Tafel himself.

Consider now the case where H(a) is fairly crowded; assume a two-dimensional array of physically identical adsorption sites, σ, of H(a) and $\sigma_r^{\ddagger}$ composed of two adjacent σ's. Let A be the assembly in which step (2.5a) is in equilibrium and A_1 or A_2 the states where a particular σ is occupied or unoccupied by H(a) respectively. The ratio of the probability of state A_1 to that of A_2 equals, according to statistical mechanics, the ratio of the partition function of A_1, $\mathfrak{Q}A_1$, to that of A_2, $\mathfrak{Q}A_2$, while $-kT \ln \mathfrak{Q}A_1$ or $-kT \ln \mathfrak{Q}A_2$ gives the free energy of A_1 or A_2, respectively.

On account of the postulated physical identity of sites, the probability of σ being occupied by δ equals the population, θ, in terms of the covered fraction or the ratio of the number of δ occupying σ over the total number of sites, either occupied or unoccupied. The probability of a site being unoccupied is similarly equal to the fraction, θ_0, the number of unoccupied sites over the total number of sites. We have in consequence

$$\theta/\theta_0 = \mathfrak{Q}A_1/\mathfrak{Q}A_2.$$

The right-hand side of the above equation is transformed as follows. Let A_0 be the assembly derived from A_1 by removing δ from the assembly to leave the σ vacant.

$$\epsilon(\delta)/N_A = -kT \ln \mathfrak{Q}A_1 - (-kT \ln \mathfrak{Q}A_0) \qquad (2.20a)$$

is now the free energy increment due to addition of δ from outside the assembly to the preliminary evacuated particular site, σ.* The free energy increment, $-kT \ln \mathfrak{Q}A_2 - (-kT \ln \mathfrak{Q}A_0)$, equals $\mu(\delta)/N_A$, on the other hand, i.e.,

$$-kT \ln \mathfrak{Q}A_2 - (-kT \ln \mathfrak{Q}A_0) = \mu(\delta)/N_A,$$

where $\mu(\delta)$ is the chemical potential of δ in the assembly. Eliminating $\mathfrak{Q}A_1/\mathfrak{Q}A_0$ and $\mathfrak{Q}A_2/\mathfrak{Q}A_0$ from the above three equations, we have

$$RT \ln (\theta/\theta_0) = \mu(\delta) - \epsilon(\delta). \qquad (2.20b)$$

Let $\delta \equiv \mathrm{H}(a)$ be the only occupant of σ. Equation (2.20b) is now

$$RT \ln[\theta/(1 - \theta)] = \mu(\mathrm{H}) - \epsilon(\mathrm{H}) \qquad (2.21a)$$

or according to Eq. (2.18H)

$$RT \ln[\theta/(1 - \theta)] = F\eta + \mu(\mathrm{H})_0 - \epsilon(\mathrm{H}), \qquad (2.21b)$$

which is reduced to Eq. (2.13a) for sufficiently small θ as compared with unity, $\epsilon(\mathrm{H})$ turning out to be the equivalent for constant $\mu(\mathrm{H})_1$. The $\theta(\sigma_r^{\ddagger}, 0)$ in Eq. (2.19) equals $(1 - \theta)^2$ according to the postulated absence of interaction, since then the probability that a site is unoccupied is independent of whether any adjacent site is occupied or not, equaling $(1 - \theta)$ invariably. Substituting θ from Eq. (2.21b) into $\theta(\sigma_r^{\ddagger}, 0) = (1 - \theta)^2$, we have

$$\theta(\sigma_r^{\ddagger}, 0) = \left[1 + \exp\{[F\eta - \epsilon(\mathrm{H}) + \mu(\mathrm{H})_0]/RT\}\right]^{-2};$$

* The $\epsilon(\delta)/N_A$ equals the corresponding energy increment exactly, provided that the dynamics of constituent particles of the electrode is independent of that of δ and the latter is in its ground state when adsorbed.

hence

$$\frac{RT}{F} \frac{\partial \ln \theta(\sigma_r^{\ddagger}, 0)}{\partial \eta} = -2\left\{1 + \exp\left[-\frac{F\eta - \epsilon(H) + \mu(H)_0}{RT}\right]\right\}^{-1}.$$

The τ_+ is given, remembering that the second term of Eq. (2.19) is zero and $n = 2$ in the present case, as

$$\tau_+ = 2\left\{1 + \exp\left[\frac{F\eta - \epsilon(H) + \mu(H)_0}{RT}\right]\right\}^{-1},$$

which states that $\tau_+ = 2$ at sufficiently small η and decreases steadily down to zero with increase of η.

The τ_+ thus passes through the experimental value, ca. 0.5, in the course of its steady decrease from 2 to 0. It is calculated by the above equation that τ_+ decreases from 0.6 to 0.4 as η increases by only 15 mV, whereas experimentally τ_+ stays around 0.5 for 150 to 800 mV.

3. τ_+ for the Catalytic Mechanism with Interactions Present

The discrepancy mentioned in the last section was dealt with by Okamoto *et al.* (1936), taking the repulsive interactions into account. It was known from quantum mechanics that atoms not bonded with each other exert mutual repulsions. The free energy increment, $\epsilon(H)$, in Eqs. (2.21) is given as its value in the absence of interaction, $\epsilon(H)_0$, plus the additional work against the repulsion. Assuming the latter is proportional to θ, we have

$$\epsilon(H) = \epsilon(H)_0 + RTu\theta, \tag{2.22a}$$

where RTu is the proportional factor, or according to Eq. (2.21b)

$$RT \ln[\theta/(1 - \theta)] + RTu\theta = F\eta - \epsilon(H)_0 + \mu(H)_0. \tag{2.22b}$$

The increment of the left-hand side of the above equation is contributed chiefly by the second term in the case where RTu is large enough and θ is not very near to zero or unity. The increase of θ with increase of η is thus retarded and approximately a linear function of η. This accounts for the experimental results of Erdey-Grúz and Volmer (1930), invalidating their argument against the catalytic mechanism referred to in Section II,B,6.

The sufficiently large magnitude of RTu postulated as above was derived by Okamoto *et al.* (1936) as a quantum-mechanical exchange repulsion among H(a). Toya (1958, 1960) has shown recently that the exchange repulsion was reinforced by repulsion due to competition for conduction electrons among the H(a), fortifying the above argument.

It is now shown semiquantitatively that α stays around 0.5 over the range of η, where θ varies slowly with η as above. The $\epsilon(\ddagger_r)$ is given similar to Eq. (2.22a) as

$$\epsilon(\ddagger_r) = \epsilon(\ddagger_r)_0 + RTu^{\ddagger}\theta, \tag{2.23a}$$

where $\epsilon(\ddagger_r)_0$ is the value of $\epsilon(\ddagger_r)$ at $\theta = 0$ and $RTu^{\ddagger}$ is the proportional factor. Consider a plane square lattice of σ, on which, at $\theta = 1$, each H(a) is surrounded by four nearest kindred ones and a critical system occupying two adjacent sites by six, respectively. Since the repulsion decays rapidly with increasing distance, the ratio of $u^{\ddagger}$ to u may be approximated by the ratio of the number of nearest neighbors to $\ddagger_r$ to that to H(a), i.e., 6:4. We have accordingly

$$\epsilon(\ddagger_r) = \epsilon(\ddagger_r)_0 + (6/4)RTu\theta. \tag{2.23b}$$

Eliminating $\epsilon(\ddagger_r)$, $\mu(I_r)$, and $RTu\theta$ from Eqs. (1.25a), (2.18d), (2.22b), and (2.23b), we have, ignoring the changes of $\ln \theta(\sigma_r^{\ddagger}, 0)$ and $RT \ln[\theta/(1 - \theta)]$ under the specified condition,

$$\ln i_+ = (2 - 6/4)F\eta/RT + \text{const};$$

hence according to Eq. (2.19), $\tau_+ = 0.5$. The additional work due to the repulsion is assumed proportional to θ as in Eqs. (2.22a) and (2.23a), which is called here the proportional approximation.

The proportional approximation has been carried out allowing for variations of $RT \ln[\theta/(1 - \theta)]$ and $\theta(\sigma_r^{\ddagger}, 0)$ by Horiuti and Kita (1964). The dependence of τ_+ on η worked out in this way is shown in Fig. 4.

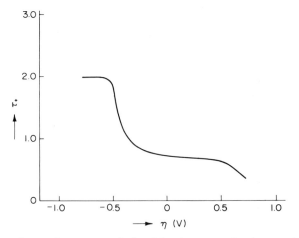

FIG. 4. Dependence of τ_+ on η, calculated by the proportional approximation on the basis of the catalytic mechanism.

The approximation was further advanced (Horiuti and Kita, 1964) by taking repulsions of H(a) individually into account instead of using the proportional approximation. The i_+ and τ_+ of (110)-, (100)-, and (111)-lattice planes of f.c.c. nickel crystals were thus worked out with the conclusion that the (111)-lattice plane contributed predominantly to the current density at higher η.

4. Saturation Current Density

It has been shown in Sections II,D,2 and II,D,3, with reference to different approximations, that τ_+ tended to vanish with an increase of η. This conclusion is arrived at from Eqs. (1.25a) and (2.21b) without referring to any particular approximation; the saturation current density is a definite property of the electrode fully covered by H(a).

Horiuti (1956) estimated theoretically the saturation current density on the hydrogen electrode of nickel at 200 A/cm^2 at 25°C. Kita and co-workers experimentally investigated the saturation current and found its lower bound to be 100 A/cm^2 for Ni (Kita and Yamazaki, 1963), 10 A/cm^2 for Ag (Yamazaki and Kita, 1965), and 20 A/cm^2 for Cu (Nomura and Kita, 1967).

5. τ_+ of the Electrochemical Mechanism

The τ_+ of the electrochemical mechanism is evaluated assuming that $\sigma_r^{\ddagger}$ and all σ for $H_2^+(a)$ are physically identical with one another and $E_p = 0$. On this basis we have from Eq. (2.20b)

$$RT \ln\{\theta(H_2^+)/[1 - \theta(H_2^+)]\} = \mu(H_2^+) - \epsilon(H_2^+), \qquad (2.24)$$

and from Eq. (1.25a)

$$i_+ = \frac{2FkT}{N_A v_r h} [1 - \theta(H_2^+)] \exp\frac{\mu(H_2^+)}{RT} \cdot \exp\left[-\frac{\epsilon(\ddagger_r) - \mu(e^-)}{RT}\right],$$

identifying $\theta(\sigma_r^{\ddagger}, 0)$ with $1 - \theta(H_2^+)$ and developing $\mu(I_r)$ as $\mu(I_r) \equiv \mu(H_2^+) + \mu(e^-)$ for the rate-determining step (2.16a). The eliminant of $[1 - \theta(H_2^+)] \exp[\mu(H_2^+)/RT]$ from the above two equations is

$$\ln[i_+/\theta(H_2^+)] = \ln(2FkT/N_A v_r h) - [\epsilon(\ddagger)_0 - \epsilon(H_2^+)_0 - \mu(e^-)]/RT, \quad (2.25)$$

where $\epsilon(\ddagger_r)_0$ and $\epsilon(H_2^+)_0$ are values of $\epsilon(\ddagger_r)$ and $\epsilon(H_2^+)$, respectively, in the absence of nonelectrostatic interactions with which $\epsilon(\ddagger_r)$ and $\epsilon(H_2^+)$ are replaced, assuming that the effects of the latter interactions upon $\ddagger_r$ and $H_2^+(a)$ compensate each other. Identifying further $\epsilon(\ddagger_r)_0$ and $\epsilon(H_2^+)_0$ approximately with relevant potential energies, i_+ is formulated as a function of η as follows. Figure 5 shows the potential energy of the initial system, $\epsilon(H_2{}^+)_0 + \mu(e^-)$, and that of the final system, $\epsilon(H_2)_0$, respectively, plotted against the distance between the two hydrogen nuclei involved in accordance with the resonance scheme (2.15). It may be noted that the potential curve, $H_2{}^+(a) + e^-$, is just insignificantly deformed from that of $H_2{}^+$ in the gas phase as worked out by Horiuti

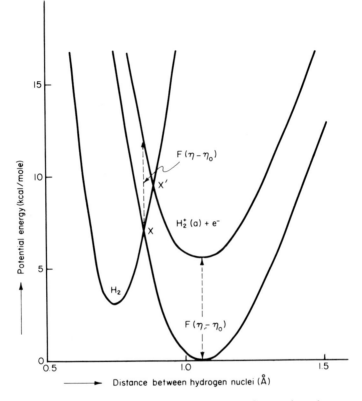

FIG. 5. Decrease of activation energy with increase of overvoltage by $\eta - \eta_0$.

et al. (1951). Step (2.16a) consists now in the decrease of the nuclear distance from the equilibrium distance of $H_2{}^+(a) + e^-$ to that of H_2 through the quantum-mechanical resonance at the intersection, X, of the two potential curves. The height of X above the minimum of curve $H_2{}^+(a) + e^-$ gives the height of the potential energy of the critical system above that of the initial system, i.e., $\epsilon(\ddagger_r)_0 - \epsilon(H_2{}^+)_0 - \mu(e^-)$.

By raising the overvoltage from a definite value, η_0, by $\Delta\eta \equiv \eta - \eta_0$, curve $H_2{}^+(a) + e^-$ shifts upwards or the potential energy of the initial system increases by $\Delta\epsilon(H_2{}^+)_0 + \Delta\mu(e^-) = F\,\Delta\eta$, whereas that of the critical system rises as the potential energy of the intersection rises, i.e., by $\alpha_E[\Delta\epsilon(H_2{}^+)_0 + \Delta\mu(e^-)] = \alpha_E F\,\Delta\eta$, where α_E is a positive proper fraction as seen in Fig. 5 and has the value $\frac{1}{2}$, particularly in the case where the inclinations of the curves around the intersection are equal in magnitude but opposite in sign. The height of the potential energy of the critical system above that of the initial system thus decreases by $(1 - \alpha_E)[\Delta\epsilon(H_2{}^+)_0 + \Delta\mu(e^-)] = (1 - \alpha_E)F\,\Delta\eta$. Hence we have from Eq. (2.25)

$$\ln i_+ = \ln \theta(H_2{}^+) + (1 - \alpha_E)[\Delta\epsilon(H_2{}^+)_0 + \Delta\mu(e^-)]/RT + \text{const}, \quad (2.26a)$$

or noting $\Delta\eta = \eta - \eta_0$,

$$\ln i_+ = \ln \theta(H_2{}^+) + (1 - \alpha_E)F\eta/RT + \text{const}. \quad (2.26b)$$

Hence according to Eq. $(1.28\tau_+)$,

$$\tau_+ = (RT/F)\,\partial \ln \theta(H_2{}^+)/\partial\eta + 1 - \alpha_E. \quad (2.26c)$$

The first term of Eq. (2.26a) is now investigated in terms of Eq. (2.24). The value of $\mu(H_2{}^+)$ is developed by using Eqs. (2.17) and (2.3b), leading to the value $\mu(H_2{}^+) = F\eta + 2\mu(H^+) + \mu(e^-)_{\text{rev}}$, while $\epsilon(H_2{}^+)$ increases with increase of $\theta(H_2{}^+)$ due to nonelectrostatic repulsion. It follows similarly to the case of H(a) in the foregoing section that $\theta(H_2{}^+)$ varies with η as $RT \ln \theta(H_2{}^+) = F\eta + \text{constant}$, or remains practically constant at a magnitude near unity according as η is sufficiently low or high, respectively. τ_+ is given by Eq. (2.26c) in the respective cases, as

$$\tau_+ = 2 - \alpha_E \quad (\eta\ll) \qquad \text{or} \qquad \tau_+ = 1 - \alpha_E \quad (\eta\gg). \quad (2.27)$$

Matsuda and Horiuti (1958) showed that Eq. (2.27) remained valid, if allowance was made for the deviation of E_p from zero on the basis of the model described in Section II,C.

Horiuti and Ikusima (1939) found that τ_+ on the hydrogen electrode of platinum in alkaline solution kept at 1.6 for η from 0 to 30 mV and at 0.6 for η from 30 mV to the extremity of observation, 100 mV, from which α_E was determined according to Eq. (2.27) as

$$\alpha_E = 0.4. \quad (2.28)$$

Mituya (1956) observed the current density around 10^{-10} A/cm^2 on the hydrogen electrode of mercury in $N/10\cdot$HCl aq. at 0°C from 95 to 270 mV overvoltage and found that $\tau_+ = 1.54$ or 0.54 below or above $\eta = 132$ mV, respectively.

The above results verify the electrochemical mechanism in both the cases of platinum and mercury. The electrochemical mechanism of the mercury electrode conforms with its $S_+(D)$ value in Section II,B,7, whereas that of the platinum electrode does not conform with $S_+(D)$ as observed at 300 mV, which was found characteristic of the catalytic mechanism. This problem has been settled as illustrated in Section II,F,5.

E. Dependence of i_+ on $\mu(H^+)$

1. General Remark

The dependence of i_+ on $\mu(H^+)$ at constant η is derived theoretically as a characteristic of the respective mechanism and compared with experiment.

We see that $\mu(I_r)$ in Eq. (1.25a) is $\mu(H^+) + \mu(e^-)$, $2\mu(H)$, or $\mu(H_2^+) + \mu(e^-)$, respectively, in the case of the slow discharge, and the catalytic or electrochemical mechanism, which equals, according to Eqs. (2.7) and (2.17), $\mu(H^+) + \mu(e^-)$, $2\mu(H^+) + 2\mu(e^-)$, or $2\mu(H^+) + 2\mu(e^-)$, respectively. Hence according to Eq. (2.3c), $\mu(I_r)$ is constant for all mechanisms at constant $\mu(H_2)$ and η. The $\theta(\sigma_r^{\ddagger}, 0)$ and $\epsilon(\ddagger_r)$ are consequently the only factors in Eq. (1.25a) which can be affected by varying $\mu(H^+)$. The $\theta(\sigma_r^{\ddagger}, 0)$ is now changed only through the change of $\theta(\delta)$, which is affected according to Eq. (2.20b) only through a change of $\epsilon(\delta)$ under the specified condition. The change of $\epsilon(\delta)$ as well as of $\epsilon(\ddagger_r)$ is brought about by a change of $\mu(H^+)$, mediated through a change of E_p in the case of charged δ or $\ddagger_r$, respectively, according to the definition given in Section II,C.

The dependence is observable in terms of i in the case of negligible i_- according to Section II,B,4. At $\eta = 0$, i_+ is reduced to i_0, which is traceable by means of a hydrogen isotope, if the specific rates of transfer of the hydrogen isotopes are in a constant ratio, so that the transfer rate of the tracer is proportional to i_0.

Conclusions are derived from the respective mechanisms.

2. Catalytic Mechanism

The $\theta(\sigma_r^{\ddagger}, 0)$ and $\epsilon(\ddagger_r)$ are not affected by change of $\mu(H^+)$, inasmuch as H(a) and $\ddagger_r$ are not charged; the neutrality of $\ddagger_r$ is plausibly admitted as it is a transient between neutral systems, 2H(a) and H_2. The i_+ remains in consequence unchanged with change in $\mu(H^+)$ at constant η and constant temperature. Figure 6 shows the slope, $\partial \log i_+/\partial pH$, which equals $-RT\,\partial \ln i_+/\partial\mu(H^+)$. The values found are fairly small, especially at lower concentration of electrolytes and differ considerably with different authors.

Horiuti and Okamoto (1938) observed the deuterium transfer from hydrogen gas to solution at $\eta = 0$ on the hydrogen electrode of nickel in $N/1000$ KOH aq. and neutral water, respectively, at $100°C$ and found its rate practically independent of pH in conformity with the inference from the catalytic mechanism. The reaction was followed by the decrease of

Electrode metal	Conc. (N) Solution	10 1 10^{-1} 10^{-2} 10^{-3}	10^{-3} 10^{-2} 10^{-1} 1 10	Conc. (N) Solution
Ag	HCl	+ − 0 1)	+0.25 2)	NaOH
Au	HCl	(~0) 3)	− 0 3)	NaOH
Bi	HClO₄	0 4)		
Ga	HCl	0 5)		
Hg	HCl	−0.09*) 6)		
Hg	HCl	0 7)		
Mo	HCl	(~0) 3)	(~0) 3)	
Ni	HCl	−0.49 −0.3 0 8)	+0.1 8)	NaOH
Ni			+0.2 9)	NaOH
Pb	H₂SO₄	−0.16 0 9)		
Ti	H₂SO₄	−0.44 ~ −0.59 10)		

FIG. 6. Table showing $\partial \log i_+/\partial$pH. ———: concentration range of observation; (~0): no systematic change of i_+ with pH; (*): estimated at $i = 0.1$ mA/cm² from Fig. 4, Ref. (4) below.

(1) Bockris, J. O'M., and Conway, B. E. (1952). *Trans. Faraday Soc.* **48**, 724.
(2) Ammar, I. A., and Awad, S. A. (1956). *J. Phys. Chem.* **60**, 1290.
(3) Pentland, N., Bockris, J. O'M., and Sheldon, E. (1957). *J. Electrochem. Soc.* **104**, 182.
(4) Palm, U. W., and Past, V. E. (1964). *Zh. Fiz. Khim.* **38**, 773.
(5) Christov, St. G., and Rajčeva, S. (1962). *Z. Elektrochem.* **66**, 486.
(6) De Béthune, A. J. (1949). *J. Am. Chem. Soc.* **71**, 1556.
(7) Post, B., and Hiskey, C. F. (1950). *J. Am. Chem. Soc.* **72**, 4203.
(8) Bockris, J. O'M., and Potter, E. C. (1952). *J. Chem. Phys.* **20**, 614.
(9) Jofa, S. (1945). *Zh. Fiz. Khim.* **19**, 117.
(10) Kolotyrkin, Ya. M., and Petrov, P. S. (1957). *Zh. Fiz. Khim.* **31**, 659.

the deuterium atomic fraction in the hydrogen gas, which would be simulated if light hydrogen were evolved, e.g., by corrosion reducing the deuterium atomic fraction of the hydrogen gas by dilution; no hydrogen evolution was detected under the latter condition (Hirota and Horiuti, 1936), although its rate, perceptible first at higher concentration of alkali, increased rapidly with alkali concentration (Horiuti and Komobuchi, 1958).

3. *Electrochemical Mechanism*

The increase of $\mu(H^+)$ by $\Delta\mu(H^+)$ gives rise to an increase of $\mu(H_2^+)$ at constant η, in accord with Eqs. (2.17) and (2.18H), which tends to increase $\theta(H_2^+)$ as shown by Eq. (2.24) at constant E_p. An increment of $\mu(H^+)$ causes on the other hand a decrease of $\mu(e^-)$ according to Eq. (2.18H) amounting to as much as $-\Delta\mu(e^-) = \Delta\mu(H^+)$, or to an increase of the electrode potential by $\Delta\mu(H^+)/F$, which tends to decrease $\theta(H_2^+)$ according to Poisson's equation at constant E_p. The $\theta(H_2^+)$ decreases now with increase of E_p according to the adsorption isotherm, Eq. (2.24), whereas it increases with an increase of E_p according to Poisson's equation. The simultaneous equations thus are satisfied by increasing E_p by $\Delta E_p = \alpha_h \Delta\mu(H^+)/F$ (Horiuti, 1954, 1957), where α_h is a positive proper fraction, which tends to unity or zero according as the ionic strength of solution approaches zero or infinity, respectively.

Curve $H_2^+(a) + e^-$ in Fig. 5 thus decreases by the amount $(1 - \alpha_h)\Delta\mu(H^+)$, so that $\epsilon(\ddagger_r)_0 - \epsilon(H_2^+)_0 - \mu(e^-)$ increases by $(1 - \alpha_E)(1 - \alpha_h)\Delta\mu(H^+)$ as inferred in the case of Section II,D,5. Noting that $\Delta\mu(H^+) = -2.30RT\,\Delta\mathrm{pH}$, we have according to Eq. (2.25)

$$\Delta \log i_+ = \Delta \log \theta(H_2^+) + (1 - \alpha_E)(1 - \alpha_h)\,\Delta\mathrm{pH}. \qquad (2.29)$$

The $\Delta \log \theta(H_2^+)$ in the above equation vanishes when $\theta(H_2^+)$ is comparable with unity and approximately constant, so that

$$\Delta \log i_+ = (1 - \alpha_E)(1 - \alpha_h)\,\Delta\mathrm{pH}, \qquad \theta(H_2^+)\ \text{const.} \qquad (2.30a)$$

The $\ln \theta(H_2^+)$ is given, eliminating $\mu(H_2^+)$ and $\mu(e^-)$ from Eqs. (2.17), (2.24), and (2.3c) as

$$RT\ln \theta(H_2^+) = RT\ln[1 - \theta(H_2^+)] + \mu(H^+) - \epsilon(H_2^+) + \tfrac{1}{2}\mu(H_2) + F\eta.$$

In the other extremity where $\theta(H_2^+) \ll 1$, $\epsilon(H_2^+) = \epsilon(H_2^+)_0$ because of the absence of nonelectrostatic repulsion, and $\epsilon(H_2^+)_0$ is in its turn given as $\epsilon(H_2^+)_0 = \epsilon(H_2^+)_0{}^0 + FE_p$, where $\epsilon(H_2^+)_0{}^0$ is the value of $\epsilon(H_2^+)_0$ at $E_p = 0$. Since $F\Delta E_p = \alpha_h \Delta\mu(H^+) = -2.30\alpha_h RT\,\Delta\mathrm{pH}$, we have from the above equation, ignoring $\theta(H_2^+)$ as compared with unity $\Delta \log \theta(H_2^+) = -(1 - \alpha_h)\Delta\mathrm{pH}$; hence from Eq. (2.29)

$$\Delta \log i_+ = -\alpha_E(1 - \alpha_h)\,\Delta\mathrm{pH}, \qquad \theta(H_2^+) \ll 1. \qquad (2.30b)$$

Equations (2.30) state that since both α_E and α_h are positive proper fractions, then i_+ increases or decreases with increase of pH according as $\theta(H_2^+) \simeq 1$ or $\theta(H_2^+) \ll 1$, respectively.

We have seen in Section II,D,5 that τ_+ of the platinum electrode in alkaline solution kept constant at 1.6 over η below 30 mV through the reversible potential. This indicates that θ is sufficiently small as compared with unity. It follows that i_0 should decrease with increase of pH. Horiuti and Ikusima (1939) observed at $\eta = 0$ that $\Delta \log i_0/\Delta\mathrm{pH} = -0.6$, which verifies the above inference.

F. SEPARATION FACTOR

1. *Formulation*

The electrolytic separation factor, $S_+(\mathrm{H}^*)$, of the hydrogen isotope is defined, inclusive of $S_+(\mathrm{D})$ referred to in Section II,B,7, as

$$S_+(\mathrm{H}^*) = \frac{i_+(\mathrm{H})/n(\mathrm{H})}{i_+(\mathrm{H}^*)/n(\mathrm{H}^*)}, \qquad (2.31)$$

where H^* stands for D and T, $i_+(\mathrm{H})$ or $i_+(\mathrm{H}^*)$ represents the current density due to transfer of H or H^*, respectively, and $n(\mathrm{H})$ or $n(\mathrm{H}^*)$ the numbers of H or H^* in solution feasible for the electrolysis, respectively.

The separation factor, $S_+(\mathrm{H}^*)$, is formulated theoretically for the simplest case, where H^* is present at such dilution among the H, that only one H^* is included at most in λ_r hydrogen atoms electrolyzed for each act of the rate-determining step. The $i_+(\mathrm{H})$ and $i_+(\mathrm{H}^*)$ are expressed by Eq. (1.25a), replacing the factor $2/\nu(r)$ with λ_r and with unity respectively as

$$i_+(\mathrm{H}) = \frac{\lambda_r F}{N_\mathrm{A}} \frac{kT}{h} N_r^{\ddagger}\theta(\sigma_r^{\ddagger}, 0) \exp\left[-\frac{\epsilon(\ddagger_r) - \mu(I_r)}{RT}\right]$$

and

$$i_+(\mathrm{H}^*) = \frac{F}{N_\mathrm{A}} \frac{kT}{h} N_r^{\ddagger}\theta(\sigma_r^{\ddagger}, 0) \exp\left[-\frac{\epsilon(\ddagger_r^*) - \mu(I_r^*)}{RT}\right],$$

where $\ddagger_r^*$ or I_r^* is the system derived from $\ddagger_r$ or I_r by replacing one of the λ_r protons to be electrolyzed by $\mathrm{H}^{*,+}$, i.e., by a deuteron or a triton. We have from Eq. (2.31) and the above two equations that

$$S_+(\mathrm{H}^*) = \lambda_r \frac{n(\mathrm{H}^*)}{n(\mathrm{H})} \exp\left[-\frac{\epsilon(\ddagger_r) - \epsilon(\ddagger_r^*)}{RT}\right] \exp\left[\frac{\mu(I_r) - \mu(I_r^*)}{RT}\right]. \qquad (2.32)$$

2. *Reduction of the General Equation*

Since $\mu(I_r) = \mu(\mathrm{H}^+) + \mu(e^-)$ or $2[\mu(\mathrm{H}^+) + (e^-)]$ according to Section II,E,1, we have, by definition of I_r^*, $\mu(I_r) - \mu(I_r^*) = \mu(\mathrm{H}^+) - \mu(\mathrm{H}_l^{*,+})$, which is transformed into $\mu(\mathrm{H}_2\mathrm{O}_{(g)}) - \mu(\mathrm{HH}^*\mathrm{O}_{(g)})$ with reference to the equilibrium of exchange reaction, $\mathrm{H}_{(l)}^+ + \mathrm{HH}^*\mathrm{O}_{(g)} =$

$H_l^{*,+} + H_2O_{(g)}$, where $H_2O_{(g)}$ is a normal or an isotopically substituted water molecule in the gas. The $\mu(H_2O_{(g)})$ or $\mu(HH*O_{(g)})$ is developed in terms of the partition function $Q(H_2O)$ or $Q(HH*O)$ of a single H_2O or $HH*O$ molecule in unit volume and its concentration in the gas, $n(H_2O_{(g)})$ or $n(HH*O_{(g)})$, respectively, to give

$$\mu(H_2O_{(g)}) = RT \ln[n(H_2O_{(g)})/Q(H_2O_{(g)})]†$$

or

$$\mu(HH*O_{(g)}) = RT \ln[n(HH*O_{(g)})/Q(HH*O_{(g)})];$$

hence Eq. (2.32) is written in the form

$$S_+(H*) = \lambda_r \exp\{-[\epsilon(\ddagger_r) - \epsilon(\ddagger_r*)]/RT\}$$
$$\times [Q(HH*O_{(g)})/Q(H_2O_{(g)})]/K/2, \quad (2.33a)$$

where

$$K = 2n(H_2O)n(H*)/n(HH*O)n(H) \quad (2.33b)$$

is the equilibrium constant of the isotopic exchange reaction, $H_{(l)}* + H_2O_{(g)} = H_{(l)} + HH*O_{(g)}$, where the suffix (l) signifies the components of solution involved in the exchange reaction. K was observed by Ikusima and Azagami (1938) for $H* \equiv D$ to be 1.076 at 18°C and 1.048 at 50°C, respectively. $Q(HH*O_{(g)})/Q(H_2O_{(g)})$ was evaluated from spectroscopic data to be 65 (Keii and Kodera, 1957) and 308 (Kodera and Saito, 1959), respectively, for $H* \equiv D$ or T at 20°C.

We now investigate the factor, $\exp\{-[\epsilon(\ddagger_r) - \epsilon(\ddagger_r*)]/RT\}$ in Eq. (2.33a). $\mathfrak{Q}A_1$ in Eq. (2.20a) is developed with reference to this case, assuming that the dynamics of the constituent particles of the electrode are independent of that of $\ddagger_r$, as $\mathfrak{Q}A_1 = \mathfrak{Q}A_0 \cdot \sum_i \exp[-\epsilon(\ddagger_r)_i/RT]$, where the latter factor is the partition function of $\ddagger_r$ inside $\sigma_r{}^\ddagger$ in the potential field of force exerted upon $\ddagger_r$ by the electrode and $\epsilon(\ddagger_r)_i$ is the ith eigenvalue of $\ddagger_r$ reckoned per mole. We have in consequence from Eq. (2.20a), $\epsilon(\ddagger_r) = -RT \ln \sum_i \exp[-\epsilon(\ddagger_r)_i/RT]$ and similarly $\epsilon(\ddagger_r*) = -RT \ln \sum_i \exp[-\epsilon(\ddagger_r*)_i/RT]$. The factor, $\exp\{-[\epsilon(\ddagger_r) - \epsilon(\ddagger_r*)]/RT\}$, depends on the shift of the eigenvalues from $\epsilon(\ddagger)_i$ to $\epsilon(\ddagger*)_i$, respectively, by the isotopic replacement. The heavier isotope thus reduces the ith eigenvalue from $\epsilon(\ddagger_r)_i$ to $\epsilon(\ddagger_r*)_i$, when substituted for protium; hence it diminishes the $S_+(H*)$ value according to Eq. (2.33a). The extent of the diminution depends on the modes of motion appropriate to the respective eigenvalue characterized by the mechanism.

† Cf., e.g., Hill (1960).

3. *The Characteristic* $S_+(H^*)$ *of the Mechanism*

The details of the evaluation of $\epsilon(\ddagger_r) - \epsilon(\ddagger^*)$ are left to the original publications and here the $S_+(H^*)$ is qualitatively reviewed as the characteristic of the mechanism.

Horiuti and Polanyi (1935) formulated step $H^+ + e^- \rightarrow H(a)$ as rate-determining of the slow discharge mechanism as mentioned in Section II,B,6; the step is the transition of a proton combined with a water molecule in H^+OH_2 to a neutral hydrogen atom covalently bonded to a metal atom, M, on the electrode surface along the line of centers of OH_2 and M. The potential curves of the respective electronic states were plotted against the distance along the combining line and the transition was considered to occur through quantum-mechanical resonance at their intersection, similarly to the transition illustrated in Fig. 5. The τ_+ near 0.5 was hence derived similarly to Section II,D,5, although the observed result that $\tau_+ > 1$ was beyond the scope of this mechanism.

The potential energy of the proton was thus described by a curve having a double minima with one maximum in-between, which was assigned to the critical system. A hydrogen nucleus has three degrees of freedom. One of them is devoted to the motion across the maximum resulting in the step, while the other two are a twofold degenerate vibration perpendicular to the combining line, which determines the partition function of the critical system.

The force acting perpendicularly on the nucleus is rather weak, being due to the distance increase from H_2O and M in the perpendicular displacement. Keii and Kodera (1957) calculated the relevant force constant and frequency, ν; hence the partition function becomes

$$\sum_i \exp[-\epsilon(\ddagger_r)_i/RT] = [2 \sinh(h\nu/2kT)]^{-2}$$
$$\times \exp(-V_0/kT) \doteqdot \exp[-(V_0 + h\nu)/kT],$$

where V_0 is the potential energy of the hydrogen nucleus in its equilibrium position; the partition function is approximated by the last member by neglecting $\exp(-h\nu/kT)$ against unity. The ν is given in terms of the mass of the nucleus, m, and the force constant, f, as $\nu = (1/2\pi)(f/m)^{1/2}$, since V_0 and f remain unchanged by isotopic replacement according to Born and Oppenheimer (1927).

The partition function, $\exp[-\epsilon(\ddagger_r^*)/RT]$, is developed similarly to $\exp[-\epsilon(\ddagger_r)/RT]$ as $\exp[-\epsilon(\ddagger_r^*)/RT] = \exp[-(V_0 + h\nu^*)/RT]$, where $\nu^* = (1/2\pi)(f/m^*)^{1/2}$ according to the above postulate and m^* is the mass of the replaced isotopic nucleus. We have in consequence

$$\exp\{-[\epsilon(\ddagger_r) - \epsilon(\ddagger_r^*)]/RT\} = \exp\{-(h\nu/kT)(1 - [m/m^*]^{1/2})\},$$

which is smaller, the greater f and, hence ν, is. $S_+(H^*)$ is thus smaller, the more tightly the nucleus is bound to the equilibrium position in the critical state.

Topley and Eyring (1934) estimated $S_+(D)$ at 25°C for the critical system of a "quasi-free hydrogen atom" at 19 as mentioned in Section II,B,7. Keii and Kodera (1957) derived the frequency, 496 cm^{-1}, for the twofold degenerate vibration using the above model; hence $S_+(D) = 12$–13 and $S_+(T) = 33$–37 (Kodera and Saito, 1959), both at 20°C. These estimates are greater than any observed value.

$S_+(H^*)$ of the dual mechanism is similarly modulated by the factor, $2 \exp\{-[\epsilon(\ddagger_r) - \epsilon(\ddagger_r^*)]/RT\}$, λ_r being 2 in this case. The two hydrogen atoms involved in the rate-determining step of the catalytic mechanism have six degrees of freedom; one of them is the transition along the reaction coordinate and five others are normal vibrations of frequencies, 936, 687, 1704, 368, and 626 cm^{-1} around the equilibrium configuration shown in Fig. 7, as worked out on the base of a distance of 3.52 Å

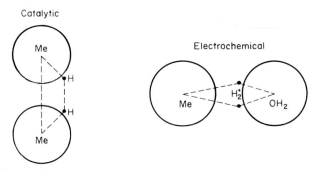

FIG. 7. Calculated configuration of the critical system for a dual mechanism (Horiuti *et al.*, 1964).

between adjacent surface metal atoms on the (110)-lattice plane of the f.c.c. nickel crystal (Okamoto *et al.*, 1936). The above factors appreciably diminish $S_+(D)$ through a number of vibrational degrees of freedom, some of them being of fairly high frequencies, down to 6.1 in accordance with experiments.* $S_+(T) \simeq 16$ was worked out on Pt on the basis of the catalytic mechanism by Horiuti and Nakamura (1950, 1951).

The above calculation was extended by Horiuti and Mueller (1968) to different $\sigma_r^{\ddagger}$ on different lattice planes of different metals and established $S_+(H^*)$ as a diagnostic tool of the catalytic mechanism.

In the case of the electrochemical mechanism, two hydrogen nuclei are situated symmetrically around the combining line as shown in Fig. 7

---

* The theoretical value increases to 7.1 when allowance is made for the quantum-mechanical tunnel effect; cf. Okamoto *et al.* (1936).

(Horiuti *et al.*, 1951); one of the modes of motion of the two nuclei is the rotation around the combining line. Four of them are vibrations of frequencies, 3338, 1516, 163, and 105 cm^{-1} and the sixth one is motion along the reaction coordinate (Horiuti *et al.*, 1951; Horiuti and Naka-mura, 1951). The critical system ought to have an exceedingly large frequency as seen above, two constituent nuclei being pinched between M and OH$_2$; the pronounced depression of the latter frequency through the isotopic replacement reduces $S_+(D)$ further below that of the catalytic mechanism to 3 (Horiuti *et al.*, 1951), in accordance with experiment.

4. *Experimental Results*

Figure 8 shows $S_+(H^*)$ summarized by Kita (1966), where $S_+(D)$ and

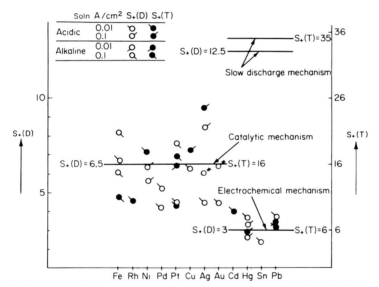

FIG. 8. Separation factors—observed (circles) and theoretical (levels); * indicates results obtained in N H$_2$SO$_4$ at $\eta = 0.35$ V (Horiuti and Okamoto, 1936).

$S_+(T)$ are graduated so as to bring $S_+(D) = 6.5$ and $S_+(T) = 16$ on the one hand and $S_+(D) = 3$ and $S_+(T) = 6$ on the other hand, respectively, upon the same levels. We see from the figure that the slow dis-charge mechanism is excluded by the theoretical value derived from it, $S_+(D) = 12–13$ or $S_+(T) = 33–37$, which is greater than any observed value and that the dual mechanism is established by the theoretical values of $S_+(H^*)$ derived from the catalytic and the electrochemical mechanisms, respectively, verified with the hydrogen electrodes of Fe, Rh, Ni, Pd, Cu, and Ag on the one hand and those of Hg, Sn, and Pb on the other hand, in line with the previous results mentioned in Section II,B,7. Some observations of Pt, Au, and Cd electrodes are found halfway; the $S_+(H^*)$ of the Pt electrode is commented upon in the next section.

5. Variation of $S_+(D)$ with η

The platinum electrode revealed the characteristics of the catalytic or electrochemical mechanism according as η was above 300 mV or around zero, as seen respectively in Sections II,B,7 or II,D,5 and II,E,3. The halfway value of $S_+(H^*)$ in Fig. 8 was observed at 40 mV η. These findings suggest a switchover of the mechanism from the electrochemical to the catalytic along with rise of overvoltage.

It is shown theoretically that the switchover is an implication of the dual mechanism. Equation (2.20b) is written respectively for coexistent $H(a)$ and $H_2^+(a)$ as

$$RT \ln[\theta(H)/\theta_0] = \mu(H) - \epsilon(H), \tag{2.34H}$$

$$RT \ln[\theta(H_2^+)/\theta_0] = \mu(H_2^+) - \epsilon(H_2^+), \tag{2.34H$_2^+$}$$

where $\theta(H)$ or $\theta(H_2^+)$ is θ of $H(a)$ or $H_2^+(a)$, respectively. Eliminating $\mu(H^+)$, $\mu(e^-)$, $\mu(H)$, and $\mu(H_2^+)$ from Eqs. (2.34), (2.7), (2.17), and (2.3b), we have

$$RT \ln \frac{\theta(H)^2}{\theta_0 \theta(H_2^+)} = F\eta + \mu(e^-)_{rev} - 2\epsilon(H) + \epsilon(H_2^+), \tag{2.35}$$

which states that, as η is raised, $H(a)$ increases at the cost of $H_2^+(a)$ and unoccupied σ, unless θ_0 is not very small, the rise of η being only partially compensated by that of E_p as shown by Horiuti (1954). The electrochemical mechanism operative at low η thus switches over to the catalytic mechanism as η is raised.

Let η be increased further until $H(a)$ covers the surface nearly completely. If η is raised further, E_p should become negative for a finite concentration of electrolyte according to the model described in Section II,D,5, no positive charge thus being allowed on the P-plane to make up for the negative E_p. Elimination of $\mu(H_2^+)$, $\mu(H)$, and θ_0 from Eqs. (2.7), (2.17), and (2.34) gives alternatively,

$$RT \ln[\theta(H_2^+)/\theta(H)] = -\epsilon(H_2^+) + \epsilon(H) + \mu(H^+),$$

which states that the H(a), covering the electrode surface nearly completely, are replaced retrograde by H_2^+(a) with further increase of η on account of the depression of $\epsilon(H_2^+)$. The electrochemical mechanism operative around $\eta = 0$ thus switches over to the catalytic mechanism and then is restored with a steady increase of η. $S_+(D)$ thus increases from ~ 3 at around $\eta = 0$ to ~ 7 and then decreases toward ~ 3.

Figure 9 shows $S_+(D)$ observed by Fukuda and Horiuti (1962) at

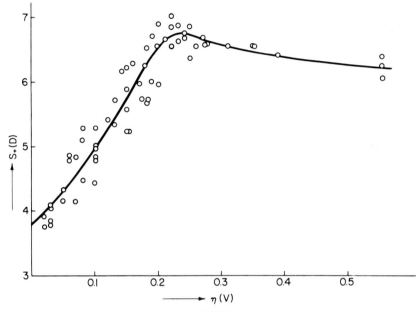

FIG. 9. Electrolytic separation factor, $S_+(D)$, dependent on overvoltage. Pt electrode in N H_2SO_4, 15°C (Fukuda and Horiuti, 1962).

different values of η in N H_2SO_4 aq. which verifies the above theoretical inference accounting for the halfway value in Fig. 8 observed on the Pt electrode at intermediate values of η.

G. The Stoichiometric Number, ν_r, of the Rate-Determining Step, r

1. General Remarks

The knowledge of ν_r referred to in Section II,B,4 enables us to discriminate between the dual and the slow discharge mechanism, for which ν_r is 1 and 2, respectively. The ν_r is given by Eq. (1.27a) by observing either i_+ or i_- as a function of η or from the latter and the direct observation of i according to Eq. (1.23).

The i_+ is, e.g., evaluated by labeling it with an isotope, provided that the isotopic composition of the unidirectional current is known, as referred to in Section II,E,1. The situation is, however, complicated inasmuch as the composition depends on $S_+(H^*)$ and the latter in its turn varies in general with η as seen in Section II,F,5. The situation is rendered exceptionally unfavorable in the very case of the hydrogen isotope, where $S_+(H^*)$ deviates widely from unity.

The process of labeling i_+ or i_- is subject to a further restriction. Let, for instance, reaction (1.6) consist of the rate-determining step (2.5a) followed by step (2.14). Isotopic exchange proceeds then quickly through step (2.14) between the solution and the hydrogen gas, bypassing the rate-determining step. The i_+ or i_- thus evaluated is apparently too large and i_+/i_- as calculated by Eq. (1.23) is too close to unity, resulting in an apparently too large value of ν_r according to Eq. (2.9). This was pointed out by Frumkin (1958) and discussed by Makrides (1962) and Matsuda and Horiuti (1962).

Equation (1.27a) has been shown invariantly applicable to hydrogen electrodes environed by isotopically mixed hydrogen gas and solution, if η was referred to the rest potential in the same environment and if i as well as i_0 represented the overall contribution of isotopes, provided that ν_r and $S_+(H^*)$ remained respectively the same as those at $\eta = 0$ (Enyo et al., 1965; Enyo and Yokoyama, 1967). Equation (1.27b) referred particularly to $\eta = 0$ thus is exact without any proviso.

2. Experimental Determination of ν_r

Horiuti and Ikusima (1939) investigated ν_r of the hydrogen electrode reaction on platinum in N KOH + 50% D_2O and normal hydrogen gas at 5°C. They derived i_0, on the one hand, from the observed value of $(\partial i/\partial \eta)_{\eta=0}$ for $\nu_r = 1$ or 2 and, on the other hand, from the observed $i_+(D)$ at $\eta = 0$ by multiplying it by the ratio, κ, of i_+ to the observed $i_+(D)$, which ratio was found constant over η from 30 to 100 mV. It was concluded that $\nu_r = 1$ from the comparison of these alternative evaluations of i_0. The validity of this conclusion rests upon the assumed invariance of κ down to $\eta = 0$. We have $\kappa = [i_+(H) + i_+(D)]/i_+(D)$ by definition or, substituting $i_+(H)/i_+(D)$ from Eq. (2.31), $\kappa = 1 + S_+(D)n(H)/n(D)$, which changes by less than 10% from $\eta = 0$ to $\eta = 30$ mV, along with the change of $S_+(D)$ by ca. 10% in the same region as read from Fig. 9. The appropriate change of κi_0 is almost within the experimental errors. This conclusion appears unaffected by the difference of solution in the experiment of Fig. 9 (H_2SO_4 aq.) from that in the above observation (KOH aq.), inasmuch as $S_+(D)$ has been found to be

4.15 in 0.1 N NaOH at $\eta = 0$ and 25°C (Enyo and Yokoyama, 1967) and to be 6–8 in 5 N KOH at $\eta > 300$ mV and 20°C (Buttler et al., 1963) in line with the result of Fig. 9.

3. Recent v_r-Determinations

Enyo and Matsushima (1969) observed i_0 quantitatively by the transfer of deuterium from pure D_2 into solution in normal water. Determining $(\partial i/\partial \eta)_{n=0}$ simultaneously, they found the v_r of different hydrogen electrodes by Eq. (1.27b) as shown below, assuming z to be 2, which verified the dual mechanism in the hydrogen evolution reaction on these electrodes.

TABLE I

EXPERIMENTAL VALUE OF v_r

Electrodes	Aqueous solution	Temperature °C	v_r
Ni	0.12 N KOH	25	1.0 ± 0.1
Pt	1.0 N H$_2$SO$_4$	12	1.1 ± 0.1
	0.5 N NaOH	12	1.05 ± 0.05
Ag	1.0 N H$_2$SO$_4$	10	1.0 ± 0.1
	0.2 N H$_2$SO$_4$	10, 50	1.0 ± 0.2
Rh	1.0 N H$_2$SO$_4$	13	1.2 ± 0.2
	0.5 N NaOH	12	1.06 ± 0.05

The v_r thus established affords, on the one hand, a sound basis for deriving i_0 from the observation of $(\partial i/\partial \eta)_{n=0}$ according to Eq. (1.27b) and excludes, on the other hand, such a mechanism as exemplified in Section II,G,1.

H. TRANSIENT PHENOMENA

Information on intermediates is derived from observations of transient phenomena responsive to change of conditions of a steady state, which are discussed in the following with reference to the underlying mechanism.

1. Galvanostatic Transient

Figure 10 shows an anodic charging curve observed on application of anodic current of constant density on a platinum electrode in 2 N H$_2$SO$_4$ aq. as observed by Breiter et al. (1955). The electrode potential is plotted against the amount of electricity supplied. The first gradual rise of potential is attributed to the consumption of the intermediates of the

HER (hydrogen intermediates) through the reverse of step (2.5a) or (2.16b), the subsequent steep rise to the charging of the double layer on exhaustion of the hydrogen intermediates, the following gradual rise to the accumulation of the product of the oxygen evolution reaction (oxygen deposit), and the further horizontal part to the steady progress of the latter reaction.

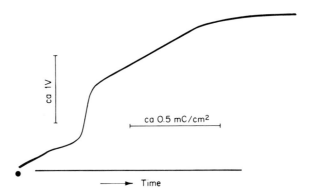

FIG. 10. Galvanostatic transient on a Pt electrode in $2 N$ H_2SO_4 aq. (Breiter *et al.*, 1955).

The transient phenomena thus are explicable by the dual mechanism, but not by the slow discharge mechanism, where the charging of the double layer is the only process before the appearance of the oxygen deposit.

2. Triangular Potential Sweep

Will and Knorr (1960a,b) developed a method of investigating inter-mediates by applying a triangular potential sweep, i.e., a linear potential increase against time followed by a linear potential decrease of the same rate, observing the responsive current density as shown in Fig. 11 for a platinum electrode. The current density above or below the abscissa was that observed in the course of the rise or fall of potential of the sweep respectively.

The humps above the abscissa with peaks around 0.2 V and the other hump are assigned to the discharge of hydrogen intermediates and to the accumulation of the oxygen deposit, respectively. During the fall of the electrode potential, reverse humps appear below the abscissa. The reverse hump at higher potential is appreciably shifted toward lower potential from the one at a higher potential above the abscissa, while the hump and its reverse at the lower potential are almost symmetrically situated with respect to the abscissa. The reverse hump at higher or lower potential is attributed to the reverse of the process assigned to the hump above the abscissa at the respective potentials. The shift of the reverse hump at higher potential is attributed to the time lag due to slow formation and consumption of the oxygen deposit.

Explanation of the phenomena is practicable using the dual mechanism but not using the slow discharge mechanism as in Section II,H,1.

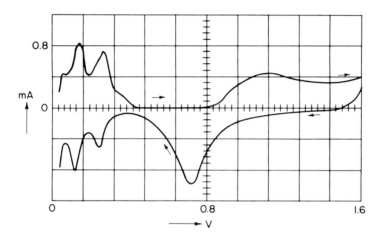

FIG. 11. Current density responsive to the triangular potential sweep on Pt electrode in 8 N H_2SO_4 bubbled with N_2 (Will and Knorr, 1960a).

3. *Capacity Measurements*

The differential capacity of an electrode per unit area, C, is given in terms of the current density, i, and the rate of overvoltage increase, $\dot{\eta}$, as $C = i/\dot{\eta}$. We see from Figs. 10 and 11 that C is extremely large in the stage where hydrogen intermediates accumulate, as compared with the C in the subsequent stage of the charging of the double layer. Breiter *et al.* (1956) observed the capacity of hydrogen electrodes of Pt, Ir, and Rh by an ac method and found the capacity of Pt as high as 1600 $\mu F/cm^2$ in 8 N H_2SO_4 at $\eta \simeq 0.1$ V, qualitatively in accord with the foregoing results.

Past and Jofa (1959) derived C for hydrogen electrodes of Ni, Fe, and Hg from the decreasing rate of η on switching off the steady current of density i as $C = -i/\dot{\eta}$, assuming that electricity flowed into the electrode from the solution at a rate, i, at the moment of switching off the steady current, which reduces η at rate i/C. Results on Ni of Past and Jofa

(1959) as well as those of Matsuda and Ohmori (1962) are shown in
Fig. 12. Both the results are, although not quantitatively coincident,
quite large as compared with the capacity assigned to the charging up of
a double layer in confirmation of the foregoing conclusions.

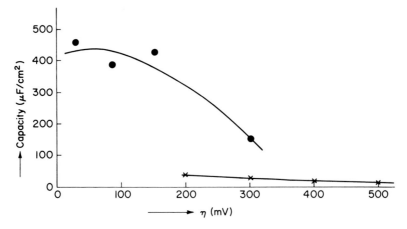

FIG. 12. Capacity of a hydrogen electrode of Ni as a function of η. NaOH aq. of
pH = 13.1 (●, Matsuda and Ohmori, 1962). Concentrated alkaline solution (× , Past and
Jofa, 1959).

I. Dependence of i_0 on Electrode Materials

1. *Periodic Dependence on Atomic Number*

Kita (1966) has shown that i_0, as derived by Eq. (2.12) from data
published over the previous fifteen years, was a periodic function of
atomic number as illustrated in Fig. 13. Its dependence on other factors

TABLE II

Nickel electrode

Solutions	1 N HCl	0.1 N HCl	0.5 N H$_2$SO$_4$	0.15 N HClO$_4$	1 M HClO$_4$
$-\log i_0$	5.4	5.87, 5.24	5.22	5.2	5.7

Mercury electrode

Solvents	H$_2$O	CH$_3$OH	C$_2$H$_5$OH	n-C$_3$H$_7$OH
$-\log i_0$	11.8	11.5	10.8	10.8

is rather insignificant, as exemplified by that of $\log i_0$ on the nickel electrode in different aqueous solutions and by that of $\log i_0$ on the mercury electrode in 0.1 M HCl in different solvents (Minc and Sobkowski, 1959) as shown in Table II.

It was further confirmed that i_0 was affected neither by preliminary treatments of electrodes nor by the state of aggregation essentially as seen, e.g., from the indistinguishable coincidence of its values on liquid and solid gallium electrodes as shown in Fig. 13 (Kita, 1966).

The periodic dependence thus established has the following aspect. The $\log i_0$ in each long period increases first with successive addition of electrons to the outermost d-orbital until $\log i_0$ attains a maximum as the d-orbital is nearly filled, then decreases sharply on completion of the d-orbital, and then increases again with further addition of electrons to the outermost p-orbital, with the latter d-orbital filled. Dowden (1957) called transition metals d-metals and the ensuing ones in the periodic table sp-metals. Ib metals are here conveniently included in d-metals.

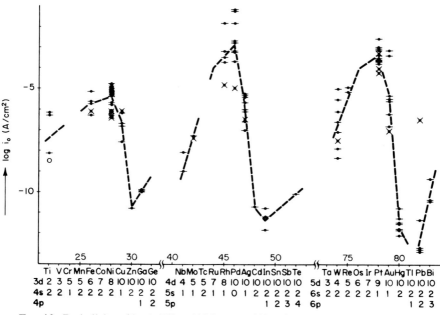

FIG. 13. Periodicity of $\log i_0$ (Kita, 1966). ●: acidic soln; ✕: alkaline soln; ○: neutral soln.

2. Dependence of i_0 on the Heat of Adsorption of Hydrogen

The $\log i_0$ of d-metals decreases approximately linearly with increase of the heat of adsorption of hydrogen as derived by the semiempirical rule of Eley (1950) and Stevenson (1955) from the work function and the heat of sublimation of adsorbents, while that of sp-metals remains constant independent of the heat of adsorption as shown in Fig. 14 (Kita, 1966).*

We see from Fig. 14 with reference to Fig. 8 or the conclusion of Section II,B,7 that d- and sp-metals are respectively relevant to the catalytic and the electrochemical mechanism.

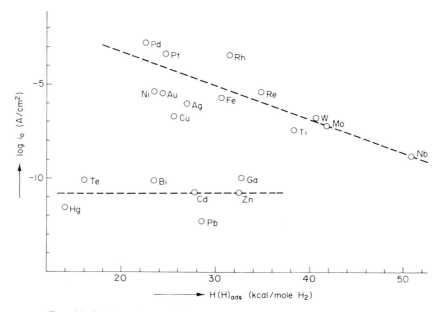

FIG. 14. log i_0 vs. heat of adsorption of hydrogen, $H(\text{H})_{\text{ads}}$ (Kita, 1966).

3. Discussion

The aspects of Fig. 14 are discussed with reference to the dual mechanism. The i_0, by premise, is given by Eq. (1.25a) for

$$\mu(I_r) = 2[\mu(\text{H}^+) + \mu(\text{e}^-)_{\text{rev}}] + \text{const}$$

in common with every electrode in accord with the equation

$$\ln i_0 = \ln \theta(\sigma_r^{\ddagger}, 0) - \epsilon(\ddagger_r)/RT + \text{const}. \tag{2.36}$$

* Rüetschi and Delahay (1955), Conway and Bockris (1957), and Bockris and Srinivasan (1965) showed that log i_0 fell into at least two groups with respect to its dependence on the heat of adsorption.

As to the catalytic mechanism, we have from Eq. (2.22b), $RTu\theta = -\epsilon(H)_0 + \text{const}$ for $\eta = 0$ over the region where θ remains approximately constant in accordance with Section II,D,3; hence substituting $RT\theta$ from the latter equation into Eq. (2.23a)

$$\epsilon(\ddagger_r) = \epsilon(\ddagger_r)_0 - (u^{\ddagger}/u)\epsilon(H)_0 + \text{const.}$$

The $\epsilon(\ddagger_r)_0$ is now expressed as

$$\epsilon(\ddagger_r)_0 = 2\alpha_c\epsilon(H)_0 + \text{const}, \qquad (2.37)$$

assuming that $\epsilon(\ddagger_r)_0$ varies linearly with $\epsilon(H)_0$ in accordance with the rule of Horiuti and Polanyi (1935), exemplified in Section II,D,5, where α_c is a positive proper fraction. We have from the above two equations

$$\epsilon(\ddagger_r) = (2\alpha_c - u^{\ddagger}/u)\epsilon(H)_0 + \text{const};$$

hence according to Eq. (2.36), noting the approximate constancy of $\theta(\sigma_r^{\ddagger}, 0)$ due to that of θ,

$$\ln i_0 = -(2\alpha_c - u^{\ddagger}/u)\epsilon(H)_0/RT + \text{const.} \qquad (2.38a)$$

Figure 12 gives $RT\,\Delta\ln i_0/\Delta H(H)_{\text{ads}} = -0.25$ for d-metals, to which the catalytic mechanism is assigned, where $\Delta H(H)_{\text{ads}}$ is the increment of heat of adsorption, $H(H)_{\text{ads}}$. Since $\Delta H(H)_{\text{ads}} = -2\Delta\,\epsilon(H)_0$, we have $2\alpha_c - u^{\ddagger}/u = -0.49$; hence $\alpha_c = 0.50$ for $u^{\ddagger}/u = 1.5$ as estimated in Section II,D,3.

The i_0 thus increases with increase of $\epsilon(H)_0$. The increase of i_0 with $\epsilon(H)_0$ does not continue, however, until θ is rendered negligible according to Eq. (2.22b) as compared with unity, when the change of $RTu\theta$ no longer outweighs that of $RT\ln[\theta/(1 - \theta)]$. In the latter case $\ln\theta(\sigma_r^{\ddagger}, 0)$ vanishes and $\epsilon(\ddagger_r)$ approximates $\epsilon(\ddagger_r)_0$ by the definition in Section II,D,3, so that we have from Eqs. (2.36) and (2.37)

$$\ln i_0 = -2\alpha_c\epsilon(H)_0/RT + \text{const}, \qquad (2.38b)$$

i.e., i_0 decreases with increase of $\epsilon(H)_0$.

The i_0 of the catalytic mechanism thus follows theoretically a so-called volcanic change, although it simply ascends so far as has been observed.

As to the electrochemical mechanism, we confine ourselves to the case of approximately constant $\theta(H_2^+)$, from which i_0 is practically extrapolated as seen from Section II,D,5. The i_0 is thus given by Eq. (2.26a), noting that $\mu(e^-) = \mu(e^-)_{\text{rev}}$, as follows:

$$\Delta\ln i_0 = (1 - \alpha_E)\,\Delta\epsilon(H_2^+)_0/RT. \qquad (2.39a)$$

The $\epsilon(H_2^+)_0$ is by definition $\epsilon(H_2^+)_0 = \epsilon(H_2^+)_0^0 + FE_p$ and $\epsilon(H_2^+)_0^0$ is approximated as $\epsilon(H_2^+)_0^0 = \epsilon(H)_0 + \text{const}$ on the basis of the resonance scheme (2.15), where the constant consists of the energy required to place H^+ as indicated in the scheme and of the resonance energy between the two alternative states, which depends little on the electrode material (Horiuti et al., 1951). The $\Delta\epsilon(H_2^+)_0$ thus is given as

$$\Delta\epsilon(H_2^+)_0 = \Delta\epsilon(H)_0 + F\Delta E_p. \qquad (2.39b)$$

The depression of $\epsilon(H)_0$ by $-\Delta\epsilon(H)_0$ hence results in the depression of $\epsilon(H_2^+)_0$ by the same amount, so that the right-hand side of Eq. (2.34H$_2^+$) is increased by the same amount, tending to increase $\theta(H_2^+)$ just as in Section II,E,3. On the other hand, the depression of $\epsilon(H)_0$ is accompanied by a decrease of the work function by approximately the amount, $-\Delta\phi = -\Delta\epsilon(H)_0/F$, as derived from the semi-empirical rule of Eley (1950) and Stevenson (1955), which conforms with the quantum-mechanical inferences by Toya (1958, 1960). The depression of the work function is now associated with the elevation of the electrode potential, E_c, by $\Delta E_c = -\Delta\phi$ on account of the chemical potential of the electrons kept at $\eta = 0$ for every electrode. This rise of electrode potential requires, according to Poisson's equation, the decrease of $\theta(H_2^+)$ at constant E_p. Summing up, the decrease of $\epsilon(H)_0$ increases both the right-hand side of Eq. (2.34H$_2^+$) as well as FE_c by $-\Delta\epsilon(H)_0$ at the same E_p, just as the increase of $\mu(H^+)$ increases them respectively by $\Delta\mu(H^+)$; the overall effect on E_p is $F\Delta E_p = -\alpha_h\Delta\epsilon(H)_0$ similar to the results $F\Delta E_p = \alpha_h\Delta\mu(H^+)$ in Section II,E,3 with identical coefficient, α_h. We have by Eq. (2.39b), $\Delta\epsilon(H_2^+)_0 = (1 - \alpha_h)\Delta\epsilon(H)_0$; hence according to Eq. (2.39a)

$$\Delta \ln i_0 = (1 - \alpha_E)(1 - \alpha_h)\Delta\epsilon(H)_0.$$

The above equation states that the dependence of i_0 on $\epsilon(H)_0$ varies with α_h and that i_0 would perceptibly increase with $\epsilon(H)_0$ as in the case of d-metals by increasing the ionic strength of the solution and thus reducing α_h according to II,E,3.

J. CONCLUDING REMARKS

The dual mechanism has been advanced from the survey of the hydrogen evolution reaction and established through the observations of i_+ as a function of η and pH, the electrolytic separation factor, the stoichiometric number of the rate-determining step, the transient phenomena, and the dependence of i_0 on $\epsilon(H)_0$ respectively as regards d- and sp-metals.

The i_0 of the catalytic mechanism increases with increase of $\epsilon(H)_0$ so far as observed, whereas the theory predicts that the increase is not interminable but is followed by a decrease *via* a maximum. This situation would be of practical importance in reducing the electric power of electrolysis.

III. Halogen Evolution Reaction

A. INTRODUCTION

The halogen evolution reaction (XER) is the evolution of a halogen from an anode in the solution of its ions, accompanying an outflow of electricity from the anode into the solution. The reaction is formulated as

$$2X^- = X_2 + 2e^-, \tag{3.1}$$

accounting for the outflow of electricity as the inflow of conduction electrons. The halide solution may be an aqueous or a molten halide.

The overvoltage, η, of the reaction is defined in accordance with Eq. (1.26) as

$$2\mu(X^-) - [\mu(X_2) + 2\mu(e^-)] = 2F\eta, \tag{3.2a}$$

where $\mu(X^-)$, etc., are chemical potentials of X^-, etc., respectively.

We have particularly in the equilibrium state of reaction (3.1)

$$2\mu(X^-) - \mu(X_2) - 2\mu(e^-)_{rev} = 0, \tag{3.2b}$$

where $\mu(e^-)_{rev}$ has a meaning analogous to that in Eq. (2.2) but appropriate to the XER. The halogen electrode under the latter condition is a *reversible halogen electrode* and the latter particularly *normal halogen electrode* in the case where the halogen and its ions in the solution are respectively of unit activity; the potential of the normal halogen electrode as referred to the normal hydrogen electrode is the *normal halogen potential*. We have from Eqs. (3.2)

$$\mu(e^-)_{rev} - \mu(e^-) = F\eta, \tag{3.3}$$

which states that η is the potential of the electrode in question as referred to the reversible halogen electrode in the same environment.

The normal halogen potentials in aqueous solution compare with the normal oxygen potential at 25°C as follows:

$\frac{1}{2} F_2/F^-$	$\frac{1}{2} Cl_2/Cl^-$	$\frac{1}{2} Br_2/Br^-$	$\frac{1}{2} I_2/I^-$	$\frac{1}{4} O_2/OH^-$
2.85	1.358	1.085	0.536	0.400 V,

where halogens and oxygen are gaseous except for iodine which is in the solid state. It is hence thermodynamically feasible that the XER in aqueous solution should be accompanied by the oxygen evolution reaction (OER), although kinetically a concurrent evolution of oxygen is hardly perceptible as was to be expected from the extremely small rate of OER separately observed. The attack of free halogen on water and the formation of ions of the type IO_3^- have to be taken into account in the determination of the normal halogen potential.

The mechanisms of XER are investigated first by formulating the characteristics of the reaction.

B. Characteristics of Halogen Evolution Reaction

The $\mu(e^-)$ in Eq. (3.2a) is connected with the potential of the electrode, E, and its work function, ϕ, as follows:

$$-F(\phi + E) = \mu(e^-). \tag{3.4}$$

Developing $\mu(X^-)$ and $\mu(X_2)$ in terms of the activity of X^- or X_2, $[X^-]$ or $[X_2]$, as

$$\mu(X^-) = \mu(X^-)_1 + RT \ln[X^-], \tag{3.5X$^-$}$$

$$\mu(X_2) = \mu(X_2)_1 + RT \ln[X_2], \tag{3.5X$_2$}$$

and eliminating $\mu(X^-)$, $\mu(X_2)$, $\mu(e^-)$, and $\mu(e^-)_{\text{rev}}$ from Eqs. (3.2b), (3.3)–(3.5), we have

$$\eta = E + (RT/F) \ln[X^-] - (RT/2F) \ln[X_2] + C_I \tag{3.6a}$$

where

$$C_I = \mu(X^-)_1/F - \mu(X_2)_1/2F + \phi \tag{3.6b}$$

is a constant for a definite halogen electrode at constant temperature and $\mu(X^-)_1$ or $\mu(X_2)_1$ is the particular value of $\mu(X^-)$ or $\mu(X_2)$ at unit activity of X^- or X_2, respectively.

We thus define the state of the halogen electrode at constant temperature with four variables of states, η, E, $[X^-]$, and $[X_2]$; hence the state of a definite halogen electrode is defined by any three of them on account of the condition imposed by Eq. (3.6a). The forward or backward unidirectional rate in terms of current density, i_+ or i_-, is thus given for a definite halogen electrode of a specified $[X_2]$ as a function of X^- and E, so that

$$d \ln i_+ = \left(\frac{\partial \ln i_+}{\partial \ln[X^-]}\right)_{E,[X_2]} d \ln[X^-] + \left(\frac{\partial \ln i_+}{\partial E}\right)_{[X^-],[X_2]} dE;$$

hence

$$\left(\frac{\partial \ln i_+}{\partial \ln[X^-]}\right)_{[X_2],\eta} = \left(\frac{\partial \ln i_+}{\partial \ln[X^-]}\right)_{E,[X_2]} + \left(\frac{\partial \ln i_+}{\partial E}\right)_{[X^-],[X_2]}\left(\frac{\partial E}{\partial \ln[X^-]}\right)_{[X_2],\eta}$$

Noting that $(\partial E/\partial \ln[X^-])_{[X_2],\eta} = -RT/F$ by Eq. (3.6a) and specifying the constant η as zero, where $i_+ = i_0$, we have from the above equation (Vetter, 1952, 1961a,b,c)

$$\gamma_+(X^-) \equiv \left(\frac{\partial \ln i_+}{\partial \ln[X^-]}\right)_{E,[X_2]} = \left(\frac{\partial \ln i_0}{\partial \ln[X^-]}\right)_{[X_2]} + \tau_+. \qquad (3.7a)$$

We have similarly

$$\gamma_+(X_2) \equiv \left(\frac{\partial \ln i_+}{\partial \ln[X_2]}\right)_{E,[X^-]} = \left(\frac{\partial \ln i_0}{\partial \ln[X_2]}\right)_{[X^-]} - \tau_+/2, \qquad (3.7b)$$

$$\gamma_-(X^-) \equiv \left(\frac{\partial \ln i_-}{\partial \ln[X^-]}\right)_{E,[X_2]} = \left(\frac{\partial \ln i_0}{\partial \ln[X^-]}\right)_{[X_2]} - \tau_-, \qquad (3.7c)$$

and

$$\gamma_-(X_2) \equiv \left(\frac{\partial \ln i_-}{\partial \ln[X_2]}\right)_{E,[X^-]} = \left(\frac{\partial \ln i_0}{\partial \ln[X_2]}\right)_{[X^-]} + \tau_-/2, \qquad (3.7d)$$

where

$$\tau_+ = (RT/F)(\partial \ln i_+/\partial E)_{[X^-],[X_2]}$$

and

$$\tau_- = -(RT/F)(\partial \ln i_-/\partial E)_{[X^-],[X_2]}$$

are special cases of those considered in Eqs. (1.28). The

$$(\partial \ln i_0/\partial \ln[X^-])_{[X_2]} \quad \text{and} \quad (\partial \ln i_0/\partial \ln[X_2])_{[X^-]}$$

are transformed, with reference to Eq. (3.6a), as

$$(\partial \ln i_0/\partial \ln[X^-])_{[X_2]} = (\partial \ln i_0/\partial E)_{[X_2]}(\partial E/\partial \ln[X^-])_{[X_2],\eta=0}$$
$$= -(RT/F)(\partial \ln i_0/\partial E)_{[X_2]} \qquad (3.8a)$$

and

$$(\partial \ln i_0/\partial \ln[X_2])_{[X^-]} = (\partial \ln i_0/\partial E)_{[X^-]}(\partial E/\partial \ln[X_2])_{[X^-],\eta=0}$$
$$= (RT/2F)(\partial \ln i_0/\partial E)_{[X^-]}, \qquad (3.8b)$$

and are shown to be connected with

$$(\partial \ln i_0/\partial \ln[X_2])_E \quad \text{or} \quad (\partial \ln i_0/\partial \ln[X^-])_E$$

according to Eq. (3.6a) as

$$(\partial \ln i_0/\partial \ln[X_2])_E = \tfrac{1}{2}(\partial \ln i_0/\partial \ln[X^-])_{[X_2]} + (\partial \ln i_0/\partial \ln[X_2])_{[X^-]} \quad (3.9a)$$

or

$$(\partial \ln i_0/\partial \ln[X^-])_E = (\partial \ln i_0/\partial \ln[X^-])_{[X_2]} + 2(\partial \ln i_0/\partial \ln[X_2])_{[X^-]}. \quad (3.9b)$$

In the case of the iodine evolution reaction, the equilibrium of reaction $X^- + X_2 = X_3^-$ is established (Vetter, 1952), so that

$$\mu(X^-) + \mu(X_2) = \mu(X_3^-);$$

developing chemical potentials in terms of relevant activities, $[X^-]$, $[X_2]$, and $[X_3^-]$, we have

$$RT \ln([X^-][X_2]/[X_3^-]) = \mu(X_3^-)_1 - \mu(X^-)_1 - \mu(X_2)_1, \quad (3.10)$$

where $\mu(X_3^-)_1$ is the particular value of $\mu(X_3^-)$ at $[X_3^-] = 1$, and eliminating $[X_2]$ from Eqs. (3.6a) and (3.10),

$$\eta = E + (3RT/2F) \ln[X^-] - (RT/2F) \ln[X_3^-] + C_{\mathrm{II}}, \quad (3.11a)$$

where

$$C_{\mathrm{II}} = 3\mu(X^-)_1/2 - \mu(X_3^-)_1/2 + \phi. \quad (3.11b)$$

The $\gamma_+(X^-)$, etc., are derived as below for the set of independent variables, E, η, $[X^-]$, and $[X_3^-]$ in place of E, η, $[X^-]$, and $[X_2]$:

$$\gamma_+(X^-) \equiv (\partial \ln i_+/\partial \ln[X^-])_{E,[X_3^-]} = (\partial \ln i_0/\partial \ln[X^-])_{[X_3^-]} + \tfrac{3}{2}\tau_+, \quad (3.12a)$$

$$\gamma_-(X^-) \equiv (\partial \ln i_-/\partial \ln[X^-])_{E,[X_3^-]} = (\partial \ln i_0/\partial \ln[X^-])_{[X_3^-]} - \tfrac{1}{2}\tau_+, \quad (3.12b)$$

$$\gamma_+(X_3^-) \equiv (\partial \ln i_+/\partial \ln[X_3^-])_{E,[X^-]} = (\partial \ln i_0/\partial \ln[X_3^-])_{[X^-]} - \tfrac{1}{2}\tau_+, \quad (3.12c)$$

and

$$\gamma_-(X_3^-) \equiv (\partial \ln i_-/\partial \ln[X_3^-])_{E,[X^-]} = (\partial \ln i_0/\partial \ln[X_3^-])_{[X^-]} + \tfrac{1}{2}\tau_-. \quad (3.12d)$$

Equations similar to Eqs. (3.8) are obtained as

$$\left(\frac{\partial \ln i_0}{\partial \ln[X^-]}\right)_{[X_3^-]} = -\frac{3RT}{2F}\left(\frac{\partial \ln i_0}{\partial E}\right)_{[X_3^-]}, \quad (3.13a)$$

$$\left(\frac{\partial \ln i_0}{\partial \ln[X_3^-]}\right)_{[X^-]} = \frac{RT}{2F}\left(\frac{\partial \ln i_0}{\partial E}\right)_{[X^-]} \quad (3.13b)$$

and those similar to Eqs. (3.9) as

$$(\partial \ln i_0/\partial \ln[\text{X}^-])_E = (\partial \ln i_0/\partial \ln[\text{X}^-])_{[\text{X}_3{}^-]} + 3(\partial \ln i_0/\partial \ln[\text{X}_3{}^-])_{[\text{X}^-]}$$

(3.14a)

or

$$(\partial \ln i_0/\partial \ln[\text{X}_3{}^-])_E = \tfrac{1}{3}(\partial \ln i_0/\partial \ln[\text{X}^-])_{[\text{X}_3{}^-]} + (\partial \ln i_0/\partial \ln[\text{X}_3{}^-])_{[\text{X}^-]}.$$

(3.14b)

C. Chlorine Evolution Reaction

Chang and Wick (1935) observed the $i(\eta)$, i.e., the current density, $i = i_+ - i_-$, of the chlorine, bromine, and iodine evolution reactions as a function of η on Pt and Ir electrodes by rotating the electrode with a view to excluding the diffusion controling the rate. Figure 15 shows the

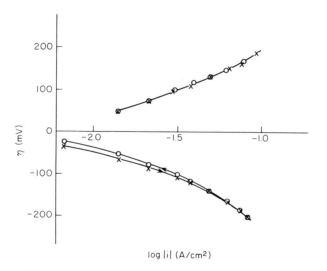

FIG. 15. η, $\log|i|$-curve of the ClER on a Pt electrode in 1 N HCl at 20°C (Chang and Wick, 1935).

observation of the chlorine evolution reaction on the Pt electrode, from which we have $\tau_+ \simeq 0.5$ and $\tau_- \simeq 0.5$; hence we have $\nu_r = 2$ according to Eq. (1.28ν_r), taking z to be 2.

The ν_r is alternatively derived at $\eta = 0$ by Eq. (1.27b) in terms of $(\partial i/\partial \eta)_{\eta=0}$ evaluated directly from experimental data as shown in Fig. 16 and i_0 approximated with i_+ extrapolated at $\eta = 0$ by the linear relation of $\log i_+$ to η; ν_r values derived by the latter method from the observation of Chang and Wick (1935) and from that of Frumkin and Tedoradse (1958) confirmed the ν_r value obtained above by Eq. $(1.28\nu_r)$. The conclusion, $\nu_r = 2$, thus established states that two identical steps have to occur for completion of an act of reaction (3.1) determining its rate.

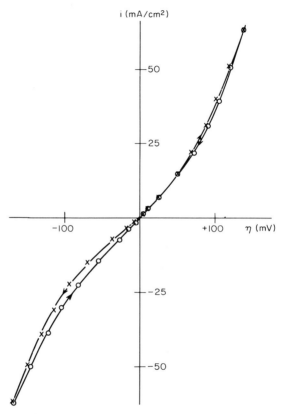

FIG. 16. η, i-curve of the ClER on a Pt electrode in 1 N HCl at 20°C (Chang and Wick, 1935).

The identical rate-determining step may be represented by $X^- + nX_a \rightarrow (n + 1)X_a + e^-$, where X_a is a constituent atom of the X_2 molecule or any species in partial equilibrium with the latter and n is the number of X_a comprised in the initial system. A rather real case of the step would be that for $n = 0$ or 1 and X_a is the adsorbed halogen atom $X(a)$; the relevant mechanism is respectively

$$X^- \xrightarrow{A} X(a) + e^-, \qquad 2X(a) \longrightarrow X_2 \qquad (3.15)$$

or

$$X_2 \longrightarrow 2X(a), \qquad X^- + X(a) \xrightarrow{A} X_2 + e^-, \tag{3.16}$$

where $\xrightarrow{A}$ signifies the rate-determining step.

The mechanism is now investigated assuming the mass action law or that i_+ or i_- is proportional to the activity of the initial or final system respectively of the rate-determining step.* The current i_+ is thus proportional to $[X^-]$, hence $\gamma_+(X^-) = 1$ according to Eq. (3.7a). The current i_- is, on the other hand, proportional to $[X_2]^{1/2}$ or $[X_2]$ according as mechanism (3.15) or (3.16) is operative, i.e., $\gamma_-(X_2) = \frac{1}{2}$ or 1, respectively, in accord with Eq. (3.7d).

The $\gamma_+(Cl^-)$ is determined by Eq. (3.7a) on the base of τ_+ as given above and $(\partial \ln i_0/\partial \ln[Cl^-])_{[Cl_2]}$ is derived from the experimental results

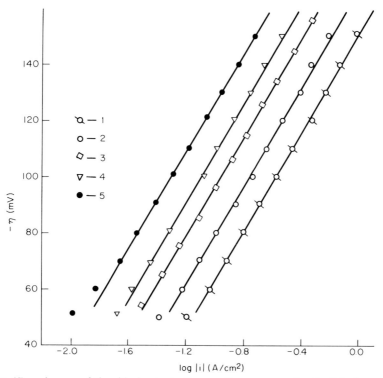

FIG. 17. η, i-curve of the chlorine ionization reaction on a rotating Pt disk electrode in 2.2 N HClO$_4$ + xN HCl (Frumkin and Tedoradse, 1958). (1) $x = 0.063$, (2) $x = 0.037$, (3) $x = 0.021$, (4) $x = 0.012$, and (5) $x = 0.0043$.

* Deviation from the mass action law occurs in these cases where the change of activity of the initial or final system accompanies the change of activity of the critical system as seen in Section II,D.

as shown in Fig. 17, approximating i_0 with i_+ extrapolated at $\eta = 0$ in accordance with the linear relation of $\log i_+$ to η. Toshima and Okaniwa (1966) obtained $\gamma_+(Cl^-) = 1$ from observations on the Pt electrode by the potentiostatic transient method, while Enyo and Yokoyama (to be published) derived $\gamma_+(Cl^-) = 1.1$ from the observations of Frumkin and Tedoradse (1958) by the rotating disc method in accordance with the above inference from the alternative mechanism. Frumkin and Tedoradse (1958) concluded, on the other hand, from their observation that $\gamma_-(Cl_2) = 1$, whereas Enyo and Yokoyama (to be published) derived $\gamma_-(Cl_2) = 0.6$–0.7 by elaborating the analysis. The latter group of authors obtained $\gamma_-(Cl_2) = 0.6$ from their own observation of $i(\eta)$, which reproduced the work of Chang and Wick referred to above.

The experimental results thus fit in with mechanism (3.15) rather than with (3.16), indicating that the former is operative in the chlorine evolution reaction on the Pt electrode.

D. Iodine and Bromine Evolution Reactions

These reactions proceed more rapidly than the chlorine evolution reaction, rendering it rather difficult to observe $i(\eta)$ without control of diffusion at the steady state, although Chang and Wick (1935) were successful in the observation of the bromine evolution reaction on an Ir electrode.

Vetter (1952) conducted an extensive investigation of the iodine evolution reaction on the Pt electrode, allowing for the equilibrium of reaction $I^- + I_2 = I_3^-$ referred to in Section III,B. He determined by the ac impedance method that $\tau_+ = 0.78$, $\tau_- = 0.21$,

$$(RT/F)(\partial \ln i_0/\partial E)_{[I^-]} = 0.78, \quad \text{and} \quad (RT/F)(\partial \ln i_0/\partial E)_{[I_3^-]} = 0.13.$$

The first two of these data lead, according to Eq. (1.28ν_r) with $z = 2$, to $\nu_r = 2$ or to the alternative mechanisms (3.15) and (3.16) as in the case of the chlorine evolution reaction. It is inferred, on the other hand, from these data using Eqs. (3.13) and (3.12) that $\gamma_+(I^-) = 1$, $\gamma_+(I_3^-) = 0$, $\gamma_-(I^-) = -0.5$, and $\gamma_-(I_3^-) = 0.5$. The first two of these data confirm the conclusion from the alternative mechanism as based on the mass action law, while it follows from the latter two that i_- is proportional to $([I_3^-]/[I^-])^{1/2}$, and hence, according to Eq. (3.10), to $[I_2]^{1/2}$; this indicates that mechanism (3.15) is operative in the evolution of iodine on the Pt electrode.

The bromine evolution reaction has been investigated on the Pt electrode by Llopis and Vazquez (1962) along the lines of Vetter (1952), with the result that $\tau_+ = 0.46$, $\tau_- = 0.54$, $(\partial \ln i_0/\partial \ln[Br^-])_{[Br_2]} = 0.56$, and

$(\partial \ln i_0 / \partial \ln[Br_2])_{[Br^-]} = 0.76$; hence according to Eqs. (3.7), $\gamma_+(Br^-) = 1.02$, $\gamma_+(Br_2) = 0.49$, $\gamma_-(Br^-) = 0.02$, and $\gamma_-(Br_2) = 0.98$.

The values of τ_+ and τ_- lead to $\nu_r = 2$ by Eq. (1.28ν_r) with $z = 2$; hence the alternative mechanisms (3.15) and (3.16) are indicated. The values of $\gamma_+(Br^-)$ and $\gamma_-(Br^-)$ fit in with either of the alternative mechanisms, whereas those of $\gamma_+(Br_2)$ and $\gamma_-(Br_2)$ accord exclusively with mechanism (3.16), contrasting the chlorine evolution reaction, indicating that the latter mechanism is operative in the bromine evolution reaction on the Pt electrode.

IV. Oxygen Evolution Reaction

A. INTRODUCTION

The oxygen evolution reaction (OER) occurs on an anode in the electrolysis of aqueous acid, base, or such salts as sulfate, which do not attack the anode appreciably on electrolysis. The OER has not been elucidated to the same degree as the HER or even as the XER, mainly because of the difficulty in reproducing the basic phenomena including the state of the electrode. A stable reversible oxygen electrode is equally difficult to obtain, perhaps because of the exceedingly minute rate of the OER as compared with the rates of concurrent electrochemical reactions, as indicated earlier in Section III,A.

The situation appears to have been improved recently by the introduction of new experimental techniques, e.g., the isotopic tracer technique and the preelectrolysis of the solution to cut down the concurrent electrochemical reactions by removing impurities.

In this section, the present status of the investigations in this field are described.

B. REVERSIBLE OXYGEN ELECTRODE

The OER is formulated as

$$4\,OH^- = O_2 + 2\,H_2O + 4e^-; \tag{4.1}$$

hence its affinity, $-\Delta G$, is

$$-\Delta G \equiv 4\mu(OH^-) - \mu(O_2) - 2\mu(H_2O) - 4\mu(e^-), \tag{4.2a}$$

where $\mu(OH^-)$, etc., are chemical potentials of OH^-, etc., respectively.

In equilibrium of reaction (4.1), the electrode is *the reversible oxygen electrode*, where $-\varDelta G = 0$ and $\mu(e^-) = \mu(e^-)_{rev}$, so that

$$4\mu(OH^-) - \mu(O_2) - 2\mu(H_2O) = 4\mu(e^-)_{rev}. \qquad (4.2b)$$

We have from Eqs. (1.26) and (4.2)

$$-\varDelta G = 4\{\mu(e^-)_{rev} - \mu(e^-)\} = 4F\eta; \qquad (4.2c)$$

the overvoltage η is thus the potential of the oxygen electrode in question as referred to the reversible oxygen electrode in the same environment.

We now investigate the potential of the reversible oxygen electrode in 1 atm O_2 as referred to the reversible hydrogen electrode in the same solution but in 1 atm H_2. Equations (2.2) and (4.2b) give, if we specify the corresponding $\mu(e^-)_{rev}$ by $\mu(e^-, H_2)_{rev}$ and $\mu(e^-, O_2)_{rev}$

$$\mu(e^-, O_2)_{rev} - \mu(e^-, H_2)_{rev}$$
$$= \mu(OH^-) + \mu(H^+) - \tfrac{1}{2}\mu(H_2) - \tfrac{1}{2}\mu(H_2O) - \tfrac{1}{4}\mu(O_2).$$

The left-hand side of the above equation is the potential of the reversible oxygen electrode as referred to the reversible hydrogen electrode, E_0, multiplied by the charge carried by 1 mole of electrons, $-F$, i.e., $-FE_0$. Assuming the electrolytic dissociation equilibrium of water in the common solution, we have $\mu(OH^-) + \mu(H^+) = \mu(H_2O)$; hence from the above equation

$$4FE_0 = \mu(O_2) + 2\mu(H_2) - 2\mu(H_2O). \qquad (4.3)$$

The right-hand side of Eq. (4.3) is the affinity of liquid water formation from its gaseous elements, which is thermodynamically evaluated with accuracy. The E_0 is hence determined for O_2 and H_2, respectively, at 1 atm to be 1.23 V at 25°C, which provides the criterion to test whether the oxygen electrode is reversible or not.* What is called below the rest potential of the oxygen electrode is the potential of the oxygen electrode of 1 atm O_2 referred to the reversible hydrogen electrode in the same solution but in 1 atm H_2.

Winkelmann (1956) found the rest potential of the oxygen electrode at 1.05 V. Bockris and Huq (1956) applied the method of pre-electrolysis to the sulfuric acid solution, with which they found the rest potential of the Pt electrode to be 1.24 ± 0.02 V; the rest potential remained steady for 1 hr and then fell but was recovered on resuming the pre-electrolysis.

* The E_0 is 0.400 V as given in Section III,A, if the solution common to the electrodes is replaced by a solution containing OH^- and O_2 respectively of unit activity around the oxygen electrode and the other solutions of H^+ and H_2 respectively at unit activity around the hydrogen electrode and the two electrodes are connected without a liquid junction potential.

The associated value of $\partial E_r/\partial \log P(O_2)$ was found to be 11 mV, which approximated the theoretical value, 14.8 mV, as follows from Eq. (4.3) by developing $\mu(O_2)$ as $\mu(O_2) = \mu(O_2)_1 + RT \ln P(O_2)$, where $\mu(O_2)_1$ is the value of $\mu(O_2)$ at $P(O_2) = 1$.

Hoare (1963) likewise resorted to pre-electrolysis and to a preliminary soak of the Pt electrode in nitric acid for more than 48 h; they thus found the rest potential to be 1.228 ± 0.010 V, which remained steady for 25 h and $\partial E_r/\partial \log P(O_2)$ gave 15 mV in agreement with the theoretical value.

Watanabe and Devanathan (1964) anodically polarized the test electrode in 1 N H_2SO_4 with a current density of 10^{-4}–10^{-2} A/cm^2 and drew it out of the solution, leaving it in moist O_2 gas over the solution for 0.5–3 h, while the solution was constantly pre-electrolyzed. They thus found that the electrode had the rest potential and $\partial E_r/\partial \log P(O_2)$, respectively, for the reversible oxygen electrode.

C. OXYGEN DEPOSIT

There is a certain range on the galvanostatic charging curve, where the electrode potential rises distinctly slowly as compared with the preceding part and this is followed by the horizontal part relevant to the steady OER or on the triangular potential sweep curve, where the responsive current markedly rises. The latter range is attributed to the accumulation of the oxygen deposit as referred to in Sections II,H,1 and 2, which may be chemisorbed oxygen atoms O(a), chemisorbed hydroxyl groups, or a thin film of oxide or hydroxide. The amount of the oxygen deposit is represented by the quantity of electricity required to complete the range.

The following facts are known about the oxygen deposit on the Pt electrode.

(i) The electrode surface is covered by at least a monolayer of oxygen when polarized by more than 1.2 V in a solution containing O_2, provided that the oxygen deposit consists of O(a), each O(a) is deposited by two elementary charges transferred from the solution to the electrode and a monolayer is completed by 10^{15} O(a) cm^{-2}.

(ii) The amount of charge required for the accumulation and consumption of the oxygen deposit, Q_A and Q_C, are independent of pH and their ratio, Q_A/Q_C, lies between 1 and 2. Q_A/Q_C was found to be 2 on Pt and Pd electrodes. This was interpreted by Vetter and Berndt (1958) as showing that O(a) is deposited by two elementary charges as mentioned above and is released on reduction by one elementary charge transferred in forming $\frac{1}{2}$ H_2O_2. Bagotzky et al. (1965) found, however, that no H_2O_2 is produced by the reduction of the oxygen deposit in the absence of

O_2 in the solution by means of the rotating disk electrode of Pt and Ag with a coaxial, coplaner ring fixed around it. The alternative explanation would be that the oxygen deposit remains partly unconsumed on reduction or that the electricity supplied in its accumulation is spent partly on some electrochemical reaction, which does not lead to an oxygen deposit.

(iii) Rosental and Veselovsky (1956) loaded a Pt electrode with ^{18}O by electrolysis to more than or less than a monolayer as estimated by the procedure described in (i). They found on electrolysis of ordinary water by the Pt electrode thus processed that the evolution of ^{18}O containing oxygen was evolved only in the former case.

(iv) The double-layer capacity of the Pt electrode loaded with the oxygen deposit leads to widely scattered results with different authors (Laitinen and Enke, 1960; Breiter, 1962; Schuldiner and ·Roe, 1963; Gilman, 1964).

D. KINETICS

1. OER

A number of authors reported the linear relation of η to $\log i$, i.e., $\eta = a + b \log i$, with constants a and b in the steady OER, where b approximated 0.12 as in the case of the HER. These constants, especially a, are, however, rather more poorly reproducible than in the case of the HER or even of the XER.

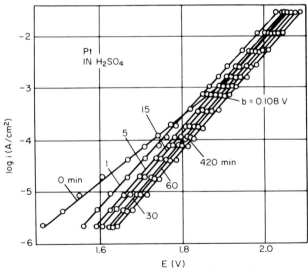

FIG. 18. $\log i$, E-relation for various periods of time of the preliminary anodic polarization with a current density of 2.13 mA/cm² (Efimov and Isgaryshev, 1956).

The E or η of the OER varies, on applying a constant current, initially rapidly with time and then more and more slowly but hardly ever attains a steady value. Efimov and Isgaryshev (1956) observed the $i(\eta)$ on a Pt electrode which had been preliminarily anodically polarized for different periods of time; the result, as shown in Fig. 18, is that b in the above equation attained a constant value for the time of polarization more than 15 min, whereas a steadily increased with the time.

The following additional facts are known about the OER.

(i) Vetter and Berndt (1958) found practically no dependence of η on pH from 0 to 12 at a constant current density on the Pt, Pd, and Au electrodes.

(ii) No H_2O_2 is produced during the steady OER (Vetter, 1961a).

(iii) $RT^2\, \partial \ln i/\partial T$ of the OER on the Pt electrode at constant η fluctuates between 12 and 23 kcal (Kodera, 1958).

2. Cathodic Reduction of Oxygen

A number of workers have reported the linear relation of the logarithm of the rate of the cathodic reduction to the electrode potential (Vetter, 1961b; Bagotzky et al., 1965; Hoare, 1966). This rate was found proportional to the function of electrode potential E, $P(O_2) \exp(-FE/2RT)$, on the Hg electrode (Bagotzky and Jablokowa, 1953), and on the Ag (Krasilischichikov, 1952) and on the Pt electrode (Winkelmann, 1956) in both acid and neutral solutions. The observed kinetics is accounted for by assuming the rate-determining step is $O_2 + e^- \rightarrow O_2^-$ (a).

The cathodic reduction has, however, a number of complicating aspects. The dependence of the rate on the electrode potential and pH varies with the pretreatment of the electrode (Bianti and Mussini, 1965; Sawyer and Day, 1963). The number of electrons transferred to reduce one O_2 molecule varies with the electrode potential (Delahay, 1950). Hydrogen peroxide is confirmed as the intermediate of the cathodic reduction of oxygen on the Hg, Pt, Au, Ag, Pd, and Rh electrodes (Vetter, 1961b; Bagotzky et al., 1965). Bagotzky et al. (1965) showed besides by means of the rotating disk electrode mentioned in Section IV,C(ii) that the electrode which is preliminarily oxidized, retarded H_2O_2 formation but accelerated its reduction on the Pt and Pd electrodes, whereas it retarded both the formation and reduction of H_2O_2 on the Rh and Ag electrodes.

Davies et al. (1959) established by means of ^{18}O that the O–O bond was not broken in the reduction of O_2 to H_2O_2 on carbon electrode.

3. ν_r

The ν_r of the OER has first been evaluated by Bockris and Huq (1956) by Eq. (1.27b) taking z to be 4; they approximated i_0 in Eq. (1.27b) with i_+ extrapolated at $\eta = 0$ from the region of η, where $i = i_+ - i_-$ equaled practically i_+; assuming the linear relationship of log i_+ to η, they determined $(\partial i/\partial \eta)_{\eta=0}$ in the same equation directly by the observation. It may be noted that the linear relationship of log i_+ to η amounts to the constancy of $\tau_+ \equiv (RT/F) \partial \ln i_+/\partial \eta$ independent of η. They thus determined ν_r to be 3.8 with a standard deviation of 1.1.

Bockris and Huq (1956) derived ν_r alternatively using Eq. (1.28ν_r) with $z = 4$, on the basis of τ_+ and τ_- derived respectively from observations of the OER and the cathodic reduction of oxygen, assuming that the current density observed in the respective regions consisted practically entirely of i_+ or i_-, respectively. This alternative method rests upon the premise, as mentioned in Section III,C, that τ_+ and τ_- are relevant to one and the same steady state, which amounts to the assumption, if admitted with τ_+ and τ_- determined at any two arbitrary steady states, that they are respectively constant throughout. They thus obtained $\nu_r = 3.5$ with a standard deviation of 0.3.

Hoare (1966) observed i_+ and i_- on the Rh electrode under 1 atm O_2 in 2 N H_2SO_4 at 25°C, which gives $\tau_+ = 0.549$ or 0.329 at the lower or higher anodic polarization and $\tau_- = 0.486$ or 0.156 at the lower or higher cathodic polarization. By combining τ_+ and τ_- at the lower anodic and cathodic polarizations according to Eq. (1.28ν_r), we have $\nu_r = 3.9$, or combining the results at the higher polarization, we have similarly $\nu_r = 8.2$. A number of workers estimated ν_r assuming the constancy of τ_+ and τ_-, which appears insecure particularly in the case of OER.

The impeccable determination of ν_r is practicable according to Eq. (1.27a), inasmuch as the unidirectional rate, i_+ or i_- is measurable by labeling it with isotopes; practically this can be accomplished without the complication met in hydrogen isotopes because of the mass ratio approximating unity. This experiment is in progress in this laboratory.

E. CONCLUDING REMARK

The OER appears to be extremely complicated and to need much further effort to elucidate the mechanism. The steady cathodic reduction of O_2 may not simply be that which would be derived from the steady OER just by reversing every constituent step of the latter, as suggested by the result that H_2O_2 is developed or not in the respective case as referred to in Section IV,D,2 or in Section IV,C(ii), respectively. It is in consequence uncertain that τ_+ and τ_- should keep respectively constant over a range of η as premised in the v_r determination.

It is desirable to practice the impeccable determination of v_r described in Section IV,D,3, which would establish a solid basis of the mechanism.

V. Oxidation of Simple Organics with Evolution of CO_2

A. Introduction

A number of works have recently been contributed to the investigation of anodic oxidation of simple organics with the evolution of CO_2 in conjunction with the practice of fuel cells. The experimental method of investigation consists generally in the observation of potential sweep transients and also of galvanostatic transients sometimes resorted to, ascribing the difference of the transients from its "blank test" referred to in Sections II,H and IV,C to the anodic oxidation. The hump due to the hydrogen intermediate mentioned in Section II,H is also improved for estimating the number of sites of adsorption on the electrode surface which is identified with the number of elementary charges composing the hump.

The potential, E, of the anode is, in what follows, the potential measured against a reversible hydrogen electrode in the same solution saturated with 1 atm H_2.

The potential sweep transients of hydrogen containing organics, e.g., CH_3OH, $HCHO$, HCO_2H, reveal three peaks at $E = 0.5\text{-}0.6$, $0.8\text{-}0.9$, and $1.3\text{-}1.5$ V as seen in Fig. 19, which potentials depend appreciably on the sweeping rate and temperature but not on the variety and concentration of the organics (Bagotzky and Vasilyev, 1964). The first peak at $0.5\text{-}0.6$ V is peculiar to the hydrogen containing organics.

The appearance of the peaks is accounted for on more or less sound grounds as follows. The first peak is attributed to the release of hydrogen atoms from the organics on its adsorption, leaving the rest adsorbed on the electrode surface, which is called below the first oxidation product. The fall of the current at E approximately equal to 0.6 V toward the end of the first peak is assigned to the blockage of the electrode surface by the first oxidation product ("self-poisoning"). The rise of the second peak at E approximately equal to 0.8 V is ascribed to the oxidation of the first oxidation product into the second oxidation product by the oxygen deposit supposed to be $O(a)$ or $OH(a)$, which begins to appear at this potential as seen in Fig. 11. The fall of the current toward the end of the second peak is caused by the excess of oxygen deposit as represented by Gilman (1963) as due to the shortage of "reaction pairs" necessary for the second oxidation, each consisting of a species of the first oxidation product and that of the oxygen deposit regarded as $O(a)$. Bagotzky and Vasilyev (1964) and Gilroy and Conway (1965) derived the shape of the second peak theoretically, assuming the blockage of the electrode surface by $O(a)$ in terms of the Frumkin–Temkin logarithmic adsorption isotherm (Frumkin and Slygin, 1935).

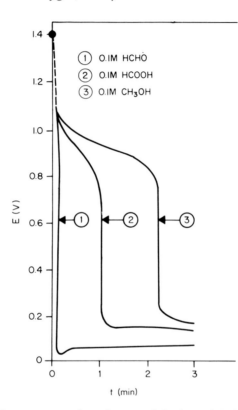

FIG. 19. Potential sweep curves of rotating smooth Pt electrode in $1\ N\ H_2SO_4$ (Bagotzky and Vasilyev, 1964).

The third peak is accepted as due to the oxidation of the organic substrate by $O(a)$ on the oxidized surface in one act or of the second oxidation product each followed by the steady OER; the fall of the current at its end in the latter case may be understood as due to the exhaustion of the second oxidation product formed at the lower electrode potential.

B. Oxidation of CO

Gilman (1963) determined the amount of adsorption, and hence the rate of adsorption, on a Pt electrode kept at $E = 0.4$ V in $1\ N$ $HClO_4$ solution saturated with CO; the amount of adsorbed CO was measured in terms of the number of elementary charges which composed the peaks attributed to the oxidation, assuming the oxidation of each CO(a) required two elementary charges. The amount of CO(a) thus determined was found to be greater than that derived from the assumption that each CO(a) occupies two of the sites and that the total number of sites equals the number of elementary charges, which compose the hump of hydrogen intermediates referred to in Sections II,H and V,A. This result was interpreted as showing that some of the CO(a) occupied only one site instead of two, i.e., their adsorption was of the linear type rather than the bridged type (Eischens and Pliskin, 1958) as accepted by other authors (Podlovchenko et al., 1963; Warner and Schuldiner, 1964; Brummer and Ford, 1965; Breiter, 1967).

The potential sweep transient on the Pt electrode in acid solution saturated with CO of known partial pressure reveals a single peak at $E = 0.9$–1.3 V (Gilman, 1963; Vielstich and Vogel, 1964), which is corroborated by the observation of Warner and Schuldiner (1964) that the galvanostatic transient of the same electrode had a plateau at $E = 1.0$–1.2 V under the same condition. On the ground that the peak appears at E is approximately 1 V, where the oxygen deposit occurs, the rate-determining step of the anodic oxidation $CO + H_2O = CO_2 + 2H^+ + 2e^-$ in acid solution is accepted to be

$$CO(a) + OH(a) \rightarrow CO_2 + H^+ + e^-$$

or

$$CO(a) + O(a) \rightarrow CO_2.$$

In alkaline solution, Roberts and Sawyer (1965) found that the rate, i_+, of the CO oxidation on an Au electrode was proportional to $[OH^-][CO]$ at constant E, and at constant $[OH^-][CO]$ the i_+ varied with E as $\tau_+ \equiv (RT/F)\, \partial \ln i_+/\partial E = 0.5$; on these grounds the latter authors proposed the following mechanism for the reaction $CO + OH^- = CO_2 + H^+ + 2e^-$ in alkaline solution:

$$CO + OH^- \longrightarrow HCO_2^-(a),$$
$$HCO_2^- \xrightarrow{A} HCO_2(a) + e^-,$$
$$HCO_2(a) \longrightarrow CO_2 + H^+ + e^-,$$

where $\xrightarrow{A}$ denotes the rate-determining step.

C. Oxidation of HCO_2H

It is known that the Pt electrode assumes the potential, when placed in a solution containing HCO_2H, which approaches zero with time as shown in Fig. 20 (Oxley *et al.*, 1964) and that the Pt electrode holds

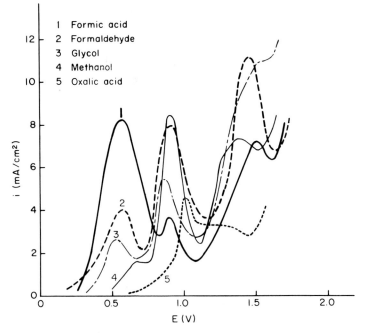

Fig. 20. Open-circuit decay curves (Oxley *et al.*, 1964).

hydrogen intermediates under the latter condition as concluded from the potential sweep analysis. On the basis of the latter experimental result Breiter (1963) suggested, admitting the ready ionization of the hydrogen intermediates, the mechanism of the steady anodic oxidation, $HCO_2H = CO_2 + 2H^+ + 2e^-$, as

$$HCO_2H \longrightarrow HCO_2(a) + H(a), \tag{5.1a}$$

$$H(a) \longrightarrow H^+ + e^-, \tag{5.1b}$$

$$HCO_2(a) \overset{A}{\longrightarrow} CO_2 + H^+ + e^- \tag{5.2}$$

with step (5.2) determining the rate. The τ_+ was found to be 0.5 along the ascending curve of the first peak of the potential sweep transient (Fleischmann *et al.*, 1964; Gottlieb, 1964) and the height of the peak was found proportional to the HCO_2H concentration (Breiter, 1963; Eckert,

1967). Breiter (1963) attributed the first peak to the reaction consisting of steps (5.1a) and (5.1b) and the second peak to the overall reaction composed of steps (5.1a), (5.1b), and (5.2). This accounts for the above experimental results as well as the appearance of the first peak at 0.5–0.6 V on the potential sweep transients, particularly in the case of the hydrogen containing organics and is in agreement with Kutschker and Vielstich (1963) and Vielstich and Vogel (1964). The latter groups of authors found the first and the second peak at $E = 0.5$ and 0.9 V respectively on the potential sweep transient using a Pt electrode in a solution of $1\,M\,HCO_2H + 1\,N\,H_2SO_4$. The third peak was found by them at $E = 1.4$ V, which they attributed to the direct attack of O(a) against HCO_2H in one act on the oxidized surface, i.e.,

$$HCO_2H + O(a) \rightarrow CO_2 + H_2O. \tag{5.3}$$

D. "Reduced CO_2"

Giner (1963) detected a certain substance, oxidizable but at different anodic potential from that of the hydrogen intermediates, when a Pt electrode was kept at $E = 0$ in acid solution saturated with CO_2. He called the substance "reduced CO_2," assuming it to be the intermediate product of the anodic oxidation of HCO_2H (Giner, 1964). Vielstich and Vogel (1964) observed the potential sweep transient of the "reduced CO_2" in comparison with that of CO, and hence concluded that these two substances were not identical in accordance with Brummer (1965).

Johnson and Kuhn (1965) and Breiter (1967) have, however, arrived at an adverse conclusion. They measured the quantity of electricity required to anodically oxidize the "reduced CO_2" as well as the amount of CO_2 produced by the oxidation, from which they inferred that two electrons were required for each molecule of CO_2 produced. Substances which fit in with this condition are CO(a), $HCO_2H(a)$, $OHC–CO_2H(a)$, etc. Johnson and Kuhn (1965) detected, however, no trace of HCO_2H and aldehydes from the solution used in their experiment, which led them to the inference that the "reduced CO_2" was nothing but CO(a). The existence of the "reduced CO_2" is still in dispute.

E. Oxidation of CH_3OH

CH_3OH is taken to be dissociatively adsorbed on the Pt electrode on the ground of the experimental result (Petry et al., 1965) similar to that in the case of HCO_2H as described in Section V,C. Adsorbed hydrogen was similarly detected by means of the galvanostatic and the potential sweep transients (Petry et al., 1965; Podlovchenko and Gorgonova, 1964;

Bagotzky and Vasilyev, 1967). They have accordingly concluded that three hydrogen atoms were released from each molecule of CH_3OH on the dissociative adsorption as

$$CH_3OH \rightarrow COH(a) + 3 \, H(a) \tag{5.4}$$

on the ground of two humps on the potential sweep transients at lower and higher potentials respectively attributable to the oxidation of the H(a) and the COH(a). On this basis, they assumed that the anodic oxidation of CH_3OH into CO_2 was completed by steps

$$H(a) \rightarrow H^+ + e^-,$$
$$COH(a) + OH(a) \rightarrow CO(a) + H_2O,$$

and

$$CO(a) + OH(a) \rightarrow CO_2 + H^+ + e^-,$$

which followed step (5.4).

Oxley et al. (1964) reported that they detected HCHO and HCO_2H from the solution during the anodic oxidation of CH_3OH but it is not yet decided whether they are intermediates or products of the anodic oxidation.

F. Oxidation of Hydrocarbons

Two peaks are observed at $E = 0.7$ and 1.25 V respectively on the potential sweep transients of saturated hydrocarbons except methane on the Pt electrode in acid solution. The first peak is attributed to the release of H(a) and the second peak to the addition of O(a) as mentioned in the Introduction (Gilman, 1966a; Tyurin et al., 1966; Brummer, 1966). In the case of methane, only one peak was observed at $E = 0.7$ V on the potential sweep transient on the Pt electrode, its mechanistic interpretation being as yet quite unknown (Niedrach, 1966).

In the case of unsaturated hydrocarbons (C_2H_4, C_2H_2) only one peak was observed at E approximately 1.25 V on their potential sweep transients (Gilman, 1966a), from which it was inferred that the addition of O(a) occurs alone without the preliminary release of H(a). It is hence probable that C_2H_4 is subject to molecular adsorption without dissociation (Bockris et al., 1965), whereas the dissociative adsorption is rendered probable by the coincidence of the potential sweep transients of C_2H_4 and C_2H_2 so far observed (Gilman, 1966b; Tyurin et al., 1966). No final conclusion is yet arrived at in this regard.

Hydrocarbons may be compared with one another in their ease of being anodically oxidized in terms of the anodic current density at a common electrode potential. The ease of oxidation has thus been found to be greater the smaller the carbon number except for methane, which is oxidized with more difficulty than C_2H_4. As regards hydrocarbons of the same carbon number, saturated hydrocarbons are oxidized more easily than unsaturated ones (Binder *et al.*, 1965).

SPECIAL REFERENCES

BAGOTZKY, V. S., and JABLOKOWA, I. E. (1953). *Zh. Fiz. Khim.* **27**, 1663.
BAGOTZKY, V. S., and VASILYEV, YU. B. (1964). *Electrochim. Acta* **9**, 869.
BAGOTZKY, V. S., and VASILYEV, YU. B. (1967). *Electrochim. Acta* **12**, 1323.
BAGOTZKY, V. S., NEKRASOV, L. N., and SCHUMILOWA, N. A. (1965). *Usp. Khim.* **34**, 1967.
BIANTI, G., and MUSSINI, T. (1965). *Electrochim. Acta* **10**, 445.
BINDER, H., KÖHLING, A., KRUPP, H., RICHTER, K., and SANDSTEDE, G. (1965). *J. Electrochem. Soc.* **112**, 355.
BOCKRIS, J. O'M., and HUQ, A. K. M. S. (1956). *Proc. Roy. Soc.* **A237**, 277.
BOCKRIS, J. O'M., and SRINIVASAN, S. (1965). *Proc. Ann. Power Sources Conf.* **19**, 4.
BOCKRIS, J. O'M., WROBLOWA, H., GILEADI, E., and PIERSMA, B. J. (1965). *Trans. Faraday Soc.* **61**, 2531.
BONHOEFFER, K. F. (1924). *Z. Physik. Chem. (Leipzig)* **113**, 199.
BORN, M., and OPPENHEIMER, J. R. (1927). *Ann. Physik* **84**, 457.
BREITER, M. W. (1962). *Electrochim. Acta* **7**, 600.
BREITER, M. W. (1963). *Electrochim. Acta* **8**, 447, 457.
BREITER, M. W. (1967). *Electrochim. Acta* **12**, 1213.
BREITER, M. W., KNORR, C. A., and VÖLKL, W. (1955). *Z. Elektrochem.* **59**, 681.
BREITER, M. W., KAMMERMAIER, H., and KNORR, C. A. (1956). *Z. Elektrochem.* **60**, 37, 119.
BRUMMER, S. B. (1965). *J. Phys. Chem.* **69**, 1363.
BRUMMER, S. B. (1966). *J. Electrochem. Soc.* **113**, 1041.
BRUMMER, S. B., and FORD, J. I. (1965). *J. Phys. Chem.* **69**, 1355.
BUTTLER, H. V., VIELSTICH, W., and BARTH, H. (1963). *Ber. Bunsenges. Physik. Chem.* **67**, 650.
CHANG, F. T., and WICK, H. (1935). *Z. Physik. Chem. (Leipzig)* **A172**, 448.
CONWAY, B. E., and BOCKRIS, J. O'M. (1957). *J. Chem. Phys.* **26**, 532.
DAVIES, M. O., CLARK, M., YEAGER, E., and HOVORKA, F. (1959). *J. Electrochem. Soc.* **106**, 56.
DELAHAY, P. (1950). *J. Electrochem. Soc.* **97**, 198, 205.
DOWDEN, D. A. (1957). "Chemisorption" (W. E. Garner, ed.), pp. 3–16. Butterworths, London and New York.
ECKERT, J. (1967). *Electrochim. Acta* **12**, 307.
EISCHENS, R. P., and PLISKIN, W. (1958). *Advan. Catalysis* **10**, 18.
EFIMOV, E. A., and ISGARYSHEV, N. A. (1956). *Zh. Fiz. Khim.* **30**, 1606.
ELEY, D. D. (1950). *Discussions Faraday Soc.* **8**, 34.
ENYO, M., and YOKOYAMA, T. (1967). *Electrochim. Acta* **12**, 1631, 1641.
ENYO, M., and YOKOYAMA, T. To be published.
ENYO, M., YOKOYAMA, T., and HOSHI, M. (1965). *J. Res. Inst. Catalysis, Hokkaido Univ.* **13**, 222.
ENYO, M., and MATSUSHIMA, T. (1969). *J. Res. Inst. Catalysis, Hokkaido Univ.* **17**, 14.

ERDEY-GRÚZ, T., and KROMREY, G. G. (1931). *Z. Physik. Chem. (Leipzig)* **A157**, 213.
ERDEY-GRÚZ, T., and VOLMER, M. (1930). *Z. Physik. Chem. (Leipzig)* **A150**, 203.
FARKAS, A. (1939). *In* "Experimental Methods in Gas Reaction," p. 153. Macmillan, New York.
FLEISCHMANN, C. W., JOHNSON, G. K., and KUHN, A. T. (1964). *J. Electrochem. Soc.* **111**, 602.
FRUMKIN, A. N. (1958). *Dokl. Akad. Nauk SSSR* **119**, 318.
FRUMKIN, A. N., and SLYGIN, A. I. (1935). *Acta Physicochim. U.R.S.S.* **3**, 791.
FRUMKIN, A. N., and TEDORADSE, G. (1958). *Z. Electrochem.* **62**, 251.
FUKUDA, M., and HORIUTI, J. (1962). *J. Res. Inst. Catalysis, Hokkaido Univ.* **10**, 43.
GIBBS, W. (1902). "Elementary Principles in Statistical Mechanics." Yale Univ. Press, New Haven, Connecticut.
GILMAN, S. (1963). *J. Phys. Chem.* **67**, 78.
GILMAN, S. (1964). *Electrochim. Acta* **9**, 1025.
GILMAN, S. (1966a). *J. Electrochem. Soc.* **113**, 1036.
GILMAN, S. (1966b). *Trans. Faraday Soc.* **62**, 466.
GILROY, D., and CONWAY, B. E. (1965). *J. Phys. Chem.* **69**, 1259.
GINER, J. (1963). *Electrochim. Acta* **8**, 857.
GINER, J. (1964). *Electrochim. Acta* **9**, 63.
GLASSTONE, S., LAIDLER, K. J., and EYRING, H. (1941). "The Theory of Rate Processes." McGraw-Hill, New York.
GOTTLIEB, M. H. (1964). *J. Electrochem. Soc.* **111**, 465.
GURNEY, R. W. (1931). *Proc. Roy. Soc.* **A134**, 137.
HABER, F., and RUSS, R. (1904). *Z. Physik. Chem.* **47**, 257.
HILL, T. L. (1960). "Introduction to Statistical Thermodynamics," pp. 79 and 158. Addison-Wesley, Reading, Massachusetts.
HIROTA, K., and HORIUTI, J. (1936). *Sci. Papers Inst. Phys. Chem. Res. (Tokyo)* **30**, 151.
HOARE, J. P. (1963). *J. Electrochem. Soc.* **110**, 1019.
HOARE, J. P. (1966). *Electrochim. Acta* **11**, 203, 549.
HORIUTI, J. (1948). *J. Res. Inst. Catalysis, Hokkaido Univ.* **1**, 8.
HORIUTI, J. (1954). *J. Res. Inst. Catalysis, Hokkaido Univ.* **3**, 52.
HORIUTI, J. (1956). *J. Res. Inst. Catalysis, Hokkaido Univ.* **4**, 55.
HORIUTI, J. (1957). *Proc. Intern. Congr. Surface Activity, and, London, 1957*, p. 280. Butterworths, London and New York.
HORIUTI, J. (1961). *J. Res. Inst. Catalysis, Hokkaido Univ.* **9**, 211.
HORIUTI, J., and IKUSIMA, M. (1939). *Proc. Imp. Acad. (Tokyo)* **15**, 39.
HORIUTI, J., and KITA, H. (1964). *J. Res. Inst. Catalysis, Hokkaido Univ.* **12**, 122.
HORIUTI, J., and KOMOBUCHI, Y. (1958). *J. Res. Inst. Catalysis, Hokkaido Univ.* **6**, 92, 197.
HORIUTI, J., and MATSUDA, A. (1959). *J. Res. Inst. Catalysis, Hokkaido Univ.* **7**, 19.
HORIUTI, J., and MUELLER, K. (1968). *J. Res. Inst. Catalysis, Hokkaido Univ.* **16**, 605.
HORIUTI, J., and NAKAMURA, T. (1950). *J. Chem. Phys.* **18**, 395.
HORIUTI, J., and NAKAMURA, T. (1951). *J. Res. Inst. Catalysis, Hokkaido Univ.* **2**, 73.
HORIUTI, J., and NAKAMURA, T. (1967). *Advan. Catalysis* **17**, 1.
HORIUTI, J., and OKAMOTO, G. (1936). *Sci. Papers Inst. Phys. Chem. Res. (Tokyo)* **28**, 231.
HORIUTI, J., and OKAMOTO, G. (1938). *Bull. Chem. Soc. Japan* **13**, 216.
HORIUTI, J., and POLANYI, M. (1933). *Nature* **132**, 819, 931.
HORIUTI, J., and POLANYI, M. (1935). *Acta Physicochim. U.R.S.S.* **2**, 505.
HORIUTI, J., KEII, T., and HIROTA, K. (1951). *J. Res. Inst. Catalysis, Hokkaido Univ.* **2**, 1.
HORIUTI, J., MATSUDA, A., ENYO, M., and KITA, H. (1964). "Electrochemistry," Proc. 1st Australian Conf. Sydney and Hobart, 1963.

IKUSIMA, M., and AZAGAMI, S. (1938). *Nippon Kagaku Zassi* **59**, 40.

JOHNSON, P. R., and KUHN, A. T. (1965). *J. Electrochem. Soc.* **112**, 599.

KECK, J. C. (1960). *J. Chem. Phys.* **32**, 1035.

KEII, T., and KODERA, T. (1957). *J. Res. Inst. Catalysis, Hokkaido Univ.* **5**, 105.

KITA, H. (1966). *J. Electrochem. Soc.* **113**, 1095.

KITA, H., and YAMAZAKI, T. (1963). *J. Res. Inst. Catalysis, Hokkaido Univ.* **11**, 10.

KODERA, T. (1958). *Shokubai (Sapporo)* **15**, 119.

KODERA, T., and SAITO, T. (1959). *J. Res. Inst. Catalysis, Hokkaido Univ.* **7**, 5.

KRASILISCHICHIKOV, A. I. (1952). *Zh. Fiz. Khim.* **26**, 216.

KUTSCHKER, A., and VIELSTICH, W. (1963). *Electrochim. Acta* **8**, 985.

LAITINEN, H. A., and ENKE, C. G. (1960). *J. Electrochem. Soc.* **107**, 773.

LLOPIS, J., and VAZQUEZ, M. (1962). *Electrochim. Acta* **6**, 177.

MAKRIDES, A. C. (1962). *J. Electrochem. Soc.* **109**, 256.

MATSUDA, A., and HORIUTI, J. (1958). *J. Res. Inst. Catalysis, Hokkaido Univ.* **6**, 231.

MATSUDA, A., and HORIUTI, J. (1962). *J. Res. Inst. Catalysis, Hokkaido Univ.* **10**, 14.

MATSUDA, A., and OHMORI, T. (1962). *J. Res. Inst. Catalysis, Hokkaido Univ.* **10**, 215.

MINC, S., and SOBKOWSKI, J. (1959). *Bull. Acad. Polon. Sci., Ser. Sci. Chim., Geol. Geograph.* **7**, 29.

MITUYA, A. (1956). *J. Res. Inst. Catalysis, Hokkaido Univ.* **4**, 228.

NIEDRACH, L. W. (1966). *J. Electrochem. Soc.* **113**, 645.

NOMURA, O., and KITA, H. (1967). *J. Res. Inst. Catalysis, Hokkaido Univ.* **15**, 35.

OKAMOTO, G., HORIUTI, J., and HIROTA, K. (1936). *Sci. Papers Inst. Phys. Chem. Res. (Tokyo)* **29**, 223.

OXLEY, J. E., JOHNSON, G. K., and BUZALSKI, B. T. (1964). *Electrochim. Acta* **9**, 897.

PAST, V. E., and JOFA, J. A. (1959). *Zh. Fiz. Khim.* **33**, 913, 1230.

PETRY, O. A., PODLOVCHENKO, B. I., FRUMKIN, A. N., and HIRA LAL (1965). *J. Electroanal. Chem.* **10**, 253.

PODLOVCHENKO, B. I., and GORGONOVA, E. P. (1964). *Dokl. Akad. Nauk USSR* **156**, 673.

PODLOVCHENKO, B. I., PETRY, O. A., and FRUMKIN, A. N. (1963). *Dokl. Akad. Nauk USSR* **153**, 379.

ROBERTS, JR., J. L., and SAWYER, D. T. (1965). *Electrochim. Acta* **10**, 989.

ROSENTAL, K. J., and VESELOVSKY, V. J. (1956). *Dokl. Akad. Nauk USSR* **111**, 637.

RÜETSCHI, P., and DELAHAY, P. (1955). *J. Chem. Phys.* **23**, 195, 1167.

SAWYER, D. T., and DAY, R. J. (1963). *Electrochim. Acta* **8**, 589.

SCHULDINER, S., and ROE, R. M. (1963). *J. Electrochem. Soc.* **110**, 332.

STEVENSON, D. P. (1955). *J. Chem. Phys.* **23**, 203.

TAFEL, J. (1905). *Z. Physik. Chem.* **50**, 641.

TOPLEY, B., and EYRING, H. (1934). *J. Chem. Phys.* **2**, 217.

TOSHIMA, S., and OKANIWA, H. (1966). *Denki Kagaku* **34**, 641.

TOYA, T. (1958). *J. Res. Inst. Catalysis, Hokkaido Univ.* **6**, 308.

TOYA, T. (1960). *J. Res. Inst. Catalysis, Hokkaido Univ.* **8**, 209.

TYURIN, V. S., PSHENICHNIKOV, A. G., and BURSHTEIN, R. KH. (1966). *Elektrokhimiya* **2**, 948.

VETTER, K. J. (1952). *Z. Physik. Chem. (Leipzig)* **199**, 22, 285.

VETTER, K. J. (1961a). *Angew. Chem.* **73**, 277.

VETTER, K. J. (1961b). "Elektrochemische Kinetik," p. 497. Springer, Berlin.

VETTER, K. J. (1961c). "Elektrochemische Kinetik," p. 370. Springer, Berlin.

VETTER, K. J., and BERNDT, D. (1958). *Z. Elektrochem.* **62**, 378.

VIELSTICH, W., and VOGEL, U. (1964). *Ber. Bunsenges. Phys. Chem.* **68**, 688.

WALTON, H. F., and WOLFENDEN, J. H. (1936). *Nature* **138**, 468.

WALTON, H. F., and WOLFENDEN, J. H. (1938). *Trans. Faraday Soc.* **34**, 436.
WARNER, T. B., and SCHULDINER, W. (1964). *J. Electrochem. Soc.* **111**, 992.
WATANABE, N., and DEVANATHAN, M. A. V. (1964). *J. Electrochem. Soc.* **111**, 615.
WIGNER, E. P. (1937). *J. Chem. Phys.* **5**, 720.
WILL, F. G., and KNORR, C. A. (1960a). *Z. Elektrochem.* **64**, 258.
WILL, F. G., and KNORR, C. A. (1960b). *Z. Elektrochem.* **64**, 270.
WINKELMANN, D. (1956). *Z. Elektrochem.* **60**, 731.
YAMAZAKI, T., and KITA, H. (1965). *J. Res. Inst. Catalysis, Hokkaido Univ.* **13**, 77.

Chapter 7

The Mechanism of Deposition
and Dissolution of Metals

JOHN O'M. BOCKRIS

AND

ALEKSANDAR R. DESPIĆ

I. Introduction 611
II. Basic Steps in Metal Deposition 613
   A. Discharge Process 613
   B. Surface Diffusion and Incorporation 638
   C. Effect of Transport in Solution on Deposition Kinetics and Properties of
      the Deposit 653
III. Crystallographic Aspects of Metal Deposition 660
   A. Morphology and Texture 660
   B. Mechanism and Kinetics of Growth of a Single Crystal 674
IV. Deposition of Alloys 697
   A. Thermodynamics of Codeposition of Metals 698
   B. Kinetics of Codeposition 701
   C. Phase Formation in Alloy Deposition 706
V. Kinetics of Metal Dissolution 708
   A. Dissolution and Deposition 708
   B. Dissolution into Solutions Free of Corresponding Ions 712
   C. Acceleration of Dissolution under Strain 716
   D. Dissolution of Alloys 720
   References 723

I. Introduction

Within the framework of electrochemistry metal deposition and dissolution are considered part of electrode kinetics, but a very special part,
distinguished by the fact that the substrate takes part in the reaction

concerned. This, of course, brings at once not only some special considerations and special theory, but also some special difficulties, because, with respect to the measurement technique, it must be recalled that the meaning of the phrase "steady-state deposition" is restrictive. Another special factor is involved which distinguishes this from other aspects of electrode kinetics. The study of metal deposition and dissolution requires at least two *types* of studies. In the first, one carries out electrodic electrochemistry and, essentially, measures the rate of passage of electrons across an interface. There are many methods of doing this, and they are of a considerable degree of sophistication, methods being available with an accuracy down to the nanosecond range.

However, knowing the rate of the transfer of electrons between the electron-conducting and ion-conducting fields does not necessarily mean that one knows at what rate the crystals are forming upon the surface, particularly under nonsteady state conditions. Thus the second aspect of the study of metal deposition and dissolution (and one in which there is an essential difference from electrodic investigations of other phenomena) is the crystallographic one. In particular, the structure of electrodeposits is affected both by crystallographic properties of the substrate and by those inherent in the depositing metal.

Another closely related subject is that of the corrosion of metals. Thus corrosion rate is determined under certain circumstances by metal dissolution, and the internal and bulk properties of metals (e.g., the spreading of cracks) can, under some circumstances, be determined by electrochemical phenomena which involve, and may be rate determined by, the anodic dissolution of the metal.

In general, fundamental research on metal deposition and dissolution did not begin until the 1930's with the work of Kossell (1927), Stranski (1928), and Erdey-Grúz and Volmer (1931a,b) [cf. Volmer (1934)]. The modern stage begins with the work of Mehl and Bockris (1957), in which sweep measurements were used for the first time to evaluate the concentration of intermediate radicals on the surface of a dissolving metal and to obtain the rate constants not only of the rate-determining, but also of the non-rate-determining, reaction. The crystallographic aspects of the theory for this new work were influenced primarily by the corresponding gas-phase work, in particular the theory of crystallization of Burton *et al.* (1949), which utilized the concept of screw dislocations in the theory of crystal growth.

At present one might say that the elementary steps up to crystallization are to some extent worked out and understood, and there is the beginning of an outline of crystallization kinetics on single crystals, with one case, that of dendritic growth, in a quantitative stage. The considerable task ahead is to make the transition from these studies of very thin deposits on single crystals to the reality of the polycrystalline situation, and then finally to understand the very complex effects of additives, which are connected with the actual situation occurring in practical electro-crystallization.

II. Basic Steps in Metal Deposition

The overall reaction of metal deposition from an aqueous solution is represented by

$$[M^{z+}(H_2O)_n]_{soln} + ze \rightarrow M_{lattice} + nH_2O, \qquad (2.1)$$

where $M^{z+}(H_2O)_n$ represents a hydrated metal ion in the bulk of solution and M is a metal atom in the bulk of the metal phase. Such a reaction may consist of the following successive or simultaneous steps:

1. Transport of the hydrated ion from the bulk of solution to a position at the outer Helmholtz plane from which it can be transferred onto the electrode surface.

2. Charge transfers between the electrode and the ion (reduction of the ion) and its partial dehydration.

3. Lateral motion to a position of incorporation into the lattice or encounter with other particles to form a crystal nucleus.

4. Incorporation of the particle into a fixed lattice position.

5. Continuation of these processes until the atom considered in steps (1)–(4) is incorporated underneath the new surface, i.e., is a bulk atom.

Essential electrode problems are connected with step 2 and, on solid metals, step 3. Step 4 is decisive for crystal growth and the structural features of the electrodeposit. There are specific cases of considerable interest in which step 1 has a pronounced effect on the outcome of electrodeposition. Step 5 may have some importance in influencing the deposition of liquid alloys (amalgams), but is of no consequence in most metal-deposition problems.

A. DISCHARGE PROCESS

Two questions of interest with definite energetic and kinetic conse-quences in this electrochemical part of the deposition process are: (1) where does the discharge of the hydrated ion take place, and (2) what is the molecular mechanism of discharge?

1. *Location of Discharge*

   a. Alternative Paths in Metal Deposition.   On solid metals two reaction paths exist as combinations of steps 2 and 3 (Fig. 1). The discharge

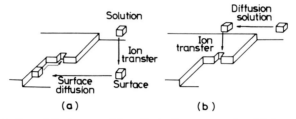

FIG. 1. Illustration of the two possible paths in the process of metal deposition. (a) Discharge followed by surface diffusion. (b) Path taken in the case when charge transfer is limited to points of incorporation.

process can take place either at specific points at the surface, i.e., the point of incorporation at which the neutral metal atom is fixed into the lattice (growth points—kinks of steps or dislocations), or else it can take place at any point at the surface away from the point of incorporation. In the first case step 3 must precede step 2, i.e., hydrated ions at the outer Helmholtz plane must move laterally until they reach the point at which discharge can take place. In the second case a complete or partial discharge accompanied by a corresponding successive release of water dipoles from the hydration sheet can take place anywhere. After the charge transfer the obtained neutral or still partly-charged and partly-hydrated metal atom—*adatom* or *adion*—must move along the surface *at the metal side* of the electrical double layer until it strikes a growth step, and along it until it reaches a kink site, at which its fixation can take place. This process is known as *surface diffusion*.

   Along this path further dehydration takes place, as depicted in Fig. 2.

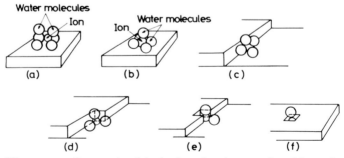

FIG. 2. The process of successive dehydration of an ion transferred from the solution (a), onto a planar surface (b), then to an edge (c), a kink (d), and finally buried into the edge (e) and the surface (f).

Conway and Bockris (1958) have investigated the likelihood of the two paths with respect to the energy changes involved in the steps.

The potential-energy diagrams could be evaluated from two types of calculations: (1) the levels of potential energy minima were evaluated from the heat contents of the initial and final states (e.g., hydrated ion in the double layer and an adion—partly hydrated and adsorbed at a surface site—respectively); (2) the potential energy as a function of distance from the minima toward the transition states were evaluated from a model in which the ion is displaced from its solvation sheet and the adion is vibrating out from its adsorbed position.

b. Heat Contents of States during the Transfer of an Ion to a Surface Site. For reaction (2.1) at equilibrium

$$\bar{\mu}_{M^{z+}} + z\bar{\mu}_e = \mu_M + n\mu_{H_2O}. \qquad (2.2)$$

At the zero-charge potential (ZCP) the electrochemical potentials $\bar{\mu}$ can be replaced by the corresponding chemical potentials corrected for the electrical work arising from the remaining interfacial potential difference due to the surface dipole layers. (The latter is assumed to be independent of potential. This assumption has obtained some approximate foundation in the work of Bockris *et al.* (1963), who estimated a change of the potential difference at a surface during a 1 V change of potential to be less than 100 mV.) If the standard states are considered and if the final state (bulk of the metal and free water) is taken as a reference state, the heat content, ΔH_i, of the initial state (metal ion in solution + z electrons in the metal) arises from the standard chemical potentials corresponding to chemical potentials of Eq. (2.2) and estimated entropies [cf. Conway (1949) and Parsons and Bockris (1951)].

To evaluate the heat content of the final state—adion at a surface site— relative to the bulk of metal and free water, first the energy of interaction of the corresponding unhydrated ion at the given type of site has to be found. This is done taking into account the reduced coordination of the particle at a surface site and thus using

$$\Delta H_f' = [1 - (2n'/n)]L_{sub}, \qquad (2.3)$$

where n is the coordination number of atoms in the bulk and n' is that of the atom (ion + electron) at the surface. Obviously, the *actual* heat content of the final state is smaller by the heat of the residual hydration. This is the effect which makes the levels of minima of the final states very different at different sites, as can be estimated by inspecting space-filling models of the ion and its solvating water molecules (cf. Fig. 2). Beside simple geometric considerations, the hydration energy has to be modified

(1) for the interaction between the water dipoles and the excess charge in the metal, which tends to reduce the heat of hydration of the adsorbed adion, and (2) for the reduction of the effective charge (cf. the Born-charging contribution to the hydration energy). In addition, the energy needed for the displacement of water molecules to make room for the adion at the surface has to be taken into account. The latter is known with accuracy only for mercury [cf. Law (1952)], but it can be inferred from determinations of the adsorption of organics at different metals, and from the fact that the energy of binding the water molecules to the surface should not differ by more than 5 kcal/mole from that on mercury (15 kcal/mole). The calculation made for, e.g., Ni^{++} ions, along the above line resulted in the following values of the energy of the adions relative to that of the atoms in the bulk (in kcal/mole):

Plane surface	-15
Edge site	-29
Kink site	-46

c. Potential-Energy Diagrams. These are obtained by intersecting the potential-energy–distance functions for the initial and final states. In the initial state the displacement energy calculated at different distances from the metal surface as the electrostatic ion–dipole interaction is added to the energy of the initial state.

The curves for vibration of the adions adsorbed at the electrode are obtained by assuming that the adsorption energy varies with distance according to some potential-energy–distance relation, e.g., a Morse function with the Morse constant *a* corresponding to that for atom–atom vibrations of the metal concerned.

In the case of the path involving the deposition onto a planar site and surface diffusion the final state of the charge transfer is only an intermediate state in the overall reaction. Hence other potential-energy–distance curves, i.e., those for the migration of the adions, have to be taken into account. The effect of potential at potentials other than the ZCP can easily be estimated by shifting the potential-energy curves of the states involving separated charges (ions + electrons in the metal) along the vertical axis by the amount of energy $zF\, \Delta\phi$, where $\Delta\phi$ is the potential of the metal with respect to the solution, relative to the ZCP.

Conway and Bockris (1958) have obtained a number of comparative diagrams, such as that shown in Fig. 3, for the deposition of cupric ions *via* cuprous ions discharging, e.g., at a kink site or at a planar surface (followed by surface diffusion).

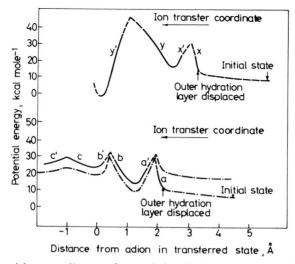

FIG. 3. Potential-energy diagrams for cupric ion discharge *via* cuprous ion as an intermediate. (a)–(c′) Deposition *via* Cu⁺ adions transferred to a plane surface site. (x)–(y′) Deposition *via* Cu⁺ adions transferred to an edge site. (a), (x) Potential-energy curves for the vibration of cupric ion inside the hydration sheet at the outer Helmholtz plane. (a′), (x′) Same as (a) and (x) but for cuprous ion. (y), (b) Same for further reduction of cuprous ion. (y′), (b′) Same as (y) and (b) but for adions (final state in the discharge reaction) at an edge and plane surface, respectively. (c), (c′) Potential-energy curves for surface diffusion of adions. [From Conway and Bockris (1958, 1960).]

d. Conclusions from the Potential-Energy Considerations. The heat of activation for metal deposition *via* different paths can be taken as equal to the energy difference between the initial state and zero-point energy level of the activated state, which can be taken to be within the rather large limits of error of the calculation of ± 5 kcal/mole, equal to the energy at the highest intersection of the potential-energy curves. The differences in the standard energies of activation for alternate paths were in many cases found to be well outside this limit, e.g., 30 kcal/mole, and significant qualitative conclusions concerning the path of deposition could be made:

1. By far the most frequently occurring charge transfers are those onto crystal *planes*, because transfer at kink sites and defects (i.e., direct deposition on growth sites) is associated with much larger activation energies, as shown in Table I.

2. The product of the charge transfer is an adion (not an adatom), since the energy of activation for the production of an uncharged species (adatom) is too large to allow appreciable transfer rates. A rough theoretical estimate indicates, e.g., for Ag⁺ adions about 50% of ionic

character. This conclusion is also supported by some other evidence. Thus Gerischer (1958, 1959), using the mean Ag–Ag distance in the lattice and an estimated double-layer thickness, deduced, from the adsorption capacitance determined as a function of Ag^+ concentration, a partial charge of 30–40%.

TABLE I

CALCULATED HEATS OF ACTIVATION AT THE ZERO-CHARGE POTENTIAL FOR DIRECT ION TRANSFER FROM THE UNDISTORTED HYDRATED ION TO SITES ON THE METAL SURFACE

Ion	Heats of activation (kcal/mole), solution to:			
	Plane surface	Edge	Kink	Surface vacancy
Ni^{2+}	130	190	> 190	190
Cu^{2+}	130	180	> 180	180
Ag^+	10	21	35	35

3. Multivalent ions are discharged in single-electron-transfer steps, since the heats of activation for simultaneous transfers of z electron are prohibitively large. Indeed, much experimental evidence has been accumulated (cf. p. 630) for the discharge of divalent ions, such as Cu^{++}, Cd^{++}, Fe^{++}, etc., in support of this conclusion.

2. Mechanism of Discharge and the Resulting Kinetics

a. Kinetics of Discharge of Univalent Ions. It has been shown in several instances that under appropriate conditions of rate-determining charge transfer (on liquid metals or on solid ones at which this process is slow, or at short time intervals where other kinds of control have not yet taken over) the potential dependence of the current density, i.e., rate of metal deposition and dissolution, follows the Butler–Volmer equation:

$$i = i_0 \left\{ \exp\left[\frac{(1 - \beta)F\eta}{RT} \right] - \exp\left[\frac{-\beta F\eta}{RT} \right] \right\}, \qquad (2.4)$$

where β is the symmetry factor, η is the overpotential, and i_0 is the exchange current density. Thus

$$i_0 = F \bar{k}_s C_{M^+}^{(1 - \beta)} C_M^{\beta}, \qquad (2.5)$$

and

$$\bar{k}_s = k_r \exp\left[\frac{(1 - \beta)F}{RT} E^0 \right] = k_f \exp\left[\frac{-\beta F}{RT} E^0 \right], \qquad (2.6)$$

where $\bar{k}_s$ is the electrochemical rate constant and k_r and k_f are potential-independent rate constants.

Thus Randles (1961) has investigated Tl^+ ion discharge on thallium amalgam using the ac impedance method and has shown a linear dependence of log $\bar{k}_s$ on potential (Fig. 4). The symmetry factor $\beta = 0.52$ was found. Similar results were obtained from Faradaic rectification (Barker et al., 1958).

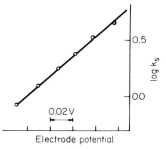

FIG. 4. The electrochemical rate constant for Tl^+ discharge on Tl-amalgam (in 1 M $HClO_4$) as a function of potential [from Randles (1961)].

Despić and Bockris (1960) have investigated the deposition of silver from silver perchlorate solutions onto helium-treated and electro-chemically-activated silver. They have shown that the galvanostatic charging curves and the steady-state data can be interpreted if the validity of Eq. (2.4) is assumed. Thus if Eq. (2.4) is extended by considering the double-layer charging current $C_{DL}(d\eta/dt)$, it can be solved for the potential time dependence at a constant current density [cf. Roiter et al. (1939a,b)]. Rearranging the solution, a linear function of time is obtained,

$$q = \log \frac{[\exp(-\eta_{A,\infty}/b') + \exp(\eta_{A,t}/b')][\exp(\eta_{A,\infty}/b') - 1]}{[\exp(\eta_{A,\infty}/b') - \exp(\eta_{A,t}/b')][\exp(-\eta_{A,\infty}/b') + 1]} = \frac{1}{2.3\tau_{DL}} t,$$
(2.7)

where $b' = RT/\beta F$ and

$$\tau_{DL} = \{b'/[\exp(-\eta_{A,\infty}/b') + \exp(\eta_{A,\infty}/b')]\}C_{DL}/i_0,$$
(2.8)

with $\eta_{A,t}$ and $\eta_{A,\infty}$ the activation overpotential at time t and after the steady-state is established, respectively. Figure 5 shows that linearity is well obeyed if $\beta = 0.5$ is assumed. That this assumption is justified was shown by analyzing the steady-state data. If Eq. (2.4) is rearranged, it is seen that a plot of

$$y = \log\left\{\left[\exp\left(\frac{F}{RT}\eta_{A,\infty}\right) - 1\right]/i\right\} = \frac{\beta F}{2.3RT}\eta_{A,\infty} - \log i_0 \quad (2.9)$$

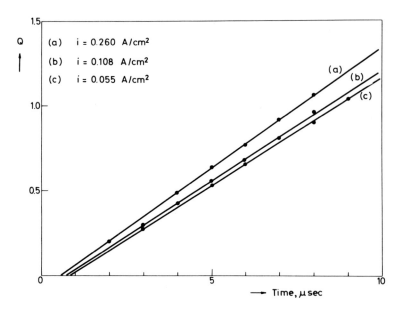

FIG. 5. Charging curves in the course of the transfer-controlled anodic polarization. Ag in 0.364 N AgClO$_4$ + 1 N HClO$_4$ solution [from Despić and Bockris (1960)].

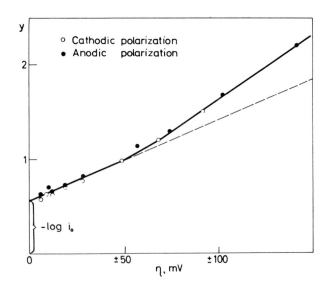

FIG. 6. Typical y–η relation for the steady-state deposition of Ag on electrochemically-activated silver [from Despić and Bockris (1960)].

should be linear over the entire range of overpotentials (contrast the Tafel plot), yielding β from the slope and i_0 from the intercept. Figure 6 shows that linearity is obeyed up to $\eta = 50$ mV. The i_0 and β values obtained are shown in Table II. A predicted concentration dependence of i_0 is also obtained [cf. Eq. (2.5)].

TABLE II

KINETIC PARAMETERS OF Ag–Ag$^+$ EXCHANGE REACTION

Electrode	C_{AgClO_4} (moles/liter)	i_0 (A/cm^2)	β
1	0.364	0.29	0.52
2	0.364	0.26	0.53
3	0.354	0.62	0.50
4	0.354	0.47	0.33
5	0.354	0.57	0.42
6	0.0724	0.26	0.49
7	0.0724	0.27	0.48
8	0.00121	0.04	0.5 ± 0.1
9	0.00121	0.04	0.5 ± 0.2

b. *Theory of Charge Transfer in Metal Deposition.* Following the procedure employed by Gurney (1931) and Butler (1936), Bockris and Matthews (1966) have developed a theory of charge transfer at electrodes (cf. that for homogeneous reactions in Chapter 8), which is applicable to metal-deposition reactions. It was shown that neither an ion transfer through the double layer to meet the electron in the metal, nor electron tunneling to an ion in its ground state at the outer Helmholtz plane can account for the observed rates of charge transfer. Instead, it is the stretching of the ion out of its hydration sheet toward the electrode which brings about the availability of quantum states in the ion having an energy equal to those near the Fermi level of the metal.

Let the energy gap to be closed by activating the ion–water bond be ΔE_0 (Fig. 7), i.e., equal to the difference between the Fermi energy of the electrons in the metal, E_I, and the maximum energy needed for transfer, E^*. The rate of discharge, i.e., the corresponding partial cathodic current density i, depends on (1) the number of electrons $\nu(E)$ with energy between E and $E + dE$ striking a unit area of the surface in unit time from inside the metal; (2) the fraction $P(E)$ of the surface at which ion–solvent bonds are sufficiently stretched so that electrons may tunnel, which is related to the number of hydrated ions $N(\epsilon)$ having an energy ϵ corresponding to the energy level E; and (3) the probability $W(E)$ of electron tunneling at the energy level E.

The first quantity, $\nu(E)$, can be found from the theory of thermionic emission [cf. Seitz (1940)] as

$$\nu(E) = (4\pi mkT/h^3) \exp[-(E - E_{\text{I}})/kT] \, dE. \qquad (2.10)$$

For hydrated ions, assuming a Boltzman distribution of the rotational-vibrational energy levels,

$$N(\epsilon) = C_{\text{M}^+} N_{\text{A}} \exp[-(\epsilon - \epsilon_0)/kT], \qquad (2.11)$$

where C_{M^+} is the surface concentration of ions (in moles/cc), ϵ_0 is the ground-state energy level of the ion, and N_{A} is Avogadro's number.

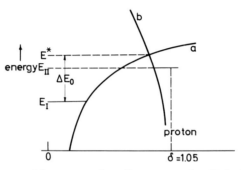

FIG. 7. Curve of potential energy against distance transfer. E_{I} is the Fermi energy of electrons in the metal and E_{II} is the ground state energy level of the final state [from Bockris and Matthews (1966)].

The energy level of the electrons E can be related to the vibrational-rotational energy levels ϵ by

$$\epsilon - \epsilon_0 = \beta(\epsilon)(E_{\text{II}} - E), \qquad (2.12)$$

where $0 < \beta < 1$ is the symmetry factor. The fraction of the surface covered by ions of suitable energy level is given by

$$P(E) = \pi r_{\text{M}^+}^2 N(\epsilon), \qquad (2.13)$$

where r_{M^+} is the radius of the ion in the hydrated state it possesses upon adsorption.

Hence, substituting (2.12) into (2.11) and this into (2.13), one obtains

$$P(E) = \pi N_{\text{A}} r_{\text{M}^+}^2 C_{\text{M}^+} \exp[-\beta(E_{\text{II}} - E)/kT]. \qquad (2.14)$$

Finally, the probability of the electron tunneling at $E = E_{\text{I}}$ is given by [cf. Moelwyn-Hughes (1961)]

$$W(E) = \exp[-(4\pi d/h)\{2m(E^* - E)\}^{1/2}], \qquad (2.15)$$

where E is the actual energy of the electron. The probability $W(E)$ depends much less strongly on E than do both $\nu(E)$ and $N(E)$. Hence, an acceptable modification is to use an average value of the tunneling probability, $\bar{W}$, for the energy range considered. Thus, the number of electrons tunneling per second is obtained by multiplying (2.10), (2.14), and (2.15) and integrating the result between the two energy levels E_I and E_{II}. The current density is obtained by multiplying this result by F/N_A. Thus

$$\vec{i} = FC_{M^+}\pi r_{M^+}^2 [4\pi m(kT)^2/h^3](1 - \bar{\beta})\bar{W}$$
$$\times \{\exp[-\bar{\beta}(E_{II} - E_I)/kT] - \exp[-(E_I - E_{II})/kT]\}, \quad (2.16)$$

where $\bar{\beta}$ is the average value of $\beta(E)$ for the energy range E_I–E_{II}. For the condition $\exp[(1 - \bar{\beta})(E_{II} - E_I)/kT] \gg 1$, Eq. (2.16) reduces to

$$\vec{i} = FC_{M^+}\pi r_{M^+}^2 \bar{W}[4\pi m(kT)^2/h^3(1 - \bar{\beta})]$$
$$\times \exp[-\bar{\beta}(E_{II} - E_I)/kT]. \quad (2.17)$$

The activated state can have two resonant forms, the one just before the electron tunneling and the one just after it. Since electron transfer becomes possible when the energies of the electron in the metal and in the ion are equal, i.e., $E_I = E_{II}$, the tunneling probability is the same in both directions. The two resonant states are thus equally probable, and if the particle immediately after tunneling were to be electrically neutral, one could say that the effective charge (i.e., the time-average charge) on the activated complex and on the metal itself is $\pm\frac{1}{2}$. These points were noted in relation to the homogeneous and electrode redox reactions by Marcus (1957a,b) (the resonant-state concept) and by Hush (1961) (the charge distribution in the activated state). Since in metal deposition the product of discharge still maintains a certain amount of charge (adions), the charge on the activated complex should deviate accordingly from the given value.

The introduction of the symmetry factor in the potential dependence of the rate, originally done by Horiuti and Polanyi (1935), has a clear physical meaning in the present theory. In the presence of a potential difference $\Delta\phi$ across the interface the gap between the energies of the electron on the two sides is changed by the amount of electrical energy associated with the transition of one mole of electrons, i.e.,

$$E = E_{II} - E_I + F\Delta\phi. \quad (2.18)$$

Accordingly, the energy of stretching the M^+—H_2O bonds to make $\Delta E = 0$ for electron tunneling must be changed, i.e. [cf. Eq. (2.12)],

$$\Delta\epsilon = \beta\,\Delta E = \beta(E_I - E_{II} + F\Delta\phi). \quad (2.19)$$

Introducing this into Eq. (2.17) yields the expected dependence of the rate of the cathodic reaction, i.e., current density–potential relationship.

The theory can be extended to calculate the anodic partial current density, and combining the two current densities, the expected from of the Butler–Volmer equation (2.4) is obtained. This theory represents an improvement over the earlier theories in that (1) the role played by the electron is clearly defined (contrast earlier ion-transfer-centered theories in which the position of the electrons was obscure); (2) the charge distribution in the activated state becomes clear; and (3) ambiguity in the meaning of the symmetry factor, or any suggestion that it is empirical, is eliminated.

c. Charge Transfer at Very High Current Densities.   It has been shown above that the symmetry factor β remains independent of potential as long as the potential energy–distance relations are linear. To investigate the possible potential dependence of β, one has to work at relatively large overpotentials, so that the potential energy–distance relations are in their nonlinear portions. To bring this about, one has to exceed the limiting current region. However, by working with transients at sufficiently short times (galvanostatic transients are preferred because of the difficulty experienced in eliminating the effect of a time-dependent pseudoohmic-polarization, i.e., ohmic drop between the electrode and the Luggin capillary), it is possible to obtain the activation-controlled dependence of rate

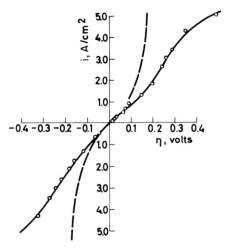

FIG. 8. Typical i–η relation in the case of an activated silver electrode in $0.364\,N$ $AgClO_4 + 1\,N\,HClO_4$ solution. Dashed curve: Butler–Volmer curve with $\beta = $ const [from Despić and Bockris (1960)].

on potential up to very high current densities (10 A/cm^2). Instead of the expected exponential dependence, however, it was shown [cf. Despić and Bockris (1960)] that in the system Ag/AgClO$_4$ at $25°$C the current density–potential relation deviates toward an apparent limiting current density (Fig. 8). The cathodic and anodic trends are symmetrical, thus indicating that transport control does not give an explanation of the results. The interpretation is that the overpotential had become large enough so that the electrical energy supplied ($F\eta$) had become comparable with the height of the activation energy barrier.

As a consequence (in contrast to what would happen with high-energy barrier processes, where a certain overpotential produces shifts of potential energies which are small compared to the barrier height), the potential-induced shifts in the potential-energy curves result in a significant change in the shape of the energy barrier. Since the symmetry factor β reflects tangents to the curves forming the barrier at their point of intersection (cf. Fig. 9), it cannot remain constant. It has been shown elsewhere (Bockris, 1954) that

$$\beta = -(\tan \gamma)/(-\tan \gamma + \tan \theta). \qquad (2.20)$$

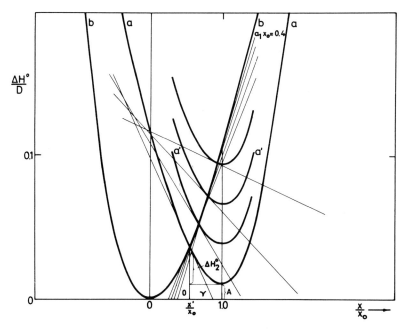

Fig. 9. The potential-energy–distance relations (Morse functions with $ax_0 = 0.4$) at different values of electrode potential [from Despić and Bockris (1960)].

If the potential-energy curves of Fig. 9 are approximated by Morse functions

$$U_1 = D_1\{1 - \exp[-a_1 x_0(x/x_0)]\}^2 \tag{2.21}$$

and

$$U_2 = D_2\big(1 - \exp\{-a_2 x_0[1 - (x/x_0)]\}\big)^2 + A \tag{2.22}$$

pertaining to solvated ions and adions respectively, with A, D, a_1, and a_2 constants, it can be deduced that

$$\beta = \frac{\exp\left[-a_2 x_0\left(1 - \dfrac{x'}{x_0}\right)\right] - \exp\left[-2a_2 x_0\left(1 - \dfrac{x'}{x_0}\right)\right]}{\exp\left[-a_2 x_0\left(1 - \dfrac{x'}{x_0}\right)\right] \exp\left[-2a_1 x_0\left(1 - \dfrac{x'}{x_0}\right)\right]} \tag{2.23}$$
$$+ \left[\frac{D_1 a_1}{D_2 a_2}\exp\left(-a_1 x_0\,\frac{x'}{x_0}\right) - \exp\left(-2a_1 x_0\,\frac{x'}{x_0}\right)\right]$$

The relative distance x/x_0 of the peak of the energy barrier can be deduced from geometric considerations as

$$\frac{x'}{x_0} = \frac{a_2 x_0}{a_1 x_0 + a_2 x_0} - \frac{1}{a_1 x_0 + a_2 x_0}$$
$$\times \ln \frac{1 - \{[\Delta H_1^{0*} - (1 - \beta)F\eta]/D_1\}^{1/2}}{1 - [(\Delta H_1^{0*} - A + \beta F\eta)/D_2]^{1/2}}. \tag{2.24}$$

Introducing (2.24) into (2.23), the potential dependence of the symmetry factor β is established. Assuming reasonable values of the constants and introducing the varying β into Eq. (2.4), the current-density–potential relationship shown in Fig. 10(b) is obtained (taking values of over-potential relative to $\Delta H°$). This is seen to follow the same pattern as that established experimentally. In this way Despić and Bockris (1960) established the concept of an activation-controlled limiting current.

  d. Discharge of Divalent Ions.   The introduction of transient methods of studying electrode kinetics, which avoid the problems of the steady state in electrodeposition of metals, stimulated intensive research in the 1950's and 1960's directly or indirectly concerned with the kinetics and mechanism of discharge of divalent ions. Reviews [e.g., cf. Tanaka and Tamamushi (1964), Jovanović (1969)] reveal a particularly large number of studies of the reduction of Cu^{++} ions (Philbert, 1943; Randles and Somerton, 1952; Hilson, 1954; Kambara and Ishii, 1961; Bockris and Enyo, 1962b; Bockris and Kita, 1962); Cd^{++} ions (Randles, 1947; Van Cakenberghe, 1951; Randles and Somerton, 1952; Gerischer,

1953a,b; Lorenz, 1953, 1954a,b; Berzins and Delahay, 1955; Lossew, 1956a,b; Vielstich and Delahay, 1957; Gerischer and Krause, 1957; Delahay, 1957; Delahay and Trachtenberg, 1958; Barker *et al.*, 1958; Bauer and Elving, 1958; Bauer *et al.*, 1960; Barker, 1961; Laitinen *et al.*, 1961; Müller and Lorenz, 1961; Randles, 1961; Kambara and Ishii, 1961; Pamfilov *et al.*, 1961, 1962; Despić and Jovanović, 1964); Zn^{++} ions (Roiter *et al.*, 1939a,b; Randles, 1947; Randles and Somerton, 1952;

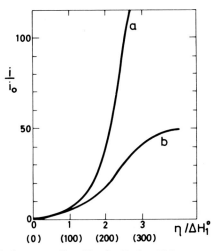

FIG. 10. The η–i relation as predicted by the Butler–Volmer equation (a) with constant β and (b) with β varying with η [from Despić and Bockris (1960)].

Gerischer, 1953a,b; Ershler and Rosental, 1953; Delahay, 1953; Lossew, 1955; Vielstich and Gerischer, 1955; Morinaga, 1955; Matsuda and Ayabe, 1955; Gerischer, 1956; Morinaga, 1956; Imai, 1958; Tamamushi and Tanaka, 1959; Matsuda and Ayabe, 1959; Imai, 1959; Sluyters and Oomen, 1960; Laitenen *et al.*, 1961; Müller and Lorenz, 1961; Gavioli and Popoff, 1961; Rius *et al.*, 1961; Kambara and Ishii, 1961; Hurlen, 1962a,b; Baticle, 1962; Behr, *et al.* 1962; Delahay and Aramata, 1962; Koryta, 1962; Dirkse, 1962; Matsuda and Ayabe, 1962; Tamamushi *et al.*, 1962; Tamamushi and Tanaka, 1963); and Fe^{++} ions (cf. Section II,A,3). Considerable discrepancies in the values of the kinetic parameters exist, due to the fact that experimental conditions were hardly ever comparable.

In several instances [cf., e.g., Bockris and Enyo (1962b)] the deposition was shown to follow a dependence corresponding to Eq. (2.4), i.e.,

$$i = i_0\{\exp(\alpha_a F\eta/RT) - \exp(\alpha_c F\eta/RT)\}, \tag{2.25}$$

where α_a and α_c are the transfer coefficients.* Assuming that all steps but one are at equilibrium, it was shown [cf. Despić (1969)] that the transfer coefficients are

$$\alpha_a = (1 - \beta)n_d + (n'/\nu_d) \qquad (2.26)$$

$$\alpha_c = -[\beta n_d + (n''/\nu_d)], \qquad (2.27)$$

where n_d is the number of electrons exchanged in the rate-determining step; n' and n'' are the numbers of electrons exchanged after and before this step, respectively; β is the symmetry factor; ν_d is the stoichiometric number (i.e., the number of acts of the rate-determining step per act of the overall reaction). Some representative kinetic parameters are listed in Table III.

Four possible mechanisms can be foreseen for the discharge of divalent ions:

$$M^{++} + 2e \rightleftarrows M, \qquad (2.28a)$$

$$M^{++} + e \rightleftarrows M^+; \qquad M^+ + e \rightleftarrows M, \qquad (2.28b)$$

$$M^{++} + M \rightleftarrows 2M^+; \qquad M^+ + e \rightleftarrows M, \qquad (2.28c)$$

$$M^{++} + e \rightleftarrows M^+; \qquad 2M^+ \rightleftarrows M^{++} + M. \qquad (2.28d)$$

Orders of reaction, transfer coefficients calculated from Eqs. (2.26) and (2.27), and stoichiometric numbers pertaining to each of these mechanisms are shown in Table IV.

Although one can argue that, except in the case of Cu^+, univalent ions are not known for the majority of divalent metals, they may exist as strongly adsorbed unstable intermediates. A direct two-electron exchange [mechanism (2.28a)] is unlikely because the high energy of activation gives a low probability of two electrons tunneling simultaneously. Of course, experimental data can be made formally to fit kinetic equations in which $\alpha_a = (1 - \alpha)n$ and $\alpha_c = \alpha n$ for $n = 2$, if any value of α is considered acceptable. Yet in the case of mechanism (2.28a) α_a and α_c must be equal to the symmetry factors $(1 - \beta)$ and β multiplied by n, respectively (since $n' = n'' = 0$) and hence α should be equal to β.

Calculations (Despić and Bockris, 1960) of β in terms of ratios of force constants showed that, using all force constants available, $0.4 < \beta < 0.6$. If $\beta = 0.5$, $(1 - \alpha)n_d = 1$ and $\alpha n_d = 1$ would be in disagreement with most experimental findings. Indeed, a number of authors have given

---

* Note that a considerable variety of definitions of transfer coefficients is found in the literature. The most often employed are $(1 - \alpha)n$ for α_a and αn for α_c. Since multiplication by the total number of electrons n has no physical justification in any other but in the case of an n-electron exchange mechanism, the general definition implied in α_c and α_a, as any factors multiplying the electrical energy term FE, is used here [cf. Despić (1969)].

TABLE III
KINETIC PARAMETERS FOR SOME MULTIELECTRON EXCHANGE REACTIONS

System	Medium	i_0^0 [a]	α_a	α_c	Ref.
Cu/0.05 M Cu^{++}	1.0 N H$_2$SO$_4$	0.51	1.47	0.49	Bockris and Kita (1962)
0.25 M Cd(Hg)/0.05 M Cd^{++}	1.0 N H$_2$SO$_4$	59	0.64	1.36	Despić et al. (1969); Lovreček and Marinčič (1966)
1.5 × 10^{-3} M Zn(Hg)/10^{-3} M Zn^{++}	NaClO$_4$	466	2.04–1.2	0.24–0.28	Lossew et al. (1967); Koryta (1962)
Fe/0.5 M Fe^{++}	0.5 M Na$_2$SO$_4$ + 10^{-2} M H$_2$SO$_4$	1.35 × 10^{-5}	0.123	0.48	Bockris et al. (1961)

[a] i_0^0 is the standard exchange current density at unit concentration of depositing ions (A cm mole^{-1}).

TABLE IV
KINETIC PARAMETERS FOR POSSIBLE MECHANISMS OF THE TWO-ELECTRON EXCHANGE REACTION [a]

Mechanism	RDS	$n_{a,ox}$	$n_{a,R}$	$n_{c,ox}$	$n_{c,R}$	α_a	α_c	ν_d
(2.28a)	I	0	1	1	0	$2(1-\beta)\approx 1$	$-2\beta\approx -1$	1
(2.28b)	I	0	1	1	0	$(2-\beta)\approx 1.5$	$-\beta\approx -0.5$	1
(2.28b)	II	0	1	1	0	$(1-\beta)\approx 0.5$	$-(1+\beta)\approx -1.5$	1
(2.28c)	I	0	1	1	1	1	0	1
(2.28c)	II	0	1	1	1	$(1-\beta)\approx 0.5$	$-\beta\approx -0.5$	2
(2.28d)	I	1	1	1	0	$(1-\beta)\approx 0.5$	$-\beta\approx -0.5$	2
(2.28d)	II	1	1	1	0	0	-1	1

[a] $n_{a,ox}$ and $n_{a,R}$ are the reaction orders for the anodic reaction with respect to the oxidized form (divalent ion) and the reduced form (metal atoms), respectively; $n_{c,ox}$ and $n_{c,R}$ are the same quantities for the cathodic reaction.

good experimental evidence that mechanisms involving univalent intermediates prevail. Thus Lossew (1955) has shown the current-density–potential relation for Zn^{++} discharge (Fig. 11) to give different cathodic and anodic slopes, as expected in the case of mechanism (2.28b) (Table IV), and suggested the existence of a mechanism involving Zn^+ ions.

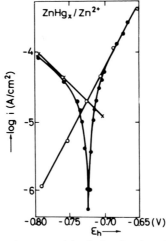

FIG. 11. The current-density–potential relation for deposition and dissolution of zinc amalgam (0.6 at. %) in O. 0.1 N $ZnSO_4$ + 5 × 10^{-5} M [$(C_4H_9)_4N]_2SO_4$. ○: Partial current densities calculated from radiotracer experiments (^{65}Zn in the amalgam); ●: experimental [from Lossew (1955)].

Mattson and Bockris (1959), Bockris and Enyo (1962b), and Bockris and Kita (1962) independently showed similar behavior in Cu^{++} ion discharge. Finally, Bockris et al. (1961) have shown the same pattern in iron deposition and suggested the existence of an FeOH species (for which spectroscopic evidence exists) (cf. Section II,A,3). This experimental evidence indicates that the slow step is that of the divalent-to-monovalent ion transition, which is in agreement with energy considerations (i.e., transfer to a heavily-hydrated divalent ion is more difficult than to a less-hydrated monovalent one).

Yet, it may be argued that the concept of the faster step being at equilibrium is not of general value. Vetter (1952, 1953a,b) has pointed out that if the two steps are similarly slow, so that both require appreciable activation overpotential, the current density is given by

$$i = 2i_{0,B'}i_{0,B''} \frac{\exp[(2 - \beta_{B'} - \beta_{B''})F\eta/RT] - \exp[-(\beta_{B'} + \beta_{B''})F\eta/RT]}{i_{0,B'}\exp[(1 - \beta_{B'})F\eta/RT] + i_{0,B''}\exp[-\beta_{B''}F\eta/RT]}$$

$$(2.29)$$

where subscripts B' and B'' signify pertaining to the first and the second step of Eq. (2.28b), respectively.

Two cases may arise here: (1) If the difference in i_0 values for the two steps is large enough, the current-density–potential relation on one side (e.g., anodic) should have three regions: the Butler–Volmer region, a Tafel region with a slope predicted by the equilibrium concept (Table IV), and another Tafel region at very high overpotentials with a slope corresponding to a single-electron exchange reaction, i.e., $RT/(1 - \beta)F$. For known i_0 values the overpotentials at which this transition takes place can be calculated from the equality of the two terms in the denominator of Eq. (2.29). Conversely, if the overpotential at which such a transition occurs is known, the ratio of the two i_0 values can be found. Since in the work on Cu^{++}, Zn^{++} (discharge on zinc amalgam), and Fe^{++} ion discharge single Tafel lines for the anodic direction with slopes of $2RT/3F$ are reported up to overpotentials of ~ 200 mV, this may be taken as an indication that the i_0 values of the second step are more than ten times larger than the i_0 values of the slow step. (2) If the difference in i_0 values is small, there is a chance for the first term in the numerator to become much larger than the second before the Tafel region of overpotentials is reached. In that case at larger overpotentials Eq. (2.29) reduces to

$$i = i_{0,B''} \exp[(1 - \beta_{B''})F\eta/RT], \qquad (2.30)$$

so that a single anodic Tafel line is obtained but with a slope of $RT/(1 - \beta)F \approx 2RT/F$, i.e., the same as for the cathodic line. Such behavior has not yet been reported.

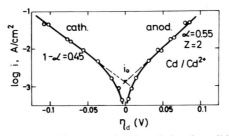

FIG. 12. Activation overpotential–current density relation for solid cadmium in 0.01 N $Cd^{++} + 0.8\ N\ K_2SO_4$ at 20°C [from Lorenz (1954b)].

Symmetrical Tafel lines have been recorded for the discharge of Cd^{++} (Lorenz, 1954a,b) and Zn^{++} (Roiter et al., 1939b) on the respective solid metals, but with slopes of RT/F (cf. Fig. 12). These cases are consistent with single-step two-electron exchange [mechanism (2.28a)] in spite of the very high activation of this step suggested by the potential-energy calculations (Conway and Bockris, 1958).

A method for obtaining the rate constants for a fast step in a deposition process was first given by Mehl and Bockris (1957), utilizing galvanostatic transients. The i_0 value of the faster step is also obtainable from Faradaic impedance measurements (Despić *et al.*, 1969). The method assumes, however, a frequency-independent double-layer capacity, and could therefore be used only for liquid metals (amalgams), since at solid surfaces this condition is not satisfied [cf. Bockris and Conway (1958), Vetter (1961)].

e. Univalent Intermediates and Pseudocapacitance Effects. When mechanisms involving univalent ions as intermediates are operative the electrode must exhibit pseudocapacitance effects in transient or ac phenomena. This is due to the fact that the system tends to follow any change in potential by adjusting the concentration of the intermediate C_{M^+} to the corresponding equilibrium value. This can be done only by exchanging charge with the electrode, i.e., by passing current. Suppose that in mechanism (2.28b) the second step is sufficiently fast and that all the ions are sufficiently available (e.g., intermediate M^+ fixed at the surface) that given any potential, the electrode can establish instantaneously the equilibrium between M^+ and M. Since at any overpotential η

$$\eta = E - E_{\text{rev}} = (RT/F) \ln(C_{M^+}/C_{M^+,0}), \tag{2.31}$$

where $C_{M^+,0}$ is the equilibrium concentration, the rate of change of the concentration C_{M^+} with time is

$$\frac{dC_{M^+}}{dt} = \frac{FC_{M^+,0}}{RT} \exp\left(\frac{nF}{RT}\right) \frac{d\eta}{dt}. \tag{2.32}$$

The current density needed to produce this concentration change is given by

$$i_{\text{PC}} = FV(dC_{M^+}/dt), \tag{2.33}$$

where V is the volume in which the concentration change is effected. Hence when, e.g., a galvanostatic pulse is applied to the electrode, at the very beginning ($\eta \to 0$) the part of the current needed to change the concentration is obtained from (2.33) and (2.32) as

$$i_{\text{PC}} = \frac{VF^2(C_{M^+,0})}{RT} \left(\frac{d\eta}{dt}\right)_{\eta \to 0} = C_{\text{PC}}\left(\frac{d\eta}{dt}\right)_{\eta \to 0}. \tag{2.34}$$

The factor with $d\eta/dt$, C_{PC}, represents the pseudocapacitance which is acting in parallel with the capacitance of the double layer. It is seen to be determined essentially by the equilibrium concentration of the intermediate species and by the volume of solution involved. If the intermediate species is fixed at the surface, V is of the order of 10^{-8} cc.

However, if the intermediate is spread into the bulk of the solution, V becomes a complex function of the diffusion properties. It is important to note that the derivation of the Eq. (2.34) does not depend on the position of the rate-determining step (i.e., the same result is obtained if it is assumed that the first step is fast and the second one is slow), but only on the condition that one of the steps is so much faster than the other that it can be considered at equilibrium.

The concentration of the univalent cuprous ion intermediate was first found by Bockris and Enyo (1962b) in the galvanostatic investigation of copper deposition.

The ac impedance method can also be used for that purpose (Despić et al., 1969) provided the double-layer capacitance can be considered independent of frequency (liquid surfaces).

f. Rate Constant as a Function of the Nature of the Substrate.   Little is known about this subject. Some changes of metal-deposition parameters are observed when the substrate is prepared in various ways. The surfaces for which comparable data exist [cf. Bockris and Kita (1962)] are those of copper prepared by quenching the metal in hydrogen or helium, by electrodeposition, dissolution, and chemical deposition, and also those surfaces partly covered by oxide film. It is shown that on all surfaces the transfer coefficients for the process are the same, and hence no change in mechanism is indicated. Two groups of the i_0 values are obtained which differ by about one order of magnitude. Electrodeposited copper and such copper covered by oxide have relatively low i_0 values, all the others being considerably higher. The effect is an indirect one, rather than being related to basic properties of the substrate, such as its work function or the effect of the substrate on the properties of Cu^+ and Cu^{++} ions in the Helmholtz layer. Bockris and Kita (1962) have shown that the difference in behavior is consistent with a higher surface coverage by adions in the first group. The fraction of free surface is reduced. Thus

$$i_0 = \overrightarrow{k}^0 C_{M^{z+}} \exp\left(-\frac{\beta F E_{rev}}{RT}\right) \exp\left(-\frac{\Delta \overrightarrow{H}^*}{RT}\right)(1 - \theta_{ad})$$

$$= \overleftarrow{k}^0 C_{M^+} \exp\left(\frac{(1-\beta)F E_{rev}}{RT}\right) \exp\left(-\frac{\Delta \overleftarrow{H}^*}{RT}\right)(1 - \theta_{ad}). \quad (2.35)$$

The surface coverage θ_{ad} was estimated to be 0.8 for the first group and 0.07 for the second group of surfaces. These differences are consistent with a tenfold difference in i_0 values, as is observed.

3. The Mechanism of the Deposition of Iron

The importance of this metal and the problems of its stability in different media appears to merit more specific attention to its electrodic behavior. The conclusions may have relevance to the deposition and dissolution of Co and Ni.

a. Facts about the Deposition of Iron. Examination of the kinetics of iron deposition involves difficulties because of the simultaneous evolution of hydrogen. In addition, hydrogen evolution produces local changes of pH at the vicinity of the electrode surface, and since the pH is known (Hoar and Hurlen, 1958) to affect the deposition kinetics, the introduction of a method of estimating the pH in the double layer as a function of the solution pH is a prerequisite for interpreting the experimental results.

In general, transition metals are found to have i_0 values in the range of $\mu A/cm^2$ or less. Hence, activation control prevails over a large range of overpotentials. Typical potential transients upon application of constant current are shown in Fig. 13. An analysis (Bockris *et al.*, 1961) shows that the cathodic galvanostatic transient should contain three plateaus. The first should be maintained by the deposition of hydrogen from H_3O^+ ions. Yet, at higher pH values and higher current densities the transition time for this reaction is so short that it is not observed in usual galvanostatic transients. The second is due to Fe^{++} ion discharge, while the third is sustained by hydrogen evolution from water. The plots of the Fe^{++} plateau overpotentials vs. log i yield Tafel lines extending in less-acidic solutions over two decades of current densities. These exhibit slopes of RT/F, characteristic of a single-step two-electron exchange mechanism (cf. Table IV). Conversely, clear Tafel sections were obtained for anodic dissolution with slopes of $2RT/3F$. The i_0 was found to be pH dependent, as shown in Fig. 14. The concentration dependencies were characterized by

$$\left(\frac{\partial \log i_0}{\partial \log C_{Fe^{2+}}}\right)_{pH} = 0.8; \qquad \left(\frac{\partial \log i_0}{\partial \log C_{OH^-}}\right)_{C_{Fe^{2+}}} = 1.$$

Anions affect the rate constants in the order $ClO_4^- > SO_4^{--} > Cl^- > Ac^- > NO_3^-$.

b. Local pH Change during Deposition. Bockris *et al.* (1961) have attributed the apparent contradiction between the Tafel slopes for cathodic and anodic processes to the effect of the local pH change in the double-layer during deposition. This can be accessed if the discharge of

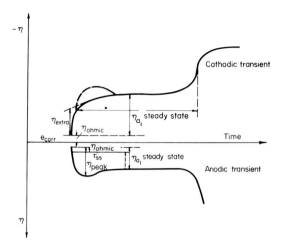

FIG. 13. Typical cathodic and anodic transients for iron in 0.05 M FeSO$_4$ of pH $= 3$. η_{extra}: Extrapolated value of overpotential at $t = 0$. η_{ohmic}: Pseudoohmic overpotential. η_{a_1}: Activation overpotential pertaining to the reaction of iron. τ_{ss}: Time necessary to establish steady state. τ: Transition time. [From Bockris *et al.* (1961).]

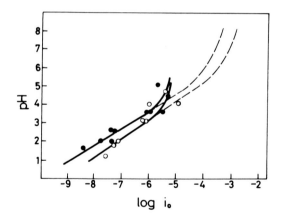

FIG. 14. pH dependence of the exchange current density for iron deposition and dissolution. Open circles: 0.5 M FeSO$_4$ + 0.5 M Na$_2$SO$_4$ solution. Darkened circles: 0.05 M FeSO$_4$ + 0.15 M Na$_2$SO$_4$ solution. [From Bockris *et al.* (1961).]

H_3O^+ is assumed to occur at its limiting current, when the total current exceeds the latter. The concentration overpotential for H_3O^+ discharge is

$$\eta_c = E_{Fe} - E_{H,\eta_c=0} = \frac{RT}{\alpha_H F} \ln \frac{(a_{H_3O^+})_s}{(a_{H_3O^+})_b} = \frac{RT}{\alpha_H F} \ln \frac{(a_{OH^-})_b}{(a_{OH^-})_s}, \quad (2.36)$$

where E_{Fe} is the actual potential of the iron electrode and $E_{H,\eta_c=0}$ is the potential the electrode would have if only the hydrogen evolution occurred upon it at the given current density and in the absence of concentration overpotential, i.e., at $\eta_c = 0$. The $(a_i)_s$ and $(a_i)_b$ are the ionic activities at the surface and in the bulk of solution, respectively.

Knowing the reaction order with respect to OH^- to be 1, the kinetic equation for the cathodic reduction of Fe^{++} ions can be written as

$$\overleftarrow{i}_{Fe} = \overleftarrow{k}_{Fe}(a_{OH^-})_s \exp(-\alpha_{c,Fe} F E_{Fe}/RT). \quad (2.37)$$

Differentiating Eq. (2.37),

$$\frac{\partial(\ln \overleftarrow{i}_{Fe})}{\partial E_{Fe}} = -\frac{\alpha_{c,Fe} F}{RT} + \frac{\partial[\ln(a_{OH^-})_s]}{\partial E_{Fe}}. \quad (2.38)$$

Considering Eq. (2.36), Eq. (2.38) can be written as

$$\frac{\partial(\ln \overleftarrow{i}_{Fe})}{\partial E_{Fe}} = -\frac{\alpha_{c,Fe} F}{RT} - \frac{\partial}{\partial E_{Fe}}(E_{Fe} - E_{H,\eta_c=0})\frac{\alpha_H F}{RT}. \quad (2.39)$$

The transfer coefficient, $\alpha_{c,Fe}$, for the iron deposition reaction (unaffected by H deposition) can be found from Eq. (2.39) if the experimental value for the Tafel slope (affected by H deposition) is taken as

$$\partial(\ln \overleftarrow{i}_{Fe})/\partial E_{Fe} = -F/RT.$$

Hence

$$\alpha_{c,Fe} = 1 - \frac{\partial}{\partial E_{Fe}}(E_{Fe} - E_{H,\eta_c=0}) = 1 - \alpha_H - \alpha_H \frac{\partial E_{H,\eta_c=0}}{\partial E_{Fe}}. \quad (2.40)$$

The concentration overpotential for H_3O^+ ions η_c can be expressed in terms of the limiting current i_L and the hydrogen deposition current i_H as

$$\eta_c = E_{Fe} - E_{H,\eta_c=0} = (RT/\alpha_H F) \ln[(i_L - i_H)/i_L]. \quad (2.41)$$

From Eq. (2.41)

$$\frac{\partial E_{H,\eta_c=0}}{\partial E_{Fe}} = \frac{1}{1 + [i_H/(i_L - i_H)]}. \quad (2.42)$$

Introducing this into Eq. (2.40), one obtains

$$\alpha_{c,Fe} = 1 - \alpha_H + \frac{\alpha_H}{1 + [i_H/(i_L - i_H)]}. \qquad (2.43)$$

As $i_H \to i_L$ the last term in Eq. (2.43) tends to zero and $\alpha_{c,Fe} = 1 - \alpha_H$. From measurements of the hydrogen evolution reaction on the same iron electrode and for the same conditions as those from which Tafel lines for iron are deduced the transfer coefficient α_H is found to be 0.51. Hence $\alpha_{c,Fe} = 0.49$ and the cathodic Tafel line would have a slope of $2RT/F$ if the pH in the vicinity of the electrode were constant with changing current density. The anodic and cathodic Tafel lines obtained experimentally and the corrected Tafel line are shown in Fig. 15.

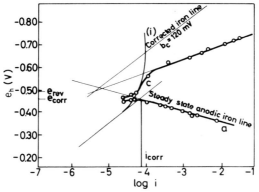

FIG. 15. (a) Anodic and (c) cathodic $\log i$ vs. potential (hydrogen scale) lines for iron deposition in 0.5 M FeSO$_4$ + 0.5 M Na$_2$SO$_4$ (pH = 4.0) and the line corrected for the local change in pH [from Bockris *et al.* (1961)].

c. Probable Mechanism of Deposition of Iron and Similar Electronegative Metals. Three facts are dominant in considering the problem of mechanism: (1) The reaction is pH-dependent and hence must involve some products of hydrolysis rather than Fe^{++} or hypothetical Fe^+ ions (calculations show that species like $FeOH^+$ are present in significant concentrations even in relatively acidic media); (2) Tafel slopes of $2RT/F$ and $2RT/3F$ are obtained in the majority of experiments carried out under conditions of high purity; (3) results indicate that small variations in experimental conditions (e.g., degree of purity) may produce significant changes in kinetic behavior, suggesting that the energy differences between various paths are small [e.g., in the results of Bockris *et al.* (1961) about 10% of the anodic Tafel lines had slopes of $RT/2F$; the same result was reported by Hoar and Hurlen (1958) for all anodic measurements].

Series of possible mechanisms have been suggested by different authors. They are represented in Table V. Table VI lists the diagnostic criteria derived for the mechanisms suggested and compares them with the experimentally obtained parameters. The latter are consistent with mechanism A only, and, with due caution necessary because of factor (3) above, this can be considered as the most probable mechanism.*

Bockris and Dražić (1962) investigated the effect of trace impurities (0.1%) in the metal phase and found no significant effect on iron dissolution kinetics (an increase in i_0 for hydrogen evolution from H_3O^+ was found to arise from 0.1% of C in iron).

B. SURFACE DIFFUSION AND INCORPORATION

1. Liquid and Solid Surfaces

While at liquid surfaces it is only charge transfer or transport which can control the deposition kinetics, at solid metals specific kinds of polarizations can arise which are due to inhibited steps in the crystal building process [cf., e.g., Lorenz (1954a,b), Mehl and Bockris (1957), Gerischer (1958), Vermilyea (1963)], in particular, the surface diffusion of adions. Good evidence for the existence of the latter is obtained by Bockris and Enyo (1962a), who compared the kinetics of deposition of gallium on liquid and solid gallium surfaces at practically the same temperature around the melting point. Considerable difference in current density at equal overpotentials is observed upon solidification in the region of low overpotentials (< 50 mV), as shown in Fig. 16. The rate

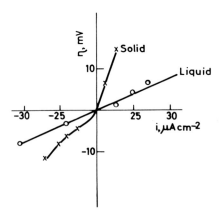

FIG. 16. Current density–overpotential relationship for deposition of gallium ions on liquid (○) and solid (×) gallium [from Bockris and Enyo (1962a)].

---

* This work has recently been confirmed by Epelboin and Morel (1968) using electrode impedance measurements.

TABLE V

POSSIBLE MECHANISMS FOR THE DEPOSITION OF IRON

Mechanism A (Bockris et $al.$, 1961)

$Fe + OH^- \rightleftarrows FeOH + e$; $\quad FeOH \xrightarrow{RDS} FeOH^+ + e$;

$FeOH^+ \rightleftarrows Fe^{++} + OH^-$

or

$Fe + H_2O \rightleftarrows FeOH^+ + H^+ + e$; $\quad FeOH \xrightarrow{RDS} FeOH^+ + e$;

$FeOH^+ + H^+ \rightleftarrows Fe^{++} + H_2O$

Mechanism B (Heusler, 1958)

$Fe + H_2O \rightleftarrows FeOH_{ads} + H^+ + e$; $\quad FeOH_{ads} + Fe \rightleftarrows [Fe(FeOH)]$;

$[Fe(FeOH)] + OH^- \rightarrow FeOH^+ + FeOH_{ads} + 2e$;

$FeOH^+ + H^+ \rightleftarrows Fe^{++} + H_2O$

Mechanism C (Bockris et $al.$, 1961)

$OH^-_{sol} \rightleftarrows OH_{ads} + e$; $\quad OH_{ads} + 2Fe \rightleftarrows Fe_2OH$;

$Fe_2OH \xrightarrow{RDS} Fe^{++} + FeOH + 2e$; $\quad FeOH \rightleftarrows Fe^{++} + OH^-_{sol} + e$

or

$H_2O \rightleftarrows OH_{ads} + H^+ + e$; $\quad OH_{ads} + 2Fe \rightleftarrows Fe_2OH$;

$Fe_2OH \xrightarrow{RDS} Fe^{++} + FeOH + 2e$; $\quad FeOH + H^+ \rightleftarrows Fe^{++} + H_2O + e$

Mechanism D (Bockris et $al.$, 1961)

$Fe + H_2O \rightleftarrows FeOH + H^+ + e$; $\quad FeOH \rightleftarrows FeOH^+ + e$;

$FeOH^+ + Fe \xrightarrow{RDS} Fe_2OH^+$; $\quad Fe_2(OH)^+ \rightleftarrows Fe^{++} + FeOH + e$;

$FeOH + H^+ \rightleftarrows Fe^{++} + H_2O + e$

Mechanism E (Bockris et $al.$, 1961)

$Fe + OH^- \xrightarrow{RDS} Fe(OH)^+ + 2e$; $\quad Fe(OH)^+ \rightleftarrows Fe^{++} + OH^-$

Mechanism F (Bockris et $al.$, 1961)

$Fe + 2 OH^- \xrightarrow{RDS} Fe(OH)_2 + 2e$; $\quad Fe(OH)_2 \rightleftarrows Fe^{++} + 2 OH^-$

Mechanism G (Kabanov et $al.$, 1947)

$Fe + OH^- \rightleftarrows FeOH + e$; $\quad FeOH + OH^- \xrightarrow{RDS} FeO + H_2O + e$;

$FeO + OH^- \rightleftarrows HFeO_2^-$; $\quad HFeO_2^- + H_2O \rightleftarrows Fe(OH)_2 + OH^-$;

$Fe(OH)_2 \rightleftarrows Fe^{++} + 2 OH^-$

Mechanism H (Hoar and Hurlen, 1958; Hurlen, 1960)

$2 Fe + OH^- \rightleftarrows 2 Fe^{++} + OH^- + 4e$

Mechanism I (Christiansen et $al.$, 1961)

$FeOH^- \rightleftarrows FeOH^+ + 2e$; $\quad FeOH^+ \rightleftarrows Fe^{++} + OH^-$

Mechanism J (Florianovich et $al.$, 1967)

$Fe + H_2O \rightleftarrows (FeOH^-)_{ads} + H^+$; $\quad (FeOH^-)_{ads} \rightleftarrows (FeOH)_{ads} + e$;

$(FeOH)_{ads} + HSO_4^- \rightarrow FeSO_4 + H_2O + e$

or

$(FeOH)_{ads} + SO_4^{--} \rightarrow FeSO_4 + OH^- + e$; $\quad FeSO_4 \rightleftarrows Fe^{++} + SO_4^{--}$

or

$FeSO_4 + H_2O \rightleftarrows FeHSO_4^- + OH^-$

TABLE VI

COMPARISON OF EXPERIMENTAL BEHAVIOR FOR IRON DEPOSITION AND DISSOLUTION WITH PREDICTIONS FROM VARIOUS MECHANISMS[a]

Quantity	Mechanism[b]								Experimental results (coefficients × 2.303)
	A	B	C	D	E	F	G	H	
$\dfrac{\partial E_{Fe}}{\partial(\ln i_a)}$	$\dfrac{2}{3}\dfrac{RT}{F}$	$\dfrac{RT}{2F}$	$\dfrac{RT}{2F}$	$\dfrac{RT}{2F}$	$\dfrac{RT}{F}$	$\dfrac{RT}{F}$	$\dfrac{2}{3}\dfrac{RT}{F}$	$\dfrac{RT}{2F}$	0.042 ± 0.008
$\dfrac{\partial E_{Fe}}{\partial(\ln i_c)}$	$-\dfrac{2RT}{F}$	$-\dfrac{RT}{2F}$	$-\dfrac{RT}{2F}$	$-\dfrac{RT}{2F}$	$-\dfrac{RT}{F}$	$-\dfrac{RT}{F}$	$-\dfrac{2RT}{F}$	$-\dfrac{RT}{2F}$	-0.116 ± 0.006
$\left[\dfrac{\partial(\ln i)}{\partial(\ln a_{Fe^{++}})}\right]_{E_{Fe}, a_{OH^-}}$	1	2	2	2	1	1	1	2	0.8
$\left[\dfrac{\partial(\ln i_0)}{\partial(\ln a_{OH^-})}\right]_{a_{Fe^{++}}}$	1	2	1	1	1	2	2	1	0.9 ± 0.05
$\left[\dfrac{\partial E_{Fe}}{\partial(\ln a_{OH^-})}\right]_{a_{Fe^{++}},i}$	$\dfrac{2}{3}\dfrac{RT}{F}$	$-\dfrac{RT}{F}$	$-\dfrac{RT}{2F}$	$\dfrac{RT}{2F}$	$\dfrac{RT}{F}$	$\dfrac{RT}{F}$	$-\dfrac{4}{3}\dfrac{RT}{F}$		
$\left[\dfrac{\partial(\ln i_0)}{\partial(\ln a_{Fe^{++}})}\right]_{a_{OH^-}}$	$\dfrac{3}{4}$	1	1	1	$\dfrac{1}{2}$	$\dfrac{1}{2}$	$\dfrac{3}{4}$	1	0.8 ± 0.1
$\left[\dfrac{\partial E_{corr}}{\partial(\ln a_{OH^-})}\right]_{a_{Fe^{++}}}$	$-\dfrac{RT}{F}$	$\dfrac{6RT}{5F}$	$\dfrac{4RT}{5F}$	$\dfrac{4RT}{5F}$	$\dfrac{4RT}{3F}$	$\dfrac{2RT}{F}$	$\dfrac{3}{2}\dfrac{RT}{F}$	$\dfrac{4RT}{5F}$	-0.060 ± 0.003
$\left[\dfrac{\partial(\ln i_{corr})}{\partial(\ln a_{OH^-})}\right]_{a_{Fe^{++}}}$	$-\dfrac{1}{2}$	$-\dfrac{2}{5}$	$-\dfrac{3}{5}$	$-\dfrac{3}{5}$	$-\dfrac{1}{3}$	0	$-\dfrac{1}{4}$	$-\dfrac{3}{5}$	-0.5 ± 0.01

[a] $\beta = 0.5$.
[b] See Table V.

expression for the case of mixed surface diffusion and activation control must take into account the change of the concentration of adions from the equilibrium value, i.e., an accumulation of adions because of slow flux of adions toward growth sites. Thus, taking the average concentration of adions to be C_{ad},

$$i = i_0\left[\frac{C_{ad}}{C_{ad,0}}\exp\left(\frac{\alpha_a F}{RT}\eta\right) - \frac{G - C_{ad}}{G - C_{ad,0}}\exp\left(\frac{\alpha_c F}{RT}\eta\right)\right], \quad (2.44)$$

where G is the total number of metal atoms on the surface (mole/cm^2), i.e., maximum possible number of adions at surface coverage equal to 1.

It was shown (Mehl and Bockris, 1957; Despić and Bockris, 1960) that at the steady state

$$C_{ad}/C_{ad,0} = 1 + [(-i)/zFv_{ad,0}], \quad (2.45)$$

where $v_{ad,0}$ is the rate constant for surface diffusion. From Eqs. (2.44) and (2.45) it follows that

$$i = \frac{i_0\{[\exp(\alpha_a F/RT)\eta] - \exp[(-\alpha_c F/RT)\eta]\}}{1 + r\exp[(\alpha_a F/RT)\eta] + \theta'r\exp[(-\alpha_c F/RT)\eta]}, \quad (2.46)$$

where $r = i_0/zFv_{ad,0}$ and $\theta' = C_{ad,0}/G - C_{ad,0}$. Damjanović and Bockris (1963) have shown that the specific diffusion rate $v_{ad,0}$ is a function of the product of the concentration of dislocations (growth sites) N and the surface diffusion coefficient D_{ad}. Equation (2.46) indicates that unless full surface coverage is reached first, then at higher overpotentials pure activation control should be established, i.e., surface diffusion is a phenomenon which is expected to affect the deposition process only close to the reversible potentials. Indeed, Fig. 17 shows that at overpotentials

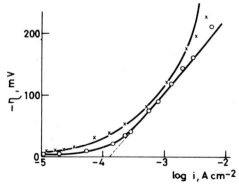

FIG. 17. Tafel plot for deposition of gallium ions on liquid (○) and solid (×) gallium. (A) Theoretical curve for charge-transfer control. (B) Theoretical curve of Bockris and Enyo (1962a).

> 50 mV the line pertaining to the deposition onto the solid approaches that for liquid gallium. The difference in state of the substrate has little influence on the velocity of the charge transfer reaction. In addition, the effect will be smaller the larger is the frequency factor (ND_{ad}) for the arrival of adions to a growth site by surface diffusion.

At still higher overpotentials the limiting current due to the saturation of the surface by adions seems to be approached, as indicated by Eq. (2.46) when the third term in the denominator becomes predominant. The situation is similar to that in which a limiting current in the hydrogen evolution reaction is observed as the surface approaches full coverage by atomic hydrogen.

2. Surface Concentration of Adions

Lorenz (1954a,b) was the first to incorporate equations concerning surface-diffusion control in the theory of electrodeposition, and did this to suggest that a determination of surface adion concentration should arise from ac measurements. Thus the steady-state ac current produced by small amplitude overvoltage $\eta = \eta' \sin \omega t$ is given by

$$= \frac{i_0 \eta' F}{RT} \left\{ \frac{\exp j\omega t}{2j} \left[\frac{j\omega}{K + j\omega} + \frac{K}{K + j\omega} \left(\frac{D}{K + j\omega} \right)^{1/2} \right. \right.$$
$$\left. \left. \times \frac{1}{x_0} \tanh \left(\frac{K + j\omega}{D} \right)^{1/2} x_0 + \text{conjugate} \right] \right\}, \qquad (2.47)$$

where $K = i_0/zFC_{ad,0}$ and $2x_0$ is the average spacing between two growth sites. Equation (2.47) can be rewritten in the form

$$i = (i_0 \eta' FE/RT) \sin(\omega t - \theta). \qquad (2.48)$$

The admittance i/η' corresponds to that of a pure resistor at low and high frequencies, and there must be a phase lead at intermediate frequencies. When the angle θ is $-45°$, $\omega = K$. Since at high frequencies the exchange current density i_0 can be found, determination of the frequency $\omega = K$ would allow the determination of $C_{ad,0}$.

Mehl and Bockris (1957) were the first to publish measurements of surface adion concentration. They use a galvanostatic transient method [cf. Gerischer (1958)]. Assuming $G \gg C_{ad}$ in Eq. (2.44) and linearizing it for $\eta_t < RT/\alpha F$, one obtains for any time t

$$-\eta_t = \frac{RT}{zF} \frac{i}{i_0} + \frac{RT}{zF} \frac{\Delta C_{n,t}}{C_{ad,0}}, \qquad (2.49)$$

where

$$\Delta C_{n,t} = C_{ad,n,t} - C_{ad,0}.$$

To relate $C_{ad,n,t}$ to time, Mehl and Bockris (1957) assumed that in the first approximation

$$dC/dt = (i_F/zF) - v, \qquad (2.50)$$

where v is the average surface diffusion flux of the adions during their passage from the point at which they are formed to that at which they meet a crystal growth site. At small departures from equilibrium it can be assumed that the flux is proportional to the difference in the actual average adion concentration and that at equilibrium (i.e., in the vicinity of the growth site where the latter is not disturbed):

$$v = v_{ad,0}\left(\frac{C_{ad,n,t} - C_{ad,0}}{C_{ad,0}}\right). \qquad (2.51)$$

Introducing Eq. (2.51) into (2.50) and integrating, one obtains

$$\frac{C_{ad,n,t} - C_{ad,0}}{C_{ad,0}} = \frac{\Delta C_{n,t}}{C_{ad,0}} = \frac{i}{zFv_{ad,0}}(1 - e^{-t/\tau}), \qquad (2.52)$$

where the time constant

$$\tau = C_{ad,0}/v_{ad,0}. \qquad (2.53)$$

Introducing Eq. (2.52) into (2.49) yields

$$-\eta_t = \frac{RT}{zF}\left[\frac{i}{i_0} + \frac{i}{zFv_{ad,0}}\left(1 - \exp\frac{-v_{ad,0}t}{C_{ad,0}}\right)\right]. \qquad (2.54)$$

At $t \to 0$, linearizing Eq. (2.54),

$$-\eta_t = \frac{RT}{zF}\left(\frac{i}{i_0} + \frac{i}{zFC_{ad,0}}t\right), \qquad (2.55)$$

or

$$\frac{d\eta}{dt} = \frac{RT}{z^2F^2}\frac{i}{C_{ad,0}} = \frac{1}{C_{PC}}i, \qquad (2.56)$$

where C_{PC} is the pseudocapacitance. As is seen, this is similar to that derived for the two-step single-electron exchange mechanism [cf. Eq. (2.34)]. It is in parallel with the double-layer capacitance, and for low activation overpotentials can be evaluated from the slopes of the charging curves at $\eta \to 0$. If the process is connected with significant activation overpotential, the slope of the charging curve at times larger than those needed for double-layer charging is the relevant one.

Alternatively, Eq. (2.54) can give

$$\eta_t - \eta_\infty = \frac{RT}{zF} \frac{i}{zFv_{\mathrm{ad},0}} \exp \frac{-v_{\mathrm{ad},0}t}{C_{\mathrm{ad},0}}, \tag{2.57}$$

or

$$d[\ln(\eta_t - \eta_\infty)]/dt = -v_{\mathrm{ad},0}/C_{\mathrm{ad},0}. \tag{2.58}$$

Also, from Eq. (2.57)

$$\eta_{t=0} - \eta_\infty = \frac{RT}{zF} \frac{i}{zFv_{\mathrm{ad},0}}. \tag{2.59}$$

Hence, from Eqs. (2.58) and (2.59), i.e., from the slope of $\log(\eta_t - \eta_\infty)$ vs. t function and from the intercept, both $v_{\mathrm{ad},0}$ and $C_{\mathrm{ad},0}$ can be determined. Figure 18 shows an experimental line obtained in the deposition of Ag^+

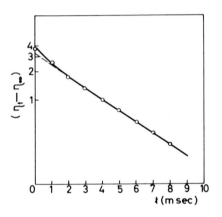

FIG. 18. Plot of $\log(\eta_t - \eta_\infty)$ against time [from Mehl and Bockris (1959)].

ions on silver. A typical set of values, as obtained in the first adion-concentration determination of Mehl and Bockris (1957), is given in Table VII.

The equilibrium surface adion concentration is sensitive to the conditions of the surface. The spread of the values, as shown in Table VIII, extends to more than an order of magnitude from a surface polarized for the first time after melting the metal and quenching it in helium to the electrode "activated" by anodic pulses.

TABLE VII

VALUES OF EXCHANGE CURRENT DENSITY (i_0), SURFACE ADION CONCENTRATION
AT EQUILIBRIUM ($c_{ad,0}$), AND SURFACE DIFFUSION FLUX ($v_{ad,0}$) FOR SILVER IN
$AgClO_4$ SOLUTIONS

$AgClO_4$ (moles/liter $\times 10^{-1}$)	i_0 (A/cm^2)	$zFv_{ad,0}$ (A/cm^2 $\times 10^{-2}$)	$C_{ad,0}$ mole (apparent cm^{-2})
2.0	0.10	1.2	8×10^{-10}
1.3	0.08	1.0	2×10^{-9}
1.0	0.07	1.4	4×10^{-10}
0.8	0.07	1.2	3×10^{-9}
0.8	0.07	1.6	6×10^{-10}
0.6	0.06	1.2	1×10^{-9}
0.6	0.06	1.1	8×10^{-10}
0.4	0.05	1.2	1×10^{-9}
0.2	0.04	1.5	6×10^{-10}
0.1	0.03	1.1	5×10^{-10}

TABLE VIII

SURFACE ADION CONCENTRATION AT THE REVERSIBLE POTENTIAL ON SILVER
ELECTRODES AS A FUNCTION OF SURFACE PREPARATION

Surface preparation	$C_{ad,0}$ (moles/cm^2 $\times 10^{-11}$)	Ref.
Quenched in H_2 atmosphere	90	Mehl and Bockris (1957)
Scraped in solution	15	Gerischer (1958)
Undefined	7	Lorenz (1958)
Quenched in He	3	Despić and Bockris (1960)
Prepared *in situ* by anodic pulse	160	Despić and Bockris (1960)

3. Surface-Diffusion Control of the Deposition Kinetics

a. Effect on the Steady-State Distribution of Adions at the Surface. A
simple model of a linear concentration gradient of adions between the
middle of a crystal plane and a growth site revealed (Section II,B,1) that
surface-diffusion control is to be expected at low overpotentials. It is of
interest to analyze more closely the adion concentration profile and
current density distribution over the crystal plane between two growth
sites when surface diffusion is slow compared to charge transfer.
Consider an infinitesimal area of the electrode (Fig. 19). Then

$$\frac{dC_{ad,x}}{dt} = -\frac{i_{a,x}}{zF} \, dx \, dy + \frac{i_{c,x}}{zF} \, dx \, dy + D \frac{\partial^2 C_{ad,x}}{\partial x^2} \, dx \, dy, \qquad (2.60)$$

where $i_{a,x}$ and $i_{c,x}$ are the local anodic and cathodic current densities, respectively, at distance x from the growth line. The two are functions of activation overpotential η_A

$$i_{a,x}/zF = (i_0/zFC_{ad,0})C_{ad,x} \exp[(1 - \beta)zF\eta_A/RT] = \vec{k}_{a,\eta}C_{ad,x} \quad (2.61)$$

and

$$i_{c,x}/zF = (i_0/zF) \exp[-\beta zF\eta_A/RT] = \vec{k}_{c,\eta}, \quad (2.62)$$

where $C_{ad,0}$ is the equilibrium adion concentration.

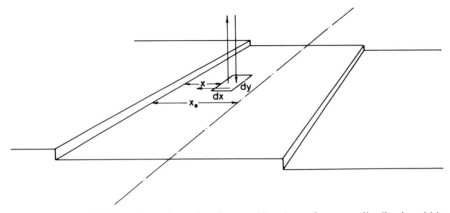

FIG. 19. Model of an electrode surface for considerations of current distribution. (↓) Cathodic current. (↑) Anodic current. (←) Surface diffusion flux. [From Despić and Bockris (1960).]

Despić and Bockris (1960) have solved Eq. (2.60) for the steady-state concentration and current-density distribution between growth sites, and Damjanović and Bockris (1963) have extended this to encompass potentiostatic transients. Thus at very short times there is a uniform adion concentration between the steps (equal to $C_{ad,0}$) and the process is activation-controlled. As deposition proceeds the concentration of adions increases the more rapidly the further the point is from the growth site, and a concentration gradient arises. The time for attaining the steady-state adion concentration depends both on the rate of surface diffusion and on the rate of the anodic partial process. If surface diffusion is rate-controlling, the time constant of the transient (the risetime) increases with cathodic overpotential, while it becomes independent of over-potential at such values of the latter at which charge transfer takes over the control of the rate. At these overpotentials it depends on the density of dislocations and increases as $v_{ad,0}$ decreases.

The steady-state distribution is obtained from Eq. (2.58) for $dC_{ad,x}/dt = 0$. Rearranging,

$$C_{ad,x} - (D/\bar{k}_{a,\eta})(\partial^2 C_{ad,x}/\partial x^2) = C_{ad,\infty}, \qquad (2.63)$$

where

$$C_{ad,\infty} = (i_c/zF\bar{k}_{a,\eta}) = C_{ad,0}\exp(-zF\eta_A/RT), \qquad (2.64)$$

i.e., $C_{ad,\infty}$ is the adion concentration at a point so distant from the growth line that $\partial^2 C_{ad,x}/\partial x^2 \to 0$.

To solve Eq. (2.63), one uses the boundary conditions

$$C_{ad,x=0} = C_{ad,0}, \qquad (2.65)$$

corresponding to the assumption that virtual equilibrium remains undisturbed at the growth line, and

$$(dC_{ad,x}/dx)_{x=x_0} = 0. \qquad (2.66)$$

Solving Eq. (2.63) with (2.65) and (2.66), one obtains

$$
\begin{aligned}
(C_{ad,x}/C_{ad,0}) = {}&\exp\!\left(\frac{-zF\eta_A}{RT}\right) + \frac{1 - \exp(-zF\eta_A/RT)}{1 + \exp[-2(i_0/zFv_{ad,0})^{1/2}p]}\\
&\times \left\{\exp\!\left[\left(\frac{i_0}{zFv_{ad,0}}\right)^{1/2}p\left(\frac{x}{x_0} - 2\right)\right]\right.\\
&\left. + \exp\!\left[-\left(\frac{i_0}{zFv_{ad,0}}\right)^{1/2}p\left(\frac{x}{x_0}\right)\right]\right\}, \quad (2.67)
\end{aligned}
$$

where

$$p = \exp[(1 - \beta)zF\eta_A/RT] \qquad (2.68)$$

and

$$v_{ad,0} = DC_{ad,0}/x_0{}^2. \qquad (2.69)$$

Damjanović and Bockris (1963) have introduced the dislocation density N as an inverse function of the growth point separation $2x_0$,

$$N = 1/(2x_0)^2. \qquad (2.70)$$

Hence

$$v_{ad,0} = 4DNC_{ad,0}. \qquad (2.71)$$

Figure 20(a,a′) shows the concentration under different conditions of overpotential at a constant ratio of activation to surface diffusion parameters $(i_0/Fv_{\mathrm{ad},0})$ and *vice versa*.

The time and position dependences of the adion concentration should be relevant for nucleation kinetics.

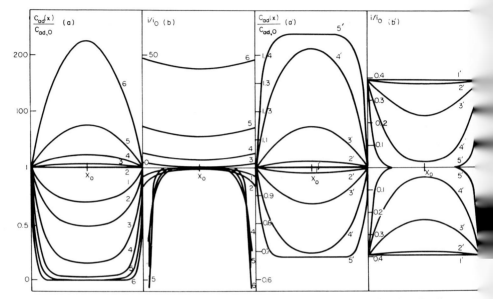

FIG. 20. Theoretical adion concentration profiles (a,a′) and current density distributions (b,b′) between two growth lines; (a) and (b) correspond to constant $i_0/Fv_{\mathrm{ad},0} = 10$ and varying η equal to: ± 10 mV (1), ± 20 mV (2), ± 50 mV (3), ± 100 mV (4), ± 150 mV (5), and ± 200 mV (6). (a′) and (b′) correspond to constant $\eta = \pm 10$ mV and $i_0/Fv_{\mathrm{ad},0}$ equal to: 0.01 (1′), 0.1 (2′), 1.0 (3′), 10 (4′), and 100 (5′). [From Despić and Bockris (1960).]

b. Rate of Arrival of Ions as a Function of the Density of Growth Steps.

The net rate of arrival of ions, i.e., the local current density at any point x, is given by

$$i_x = i_0\{(C_{\mathrm{ad},x}/C_{\mathrm{ad},0}) \exp[(1 - \beta)zF\eta_{\mathrm{A}}/RT] - \exp[-\beta zF\eta_{\mathrm{A}}/RT]\}. \quad (2.72)$$

By introducing Eq. (2.67) into (2.72) the distribution of the current density can be calculated as a function of the position x/x_0 on the electrode surface, i.e.,

$$i_x = i_0\Big\{\exp[(1 - \beta)zF\eta_{\mathrm{A}}/RT] - \exp[-\beta zF\eta_{\mathrm{A}}/RT]$$
$$\times \frac{\cosh(m/D^{1/2})}{\cosh[m/2(ND)^{1/2}]}\Big\}, \quad (2.73)$$

and

$$m = (i_0/zFC_{\mathrm{ad},0})^{1/2}p^{1/2}.$$

Figure 20(b,b') shows the current density distribution at different overpotentials and at a varying ratio of i_0 to $v_{\mathrm{ad},0}$, which contains the dislocation density [cf. Eq. (2.71)].

It is seen that when surface diffusion is the rate-controlling step the rate of charge transfer at the midpoint between the growth steps is small compared to that near the growth steps themselves, i.e., *a large part of the current flows to the vicinity of growth steps.* This, however, does not imply direct deposition from solution to kink sites at steps, which is shown to be associated with a prohibitively large activation energy (see Section II,A,1).

c. Average Current Density during Cathodic Polarization. Fleischmann and Thirsk (1960) have solved Eq. (2.60) for the potentiostatic transient conditions in terms of the average current density over the whole surface, which is measurable experimentally. They introduced at the outset

$$i = (zF/x_0) \int_0^{x_0} [\bar{k}_{c,\eta} - \bar{k}_{a,\eta}C_{\mathrm{ad},x}] \, dx. \tag{2.74}$$

Integral transforms were used and the solution obtained as

$$i = zF(\bar{k}_{c,\eta} - \bar{k}_{a,\eta}C_{\mathrm{ad},0})\left(\frac{D}{x_0^2\bar{k}_{a,\eta}}\right)^{1/2} \tanh\left(\frac{x_0^2\bar{k}_{a,\eta}}{D}\right)^{1/2}$$

$$+ \sum_{n=0}^{\infty} \frac{32zF\bar{k}_{a,\eta}x_0^2(\bar{k}_{c,\eta} - \bar{k}_{a,\eta}C_{\mathrm{ad},0})}{(2n+1)^2\pi^2[4\bar{k}_{a,\eta}x_0^2 + (2n+1)^2\pi^2D]} \cdot \tag{2.75}$$

From the time-dependent term it can be deduced that the order of magnitude of the relaxation time is

$$\tau \approx 4x_0^2/(\pi^2D + 4\bar{k}_{a,\eta}x_0^2), \tag{2.76}$$

and for the probable values of x_0 it should not be larger than 0.1 sec.

The attempt to obtain an expression for the galvanostatic transient resulted in a complex expression which could not be explicitly solved for the η–t dependence [cf. Fleischmann and Thirsk (1960)].

The first term in Eq. (2.75) represents the steady-state current density as a function of overpotential. Similar equations were obtained by Despić and Bockris (1960) and Damjanović and Bockris (1963) upon averaging the current density of Eq. (2.73) over the entire surface.

Figure 21 shows the η–i relation at different values of $D/x_0{}^2 = 4ND$.

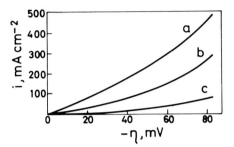

FIG. 21. Calculated average current density at surface-diffusion-controlled steady-state deposition as function of overpotential. ND equal to (a) 10^4, (b) 10^2, and (c) 10 sec^{-1}. $i_0 = 100 \text{ mA/cm}^2$ [from Damjanović and Bockris (1963)].

At very high values of $4ND \gg \bar{k}_{a,\eta}$ the factor

$$(D/x_0{}^2\bar{k}_{a,\eta})^{1/2} \tanh(x_0{}^2\bar{k}_{a,\eta}/D)$$

tends to 1. Hence the steady-state current term of Eq. (2.75) tends to

$$i = i_0\{\exp[-\beta z F\eta/RT] - \exp[(1-\beta)z F\eta/RT]\}, \qquad (2.77)$$

which is a pure activation-controlled current. This accounts for the transition from rate-controlling surface diffusion at low cathodic overpotentials (large $\bar{k}_{a,\eta}$), to rate-controlling charge transfer (small $\bar{k}_{a,\eta}$) as shown in Fig. 22. The transition overpotential obviously depends on the value of dislocation density N.

Fleischmann and Thirsk (1963) have stressed that these equations are valid for a model of x_0, i.e., N, independent of overpotential, and that in another case a more complex relation between i and η is expected.

Correspondingly, Kita et al. (1961) considered the potential dependence of the activity of the dislocations. It is not correct to assume that all the dislocations are active and to base the consideration of what determines the mean distance for diffusion on their total number. This is because the screw dislocations rotate. However, they will not rotate unless there is an overpotential which allows the radius of the rotation to be more than a critical value. The above authors showed that only a certain fraction of dislocations are active. This fraction is

$$\exp[-(2A\gamma/\beta z F\eta)^2 N_{\text{dis}}], \qquad (2.77a)$$

where A is the atomic weight, γ the surface free energy of the edge, β the density, and N_{dis} the number of dislocations.

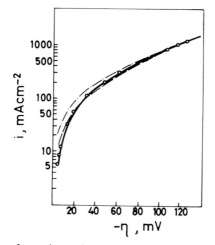

FIG. 22. Log i–η plot of experimental results (solid curve) of Mehl and Bockris (1957) on silver; dash-dot curve: transfer-controlled current; dashed curve: surface-diffusion-controlled current at low overpotentials. $i_0 = 120$ mA/cm^2; $C_{ad,0} = 10^{-10}$ mole/cm^2; $ND = 3.4 \times 10^3$ sec^{-1}. [From Damjanović and Bockris (1963).]

d. *Activity of Metal Surfaces in Solution.* The observed scatter of experimentally determined kinetic parameters for metal-deposition processes and the change of these parameters with time in one and the same experiment has resulted in the use of the vague term "activity of the surface." There seems to be a general assumption that the most active surface should be a freshly prepared one in contact with solution of extreme purity. Bockris and Kita (1962) reported a series of experiments in which fresh and active copper surfaces were produced by anodic dissolution of a predetermined number of atomic layers from an old surface, and then i_0 of the deposition process was determined as a function of time by imposing cathodic transients. Figure 23 shows a typical result.

Several reasons may exist for the loss of activity: (1) Smoothing of the surface during multilayer deposition; (2) adsorption of impurities; and (3) tendency to form crystal faces of lower indices.

Reason (1) seems an unlikely explanation of the extent of change. The adsorption of impurities, reason (2), seems to be a possible cause. It is reasonable to assume that the number of available growth steps N should fall with time after cessation of dissolution because impurities adsorb on them, i.e.,

$$N_t = N_0\left(1 - \frac{N_\infty'}{N_0}\frac{N_t'}{N_\infty'}\right),\tag{2.78}$$

where N_t' and N_∞' are the number of inactivated sites after time t and after very long time, respectively. The diffusion of impurities should lead [cf. Delahay and Trachtenberg (1957)] to

$$N_t'/N_\infty' = 1 - \exp(D't/k^2)\,\mathrm{erfc}(D't/k^2)^{1/2}, \qquad (2.79)$$

where D is the diffusion coefficient of the impurity in solution and k is a constant characteristic of the adsorption isotherm. In the region of complete surface-diffusion control, i.e., at $zFv_{\mathrm{ad},0} \ll i_0\exp(-\beta zF\eta/RT)$,

$$i = 8zFDC_{\mathrm{ad},0}N_t[\exp(-zF\eta/RT) - 1]. \qquad (2.80)$$

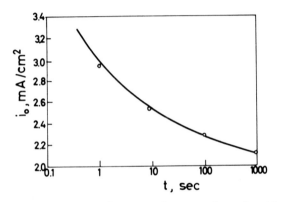

FIG. 23. Change of i_0 with time of contact of copper electrode with solution after activation by anodic dissolution [from Bockris and Kita (1962)].

Introducing Eqs. (2.78) and (2.79) into (2.80) and linearizing for small η values, one obtains

$$\eta_t = \eta_\infty\{[1 - (N_0'/N_0)] + (N_\infty'/N_0)\exp(D't/k^2)\,\mathrm{erfc}(D't/k^2)^{1/2}\}^{-1}, \quad (2.81)$$

where $\eta_\infty = RT/8(zF)^2DC_{\mathrm{ad},0}N_0$. This time dependence seems to reproduce reasonably well the experimental observation, as shown in Fig. 24 (curve a).

Reason (3) above assumes recrystallization of the surface in solution. Higher activity of some crystal faces should also result in a higher adion concentration and faster charge transfer kinetics. This can be given a quantitative account [cf. Bockris and Kita (1962)], but the resulting curves (Fig. 24, curves b and c) do not seem to follow the experimentally recorded changes better than the impurity hypothesis.

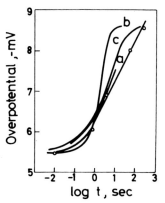

FIG. 24. Change of overpotential with time of contact with solution after activation by anodic dissolution. Calculated curves: (a) using Eq. (2.81) with $D/k^2 = 1$, $N_\infty/N_0 = 0.7$; (b) and (c) using the concept of recrystallization. [From Bockris and Kita (1962).]

C. EFFECT OF TRANSPORT IN SOLUTION ON DEPOSITION KINETICS AND PROPERTIES OF THE DEPOSIT

It has been realized for some time [cf. Kasper (1940a,b,c)] that a non-uniform current distribution over the surface can cause an uneven rate of metal deposition. When this situation is caused by a difference in ohmic resistance in the electrolyte between some points of the substrate (e.g., on peaks and recesses at the surface), and the counter-electrode, it is known as primary current distribution [cf. Wagner (1951, 1961)]. This tendency is usually modified by the resistance to the electrochemical reaction at the surface (Faradaic resistance), i.e., a kinetic factor, and the change produced relative to the primary current distribution is termed secondary current distribution. Two opposing situations can arise here, resulting, after prolonged deposition, in different structure of the obtained electrodeposit. For deposition from pure ionic solutions roughening of the surface tends to occur ("bad microthrowing power") with further effects on the type of deposit obtained. When appropriate additives are present in solution smoothening tends to occur ("good microthrowing power"), with leveling of surface irregularities and brightening. Both effects are connected with some limitations in transport from the bulk of solution. They are analyzed below in more detail.

1. Diffusion-Controlled Deposition and Surface-Roughness Initiation of Dendritic Growth

The quantitative theory of surface roughening and initiation of dendritic growth was developed by Despić et al. (1968a). It was assumed that (1) prior to the commencement of deposition any surface possesses some degree of roughness which could be described by some function

$$y_0(x) = f(x) \qquad (2.82)$$

(cf. Fig. 25) which need not be precisely known for the subsequent

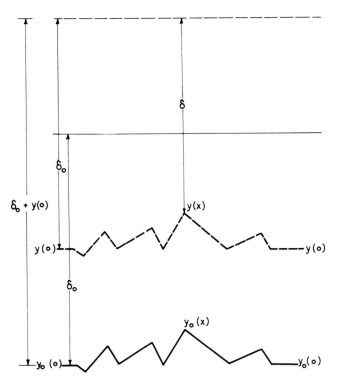

FIG. 25. Model of an electrode surface with microroughness much smaller than diffusion-layer thickness [from Despić *et al.* (1968a)].

argument; (2) hydrodynamic conditions in the solution are such that a diffusion layer is established of constant thickness δ_0 relative to a point $y_0(0)$ representative of a flat surface; (3) at any x the amplitude $y_0(x) \ll \delta_0$. The rate of growth at any point at the surface is related to the current density at that point as

$$dy(x)/dt = (M/\rho z F)i, \qquad (2.83)$$

where M is the atomic weight and ρ is the density of the depositing metal.

Suppose that at any point the deposition occurs under mixed activation and diffusion control. The current density is given by

$$i = i_0\left\{\left(\frac{i_L - i}{i_L}\right) f_c(\eta) - f_a(\eta)\right\},\tag{2.84}$$

where

$$f_c(\eta) = \exp(\alpha_c F\eta/RT),\tag{2.85}$$

$$f_a(\eta) = \exp(\alpha_a F\eta/RT),\tag{2.86}$$

and i_L is the limiting current density at that point. Rearranging Eq. (2.84) for i explicitly yields

$$i = \frac{i_L[f_c(\eta) - f_a(\eta)]}{(i_L/i_0) + f_c(\eta)}.\tag{2.87}$$

The limiting current density is given by

$$i_L = zFDC_0/\delta_{x,t} = (i_L)_0/\delta_{x,t},\tag{2.88}$$

where D is the diffusion coefficient, C_0 is the concentration of depositing ions in the bulk of solution, $\delta_{x,t}$ is the diffusion layer thickness at this point and at time t after the beginning of the deposition process, and $(i_L)_0$ represents the limiting current density reduced to unit diffusion-layer thickness. Since the outer boundary of the diffusion layer is also advancing during deposition, the value of $\delta_{x,t}$ is given by

$$\delta_{x,t} = \delta_0 + y(0) - y(x),\tag{2.89}$$

i.e., is dependent on the position of the point at the surface. This non-uniform effective diffusion layer-thickness was found experimentally by Beacom (1959), and introduced into theory by Ibl (1961, 1963a,b) in considering the problem of microthrowing power.

Introducing Eq. (2.89) into (2.88), (2.87), and (2.83) and rearranging, one obtains

$$\frac{dy(x)}{dt} = \frac{M}{\rho z F} \frac{(i_L)_0\{1 - [f_a(\eta)/f_c(\eta)]\}}{[(i_L)_0/i_0 f_c(\eta)] + \delta_0 + y(0) - y(x)}.\tag{2.90}$$

Since one is interested in the propagation at point x *relative* to the advancement of the average surface, as represented by $y(0)$, a new variable is introduced

$$y = y(x) - y(0).\tag{2.91}$$

Replacing this into Eq. (2.90) and rearranging,

$$\frac{dy}{dt} = \frac{M}{\rho z F} \frac{(i_L)_0\{1 - [f_a(\eta)/f_c(\eta)]\}}{[(i_L)_0/i_0 f_c(\eta)] + \delta_0} \left\{\frac{y}{[(i_L)_0/i_0 f_c(\eta)] + \delta_0 - y}\right\}$$

$$= \frac{A}{B}\left[\frac{y}{B - y}\right]. \tag{2.92}$$

Integration of Eq. (2.92) between y and y_0 gives

$$B(\ln y) - y = (A/B)t + B(\ln y_0) - y_0. \tag{2.93}$$

At the early stage of deposition $y \ll \delta_0$, and hence also $y \ll B$, so that it can be neglected in Eq. (2.92). With this approximation one obtains

$$\ln y = (A/B^2)t + \ln y_0, \tag{2.94}$$

and hence

$$y = y_0 \exp(t/\tau) \tag{2.95}$$

with the time constant

$$\tau = \frac{\rho z F}{M} \frac{\{[(i_L)_0/i_0 f_c(\eta)] + \delta_0\}^2}{(i_L)_0\{1 - [f_a(\eta)/f_c(\eta)]\}}. \tag{2.96}$$

Since $y_0 = f(x)$, this means that the function y will reproduce the substrate with a factor *exponentially increasing with time*, i.e., the original surface roughness tends to become amplified, and because of the exponential nature of such a process it should "suddenly" become visible. Figure 26 represents an idealized model of the surface at different constant time intervals with the values of the relative time $t/\tau = 1, 2, 3, 4, \ldots$ and a dependence of the height of one of the peaks on time. This model appears to rationalize the appearance of protrusions. The latter may develop into a dendrite or a whisker if some additional conditions are satisfied (cf. Section III,B,3). To check this theory against experiments, its consequences for the apparent current density (per unit geometric surface area) may be analyzed.

Differentiating Eq. (2.95) and introducing the rate obtained into Eq. (2.93), one obtains

$$i = \frac{(i_L)_0\{1 - [f_a(\eta)/f_c(\eta)]\}}{[(i_L)_0/i_0 f_c(\eta)] + \delta_0} \left\{1 + \frac{y_0}{[(i_L)_0/i_0 f_c(\eta)] + \delta_0} \exp\left(\frac{t}{\tau}\right)\right\}. \tag{2.97}$$

To obtain the average current density, one integrates Eq. (2.97) over the entire surface. Since y_0 is the only variable dependent on x and z, this results in an equation analogous to Eq. (2.97) with a somewhat changed value of the constant with the exponential term in the equation.

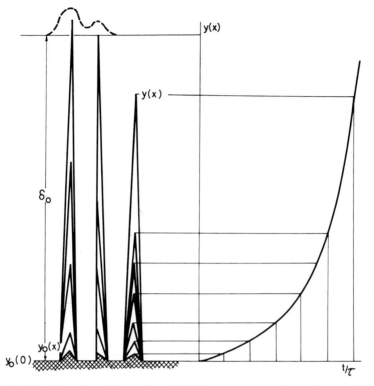

FIG. 26. The time dependence of peak heights at constant electrode potential [from Despić *et al.* (1968a)].

Two significant conclusions can be drawn:

1. The current density should be an exponential function of time with a time constant dependent on the concentration of depositing ions. This is found experimentally, e.g., for the deposition of zinc from alkaline zincate solutions (Despić *et al.*, 1968a), as shown in Fig. 27.

2. The exponential increase in current with time, as well as the growth of protrusions causing it, does not necessarily require complete, or even predominant, diffusion control of the deposition. In the case of very low i_0 values or high overpotentials, or both, it may be that

$$[(i_L)_0/i_0 f_c(\eta)] \gg \delta_0,$$

so that in this case the diffusion-layer thickness obviously loses its significance in the preexponential factor and in the time constant. The

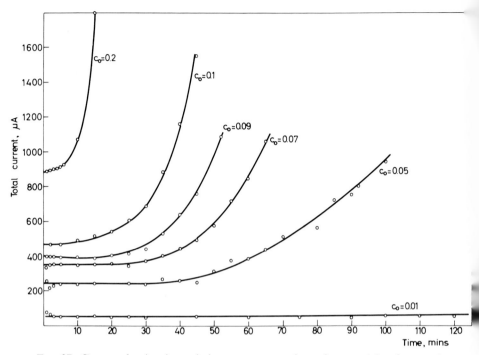

Fɪɢ. 27. Current density time relations at constant electrode potential and at varying concentrations of zincate ions in solution [from Despić *et al.* (1968a)].

diffusion coefficient remains in both factors, indicating that the basic reason for the phenomenon is still the difference in the rates of transport to different points at the electrode surface.

2. *Diffusion-Controlled Adsorption of Additives. Mechanism of Leveling and Brightening*

The reaction resistance at the electrode may be a function of the concentration of additives in solution. This should be the case if they adsorb at the surface and thus block sites for incorporation of metal atoms into the lattice, or inhibit the surface diffusion as adions. If they are consumed during the deposition process (e.g., buried into the deposit or chemically changed by reduction and desorbed), a diffusion flux of the additive from the bulk of solution can determine the reaction resistance at the surface. This, in turn, should cause a secondary current distribution with the tendency to level the surface irregularities. Kardos (1956) proposed a mechanism of leveling based on the following assumptions:

1. The concentration of the metal ions is sufficiently high so that the deposition is activation-controlled over the whole surface.

2. At a given current density an increase in coverage by the additive—the leveling agent—results in a strong decrease of the rate constant, i.e., at a given electrode potential the current density decreases.

3. The surface coverage by the additive is not the equilibrium one, but is controlled by the rate of diffusion of the additive toward the electrode.

Foulke and Kardos (1956) have shown the similarity between the ratio of the thickness of a deposit in the recess, h_r, to that at a peak, h_p, and the ratio of the current at a flat surface in a less-agitated solution to that of a more-agitated one. The similarity is based on the fact that the thickness of the deposit is proportional to the rate of deposition. If the process is controlled by diffusion, its rate is a function of the diffusion-layer thickness. Hence the thinner layer in the agitated solution corresponds to the situation at the peak of the microroughness, while that in less-agitated solution to the situation in the recess.

Figure 28 shows the experimentally recorded current densities at

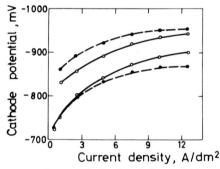

FIG. 28. Polarization curves (vs. standard calomel electrode) in nonagitated (open circles, solid curves) and agitated (darkened circles, dashed curves) Watts' nickel (50°, pH = 3.0) in the absence of addition agents (lower pair) and in presence of 0.4 g/liter of 2 butyne-1,4-diol (upper pair) [from Foulke and Kardos (1956)].

different potentials in the absence and in the presence of 2-butyne-1,4-diol as a leveling agent in a nickel-containing solution. It is seen that in the absence of the additive the deposition occurs at less negative potentials and is predominantly activation-controlled. Some positive effect of stirring observed at higher current densities indicates that diffusion of the depositing ions has taken a part of the rate control. In the presence of the additive, however, the process is inhibited, and this shift in the negative direction is stronger the higher is the flux of the additive.

Kruglikov *et al.* (1963) have used the rotating-disk technique to show the quantitative correlation between the rate of deposition and the diffusion-layer thickness, which is proportional to the square root of the angular velocity (Fig. 29). The negative slopes of the straight lines indicate that the rate of deposition is controlled by the diffusion of the inhibitor. The direct dependence of the inhibition on the concentration of additive was demonstrated by Rogers and Taylor (1963, 1965) under conditions of constant diffusion-layer thickness obtained at the rotating disk with constant rotation rate (Fig. 30). Finally, Kruglikov *et al.* (1964, 1965) have shown that under otherwise identical conditions the same values of $C_0\sqrt{\omega}$ produce, over a wide range of values of C_0 and ω, the same polarization curves in nickel baths containing coumarin, thiourea, quinoline, quinaldine, and chloral hydrate as leveling agents.

However, at increasing concentration of the leveling agent maxima in leveling effect were observed in several instances, past which the leveling becomes less pronounced again [cf. Watson (1960), Watson and Edwards (1957)]. This is in agreement with the tendency of the deposition current in Fig. 30 to level out at increasing concentration of the additive as the diffusion control over the adsorption of the latter starts to break down. Thus, leveling maxima confirm that leveling is due to the *difference* in inhibition between micropeak and microrecess points; if this difference disappears, so does the leveling, although inhibition remains high.

The phenomenon of brightening is more complex and less well-defined than that of leveling. It depends not only on diffusion, but also on some structural features of the deposit. Yet, to the extent that it is dependent on the smoothing of microroughness, the same factors are operative as those in leveling.

III. Crystallographic Aspects of Metal Deposition

A. Morphology and Texture

The crystal structure of electrodeposits depends on the basic crystallographic properties of the depositing metal. The morphological properties and texture, however, depend largely on the conditions of electrocrystallization. Close analogies exist between the latter and the vapor-phase crystallization on the one side and ordinary crystallization from solutions and melts on the other. Hence, many of the basic concepts such as surface diffusion, growth on planes of high index, two-dimensional nucleation, and screw-dislocation mechanisms of growth were taken from these fields. However, considerable differences should also be appreciated arising from such facts as the following:

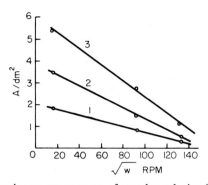

FIG. 29. Current density vs. square root of angular velocity (rpm) obtained at three electrode potentials (-700, -725, and -750 mV vs. normal hydrogen electrode) in a Watts' nickel bath (50°, pH $= 4.5$) containing 0.15 g/liter coumarin [from Kruglikov *et al.* (1963)].

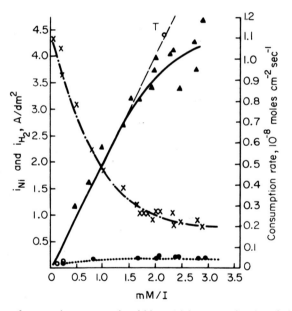

FIG. 30. Rate of coumarin consumption (▲), partial current density of nickel deposition i_{Ni} (×), and hydrogen evolution i_{H_2} (●), vs. coumarin concentration in a Watts' nickel bath at -960 mV vs. standard calomel electrode at 980 rpm, 48.5°C, pH $= 4.0$. Point T is calculated from the Levich equation for the rotating disc [from Rogers and Taylor (1965)].

1. The presence of adsorbed layers, e.g., of anions or of water molecules and of solvated metal adions (instead of adatoms).

2. Existence of the double-layer field.

3. Difference in the nature of the interaction of particles adsorbed at the surface with the substrate prior to incorporation into the lattice. Thus, not only are there adions instead of adatoms in the electrochemical case, but there is interaction with the solvent in the case of a metal.

4. Difference in the rate of arrival of particles to the substrate— diffusion of ions in solution is slower than that of atoms in the gas phase; hence there will be a greater tendency to control the rate by diffusion.

The much more varied morphology and texture which are observable with electrodeposits are due to the effect of potential on surface free-energy and the contact adsorption of anions, of which there are no parallels in deposition from the gas phase.

1. Descriptive Morphology

Early work on metal deposition was overwhelmingly preoccupied with the qualitative description of crystal growth forms. The first observations under a microscope were recorded by Huntington in 1905. In subsequent developments the introduction of Nomarsky interference-contrast microscopy and polarized interferometry has proven more fruitful than the introduction of electron microscopy. The latter is not applicable for observations *in situ* because of absorption, whereas the former allows vertical growth of about 2000 Å to be observed, as well as contours and objects as large as $\lambda/2$.

a. Growth Forms. In analyzing the shapes of individual crystals the following forms are encountered in numerous communications found in the literature:

1. *Layers* [Fig. 31(a)] become visible at an average height of the steps of about 500 Å [cf. Volmer (1922), Kohlschütter and Übersax (1924), Erdey-Grúz and Volmer (1931a,b), Kohlschütter and Torrichelli (1932), Vagramyan and Gorbunova (1937), Wranglén (1955), Sroka and Fischer (1956), Wilcock (1956–1957), Howes (1959), Sato (1959), Seiter *et al.* (1960), Barnes *et al.* (1960), Wranglén (1960), Pick *et al.* (1960), Economou *et al.* (1960), Economou and Trivich (1961), Giron and Ogburn (1961), Damjanović *et al.* (1965a,b)]. The layers themselves contain numerous microsteps [cf. Pick *et al.* (1960), Howes (1959), Damjanović *et al.* (1965a,b)].

2. *Pyramids* [Fig. 31(b)] [cf. Seiter and Fischer (1959), Seiter *et al.* (1960)] are usually obtained at low current densities and only on surfaces

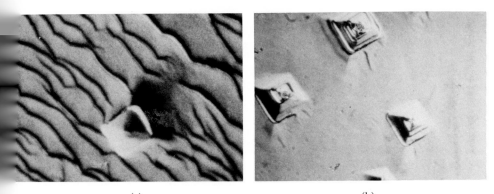

(a)   (b)

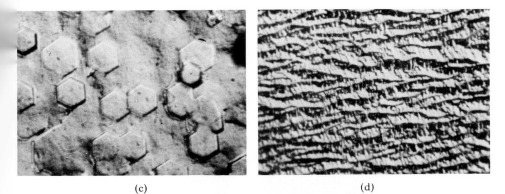

(c)   (d)

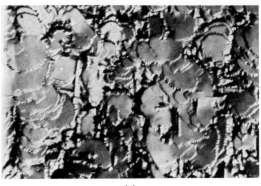

(e)

FIG. 31. Growth forms in various systems. (a) Top left, layers; (b) top right, pyramids; (c) middle left, blocks; (d) middle right, ridges; (e) bottom, cubic layers.

FIG. 31(f), spirals.

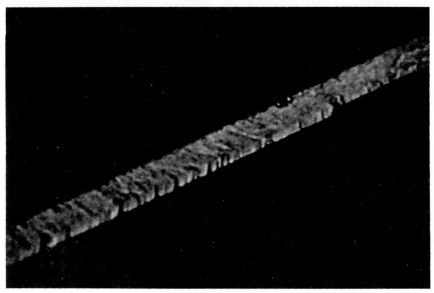

FIG. 31(g), whiskers.

FIG. 31(h), dendrites.

which approximate in their orientation certain specific crystallographic planes [Damjanović et al. (1965a,b)]. The symmetry of the pyramids is related to that of the substrate [Economou et al. (1960)]. The side faces do not seem to be planes of higher indices, i.e., smooth on the atomic level. Instead, they appear to be composed by the piling up of macrolayers [cf. Wranglén (1955, 1960), Economou et al. (1960)].

3. *Blocks* [Fig. 31(c)] [cf. Damjanović et al. (1965a,b)] can be considered as truncated pyramids. This form is particularly sensitive to the purity of the solution (Barnes et al., 1960; Turner and Johnson, 1962).

4. *Ridges* [Fig. 31(d)] (Pick et al., 1960; Damjanović et al., 1965a,b) are a special kind of layer growth which appear in the presence of adsorbed impurities. They may develop on an already-formed layer structure if small amounts of surfactants are added.

5. *Cubic layers* [Fig. 31(e)] (Seiter et al., 1960) are an intermediate structure between blocks and layers.

6. *Spirals* [Fig. 31(f)] (Streinberg, 1952; Kaishev et al., 1955; Pick, 1955; Seiter and Fischer, 1959; Seiter et al., 1960) appear as pyramids with pronounced layers in a spiral arrangement winding toward the top. The height of the observable steps may be as low as 100 Å. The step separation is of the order of 1–10 μ and increases with decreasing current density (Kaishev et al., 1955; Seiter et al., 1960).

7. *Whiskers* [Fig. 31(g)] are threadlike long single crystals [cf. Aten and Boerlage (1920), Gorbunova and Zhukova (1949), Gorbunova and Dankov (1949), Graf and Morgenstern (1955), Van der Meulen and Lindstrom (1956), Ovenston et al. (1957), Price et al. (1958)]. They form at fairly high current densities, and the presence of organic solutes is important (Price et al., 1958).

8. *Dendrites* [Fig. 31(h)] are needlelike and pine-treelike deposits [cf. Kohlschütter and Übersax (1924), Kohlschütter and Good (1927), Wranglén (1955, 1960), Matthew et al. (1961), Barton and Bockris (1962), Despić et al. (1968a)]. They are likely to form from simple salt solutions and melts at low cation concentrations. The stalk and branches are parallel to low-index direction in the lattice. Hence angles between the stalk and branch are defined. They may be of a two-dimensional (fernlike) or three-dimensional variety. They appear to contain numerous layers of pyramidal growth forms (Wranglén, 1955, 1960).

This description is somewhat arbitrary, being based on the visible surface topography.

b. The Effect of Potential on the Growth Forms. Seiter and Fischer (1959) showed some correlation between growth forms and current density, as seen in Fig. 32. Although the current density is the variable used in Fig. 32, this is related to overpotential, as shown in the same figure, and, indeed, it is the overpotential which would be expected to affect the morphology on several grounds. Thus:

1. The adion concentration and the local current-density distributions are potential-dependent (Despić and Bockris, 1960; Damjanović and Bockris, 1963). If local supersaturation is reached, nucleation should occur. The supersaturation is shown (Vermilyea, 1957; Mehl and Bockris, 1959; Fleischmann and Thirsk, 1960) to be potential-dependent.

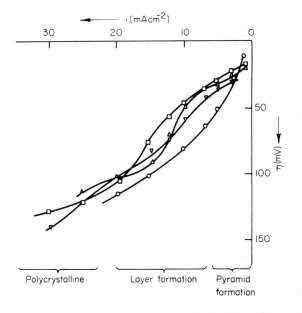

FIG. 32. Correlation between polarization curves obtained by various authors and the observed growth forms in copper electrocrystallization [from Seiter and Fischer (1959)].

2. The fraction of steps active in the growth process depends on the radius of the critical nucleus, which in turn is a function of potential [Vermilyea, 1957; Kita *et al.*, 1961; cf. Eq. (2.77a)].

3. Potential-dependent adsorption of different species present in solution, particularly organic substances, is of decisive importance for the growth form (see below).

4. Some secondary effects of potential may arise, e.g., *via* the effect of potential on interfacial tension.

c. Impurity Effects. Very small amounts of impurities profoundly modify the growth morphology. Direct effects have been shown and correlations with amounts in solution obtained by Economou *et al.* (1960) and Barnes *et al.* (1960). Using known adsorption isotherms for certain organic compounds on copper (Bockris and Swinkels, 1964), Damjanović *et al.* (1965a,b, 1966) were able to relate the growth forms to the quantities of adsorbed materials. Thus, the distance between steps on a {100} face was observed to change compared to that in pure solution at the same current density at a surface coverage by *n*-decylamine as low as 10^{-4}. At surface coverage 10^{-2} the deposit was of a ridge type. Addition of *n*-decylamine made pyramids become more readily truncated (at lower current densities) and transform into blocks (Turner and Johnson, 1962).

When purification of solution consists only of the recrystallization of salt and pre-electrolysis the distance between steps on layer-type structures ({100} face) is noticeably shorter than if strong adsorbents (activated alumina and charcoal) are used in purification.

These factors and many others, as reviewed by Fischer (1960), account for the noteworthy difficulties in reproducing defined morphology under nominally identical conditions and in finding definite correlations between kinetic conditions and morphological properties of the deposit.

2. *Polycrystalline Deposits*

a. Problems of Epitaxy.   In an early stage of the electrodeposition of metal ions onto a corresponding metal substrate there is a tendency to preserve the orientation of the substrate by extended incorporation into existing lattice sites. Even at electrodeposition onto single crystals of a different metal a remarkable degree of epitaxial order can initially be achieved. Parallel lattice orientation is found to occur when the lattice spacings of the two metals in parallel directions differ by less than 15% (Thirsk, 1939). The problem of epitaxy was treated in a static way by Frank and Van der Merve (1949a,b). They calculated the equilibrium configuration of a deposited layer when constrained by the periodicity of the substrate surface. If the amplitude of the first harmonic in a Fourier series representing the potential field of a substrate of wavelength a is $\frac{1}{2}W$, it was shown that a critical value of misfit up to which a lattice of a natural spacing b and force constant γ can follow the substrate is given by

$$1/P = [(b/a) - 1] = 2/\pi m, \qquad (3.1)$$

where $m = (\gamma a^2/2W)^{1/2}$. The calculations using a Lennard-Jones potential give $m \approx 7$, and the critical misfit comes to about 9%.

Larger misfits produce the source of dislocations. There is an activation energy for the formation of dislocations, and at sufficiently low temperatures a metastable state can exist up to 14% misfit.

Although the static concept of small misfit values may have been given more credit than is justified by the experimental facts [cf. Pashley (1956)], purely geometric restrictions are certainly some, if not the only, factors which determine the chances of the initial epitaxy. In cases where the induced stress is light or absent, or where the symmetry of the potential field matches that of the preferred habit of growth of the material, epitaxy can be reasonably controlled.

b. Randomly Oriented Deposits. Still, the orientational influence extends only to a certain limit. The departure is first reflected in the formation of a number of twins in the deposit [cf. Kumar and Wilman (1957), Poli and Bicelli (1959), Barnes (1961), Vaughan (1961)]. Eventually, the deposit turns polycrystalline, with randomly oriented crystallites. In prolonged electrodeposition, regardless of the type and nature of the starting substrate, a polycrystalline deposit always forms. However, the orientational influence of a polycrystalline substrate with small crystallites (of the order of 1000 Å) ceases at considerably earlier stages of deposition than that of the substrate consisting of large single crystals (50,000 Å). Any factor causing dislocations (misfit, surface-active substances, impurities, or inhibitors) makes this transition occur at an earlier stage (Wilman, 1955).

In general, two kinds of deposits, compact or loose, are obtained. Loose deposits, mossy or dendritic, are usually obtained from pure solutions at metals with high exchange-current densities and at low ionic concentrations. Conversely, compact deposits result in systems with low exchange-current densities at high concentrations and in the presence of surface-active agents. The average grain size depends to a considerable degree on the concentration of the latter. Large molecules block the surface by adsorption, and nucleation is continuously promoted. Hence smaller-size crystallites result [cf. Vermilyea (1959), Fischer (1960)]. This is more pronounced the lower is the current density, since at high current densities the molecules of the agent do not adsorb fast enough to prevent growth.

3. *Texture*

In the advanced stage of polycrystalline growth electrodeposits tend to develop a preferred crystal orientation. A definition of texture can be made in terms of the concept of degree of orientation (Finch *et al.*, 1947): in a polycrystalline deposit the orientation of each grain can be specified by giving the angles formed by the crystallographic directions of the crystal and the axes of a reference system fixed with respect to the macroscopic substrate. In *random orientation* all the three axes are randomly oriented; *one-degree orientation* or texture is achieved as one of the axes becomes fixed relative to the substrate, the other two remaining randomly disposed. Two-degree and three-degree orientations can in principle also be obtained. The latter case corresponds to a single crystal built upon a substrate.

A one-degree orientation in a direction perpendicular to the substrate is found in electrodeposits by electron diffraction (Finch *et al.*, 1935; Finch and Sun, 1936; Finch and Williams, 1937; Finch *et al.*, 1947; Sato, 1959; Pangarov and Velinov, 1966; Pangarov and Vitkova, 1966a,b), by X-ray analysis (Glocker and Kaupp, 1924; Pangarov and Rashkov, 1960a,b; Pangarov and Dobrev, 1962; Pangarov and Michailova, 1964; Pangarov and Vitkova, 1966a,b), and by electron micrography (Pangarov and Velinov, 1966).

Finch and co-workers stressed two factors as dominant in determining texture: (1) the influence of base metal, and (2) the influence of the electrolyte used. A number of investigations were done using single crystals as substrates (Cochrane, 1936; Gorbunova, 1938; Evans, 1952–1953; Takahashi, 1952; Setty and Wilman, 1955; Grechukina, 1956; Setty, 1957; Reddy and Wilman, 1959; Poli and Bicelli, 1959a,b; Fukuda, 1959). The effect of electrolyte involves composition, pH, surface-active substances and colloidal additives, current density, and temperature (Glocker and Kaupp, 1924; Finch and Sun, 1936; Finch *et al.*, 1947; Hirata *et al.*, 1939; Arkharov, 1936; Kochergin, 1953; Wilman, 1955; Banerjee and Goswami, 1955, 1957–1959; Washi, 1957; Gorbunova and Sutyagina, 1958; Matsunaga, 1960).

Finch *et al.* (1947) pointed out that two types of oriented crystal growth prevail: "lateral growth," when the most densely populated net plane is parallel to the surface, and "outward growth," when the most densely populated lattice row is normal to the surface. Yet, Pangarov (1962) pointed out that in some cases there is a strong texture which does not belong to either of these types, as e.g., with cobalt (Pangarov and Rashkov, 1960a,b), hexagonal nickel (Banerjee and Goswami, 1955, 1957), and hexagonal silver (Layton, 1952), which tend to have the ($10\bar{1}0$) plane parallel to the surface. Pangarov and Rashkov (1960a,b) further showed that the orientation of the crystallites also depends upon the overvoltage at which deposition is taking place.

Several attempts have been made to find a mechanism by which the crystallites growing on a randomly oriented substrate develop texture.

The early theory (Bozorth, 1925) attempted to explain the texture in terms of internal stress which develops in crystals during deposition, resulting in a plastic deformation, which in turn causes preferred orientation. As a consequence, a relation between texture and hardness was sought by many authors, since internal stress would be related to the latter. Little success was achieved [cf. Wood (1935)].

Gorbunova *et al.* (1957) suggested that the experimental results on the texture of electrodeposited zinc can be explained on the basis of the work of Kaishev and Bliznakov (1949) on the orientation of two-dimensional nuclei. These authors put forward the idea that the crystallization of homopolar crystals on a structureless substrate should start with those two-dimensional nuclei for which (depending on supersaturation) the work of formation has the smallest value.

Pangarov and co-workers (1960–1966) applied this concept quantitatively to growth on polycrystalline substrate, maintaining that repeated nucleations at such a surface, governed by this minimum-energy principle, must lead to texture of the deposit as a whole.

The Stranski–Kaishev method (1934a,b,c, 1935a,b) gives the work of formation of two-dimensional nuclei, W_{hkl}, for an *hkl* plane as

$$W_{hkl} = \frac{B_{hkl}}{(1/mN)(\mu - \mu_0) + \psi_0 + C_{hkl}}, \qquad (3.2)$$

where μ and μ_0 are chemical potentials of the vapors in equilibrium with the nucleus and with an infinitely large crystal, respectively, ψ_0 is the work needed for extracting an atom from the substrate, m is the number of atoms in a gas molecule, and B_{hkl} and C_{hkl} are constants which depend on the type of nucleus and can be calculated in relative units (Pangarov, 1964), taking as a basis the work in breaking a bond between the first two neighbors ψ_1. In the most general case $B_{hkl} = b\psi_1{}^2$ and $C_{hkl} = c\psi_1$ (b and c are constants which can be determined from model investigations).

The work W_{hkl} can be represented graphically as a function of supersaturation, $(1/mN)(\mu - \mu_0) + \psi_0$, expressed in relative units. Figure 33 shows this work for different planes of (a) a face-centered-cubic lattice, (b) a body-centered-cubic lattice, (c) a hexagonal-close-packed lattice, and (d) a tetragonal lattice of tin. The fact that the curves intersect indicates that at different degrees of supersaturation different preferred orientations should develop from the two-dimensional nuclei having the lowest energies of formation. In metal deposition the chemical potential should be replaced by electrochemical potential $\bar{\mu}$, i.e.,

$$\bar{\mu} = \mu + zF\phi, \qquad (3.3)$$

where ϕ is the absolute electrode potential. In such a case

$$\bar{\mu} - \bar{\mu}_0 = zF(\phi - \phi_{\text{rev}}) = zF\eta, \qquad (3.4)$$

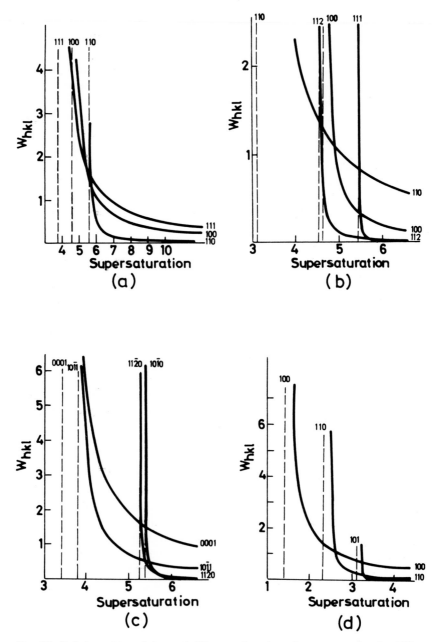

Fig. 33. Relative values of the work W_{hkl} as a function of supersaturation for different planes of an (a) fcc lattice, (b) bcc lattice, (c) hcp lattice, (d) tetragonal lattice [from Pangarov (1964)].

and hence from Eq. (3.3) it follows that

$$W_{hkl} = \frac{B_{hkl}}{ze_0\eta + \psi_0 - C_{hkl}},\tag{3.5}$$

i.e., the work of formation for each particular crystal plane is a different function of overpotential. As the overpotential is increased a change of texture can arise as the work of formation of, say, a {110} nucleus, becomes smaller than that of formation of the {100} nucleus. This is indeed observed experimentally [cf. Pangarov and Rashkov (1960a,b), Pangarov and Vitkova (1966b)] in the case of thick cobalt deposits crystallizing in hexagonal array. Thus with increasing overpotential the following axes should appear: {0001}, {10$\bar{1}$1}, {11$\bar{2}$0}, {10$\bar{1}$0}. This is confirmed by experiment. Numerous other data are also in agreement with the predictions arising from the calculations [cf. those for body-centered-cubic lattices of α-Fe and α-Cr (Glocker and Kaupp, 1924; Arkharov, 1936), as well as those in which two axes of preferred orientation were predicted and observed, as in the case of the tetragonal lattice of tin (Pangarov and Michailova, 1963)].

Hence this theory appears to offer fairly comprehensive coverage of the phenomena noted so far. Yet it can be criticized on the grounds that it assumes nucleation to be a basic and unavoidable step in crystal growth. Much of the experimental evidence contradicts this, and makes other mechanisms of growth seem more relevant. Hence other reasons for the appearance of texture upon prolonged deposition must be considered, at least as additional factors. Besides the possible effects of internal stress favored in the early theory, Reddy (1963) has developed a theory of the effect of microscopic kinetics on macroscopic morphology. If different crystal faces offer different kinetic parameters for metal deposition, after some time of growth the crystal will develop well-defined, usually low-index, slow-growing faces aligned according to fixed crystallographic planes. In addition, different rates of growth can be expected in a direction normal to the surface and in a direction parallel to it because of the influence of the electric field and concentration profile in the near vicinity of the growing substrate when an outward growth mode predominates.

A correlation can then be established between certain crystallographic axes and a set of axes referred to the substrate: the slowly growing faces will align themselves perpendicularly to the surface, producing a preferred orientation. Although less specific in its predictions, this theory is an attractive one as an addition to the theory of Pangarov, inasmuch as it covers the influence of additives, codeposition of hydrogen, etc. These influences can be explained (Sato, 1959; Reddy, 1963) by considering the effect they have on the kinetic parameters for different crystallographic faces. If the presence of an additive considerably slows down the growth of a certain plane, a change in texture should result when it is added.

B. Mechanism and Kinetics of Growth of a Single Crystal

1. *Mechanisms of Crystal Growth*

The present concepts of the mechanisms of growth originate with the work of Volmer (1921), Kossel (1927), Stranski (1928), and Becker and Döring (1935). These early theories postulated a scheme according to which atoms and molecules strike the metal surface at any point and, by means of surface diffusion, reach the monoatomic steps. Along the steps they reach the "half-crystal positions" or "kinks," where they become incorporated into the lattice. This type of model requires the surface to have at least one high crystallographic index, which would contain atomic or microsteps or even microscopically observable macrosteps. After the steps originally present grow out and disappear and a low-index surface forms, the growth can be perpetuated by one of the two alternatives: (1) two-dimensional nucleation recognized by the early theory, (2) a screw-dislocation mechanism introduced by Burton et al. (1949). The three modes of growth are represented in Fig. 34.

a. Growth on Planes of High Index.   High-index planes arise from misorientations of the surface plane with respect to a crystallographic plane in the cutting of a given single crystal. Even if the crystal surface was originally optically smooth, i.e., consisting of invisible microsteps only, visible macrosteps frequently appear which move across the surface, increase in height, and sharpen, or sometimes fade out.

These phenomena of bunching and dispersing of microsteps have found several mechanistic explanations. Thus Bunn and Emmett (1949) have advanced a *crystal-morphology theory* which is based on a well-established tendency for crystals to grow in such a way as to be bounded by planes of low indices. According to this theory, monoatomic microsteps would tend to join during the growth into larger steps, the surface between the latter corresponding to a low-index plane with lower surface energy than the original surface (cf. Fig. 35).

Fischer (1954) has suggested that *local solution exhaustion* may be the reason for stepwise growth. At some places on the electrode successive nucleations of monoatomic layers on top of each other may take place, followed by a lateral growth of these nuclei. Yet, intensive growth must result in a local depletion of adions, so that the supersaturation falls below a critical value and further nucleation stops (cf. Section III,B,1b). After the macrostep formed thus far has moved away the original degree of supersaturation is reestablished, and conditions for the formation of a new step are created once more. In this way the periodic appearance of visible steps some distance apart and their wavelike motion across the crystal surface can be qualitatively understood. This mechanism can be criticized for containing the assumption that nucleation is an essential precursor of the appearance of macrosteps. Yet it can be argued that a

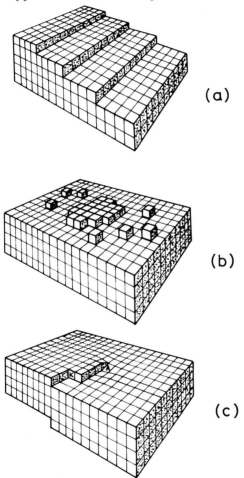

(a)

(b)

(c)

FIG. 34. Models of different sources of microstep on a surface. (a) Misorientation of the surface with respect to the ideal low-index plane. (b) Two-dimensional nucleus. (c) Emergent screw dislocation [from Bockris and Razumney (1967)].

similar periodic sequence of events can arise if the growth originates from, e.g., a pair of screw dislocations forming a loop (cf. Section III,B,1c), since such a structure resembles a two-dimensional nucleus and also requires a certain adion concentration before it may develop into a spreading layer.

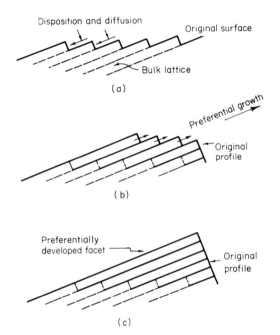

FIG. 35. Successive stages in preferential step or edge growth to produce surfaces of lower index [from Bunn and Emmett (1949)].

A kinematic theory of macrostep formation was formulated by Frank (1958), Cabrera and Vermilyea (1958), and Chernov (1960). According to these authors clustering of monoatomic steps is a purely kinetic phenomenon due to differences in rates of advance of different layers. As shown below (cf. Section III,B,2a) a motion of invisible incipient bunches of microsteps should first result in the appearance of a visible sharp edge, and after some time this step should fade away again.

Finally, *adsorption of impurities* also seems to be a good reason for macrostep formation. Even if the surface originally consisted of microsteps only, when these begin to move across the surface they would frequently come across adsorbed impurity molecules and stop moving or move at a lower rate. The next layer growing on top of the one considered would continue moving until it reaches the edge of the blocked layer.

This process continues until so many steps bunch on top of each other that a macrostep is formed. Eventually, the macrostep may start moving as a whole if the impurity molecules vanish, e.g., by being buried into the lattice.

The capacity of a strongly adsorbed molecule to prevent movement of a step arises from the fact that going around it requires bending the edge of the step. If two such impurity molecules are sufficiently close together, e.g., a distance closer than a critical value similar to that governing nucleation (cf. Section III,B,3), the probability of "squeezing" between them decreases considerably because of the large edge energy required for that process. This mechanism of bunching is shown schematically in Fig. 36.

All these theories seem to complement rather than to exclude each other. Thus a crystal-morphology theory seems to give a reasonable explanation for the experimental fact (Damjanović et al., 1965a,b) that the distance between the steps increases as the deposit grows thicker (cf. Fig. 37), i.e., prolonged deposition results in fewer steps. The kinematic

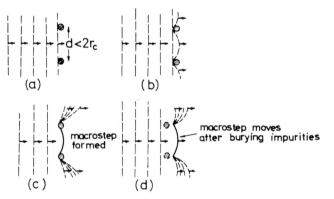

FIG. 36. Schematic representation of the effect of adsorbed impurities on the propagation of microsteps (dotted lines) [from Bockris and Razumney (1967)].

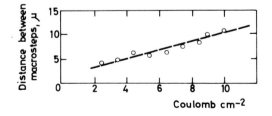

FIG. 37. Variation of average distance between macrosteps with thickness of deposit for copper electrocrystallization [from Damjanović et al. (1965a,b)].

theory is attractive inasmuch as it gives a single reason both for the appearance of visible macrosteps and for their fading away. Yet, according to this theory macrosteps are inherently unstable, and there is no obvious reason why macrosteps should appear at fairly equal distances or why that distance should increase as the deposition proceeds. The local-exhaustion theory gives a good explanation for periodicity in the appearance of microsteps, while impurity-adsorption theory accounts best for their stability. Hence it appears that in any development of a macrolayer structure a suitable combination of the explanation cited should be sought in an attempt at rationalization.

b. Two-Dimensional Nucleation. If a crystal plane lacks steps and kinks, i.e., points of growth, or if growth at these sites is sufficiently inhibited so that a large concentration of adions builds up compared to the equilibrium concentration, the probability increases that new growing centers will form in the form of two-dimensional nuclei. Very convincing illustrations of such a situation were made by Budevski *et al.* (1966a,b) after a remarkable achievement in preparing metal surfaces free of any dislocations. At constant current, polarizations much larger than those upon ordinary planes of such a metal are obtained. Moreover, they are distinguished by periodic oscillations. Similar oscillation of potential were recorded by Despić *et al.* (1966) when nickel electrodes were shot at high speeds (300 m/sec) into silver nitrate-containing solutions (Fig. 38). These phenomena are ascribed to fluctuations in the formation of two-dimensional nuclei. Under potentiostatic conditions a current can be observed on a dislocation-free surface only at overpotentials exceeding 8–12 mV [cf. Budevski (1966)], whereas at lower overpotentials the cell is electrically cut off. When a short voltage pulse in excess of this value is applied to the cell in the cut-off condition supersaturation by adions is achieved and a nucleus of a new lattice net is formed. The propagation of the step produced is accompanied by a certain current flow. When the new layer has spread out over the whole surface the current again drops to zero, since the steady-state potential is insufficient to form a new nucleus.

The current–time curves for a series of successive voltage pulses are shown in Fig. 39. Although the curves vary in form, the integral of the current over time has the same value in all cases. The amount of electricity given by this integral corresponds exactly to the amount required for the completion of a monoatomic layer over the cubic plane.

c. Screw-Dislocation Mechanism. This represents a way of perpetuating the growth without repeated nucleation, i.e., at low supersaturation, where the latter cannot occur. A simple way to visualize a dislocation

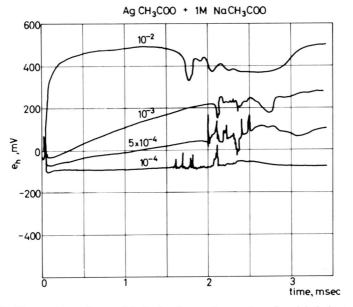

FIG. 38. Fluctuations of potential obtained upon immersion of a nickel electrode into a silver-ion-containing electrolyte [from Despić *et al.* (1966)].

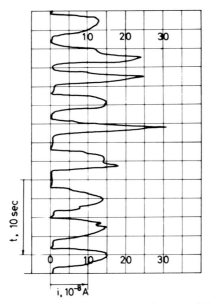

FIG. 39. Current–time curves following a voltage pulse on a dislocation-free plane in a cut-off condition [from Budevski (1966)].

providing for a screwlike development of a crystal plane is shown in Fig.
40. The step promotes the growth by rotating itself with one end fixed
at the point where the screw dislocation emerges. A gradual sloping of
the crystal surface occurs, so that at no point is there a discontinuity
between the lattice layers. The two directions of incorporation of adatoms
into the lattice—radial and circular—result in a radial propagation of
each layer and the emergence of new layers. Thus a spiral develops, as
shown by Frank (1949a,b) (Fig. 41).

The above hypothetical growth by an isolated screw dislocation can be
expanded to a less hypothetical one where more than one screw dislocation emerges on the surface of a crystal (Burton *et al.*, 1951; Frank,
1949a,b; Cabrera and Burton, 1949). In the simplest case of a pair of
dislocations emerging sufficiently far apart and being of different sign
(one rotating clockwise and one rotating counterclockwise) pyramidal
growth results from merging of the two, as can be seen from Fig. 42.

With random distribution of an equal number of screw dislocations of
both signs a complex pattern of the joining of neighboring outgrowths
results, at low supersaturations, in a relatively smooth and even deposit
[cf. Verma (1953)]. Much experimental evidence qualitatively supports
the screw-dislocation mechanism of crystal growth. Spiral forms of
growth are observed in crystals deposited from the vapor phase (Verma,
1951; Forty and Frank, 1952, 1953), as well as in those crystallized from
solution (Forty and Frank, 1951; Amelinckx, 1952; Kozlovskii and
Lemmlein, 1958), by electrolysis from solutions (Kaishev *et al.*, 1955;
Kaishev, 1961; Pick, 1955; Seiter *et al.*, 1958; Wranglén, 1955; Sato,
1959), and from fused salts (Streinberg, 1952).

2. *Kinetics of Step Propagation*

a. Macrostep Movements.   In layer growth the steps all propagate in
the same direction, and this does not depend upon the direction of movement of solution over the electrode surface. The average rate of step
propagation is proportional to current density (Damjanović *et al.*,
1965a,b).

Knowing the rate v of macrostep propagation, the average height h and
the average density of macrosteps, i.e., distance l between them, it is
possible to calculate the average rate of metal deposition at these macrosteps [cf. Damjanović (1965)]:

$$i_{\text{step}} = \frac{zFg}{M}\frac{h}{l}v, \tag{3.6}$$

where g is the specific and M the atomic weight of the metal.

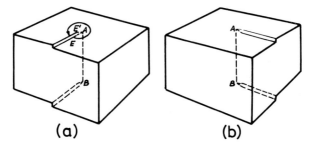

FIG. 40. Scheme of (a) right-handed and (b) left-handed dislocations emerging on a surface [from Bockris and Damjanović (1964)].

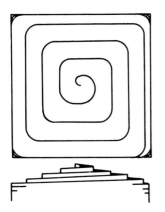

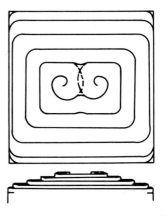

FIG. 41. Simple spiral with straight turns (above) and pyramidal growth resulting from it (below) [from Frank (1949a,b)].

FIG. 42. Formation of pyramidal growth due to a pair of screw dislocations [from Frank (1949a,b)].

It was found (Damjanović *et al.*, 1965a,b, 1966) that only a few per cent of the total deposited metal atoms are incorporated into the metal lattice at these macrosteps. Obviously, the balance must be deposited on the optically-smooth portions between the macrosteps. This can occur either (1) by the self-perpetuating screw-dislocation mechanism forming sub-optical spirals on the surface; or (2) by the growth across the surface of monoatomic steps formed at the bottom of macrosteps, as the most favorable position for incorporation, i.e., transfer of microsteps from the bottom edge of one macrostep to the top edge of another.

The kinetics of the step propagation can be discussed in terms of the kinematic theory of macrostep formation. This assumes that at the start the monoatomic steps at an optically-smooth surface are *not* uniformly spaced. Then as the deposition begins the steps must travel across the surface with various velocities, depending principally on the proximity of other steps. The higher the density (number per unit length) of steps k, the slower should the steps travel, as the same amount of material deposited per unit area is distributed over the larger total length of steps. The "flux of steps" q can be defined as the number of steps passing a point x per unit time. Then

$$q = f(k). \tag{3.7}$$

At any point on the surface the law of conservation of monoatomic steps requires that

$$\partial q/\partial x + \partial k/\partial t = 0, \tag{3.8}$$

i.e., the flux gradient in the direction of the step propagation equals the rate of change of step density with time at the same point. In view of Eq. (3.7), Eq. (3.8) can be written as

$$\frac{dq}{dk}\frac{\partial k}{\partial x} + \frac{\partial k}{\partial t} = 0, \tag{3.9}$$

or

$$dq/dk = C(k) = dx/dt. \tag{3.10}$$

Frank (1958) has assumed a reasonable relationship between q and k, as shown in Fig. 43. It follows from Eq. (3.10) that for each value of k there

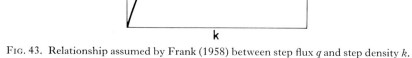

Fig. 43. Relationship assumed by Frank (1958) between step flux q and step density k.

is a definite value of the velocity of step propagation, i.e., the slope of the q–k relation.

Because dx/dt is a constant for a given value of k, there must be a straight line in the (x, t) plane along which k is constant, i.e., the density of steps remains the same, or, on the surface of the metal there are regions of constant step-density which travel at the same rate. Correspondingly, there are regions of different step-density which travel at different velocity, called the kinematic wave velocity $C(k)$.

The line of constant k in the (x, t) plane is called the "characteristic." The slope of the characteristic, dx/dt, will be smaller the larger is k, as seen from Fig. 43.

The way the step-density profile changes with time is illustrated in Fig. 44. Initially, at time t_0, it is assumed that the density of steps along

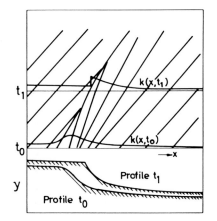

FIG. 44. Characteristics of arbitrary step density at time t_0, $k(x, t_0)$; density profile at time t_1, $k(x, t_1)$; surface profiles at t_0 and t_1 [from Frank (1958)].

the x axis is even except for a symmetrical hump representing a region of higher density of steps.

Characteristic lines for regions with the same densities of step have the same slope, $dx/dt = dq/dk$. Steps (and the hump) propagate from left to right. On the right-hand side of the hump the characteristics diverge [since as k decreases, dq/dk increases (see Fig. 43) and hence dt/dx decreases], while on the left-hand side they converge and, at some time greater than t_0, will meet. When the characteristics meet the intervening values of the step density disappear and a discontinuity in step density, a shock wave, develops. This discontinuity travels across the surface following a trajectory of the slope

$$dx/dt = (q_2 - q_1)/(k_2 - k_1), \qquad (3.11)$$

where q_1 and k_1, q_2 and k_2 are fluxes and densities of steps at places adjacent to the discontinuity. At some later time t_1 the density of steps is as shown in Fig. 44. The discontinuity in the density of steps occurs at the rear of the advancing step hump, or "cluster" of steps, as it may be called. However, it may occur on either side of the cluster. When dq/dk decreases with increasing density of steps, or, in other words, when

d^2q/dk^2 is negative, a discontinuity occurs at the rear of the advancing cluster. The condition for its occurrence at the front of the advancing cluster is an increase of dq/dk with increasing density of steps ($d^2q/dk^2 > 0$), so that the characteristics in Fig. 44 converge at the front of the advancing step-density hump and diverge at the rear of the hump. Such conditions may be created, e.g., if the movement of steps is affected by adsorbed impurities. These will reduce the velocity of propagation of a step. An advancing step creates behind itself a clean surface on which adsorption can take place. The density of adsorbed species will gradually increase with time and will be greater the smaller is the flux, providing the concentration of impurities is low and no equilibrium concentration of adsorbed species is reached by the time the next step is swept across. If a following step is further apart than a preceding one, it will meet a surface with a higher density of adsorbed impurities than if the steps follow each other uniformly. Thus the velocity of steps and their rate of flow q will be reduced more for the smaller values of step density k than for the larger values. For low values of k, dq/dk increases with k and d^2q/dk^2 becomes positive. Characteristic lines in Fig. 44 at places of higher step densities will then converge on the right-hand side of the density hump, and eventually a discontinuity will develop at the front of the advancing hump. In this latter part of the discussion the invocation of the effect of impurities is equivalent to the assumption (in contrast to the former part) that the flux of steps is not only a function of their density at the point considered, but also of densities in advance of this point, or, in other words, also of the dk/dx value.

Profiles of the growing surfaces, which can be obtained by integrating step densities over distance, are given in Fig. 45. As can be seen, if the step flux depends on step density only, the discontinuity in step density and in the surface-profile slope, once developed, will gradually decrease with time as the difference in the slope of characteristic lines which meet in subsequent periods of time decreases. Eventually, the discontinuity will completely fade away. Such a fading away of steps with time has been observed by Damjanović et al. (1965a,b).

b. Overlap between Growing Centers. Layer growth can attain a steady state in lateral propagation as the steps flow across the surface in one direction, appearing on one end and vanishing on the other. Yet, this situation is relatively rare. More often layers spread from point centers, as in the case of growth from two-dimensional nuclei or screw dislocations. In this case after some time the neighboring layers must start affecting each other by competing for the depositing material as well as for the space in which to grow. No steady state with respect to lateral growth can exist. It can be established in the outward direction perpendicular to the surface only after the basic layer is completely filled (Fleischmann and Liler, 1958; Armstrong et al., 1966).

The problem of overlap of growth was reveiwed in detail by Fleischmann and Thirsk (1963).

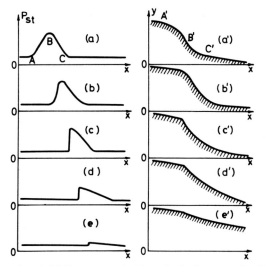

FIG. 45. Progress of an initially symmetrical bunch of microsteps (a, a′) according to the kinematic theory. (a)–(e) Variation with time of the distribution of density of microsteps. (a′)–(e′) Corresponding variation of surface profiles [from Bockris and Razumney (1967)].

In a number of simple cases it is possible to evaluate the consequences on the total rate of growth of the overlapping of the growing areas. Suppose that N_0 growing centers are evenly distributed over a unit of surface area, and that discharge takes place at or quite near the edges of the growing two-dimensional crystal planes, e.g., because surface diffusion is heavily inhibited. The total rate of discharge per unit geometric surface should in this case be proportional to the total length of edges l,

$$i = i_1 l, \tag{3.12}$$

where i_1 is the local current density at the edge and is given by a kinetic equation similar to Eq. (2.4) with rate constants referring to unit length rather than unit area. The increase in linear dimensions of a layer of height h growing in two dimensions over the surface is given by

$$dr = \frac{i_1}{zF} \frac{M}{h\rho} dt. \tag{3.13}$$

If all the centers are assumed to have started growing at the same time, at any time t after the beginning of the current pulse the average radius of the growing centers should be

$$r = \frac{i_1}{zF}\frac{M}{h\rho}t, \tag{3.14}$$

and consequently the total length of all the edges should be

$$l = 2r\pi N_0 = (2\pi N_0 M i_1/zF\rho h)t. \tag{3.15}$$

Introducing Eq. (3.15) into (3.12), one obtains for the total current density

$$i = (2\pi N_0 M/zF\rho h)i_1^2 t, \tag{3.16}$$

i.e., the current density should increase continuously with time.

However, as the circular growing layers touch each other and start overlapping the situation changes, and the limitations in their further expansion have to be taken into account.

Avrami (1939–1941) has approached this problem by defining the "extended" sizes of the regions in which more than some minimum number of centers overlap. Thus in Fig. 46, the area of the extended

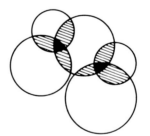

Fig. 46. Overlap of two (shaded region) and three (dotted region) cylindrical centers growing on a surface [from Fleischmann and Thirsk (1963)].

regions $S_{1,\text{ex}}$, is the area of the regions occupied by at least one layer, counted for each center of growth separately. Hence if S_1 is the area of all the regions of growth that are not affected by overlap, S_2 the area covered by two regions, S_3 the area covered by three centers, etc., the extended size $S_{1,\text{ex}}$ is

$$S_{1,\text{ex}} = S_1 + 2S_2 + 3S_3 + \cdots + mS_m. \tag{3.17}$$

Similarly, the extended size of the regions covered by two or more overlapping centers (overlap of the first or higher order) is

$$S_{2,\text{ex}} = S_2 + 3S_3 + 6S_4 + \cdots + \frac{m(m - 2 + 1)}{2!} S_m, \quad (3.18)$$

or, in general,

$$S_{k,\text{ex}} = \frac{1}{k!} \sum_{m=k}^{\infty} [m(m - 1) \cdots (m - k + 1)] S_m. \quad (3.19)$$

The actual size of the region affected by growth is

$$S = S_1 + S_2 + S_3 + \cdots + S_m. \quad (3.20)$$

Using the set of equations of the type of Eq. (3.19) to extract the size of regions S_1, S_2, etc., one obtains from Eq. (3.20)

$$S = S_{1,\text{ex}} - S_{2,\text{ex}} + S_{3,\text{ex}} + \cdots + (-1)^{(m+1)} S_{m,\text{ex}}. \quad (3.21)$$

The calculation of the actual size S is therefore reduced to the statistical evaluation of the extended sizes.

To evaluate the rate of growth for the simple case of two-dimensional growth from uniformly distributed growing centers, it is simplest to calculate the extended surface of overlap. Suppose that the extended size $S_{1,\text{ex}}$ expands by $dS_{1,\text{ex}}$. The total extended overlap $dS_{2,\text{ex}}$ created by this expansion is clearly $dS_{1,\text{ex}}$ multiplied by the local density $S_{1,\text{ex}}$, i.e.,

$$dS_{2,\text{ex}} = S_{1,\text{ex}} \, dS_{1,\text{ex}}, \quad (3.22)$$

and similarly for other extended regions,

$$dS_{m+1,\text{ex},q} = S_{m,\text{ex},q} \, dS_{1,\text{ex},q}. \quad (3.23)$$

Integrating these equations successively and substituting in Eq. (3.21), one obtains

$$S = S_{1,\text{ex}} - \frac{1}{2} S_{1,\text{ex}}^2 + \frac{1}{2 \cdot 3} S_{1,\text{ex}}^3 \cdots = 1 - \exp(-S_{1,\text{ex}}), \quad (3.24)$$

which is the most general equation for overlap in this particular case.

Now, $S_{1,\text{ex}}$ can be obtained from Eq. (3.14), taking into account that $S_{1,\text{ex}} = r^2\pi$. Hence the law of the growth of the surface area is obtained from Eq. (3.24) as

$$S = 1 - \exp[-\pi(i_1 M/zFh\rho)^2 t^2], \quad (3.25)$$

i.e., S changes from 0 to 1 for t between 0 and ∞.

Fleischmann and Thirsk (1963) have reviewed the general formulation of $S_{m,ex}$. Such an evaluation of S, however, meets with great difficulties, and in the few cases which have been treated—in the field of the electro-crystallization of substances other than metals—many simplifying assumptions have had to be made [cf. Fleischmann and Thirsk (1955), Bewick (1961), Dugdale et al. (1961)].

3. Nucleation

a. *Conditions of Nucleation.* Nuclei tend to form when the adion concentration exceeds the equilibrium value. However, stable formations require a certain critical degree of supersaturation. At sufficiently high rates of discharge, and if the rate control is such that the adion concentration increases with increasing negative potential, so does the probability of nucleation.

The classical approach to the problem of the critical nucleus was best reviewed by Frenkel (1946). This approach was developed primarily with reference to the formation of liquid droplets from the vapor phase and is based on the same simple model as that used for deriving the Kelvin relation. Nevertheless, it resulted in definite relationships between current density and overpotential for the case of rate-controlling nucleation (Erdey-Grúz and Volmer, 1931a,b):

$$\ln i = A - B\eta^{-2} \qquad (3.26)$$

and

$$\ln i = A' - B'\eta^{-1}, \qquad (3.27)$$

for three-dimensional and two-dimensional nucleation, respectively.

A consideration of this problem more directly reflecting the conditions existing in electrodeposition of metals is due to Damjanović and Bockris (1963). They assumed that the free enthalpy of formation of an embryo from adions, ΔG, should consist of two parts: the "condensation" energy of adions in the inner part of the embryo, and the energy of formation of its edge.

Thus

$$\Delta G = \left(\frac{2\pi r^2}{\sqrt{3}\, d^2} - \frac{2r - d}{d} \right) \Delta G_1 + \left(\frac{2r - d}{d} \right) \pi\, \Delta G_2, \qquad (3.28)$$

where ΔG_1 (< 0) is the free-energy change when a free adion joins the bulk of the cluster, and ΔG_2 (> 0) that of an adion entering the edge of the cluster; d is the interatomic distance in the embryo; and r is the radius of the embryo.

If r exceeds a certain value, ΔG becomes negative, i.e., the nucleus is stable and can continue to grow. From $[\partial(\Delta G)/\partial r] = 0$ one obtains

$$r_c = -\frac{\sqrt{3}\, d(\Delta G_2 - \Delta G_1)}{2\,\Delta G_1} = -\sqrt{3}\,\frac{d}{2}\frac{\Delta G_3}{\Delta G_1}, \tag{3.29}$$

where ΔG_3 is the free-energy change for an adion transferred from the bulk of the nucleus into its edge.

The two free-enthalpy changes can be expressed as

$$\Delta G_1 = -\frac{L}{2} - kT\left(\ln\frac{n}{N-n}\right) + \Delta G_4 \tag{3.30}$$

and

$$\Delta G_3 = (2.5/12)L + \Delta G_5, \tag{3.31}$$

where L is the heat of sublimation of the metal, n the number of adions per cm², N the number of available sites per cm² ($\sim 1.5 \times 10^{15}$), ΔG_4 the free-enthalpy difference between the enthalpy of hydration of the ion in the two-dimensional nucleus and that of the adion at the surface, and ΔG_5 the free-enthalpy difference between the ion in the edge and in the nucleus. The factor of 2.5 in ΔG_3 arises from the difference in coordination numbers of an adion in the edge of a nucleus (6–7) and in the bulk of the nucleus (nine). For a stable nucleus at equilibrium with adions at the surface the free-enthalpy change $\Delta G_1 = 0$, and the equilibrium number of adions per cm², n_0, can be calculated from Eq. (3.30). If this is replaced back into Eq. (3.30) for the nonequilibrium case, one obtains

$$\Delta G_1 = -kT\ln\frac{n(N-n_0)}{n_0(N-n)}. \tag{3.32}$$

Assuming $N \gg n > n_0$, and introducing Eq. (3.32) into (3.29),

$$r_c = \frac{\sqrt{3}}{2}\frac{d^2\gamma}{kT\ln(n/n_0)} = \frac{\sqrt{3}}{2}\frac{N_0\gamma d^2}{\eta} \tag{3.33}$$

where γ, the edge energy in ergs/cm of the circumference of the nucleus, is defined as

$$\gamma = \Delta G_3/d, \tag{3.34}$$

and η is the overpotential due to the supersaturation by adions, as represented by the ratio of the actual concentration of adions to the equilibrium one n/n_0.

This can be numerically obtained provided the hydration energy ΔG_5 can be estimated [cf. Conway and Bockris (1958, 1960)].

b. Frequency of Nucleation.   An embryo of a critical radius r_c becomes a stable nucleus after a further adion is incorporated into it. This should occur after the elapse of time

$$\tau = 1/(2\pi r_c v n), \tag{3.35}$$

where v is the average velocity of adions in a given direction. If N_c is the number of embryos per cm^2 having the critical radius and if all of them become stable nuclei in τ sec, the number formed in 1 sec, i.e., the rate of nucleation, is given by

$$R = N_c[1/(2\pi r_c v n)]^{-1} = 2\pi r_c N_c n (kT/2\pi m)^{1/2}, \tag{3.36}$$

if $v = (kT/2\pi m)^{1/2}$ is taken from the kinetic theory. The number of nuclei N_c can be obtained as

$$N_c = (n/n_c) \exp(-\Delta G_c/kT), \tag{3.37}$$

where n_c is the number of atoms in the critical nucleus and ΔG_c is the free-enthalpy change for the formation of a nucleus of critical size. The latter can be obtained from Eq. (3.28) when r is replaced by r_c as

$$\Delta G_c = \pi \Delta G_3[(r_c/d) - 1]. \tag{3.38}$$

If H is the surface area taken up by an adion in the nucleus

$$n_c = r_c^2 \pi / H. \tag{3.39}$$

Introducing (3.38) and (3.39) into Eq. (3.37), also allowing for the energy of activation of surface migration, E, and introducing Eq. (3.37) into Eq. (3.36), the rate of nucleation just after the latter begins is

$$R = \frac{2Hn^2}{r_c} \left(\frac{kT}{2\pi m}\right)^{1/2} \exp - \frac{[(r_c/d) - 1]\pi \Delta G_3 + E}{kT}. \tag{3.40}$$

Considering the dependence of n and r_c on overpotential, it is seen that a complex dependence of the rate of nucleation, i.e., of the corresponding current density on overpotential, is obtained.

Numerical calculations using Eq. (3.40) show that significant rates of nucleation can be expected only at overpotentials of the order of 80–100 mV (Damjanović and Bockris, 1963). Indeed, experimentally it is found that at copper whiskers, which are supposed to be bounded by ideally flat surfaces, with no steps and growth sites on them there is no deposition current unless overpotentials of the order of 100 mV are reached (Vermilyea, 1957). However, much lower overpotentials seem to be needed for the initiation of growth on a dislocation-free surface of silver [cf. Budevski (1966)].

In addition, Eq. (3.35) for the rate of growth, and, for that matter, Eqs. (3.26) and (3.27), are difficult to verify experimentally, since a case of prolonged nucleation control of electrodeposition is difficult to achieve. Instead, as soon as a number of stable nuclei are formed at sufficiently high overpotentials some other step in deposition (e.g., charge transfer or surface diffusion) takes over the control of the process.

The probability of nucleation on a stepped surface is less than on the dislocation-free surface, since the adion concentration, even at the midpoint between steps, is smaller. Analysis shows that on a surface with a high dislocation density ($\sim 10^{10}$ cm^{-2}) no nucleation is to be expected unless cathodic overpotentials are over 150 mV.

4. Dendritic and Whisker Growth

After protrusions have been formed during electrodeposition by a diffusion-controlled amplification of surface roughness (cf. Section II,C,1) two kinds of further outward growth tend to develop, depending primarily on the properties of the electrolyte: dendrites and whiskers. The former are usually obtained from pure solutions or melts of simple salts. Conversely, impurities or additives seem to be an essential condition for the appearance of whisker growth.

a. Appearance and Propagation of Dendrites. Phenomenological explanations for the appearance of dendrites upon the crystallization of melts were discussed by Kirkaldy (1959). These are based on the principle of minimum rate of entropy production as introduced by de Groot (1952) and Prigogine (1955).

In a suitable formulation this method could be applied to dendrite formation by electrocrystallization from solution to account for the morphological aspects of dendritic deposits, the dendrite spacings, and the branch spacing in a single dendrite. The interesting implication of the derived equations, which seems to be applicable to this case as well, is that a diffusive entropy production plays a dominant role in the determination of interface morphology, since no mechanism is available to create the required chemical free energy from thermal energy at the rate required by the derived equations.

That the nonspecific diffusional effects are the major dendrite formation and propagation factor was also shown in the mechanistic theory derived by Barton and Bockris (1962) in order to explain phenomena observed in electrocrystallization of silver from melts of $AgNO_3$–KNO_3–$NaNO_3$ at 300°C.

Two major characteristics of such a growth were established: (1) a certain critical overpotential must be exceeded at the electrode before dendrites could appear; and (2) there was an initiation time for the commencement of the growth.

At a given controlled overpotential η a number of dendrites grow at a characteristic rate which increases with increase of overpotential, becoming roughly linear to it at overpotentials above a few millivolts. Electron-microscope photographs indicated a parabolic shape of dendrite tips, with a radius of curvature of about 10^{-5} cm. The tip radius tends to remain constant, i.e., the dendrite does not change its lateral dimensions appreciably.

It was clear that the rate of deposition of Ag$^+$ ions on the surface of the electrode was diffusion-controlled, because of the very high value of i_0 for the charge transfer in silver deposition from a molten salt.

As a protrusion appears at such a surface the rate of transport of the depositing ions to its tip becomes larger than that at the rest of the surface because conditions of spherical diffusion are established and the rate of deposition becomes inversely proportional to the radius of curvature. This provides for the observation of much faster growth of the dendrite as a whole. In this respect the rate of growth at constant overpotential, if only diffusion-controlled, would tend to infinity as the radius of the dendrite tends to zero. However, as the radius decreases, the reversible potential is shifted from the usual value characteristic of a flat surface (infinite radius of curvature) because the free energy of the surface changes with curvature in the sense that the smaller the radius of curvature, the greater is the tendency to dissolve (cf. the Kelvin vapor pressure equation). Also, as diffusion becomes very fast activation control should eventually take over. Taking these limitations into account, the total over-potential (for $\eta < |20|$ mV) could be considered to be composed of three terms,

$$\eta = \frac{irRT}{DC_\infty F^2} + \frac{2\gamma V}{Fr} + \frac{i}{i_0}\frac{RT}{R}, \tag{3.41}$$

representing, respectively, the diffusional effect and the effects of surface energy and of activation control. This equation contains the explanation of the tendency of dendrites to grow with a constant tip radius and velocity (i.e., local current density i) at a given potential. These are defined by $di/dr = 0$, i.e., the maximum of the function (3.41). The theory also gave a quantitative account of the observed rates and tip radii.

This theory was supported by further experimental evidence of Reddy (1966) and Hamilton (1963).

An extension of this theory was derived by Despić et al. (1968a) in the attempt to explain additional facts obtained from a study of dendrite growth on zinc electrodeposited from alkaline zincate solutions at room temperature. The existence of a critical initiation overpotential below which no dendrite forms was confirmed here as well. For this system it was of the order of 70–80 mV. The rates of propagation were much lower, and clusters of dendrites, rather than individual needles (as for silver from silver nitrate), were observed.

Since at these overpotentials linearization made in the Barton–Bockris theory was no longer justified, the essential step in the extension was to assume a complete relationship between the current density at the tip and overpotential as given by a Butler–Volmer type of equation applied for the probable two-step single-electron exchange mechanism (cf. Section II,A,2d)

$$i = i_0 \left\{ \frac{C_{\mathrm{Zn(OH)_4^-}}}{(C_{\mathrm{Zn(OH)_4^-}})_0} \exp\left(-\frac{\beta F}{RT}\eta\right) - \frac{a_{\mathrm{Zn}}}{(a_{\mathrm{Zn}})_0} \exp\left[\frac{(1+\beta)F}{RT}\eta\right] \right\}. \quad (3.42)$$

Then

$$\frac{C_{\mathrm{Zn(OH)_4^-}}}{(C_{\mathrm{Zn(OH)_4^-}})_0} = \frac{i_{\mathrm{L,s}} - i}{i_{\mathrm{L,s}}}, \quad (3.43)$$

where $i_{\mathrm{L,s}}$ is the spherical diffusion limiting current at the dendrite tip,

$$i_{\mathrm{L,s}} = nFDC_0/r \quad (3.44)$$

and

$$a_{\mathrm{Zn}}/(a_{\mathrm{Zn}})_0 = \exp(2\gamma V/RTr), \quad (3.45)$$

considering that the surface energy affects the activity of adions as reactants in the anodic reaction. Introducing (3.44) and (3.45) into (3.43) and solving for i, one obtains

$$\frac{i}{K} = \frac{f_c(\eta) - f_a(\eta)\exp(1/r_n)}{(K/i_0) + f_c(\eta)r_n}, \quad (3.46)$$

where

$$f_c(\eta) = \exp(-\beta F\eta/RT) \quad (3.47)$$

and

$$f_a(\eta) = \exp[(1+\beta)F\eta/RT], \quad (3.48)$$

and

$$r_n = r/(2\gamma V/RT) \quad (3.49)$$

is a normalized tip radius.

The normalization constant for the current density is given by

$$K = RTnFDC_0/2\gamma V. \qquad (3.50)$$

Figure 47 shows a set of curves of the tip current density as a function of dendrite radius calculated using Eq. (3.46) for an independently obtained i_0 value of $12\ \mathrm{mA/cm^2}$ in $0.1\ M$ zincate solution and different overpotentials.

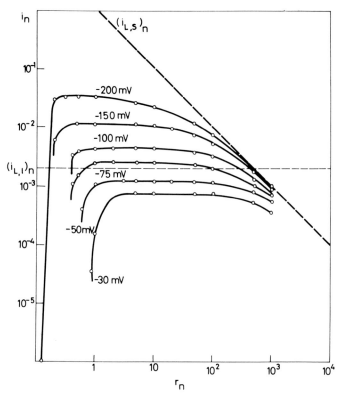

FIG. 47. Dependence of the current density at the dendrite tip on the tip radius at different constant overpotentials for the deposition of divalent metals. $(i_{L,l})_n$ is the value of the limiting current density of linear diffusion; $(i_{L,s})_n$ is that of the spherical diffusion. [From Despić et al. (1968a).]

On the same figure a horizontal line is drawn representing the value of the normalized limiting current density for the diffusion onto the flat surface. An important conclusion is indicated by the observation that curves for $\eta < |50|$ mV lie below that line, while those for $\eta > |75|$ mV lie above it: the dendrites will appear when the overpotential is such that the i_n–r_n curve crosses over the linear diffusion limiting-current density line $i_{L,1}$ for any value of r, since only under those conditions would a protrusion of the radius r grow faster than the rest of the surface. Equation (3.46) can be solved explicitly for the critical overpotential value at which $i = i_{L,1}$. For $f_a(\eta) \ll f_c(\eta)$

$$-\eta_c = \frac{RT}{\beta F} \ln \frac{i_{L,1}}{i_0} = \frac{RT}{\beta F} \ln\left(\frac{nFD}{\delta_0} \frac{C_0}{i_0}\right). \tag{3.51}$$

The value of η_c obtained for zinc deposition agreed reasonably well with that observed experimentally.

The same theory was applied to the deposition of silver (taking into account the difference in the mechanism of discharge). Figure 48 resulted, which indicated the critical overpotential of a few millivolts, in agreement with experiment.

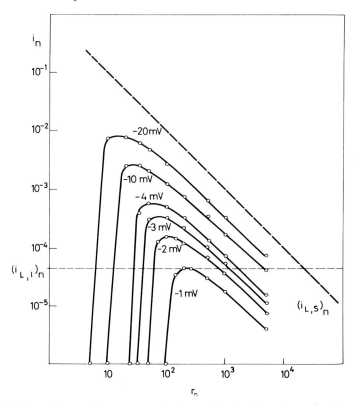

FIG. 48. Dependence of the current density at the dendrite tip on the tip radius at different constant overpotentials for the case of silver deposition [from Despić *et al.* (1968a)].

A comparison of Figs. 47 and 48 also reveals the reasons for the difference in the type of dendritic growth in the two systems: the silver lines are seen to have sharper maxima than in the case of zinc, indicating a narrow region of dendrite radii for which the conditions of growth would be satisfied. Since the probability of finding protrusions of desired radii increases with increasing region of allowed radii, the number of dendrites growing simultaneously on zinc is expected to be much larger than that on silver. Qualitatively, dendrite formation on zinc would tend to occur over a greater range of conditions than that on silver.

According to these theories the appearance and growth of dendrites are determined solely by electrodic parameters, and no specific effects of crystal structure have been incorporated. Yet these do seem to have some effect on the orientation of propagation and on the appearance of branching. Crystallographic studies of dendrite formation have been quite frequent over the last few years, notably that due to Wranglén (1960), in which the dendrites were identified as single crystals whose plane of growth corresponded to the closest-packed lattice. Silver dendrites grown by Yang et al. (1959) were examined by Faust and John (1961) and identified not as single crystals, but as twinned structures. This twinning was proposed to be an integral part of dendritic formation [cf. Reddy (1966)]. The presence of twinned dendritic structures was further supported by the work of Giron and Ogburn (1961), Ogburn et al. (1964, 1965) and Hamilton and Seidensticker (1960) and further work by Faust and John (1963). It should be mentioned, however, that this twinning may be the result of very fast growth of dendrites [cf. Cohen and Weertman (1963)] and therefore not be an essential part of the growth mechanism. These aspects of dendrite growth await conclusive results.

b. Whiskers. Those grow on tips only, the side faces being blocked by the adsorption of impurities. The current density at the tips is usually high (in the amp/cm² region) and increases with increasing concentration of organic additives such as gelatine, oleic acid, etc. (Gorbunova and Zhukova, 1949; Gorbunova and Dankov, 1949; Price et al., 1958). Price et al. derived the dependence of the current density on the concentration of additives: impurity molecules are first adsorbed on the tip of the whiskers and then buried into a growing whisker. If the rate of burial is smaller, the adsorbed molecules should accumulate at the surface and block the tip. Hence the growth should stop. At the steady state the rates of the adsorption and burial processes are equal.

The rate of spherical diffusion to the tip of radius r is given by

$$(dn/dt) = D_{imp}C_{imp}/r, \qquad (3.52)$$

where D_{imp} and C_{imp} are the diffusion coefficient and concentration of impurities, respectively.

The rate of burial of the adsorbed species should be

$$(dn_{imp}/dt) = (iN/zFn_0)(C_{ad,imp}/h),\qquad (3.53)$$

where $C_{ad,imp}$ is the surface concentration of impurities, n_0 is the number of atoms in 1 cm² of a monolayer, and h is the number of monolayers required to cover one molecule of the adsorbed impurity.

The concentration of adsorbed impurities on a growing surface (tip of the whisker) cannot be larger than

$$C_{ad,max} = 1/N(2r_c)^2,\qquad (3.54)$$

where r_c is the radius of the two-dimensional critical nuclei. The meaning of this relation is that an advancing step (e.g., of a screw dislocation) providing the growth of the whisker cannot proceed if the impurity atoms are closer than $2r_c$ (cf. Section III,B,1a and Fig. 36). With r_c related to overpotential [cf. Eq. (3.33)] and the latter to the current density, the maximum concentration of adsorbed impurities, as given by Eq. (3.54), is

$$C_{ad,max} = K(i)^2.\qquad (3.55)$$

The critical current density below which impurities (or additives) prevent the growth of whiskers is obtained by setting Eqs. (3.52) and (3.53) equal and replacing $C_{ad,imp}$ by Eq. (3.55), i.e.,

$$i_c = K(C_{imp}/r)^{1/3}.\qquad (3.56)$$

It is seen that the critical current density is strictly dependent on the concentration of adsorbing matter. It was assumed that a whisker starts to grow from a three-dimensional nucleus on which all but one of the crystal faces have been blocked for growth by the adsorption of impurities. The growing whisker then adjusts the current density at its tip and the tip radius so that the product $i_c^3 r$ matches the impurity concentration.

With reasonable choice of constants Price *et al.* have shown that the above hypothesis is in a fair agreement with experimental results obtained in the growth of silver whiskers from solutions containing gelatin and oleic acid. However, the mechanism of the above-mentioned adjustment remains rather obscure.

IV. Deposition of Alloys

Metals can be codeposited from solutions containing mixtures of their ions. A recent review of the phenomenon is due to Gorbunova and

Polukarov (1967), while an extensive survey of the field was made by Brenner (1963).

The basic condition of codeposition is obtained from the fact that the entire surface of a metallic substrate at which codeposition takes place is always equipotential with respect to the bulk of solution. Thus for metals A and B the codeposition should occur if their potential differences at the double layer with respect to their ions in solution are related by the equality

$$\Delta\phi = (\Delta\phi_{rev})_A + \eta_A = (\Delta\phi_{rev})_B + \eta_B, \qquad (4.1)$$

where the $\Delta\phi_{rev}$ are the reversible potentials and η the overpotentials for the deposition processes.

This reveals a thermodynamic and a kinetic aspect of the codeposition of metals.

A. THERMODYNAMICS OF CODEPOSITION OF METALS

An early thermodynamic analysis of the problem is due to Foerster (1931). He has shown that considerable shifts in the reversible potentials of metals in alloys with respect to those in pure metals are possible. Some metals may codeposit which otherwise cannot be electrodeposited in the pure state from aqueous solutions.

The reversible potential of any one of the metals undergoing reactions $A^{z_A+} + z_A e \rightarrow A$ or $B^{z_B+} + z_B e \rightarrow B$ is defined by

$$\Delta\phi_{rev} = \Delta\phi_x{}^\circ + (RT/zF) \ln a_{M^{z+}}, \qquad (4.2)$$

i.e., by a term reflecting the basic properties of the metal, $\Delta\phi_x{}^\circ$, and a term reflecting the activity of metal ions in solution. Of course, $\Delta\phi_x{}^\circ$ is the standard potential corresponding to the given metal *in the given alloy* rather than in its pure state. It can be represented formally by

$$\Delta\phi_x{}^\circ = \Delta\phi^\circ - \frac{RT}{zF} \ln a_M = \Delta\phi^\circ - \frac{RT}{zF} \ln f_M - \frac{RT}{zF} \ln X_M, \quad (4.3)$$

where $\Delta\phi^\circ$ is the standard potential of the pure metal, a_M is its activity in the alloy, defined by X_M, the mole fraction of the metal in the crystallite of the alloy containing it, and by f_M, the activity coefficient of the metal in the alloy phase.

The latter reflects the effect of the changes in the environment of an atom in the alloy lattice, relative to that in the pure metal phase, on the standard free energy of that species, i.e.,

$$-\frac{RT}{zF} \ln f_M = -\frac{\Delta G^\circ_{M,All}}{zF} = \Delta\phi^\circ_{M,All}, \qquad (4.4)$$

where $\Delta G^\circ_{M,All}$ is the change in the standard free energy and $\Delta \phi^\circ_{M,All}$ is the change in the standard potential of a mole of metal M when introduced into a very large amount of the alloy of the given composition.

1. *Change of Free Energy of the Components in Alloy Formation*

In general, three different situations can arise in the formation of an alloy: (1) a eutectic (mechanical) mixture of pure metal crystallites can be formed, (2) the components can form a homogeneous solid solution, and (3) the components can form intermetallic compounds. Combinations of the above situations are also encountered.

a. Free Energy of Components in the Eutectic Mixture. The mole fraction of each component, X_M, in Eq. (4.3) should be considered as 1. The standard free energy could be expected to be the same as that of pure metals.

A Kelvin-type effect of the crystallite size would lead to a somewhat increased standard free energy, with a resulting shift of the standard potential, $\Delta \phi^\circ_{M,All}$, in the negative direction. However, in the electrodeposition of alloys of this type (e.g., Cu–Pb and Cu–Tl) the reduction of the less-noble metal occurs at more positive values of potentials than estimated on the basis of standard potentials of the pure metal phase and Eq. (4.3) (Polukarov *et al.*, 1962a,b). These potential shifts can be of the order of several tens of millivolts. It appears that in such cases the crystallites of the eutectic are supersaturated solid solutions of one component in another, rather than pure metal phases. The situation is then essentially a nonequilibrium one.

b. Free Energy of Components in Solid Solution. In the case of ideal solutions, f_M in Eq. (4.3) should be equal to 1, and the shift in the standard potential should be linearly proportional to the log of the mole fraction. However, deviations are often encountered leading to values of potential more positive than that expected, indicating a decrease in the standard free energy of the component in the solution compared to that of a pure metal phase. This primarily reflects the change of the bonding energy of atoms leaving the lattice of a pure metal and entering that of an alloy.

c. Free Energy of Components Forming Intermetallic Compounds. Application of Eq. (4.3) would have to be only formal, since in such a case one could not consider any one of the metals as separate species, and the activity of each component no longer has physical meaning.

However, the standard reversible potential for the deposition of the metal into the alloy can be related to that of the pure metal phase if the standard free energy of formation of the compound, $\Delta G^{\circ}_{A_nB_m}$, is known. If the formation reaction is

$$nA + mB \rightarrow A_nB_m,$$

then, e.g., for the metal A

$$\Delta\phi_x^{\circ} = \Delta\phi^{\circ} + \frac{1}{nzF}\Delta G^{\circ}_{A_nB_m}. \tag{4.5}$$

2. The Role of Ionic Activity in Solution

The ability to perform a *reversible* deposition of alloys depends on the possibility of making the reversible potentials in the alloy equal for both metals. In principle, this can be achieved by adjusting the ionic activities in Eq. (4.2) to compensate for the difference in standard electrode potentials. In practice, however, too large a gap cannot be overcome, because of the logarithmic nature of the dependence (4.2). The concentration of the species having a more positive standard potential would in most cases have to be below the practically attainable limits ($< 10^{-6}$ mole/liter). One way of overcoming this difficulty that is sometimes possible is to complex the ions by a suitable complexing agent with a higher affinity for the species with a more-positive standard potential than for the other. This amounts to an additional change of the standard potential with respect to that of a pure metal in solution in the form of uncomplexed hydrated ions. If K_{MX} is the thermodynamic stability constant of a complex between a metal and a ligand,

$$M^{z+} + nX \rightleftarrows MX_n^{z+},$$

i.e.,

$$K_{MX} = \frac{a_{MX}}{a_M a_X{}^n}, \tag{4.6}$$

replacing a_M from (4.6) into Eq. (4.2), one obtains

$$\Delta\phi_{rev} = \Delta\phi_x^{\circ} - \frac{RT}{zF}\ln K_{MX} + \frac{RT}{zF}\ln\left(\frac{a_{MX}}{a_X{}^n}\right)$$

$$= \Delta\phi_{MX}^{\circ} + \frac{RT}{zF}\ln\left(\frac{a_{MX}}{a_X{}^n}\right), \tag{4.7}$$

where $\Delta\phi^\circ_{MX}$ is the standard electrode potential for the new electrode reaction

$$MX_n^{z+} + ze \rightleftharpoons M + nX,$$

and is seen to be more negative than $\Delta\phi_x^\circ$ for reasonably high activities of the ligand X in solution.

The complexing effect of the ligand on the other metal also has to be taken into account. It can be shown that a suitable complexing agent is one which has a difference between the standard free energies of complexing for the two metals approximately equal to

$$\Delta(\Delta G^\circ_{MX}) = (\Delta G^\circ_{AX} - \Delta G^\circ_{BX}) \approx z_B F(\Delta\phi_x^\circ)_B - z_A F(\Delta\phi_x^\circ)_A. \quad (4.8)$$

This is the condition by which the inequalities in the reversible potentials can be compensated by an appropriate selection of solution composition.

B. KINETICS OF CODEPOSITION

As seen from Eq. (4.1), the situation with respect to codeposition is of course influenced by kinetic conditions as well, i.e., if the difference in the reversible potentials of the depositing ions is compensated by appropriate adjustments in overpotentials. Exchange currents will obviously be affected by alloying. The effects depend on the type of depositing alloy.

1. *Kinetics of Deposition of Binary Systems with Eutectic*

It is observed experimentally, e.g., in the deposition of tin–lead and tin–zinc alloys from pyrophosphate solutions (Rama-Char and Vaid, 1961a,b, 1962), that partial current densities for each metal at a given potential are lower than in deposition onto corresponding pure metals.

Since in this case separate crystallites of pure metals are formed, one may assume, to a first approximation, that the deposition kinetics of one metal are unaffected by the deposition of the other. Hence the current density–potential relation for a microsurface area of each crystal may be taken to be the same as that for macrodeposition of the metal concerned (Fig. 49). Since the effective surface area for the deposition of each metal is less than in the same external geometric area for a pure metal, the apparent current density (per unit geometric surface area) of its deposition for a given overpotential is

$$i_A' = i_A[S_A/(S_A + S_B)], \quad (4.9)$$

where S_A and S_B are the surface areas covered by metals A and B, respectively (Gorbunova and Polukarov, 1967). Since in the competition for the surface, each metal will cover a portion in proportion to the real current density of its deposition,

$$\frac{S_A}{S_A + S_B} = \frac{i_A(\epsilon_A/\rho_A)}{i_A(\epsilon_A/\rho_A) + i_B(\epsilon_B/\rho_B)},$$ (4.10)

where the ϵ's are the electrochemical equivalents and the ρ's are the densities of the metals. Introducing Eq. (4.10) into Eq. (4.9), one obtains for the experimentally obtainable current density

$$i_A' = \frac{i_A^2}{i_A + (\epsilon_B\rho_A/\epsilon_A\rho_B)i_B}.$$ (4.11)

The current density–potential relationships for the two metals obtained using Eq. (4.11) are shown in Fig. 49 together with the curve for the total current density which is the sum of those for the two metals.

Thus, although the deposition kinetics remains unchanged, there is a quasidecrease in the rate of individual electrochemical reactions as compared with the process of pure metal deposition.

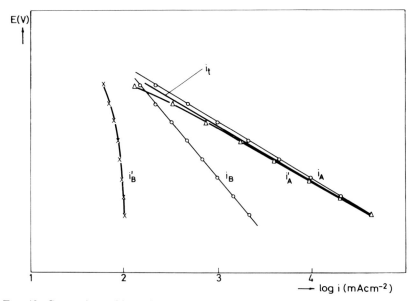

FIG. 49. Comparison of hypothetical polarization curves for deposition of metals onto pure metal phases (i_A and i_B) with those that should be obtained for the deposition on a eutectic-type alloy according to Eq. (4.11). i_t: The total current density line. [From Gorbunova and Polukarov (1967).]

There is, however, a limiting assumption implicit in Eq. (4.11): discharge of a metal can take place on the surface of its corresponding metal phase only. This can be true at relatively low overpotentials only, where nucleation processes are relatively slow. Otherwise, and if the discharge step is rate-controlling, there is little reason for such a situation, and other explanations should be sought [cf. Rama-Char and Vaid (1961a,b, 1962)]. Even at low overpotentials the partial current densities should be larger than calculated according to Eq. (4.11), because additional adatoms come to the corresponding crystals from deposition sites on the other metal by surface diffusion.

Some change in deposition kinetics in dilute solutions may also be due to a change in double-layer structure, inasmuch as the relevant concentration of each of the depositing ions in the double layer is lower than if they were the only ones present in the electrolyte (Vagramyan and Fatueva, 1960). Thus the Frumkin effect is not as pronounced in the deposition of alloys as it is in the deposition of pure metals.

2. Codeposition into Solid Solutions

The standard free energy of a metal in the solid solution, G°_{All}, usually differs significantly from that in the pure metal phase, $G_M{}^\circ$. The changes in free-energy values for the deposited metal atoms from those which would exist in the corresponding pure metal, $\varDelta G^\circ_{M,All}$, can be expected to affect the activation energy barrier of the process by $\beta\,\varDelta G^\circ_{M,All}$. Exceptions to this would occur when there is an intermediate state between the transition state and the final state, unaffected by alloying. For example, if in the discharge of divalent ions the second step is fast and if the standard free energy of the univalent intermediate is not affected by alloying, there should be no effect of alloying on the activation energy barrier either. Hence, except for such cases, if $\hbar_M$ is the electrochemical rate constant for the deposition of a metal into the pure phase and $\hbar_{All}$ is that for deposition into the alloy, the latter should be different from the former (cf. Fig. 50) by

$$\hbar_{All} = \hbar_M \exp(-\beta\,\varDelta G^\circ_{M,All}/RT). \qquad (4.12)$$

However, the rate constants refer to a constant absolute value of the potential difference across the double layer. It can be shown, on a model which accepts the vertical shifting of the potential-energy curves, that if the rates are compared at constant overpotential, or if the chemical rate constants or i_0 values are compared at $\eta \to 0$, they should be equal for both kinds of deposition and independent of the energies of mixing.

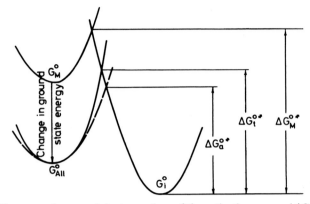

FIG. 50. The approximate and the true values of the activation energy ($\Delta G_a°$ and $\Delta G_t°$, respectively) obtained upon simple vertical shifting of the potential-energy curve (dotted line) and upon taking into account the effect of the shape of the curve.

Thus the electrochemical rate constants in Eq. (4.12) can be written as

$$k_{\text{All}} \exp \frac{-\beta F\phi}{RT} = k_{\text{M}} \exp \frac{-\beta F\phi}{RT} \exp \frac{\beta \Delta G°_{\text{M,All}}}{RT}. \qquad (4.12a)$$

If the reversible potentials

$$(\phi_{\text{rev}})_{\text{M}} = -(G_{\text{M}}° - G_{\text{i}}°)/zF \qquad (4.13)$$

and

$$(\phi_{\text{rev}})_{\text{All}} = -(G°_{\text{All}} - G_{\text{i}}°)/zF \qquad (4.14)$$

(where $G_{\text{i}}°$ is the energy level of the initial state in reduction) are introduced to the two sides of Eq. (4.12a), one obtains

$$k_{\text{All}} \exp \frac{\beta(G_{\text{M}}° - G_{\text{i}}°)}{RT} \exp \frac{-\beta F\eta}{RT}$$

$$= k_{\text{M}} \exp \frac{\beta(G°_{\text{All}} - G_{\text{i}}°)}{RT} \exp \frac{-\beta \Delta G°_{\text{M,All}}}{RT} \exp \frac{-\beta F\eta}{RT}. \qquad (4.15)$$

The first exponential term on the right-hand side can be written as

$$\exp \frac{\beta(G°_{\text{All}} - G_{\text{i}}°)}{RT} = \exp \frac{(G - \beta_{\text{All}} G_{\text{M}}° + G_{\text{M}}° - G_{\text{i}}°)}{RT}$$

$$= \exp \frac{\beta \Delta G°_{\text{M,All}}}{RT} \exp \frac{\beta(G_{\text{M}}° - G_{\text{i}}°)}{RT}. \qquad (4.16)$$

Introducing this into Eq. (4.15), one can see that all the exponential terms cancel and

$$k_{All} = k_M. \qquad (4.17)$$

In a similar manner no direct effect on deposition kinetics is expected from the change in zero-charge potential from that for the pure metals to that of the alloy [cf. Frumkin (1965a,b)]. Yet experiments show that there is some effect, in the sense that the reduction rate decreases for codeposition, relative to the deposition of a pure metal, for the component with a more-positive zero-charge potential, whereas the opposite is true for the other component (Stabrovsky, 1951; Persiantseva and Titov, 1958; Vagramyan and Fatueva, 1959; Lainer and Yu, 1963).

Gorbunova and Polukarov (1967) drew attention to the fact that the simple vertical shift of the potential-energy curves represents an oversimplification. If they are adequately represented by an equation of the Morse type [cf. Eqs. (2.21) and (2.22)], then any change in the parameter D (reflecting the change in the ground-state energy) results not only in vertical shifting, but also in a change of slope. This produces a change of the activation energy which adds to that produced by shifting (Fig. 50).

Some other effects resulting from the change in the zero-charge potential are also possible. For example, adsorption of some surface-active substances depends on this change. This in turn significantly affects the kinetics of reduction as well as the phase composition of the deposited alloy (Loshkarev and Kryukova, 1948, 1949; Martirosyan and Kryukova, 1953).

In addition, the Frumkin effect of the double-layer structure can be significant in dilute solutions.

3. *Special Cases of Codeposition*

A special case seems to exist in codepositions with chromium from chromate solutions. The presence of a surface film hinders the reduction of ions normally reduced at more positive potentials (Kasper, 1932). In some cases, in which the other metal constitutes the complex anion, it becomes a part of the surface film and is reduced directly from it onto the metal substrate. In this way alloys of manganese, rhenium, and selenium with chromium are obtained (Vagramyan and Usachev, 1954; Vagramyan et al., 1961). In this case the deposition rate is related to the ionic concentration in the film rather than in the bulk of the electrolyte.

In addition, passivating films may have such retarding effects on the deposition of some ionic species as to change the properties of deposited alloy. Vagramyan (1959) has shown a hydroxide film and a layer of adsorbed hydrogen to be explanations for a changing rate of deposition of iron during the codeposition of cobalt or nickel.

C. PHASE FORMATION IN ALLOY DEPOSITION

In a number of cases, e.g., the copper–zinc system [cf. Raub and Krause (1944)], the electrodeposited alloy has the phase structure predicted by the equilibrium diagram. However, in the majority of cases considerable deviations are encountered. These are mainly in the direction of the formation of supersaturated solid solutions, the compositions of which are beyond the solubility limits for equilibrium systems [Raub and Engel (1943), for silver–lead alloys, or Lainer (1950) for tin–copper deposits]. In some cases high-temperature phases are formed instead of the equilibrium ones (Raub and Wullhorst, 1947).

According to the zone theory of metals as applied to alloys [cf. Jones (1950), Mott (1952)] the limits of existence of different phases are related mainly to the number of free electrons per atom of the alloy. Gradual filling of the Brillouin zones results in a change of the phase energy to an extent which makes it unstable compared to other possible phases. Thus Raynor (1949) and Hume-Rothery and Raynor (1956) have found in a series of alloys of metals codeposited with copper and silver that successive phase transitions take place at definite electron concentrations. However, in the majority of cases the equilibrium solubilities are lower because of an additional effect of the difference in the volumes of atoms that should build a common lattice. Atoms of larger volumes cause considerable distortion of lattices built of atoms of smaller radii, and the energy of distortion adds to other factors in the phase energy in determining the equilibrium properties.

During electrodeposition the influence of the volume factor seems to be negligibly small, and this seems to be the main cause of increased solubility. The latter never exceeds the limits determined by the zone theory.

1. *The Relation between the Phase Structure and Deposition Kinetics*

The formation of nonequilibrium phases, particularly those which are normally found at higher temperatures only, was accounted for by Banerjee (1952) and Fischer (1955) in terms of the formation by the reduced ions of a layer of atoms which are in a similar state and energy content as those in the melt, which subsequently solidifies into a solid phase corresponding to that liquid state.

The increase in solubility of a less-noble metal in the more-positive metal can be explained in terms of the energetics of the deposition. The partial molar free-energy of the more electronegative component (e.g., lead or thallium in copper) relative to that of the pure metal phase may be written as

$$\Delta \bar{F} = R[T + \Delta T f(\eta)] \ln x - (1 - x)^2 (U_{\text{All}} + \gamma z F \eta), \quad (4.18)$$

where x is the mole fraction of the less-noble component. The first term represents the change in entropy, and its overpotential dependence is not explicitly known. The energy U_{All} can be found as

$$U_{\text{All}} = \sigma N_0 [U_{AB} - \tfrac{1}{2}(U_{AA} + U_{BB})], \quad (4.19)$$

where the U's are the bond energies between the atomic species denoted in the indices and σ is the coordination number. The coefficient γ in (4.18) is a rate-dependent quantity and tends to 1 at high rates.

The change in partial molar free-energy, $\Delta \bar{F}$, determines the extent to which the equilibrium potential for the deposition of the less-noble metal is shifted to the positive side and its deposition made possible. Moreover, it indicates that, other conditions being equal, the increased overvoltage should lead to increased solubility, or supersaturation, with respect to the state at zero overpotential.

For the appearance of the second metal phase—that of the less-noble metal—the potential of the electrode should be made more negative than the equilibrium potential of the latter. In addition, a nucleation over-potential may be required, the more so the larger the difference in atomic radii between the two components. This is in accordance with the obser-vations (Lustman, 1943; Rama-Char and Vaid, 1961a,b; Sree and Rama-Char, 1961a,b; Sree et al., 1961; Polukarov et al., 1962a,b) of potential–current relations leading to some kind of limiting currents, the overcoming of which coincides with the observed phase transformations.

During the simultaneous growth of the two phases the distribution of the less-noble metal between them will be in proportion to the growth rates of the two phases. In general, the higher the overpotential, the higher will be the degree of supersaturation of the solid solution phase, i.e., the higher the concentration of the less-noble metal in it.

3. Effect of Additives on Phase Formation

In general, surface-active substances have a profound effect on deposi-tion processes, changing both the kinetics of deposition (increasing over-voltage) and the morphology of the deposits. In the deposition of alloys these complex effects result in changes in composition of phases and sometimes affect the phase structure of the deposit.

The change in the composition of solid solutions is a consequence of increased overvoltage, which is seen to increase the solubility of the less-noble component in the more-noble one.

The change in phase structure can occur when the surface-active substance is adsorbed differently at different surfaces. Such a situation can arise even in the absence of specific adsorption forces if the surface-active substance is undergoing electrosorption. The latter depends on the difference between the actual electrode potential and the point of zero charge. Since the potential of zero charge is different at different phases, the growth will be more inhibited or even prevented for those phases on which strong adsorption occurs. The differences will be particularly pronounced when the two phases have a different sign of charge at the given potential.

Both kinds of effect are found in the codeposition of lead with copper (Gorbunova and Polukarov, 1967). In the presence of thiourea as a surface-active substance the overpotential is increased and there is a proportional increase in the lead content of both the alloy as a whole and the solid solution as one of its phases (Fig. 51, curves 1a and 2a), but two phases persist at any concentration of the additive. However, if the anodic compartment is not separated from the cathodic one, so that the oxidation product of thiourea (disulfide formamide) is present, which is strongly adsorbed on lead, one phase vanishes and only solid solution remains, so that the lead content in the alloy and in the solid solution phase coincides (Fig. 51, curves 1b and 2b). The solubility of lead is seen to show a strong increase at higher overpotentials. The current density–potential relation shows a correspondingly strong increase in inhibition of the process (Fig. 52).

Quantitative treatments of this aspect of alloy deposition are lacking.

V. Kinetics of Metal Dissolution

A. Dissolution and Deposition

1. *Microscopic Reversibility and the Overall Process*

Dissolution should follow the same reaction path as the deposition process, i.e., the overall reaction of formation of a metal atom incorporated in the bulk of the lattice from a hydrated ion in solution and electrons in the metal should be reversible in each unit step and thus also

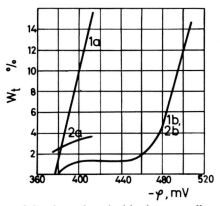

FIG. 51. Lead content of the electrodeposited lead–copper alloy as a function of over-voltage. Curves 1a,b: The total lead content; curves 2a,b: Lead content in the solid solutions. (a) In the presence of thiourea. (b) In the presence of disulfide formamide [Gorbunova and Polukarov (1967).]

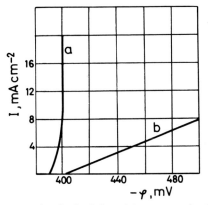

FIG. 52. Polarization curves for the lead deposition process in the presence of (a) thiourea and of (b) disulfide formamide [from Gorbunova and Polukarov (1967)].

as a whole. However, this microscopic reversibility is limited to conditions of rate and potential not far from the equilibrium ones. There always exists the possibility of the reaction assuming an alternative parallel path and this may be favored by the potential being made anodic.

Irreversibility often occurs in the lattice-building stage. Thus centers of most-active growth by no means need also be centers of most-active dissolution.

For this reason application of alternating current to equilibrated metal electrodes seldom remains without effect on the surface structure. Depending on conditions, roughening (pitting) or smoothing action is observed.

A significant role is played here by surface-active substances which undergo electrosorption. Since the coverage is potential-dependent, they may inhibit the deposition process at the required negative potentials while desorbing and not affecting the anodic dissolution, and *vice versa*.

Knowledge of the extent of irreversibility and the effects of various factors is still very limited and qualitative in nature. Hence the application of cycling or superimposed alternating current in practical metal deposition for achieving different effects (e.g., prevention of dendritic growth) is largely empirical.

2. *Dissolution with the Formation of Insoluble Products*

This covers a large area of the electrocrystallization of substances other than metals [cf. the review by Fleischmann and Thirsk (1963)], and in one particular aspect, i.e., that of special film formation, also the phenomenon of passivity.

Electrocrystallization can be a secondary effect arising from supersaturation of the solution in the vicinity of the electrode by the dissolution products. However, instead of reprecipitation, a direct formation of insoluble products at the electrode surface as a part of the electrodic reaction is also possible.

A detailed study of the system is required to distinguish between the two mechanisms. A very illuminating investigation has been carried out by Bockris *et al.* (1964) in an attempt to solve the puzzling mechanism of the calomel electrode. It has been realized for some time (Hills and Ives, 1951) that the concentration of mercurous ions in equilibrium with solid calomel is too small ($10^{-18}\ M$) to sustain the extremely high exchange-current density of the calomel electrode ($\sim 1\ A/cm^2$). A role is given to unstable chloro-mercurous ions, Hg_2Cl^+, and also to "chloro-mercury," a covalently-bound two-dimensional layer of chlorine, as a precursor to calomel formation.

Bockris *et al.* (1964) have used a combination of galvanostatic and elipsometric techniques to compare the charge–potential relationship with an optical detection of the appearance of any film on the mercury surface.

A typical observation is shown in Fig. 53. Thus it was proved that upon application of an anodic current pulse a film appears long after the polarizable mercury interface has become a nonpolarizable one, because of the presence of a potential-determining species. Also, as the anodic current is cut off the ellipsometric recording indicates further film growth. Thus it is concluded that the anodic current is not the direct cause of the growth of the calomel film. Rather, the film growth is due to precipitation from solution. This is further supported by the fact that the induction time was shown to depend on the degree of stirring of the electrolyte. This can be explained only by a diffusion-controlled process of accumulation of the species in the solution to a value exceeding the solubility. The assumption of a dissolution–precipitation mechanism of

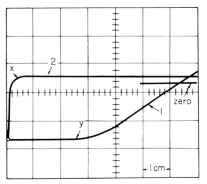

FIG. 53. Elipsometric and galvanostatic transients for anodic polarization of mercury in 1 M KCl. Curve 1: Ellipsometer photomultiplier output against time transient. Curve 2: Potential against time transient. [From Bockris *et al.* (1964).]

calomel formation involving the chloro-mercurous ion as the carrier of mercury away from the electrode led to the following expression for the induction time in unstirred solutions

$$i\sqrt{\tau_i} = (D\sqrt{\pi}/2)(1 + K_F)C^* + (K_F\sqrt{\pi}/2\sqrt{z})i\,\mathrm{erf}(z\sqrt{\tau}), \quad (5.1)$$

where $K_F = k^-/k^+$ and $Z = k^- + k^+$, while k^+ and k^- are the rate constants for a homogeneous chemical reaction of disproportionation of the Hg_2Cl^+ ion.

The experimental plot of $i\sqrt{\tau_i}$ vs. i (Fig. 54) is in complete accordance with Eq. (5.1). A calculation based on this equation rendered a critical

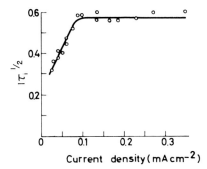

FIG. 54. Plot of $i\sqrt{\tau_i}$ against galvanostatic current density [from Bockris *et al.* (1964)].

concentration C^* of Hg_2Cl^+ ions for the formation of calomel of 10^{-4} M. Such a concentration value supports the basic argument of Hills and Ives (1951) that chloro-mercurous ions are generated in sufficient concentration to sustain the high exchange–current density of the calomel electrode. However, the ellipsometric observations, which can detect films of monomolecular thickness, revealed no appearance of a film of chloro-mercury at the electrode surface.

In a general case the choice between direct precipitation and dissolution–reprecipitation mechanisms depends mainly on the rate-determining step of the overall process of formation of the insoluble product. If the latter is controlled by the discharge process, i.e., ionization of the metal, the chances are that the product will be formed in the double-layer region. Conversely, if diffusion from the bulk of solution of the other component forming the product is rate-controlling, or if an intermediate of considerable solubility is formed, crystallization will occur independently of the discharge process and no epitaxial formations are to be expected.

B. Dissolution into Solutions Free of Corresponding Ions

This special case, which is rarely met in metal deposition (e.g., deposition of metals into a stream of pure mercury), is of interest because of its consequences for the corrosion of metals.

1. The Concept of the "Floating" Electrode and the Dissolution of Univalent Metals

The problem of a "floating" electrode, i.e., "a piece of metal simply immersed in a solution, with no control over its electrical potential," was treated in a rather general way by Kimball (1940).

In the case of a one-electron exchange reaction the current density is given by

$$i = \overrightarrow{i}_M - \overleftarrow{i}_{M^+}. \tag{5.2}$$

With the usual current density–potential relationship one can write

$$i = \overrightarrow{k}_M{}^0 C_M f_a(E) - \overleftarrow{k}_{M^+}^0 C_{M^+} f_c(E), \tag{5.3}$$

where

$$f_a(E) = \exp[(1 - \beta)FE/RT] \tag{5.4}$$

and

$$f_c(E) = \exp[-\beta FE/RT], \tag{5.5}$$

and E is a relative potential of the electrode with respect to a selected reference electrode. The electrochemical rate constants $\vec{k}_M{}^0$ and $\overleftarrow{k}_{M^+}^0$ are seen to be equal to the specific rates at unit concentrations and *at the potential equal to the reference electrode potential*. This reference point is selected arbitrarily, since the system lacks a defined reversible potential with respect to which the rates are usually related. If the concentration of ions outside a diffusion layer of a defined thickness δ is maintained equal to zero, the material balance at the outer Helmholtz plane can be written as

$$\frac{dC_{M^+}}{dt} = \frac{\vec{k}_M{}^0 C_M f_a(E)}{F} - \frac{\overleftarrow{k}_{M^+}^0 C_{M^+} f_c(E)}{F} - \frac{D_{M^+} C_{M^+}}{\delta} = 0. \quad (5.6)$$

Hence

$$C_{M^+} = \frac{\vec{k}_M{}^0 C_M f_a(E)}{\overleftarrow{k}_{M^+}^0 f_c(E) + (F D_{M^+}/\delta)} \quad (5.7)$$

and replacing this into Eq. (5.3) one obtains

$$i = \vec{k}_M{}^0 C_M f_a(E) \frac{F D_{M^+}/\delta}{\overleftarrow{k}_{M^+}^0 f_c(E) + (F D_{M^+}/\delta)}. \quad (5.8)$$

Equation (5.8) indicates that at a constant flow rate (constant δ) two regions of current–potential relation are to be expected, both of the Tafel type:

1. In the region of relatively negative potentials, where $\overleftarrow{k}_{M^+}^0 f_c(E) \gg F D_{M^+}/\delta$, one obtains from Eq. (5.8)

$$\ln i = [\ln(\vec{k}_M{}^0/\overleftarrow{k}_{M^+}^0)(F D_{M^+}/\delta) C_M] + (F/RT)E, \quad (5.9)$$

i.e., a linear $\ln i$ vs. E relationship with a slope of F/RT and an intercept which is a complex quantity containing the equilibrium constant for the electrode reaction and the diffusion parameters. This is essentially a region where diffusion is controlling the rate of the process. Correspondingly, in the anodic process a logarithmic relationship is maintained and there is no tendency to limiting current.

2. As the potential is made positive, a condition arises at which $\overleftarrow{k}_{M^+}^0 f_c(E) \ll F D_{M^+}/\delta$. Hence Eq. (5.8) reduces to

$$\ln i = (\ln \vec{k}_M{}^0 C_M) + \frac{(1-\beta)F}{RT} E, \quad (5.10)$$

i.e., the process becomes activation controlled, with a slope of the Tafel relation equal to $(1 - \beta)F/RT$ and an intercept which represents a kind of standard current density obtained at the potential equal to the reference-electrode potential. If the reference electrode is of the same metal in a solution of ions of unit activity, this is equal to the standard exchange–current density.

2. Dissolution of Divalent Metals

Despić et al. (1969) have shown, by applying a treatment essentially similar to that of Kimball (1940) to the dissolution kinetics of divalent metals, that some new diagnostic criteria can be found in this case which are useful for deriving the detailed mechanism of two-electron exchange reactions.

With regard to the possible unit steps which could appear in such a reaction (cf. Section II,A,2d) it was assumed that the disproportionation of univalent intermediate ions can be neglected in the reaction layer at the electrode surface. Thus the probability of electron transfer between the positively charged metal substrate and a hydrated ion is greater than that between two hydrated ions.

An additional possibility was taken into account, i.e., a univalent intermediate species can exist but can be so strongly adsorbed that it is never discovered in solution (e.g., univalent ions of Ni, Cd, etc.), i.e., that its concentration at the plane $x = 0$, relevant to diffusion, is always zero. In a first approximation this can be described by the linear portion of a Langmuir isotherm,

$$C_{M^+}/(C_{M^+})_{x=0} = K. \tag{5.11}$$

For a direct two-electron exchange reaction an equation identical to Eq. (5.8) should be applicable. In the two one-electron step paths the net current density at the electrode is

$$i = \overrightarrow{i}_M - \overleftarrow{i}_{M^+} + \overrightarrow{i}_{M^+} - \overleftarrow{i}_{M^{++}}$$
$$= \overrightarrow{k}_1{}^0 C_M f_a(E) - \overleftarrow{k}_1{}^0 C_{M^+} f_c(E) + \overrightarrow{k}_2{}^0 C_{M^+} f_a(E) - \overleftarrow{k}_2{}^0 C_{M^{++}} f_c(E), \tag{5.12}$$

assuming both steps have the same value of the symmetry factor β.

At the steady state the concentration of M^+ and M^{++} ions at the plane $x = 0$ are constant, and from the material balance equations, similar to

Eq. (5.6), the concentrations C_{M^+} and $C_{M^{++}}$ are found. Replacing these into Eq. (5.12), one obtains

$$i = \frac{\overset{\rightarrow}{k_1}{}^0 C_M f_a(E)\{P_1[\overset{\leftarrow}{k_2}{}^0 f_c(E) + P_2] + 2\overset{\rightarrow}{k_2}{}^0 f_a(E) P_2\}}{\overset{\leftarrow}{k_1}{}^0 \overset{\rightarrow}{k_2}{}^0 [f_c(E)]^2 + P_1 \overset{\rightarrow}{k_2}{}^0 f_c(E) + P_2\{\overset{\leftarrow}{k_1}{}^0 f_c(E) + \overset{\rightarrow}{k_2}{}^0 f_a(E) + P_1\}}$$

(5.13)

where

$$P_1 = F D_{M^+} / K \delta_{M^+}$$ (5.14)

and

$$P_2 = F D_{M^{++}} / \delta_{M^{++}}.$$ (5.15)

If disproportionation inside the diffusion layer is negligible, the diffusion-layer thicknesses δ_{M^+} and $\delta_{M^{++}}$ are equal to each other and to the hydrodynamic diffusion-layer thickness. If this is not the case, however, apparent diffusion-layer thicknesses are determined by a complex relationship between the rates of diffusion and the rate of the chemical reaction, as, e.g., in deposition and dissolution of copper [cf. mechanism (2.28b) in Section II,A,2d].

A number of situations can arise if the two-step mechanism is operative, which can be distinguished by the resulting kinetic consequences, in particular, by the slopes of the potential–current density relations in different regions of potential. These situations are:

1. Both steps are fast compared to the diffusion of univalent intermediate away from the electrode, but the second step is slower than the first.
2. Both steps are fast compared to diffusion of univalent intermediate, but the first step is slower than the second.
3. The first step is fast, while the second is very slow and slower than the diffusion of the univalent intermediate.

Table IX lists the Tafel slopes for the cases cited. It is seen that they all can be experimentally distinguished if the kinetics is followed over a sufficiently large region of potentials. Such transitions of slopes were observed in the systems investigated experimentally, as, e.g., in the anodic dissolution of copper and nickel into hydrochloric acid (Bockris *et al.*, 1969).

TABLE IX

Tafel Slopes at Different Situations of Dissolution of Divalent Ions into Electrolytes free of Corresponding Ions ("Floating" Electrode)

Mechanism	Situation[a]	Tafel slope		
		Low potentials	Highly positive potentials	
Single-step, two-electron exchange	—	$RT/2F$	$RT/2(1 - \beta)F$	RT/F
Two-step, one-electron exchange	(1)	$RT/2F$	$RT/(2 - \beta)F$	$2RT/3F$
	(2)	$RT/2F$	$RT/(1 - \beta)F$	$2RT/F$
	(3)	RT/F	$RT/(1 - \beta)F$	$2RT/F$

[a] See text for explanations of situations 1, 2, and 3.

C. Acceleration of Dissolution under Strain

1. *Phenomenology*

A number of authors have noted that straining a metal produces some change either in the steady-state potential attained in a corrosive medium (Gerischer and Rickert, 1955) or in the current recorded when the metal is maintained at a constant potential (Hoar and West, 1958; Raicheff *et al.*, 1967).

While the metal is strained in the elastic region the effects are small and no acceleration of the dissolution process could be observed. The recorded changes could be ascribed to the changes in the rate of the depolarization reaction of the corrosion couple (Despić *et al.*, 1968b).

Yielding of the metal beyond the plastic-deformation-point results, however, in very strong mechanoelectrochemical effects. An increase of many times in the metal dissolution currents is obtained. Figure 55 shows a typical simultaneous recording of the stress–strain and current–strain relationships for iron for straining done at a constant rate. Molybdenum exhibited a similar behavior, the current passing through a maximum before the breaking point. The effect was much less pronounced in the case of nickel and copper, and the current–strain relation was of a different character.

Characteristic of these relationships is a linear dependence of the maximum current attained on the rate at which the metal is strained. As seen in Fig. 56, the slope of this relationship depends on the metal, and was highest in the case of iron. For a given metal the slope was found to be strongly potential-dependent.

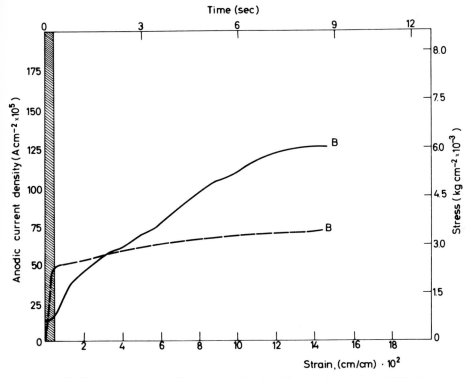

FIG. 55. Simultaneous recording of stress (dashed line) and current (solid line) at an iron wire as a function of strain at a constant strain rate up to the breaking point B. Shaded area corresponds to the elastic region. [From Raicheff *et al.* (1967).]

2. *Theories of Accelerated Dissolution*

Several concepts of accelerated dissolution of metals under strain have been developed. Logan (1952, 1958, 1961) suggested protective-film rupture to be a dominant phenomenon in the acceleration of dissolution. However, the same phenomenon was observed under conditions in which the existence of such a film was excluded.

Gerischer and Rickert (1955) attributed the increase in the dissolution rate to the effect of "loosely bound" atoms at deformed grain boundaries. Bakish and Robertson (1955, 1956) observed that dissolution of Cu–Au alloys immersed in $FeCl_3$ solution occurs preferentially along the surface traces of slip bands of plastically deformed Cu_3Au phase. Raicheff *et al.* (1967) were the first to suggest that this should be the case whenever high-index planes are developed on metal surfaces as slip planes upon

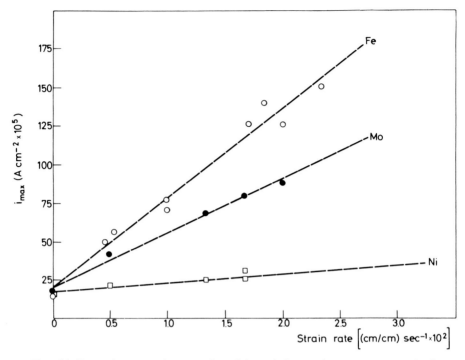

FIG. 56. Dependence on the rate of straining of the maximum current attained at constant potential for the straining of wire samples of different metals [from Despić *et al.* (1968b)].

yielding, since the exchange current density and corresponding discharge activities at a given potential should be higher at less-closely packed surfaces.

Despić *et al.* (1968b) confirmed the value of this concept by obtaining the cited difference in the effect of strain on the metals which are known to develop high-index planes, such as iron and molybdenum, compared to those which slip along planes of low index, as do copper and nickel.

Despić *et al.* (1968b) developed a quantitative theory of this effect. According to the model assumed, at the yielding point slip edges appear all along the metal surface and slip planes start emerging in succession to one another in discontinuous increments rather than simultaneously and continuously. As the strain increases continuously more of the new planes emerge, and since they provide an increased dissolution activity, the overall rate of dissolution increases continuously with increase of strain. Yet these new active planes dissolve preferentially, and the surface in contact with the solution tends to reacquire the more stable planes having rate constants for dissolution similar to that of the surface of the un-strained wire. There is thus a lifetime for the activity of each developed high-index slip plane, after which it becomes inactive, i.e., only as active as the initial surface of the wire. The process is represented in Fig. 57.

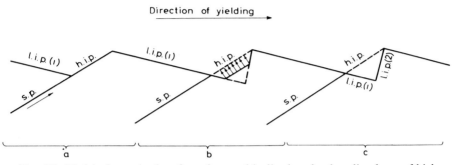

FIG. 57. Model of a strained surface of a metal inclined to develop slip planes of high index. (a) Appearance of slip planes. (b) Dissolution of the plane of high index. (c) End of the dissolution process.

Such a model resulted in the following equation for the net increase in current density in the plastic region:

$$\Delta i = Q + (P + B)t - At^2, \tag{5.16}$$

where

$$Q = K_{DL}(L_0/S_0)(\partial S/\partial L)\,\Delta E(\partial\epsilon/\partial t) + N_e y^0(L_0/S_0)i_e, \tag{5.17}$$

$$P = i_{ns}(L_0/S_0)(\partial S/\partial L)(\partial\epsilon/\partial t)f, \tag{5.18}$$

$$B = K(L_0/S_0)s_0 i_{hip}(\partial\epsilon/\partial t), \tag{5.19}$$

and

$$A = (K/zFn)(L_0/S_0)y^0 i_{hip}[i_{hip} - i_{lip,1}\cos\alpha - i_{lip,2}\sin\alpha]$$
$$\times (\cot\alpha + \tan\alpha)(\partial\epsilon/\partial t), \tag{5.20}$$

with K_{DL} the integral double-layer capacity; $\partial S/\partial L$ the newly created surface per unit elongation; L_0 and S_0 the original length and surface area of the metal sample (wire), respectively; ΔE the value of the potential of the sample relative to the ZCP; $(\partial\epsilon/\partial t)$ the strain rate; N_e the number of slip lines appearing per unit length of the wire; y^0 the average length of a slip line; i_e the anodic current per unit length of the slip line (edge); i_{ns} the current density at the newly formed surface; f the roughness factor; K the frequency of occurrence of slips per unit elongation;

and i_{hip}, $i_{lip,1}$, and $i_{lip,2}$ the current densities at the three planes shown in
Fig. 57. The equation was shown to describe adequately the experi-
mentally observed change in current with time (or strain) up to the
current maximum. The maximum current is found as

$$\Delta i_{max} = Q + 2A - [4A^3/(P + B)^2] = M + N(\partial\epsilon/\partial t), \quad (5.21)$$

i.e., the linear relationship obtained between the maximum current and
the strain rate (cf. Fig. 56) is theoretically justified. With assumption of
reasonable values of some constants a number of parameters could be
obtained from slopes and intercepts of the derived relationships. The
current density at the emerging high-index planes, i_{hip}, is found to be
50–100 times larger than that at the normal surface, which is in the
expected range.

The theory shows that the high-index-plane effect is not the only effect
resulting in accelerated dissolution. There is also a significant contribution of the emerging edges, and this explains the fact that some acceleration is also observed in metals which do not develop high-index planes
upon slipping.

This theory does not *exclude* a mechanism of acceleration based on
protective film rupture. The latter can, however, be distinguished from
the former by the fact that it produces effects, of such an extent, which
cannot be accounted for by the high-index-plane mechanism.

Raicheff and Despić (1969) have shown on the same metal in constant
conditions of environment (iron in aerated nitrate solution) that in the
active region of potentials the increase in current upon straining is of the
order of magnitude expected on the basis of the high-index-plane theory,
while there is a 1500–2000-fold increase in current in the passive region,
which can be explained only in terms of a film-rupture mechanism.

Additional support is found in the fact that in the latter case the increase in current starts *before* the yielding of the metal.

Thus the two mechanisms seem to supplement rather than to exclude
each other.

D. DISSOLUTION OF ALLOYS

1. *Basic Kinetic Relations*

Before anodic dissolution of binary alloys takes place the concentrations
of the constituents in the surface should be the same as in the bulk of the
alloy. As soon as the dissolution starts, however, the surface concentrations as a rule depart from their bulk values. This is due to the fact that
at the steady state the ratio of the partial current densities of individual

components is fixed, determined solely by alloy composition and independent of its other properties, i.e.,*

$$i_A/i_B = x_A/x_B, \tag{5.22}$$

where x_A and x_B are the molar ratios of the components A and B, respectively.

Since the rate constants for the dissolution of the two components normally differ from each other, the fixed relation between the partial current densities can be achieved only by an appropriate adjustment of the surface concentrations, which will then differ in ratio from those in the bulk. To obtain the new surface concentrations, one must, on one side, take into account the proportionality between partial currents and the corresponding surface concentrations and the relation between the molar concentrations and the molar ratios. Equation (5.22) can then be shown to be

$$\frac{\bar{k}_A C_A{}^\sigma}{\bar{k}_B C_B{}^\sigma} = \frac{M_B C_A{}^\sigma[\rho + (M_A - M_B)C_B{}^\sigma]}{M_A C_B{}^\sigma[\rho + (M_B - M_A)C_A{}^\sigma]}, \tag{5.23}$$

where $\bar{k}_A$ and $\bar{k}_B$ are the electrochemical rate constants for dissolution of the components A and B, respectively, M_A and M_B are the respective atomic weights, and ρ is the density of the alloy.

On the other side, from $x_A + x_B = 1$ it follows that

$$\frac{M_B C_A{}^\sigma}{\rho + (M_B - M_A)C_A{}^\sigma} + \frac{M_A C_B{}^\sigma}{\rho + (M_A - M_B)C_B{}^\sigma} = 1. \tag{5.24}$$

From Eqs. (5.23) and (5.24) the surface concentrations are found as

$$C_A{}^\sigma = \rho \bar{k}_B x_A / (M_A \bar{k}_B x_A + M_B \bar{k}_A x_B) \tag{5.25}$$

and

$$C_B{}^\sigma = \rho \bar{k}_A x_B / (M_A \bar{k}_B x_A + M_B \bar{k}_A x_B). \tag{5.26}$$

These relations were first derived by Scorcheletti et al. (1958) based on somewhat different reasoning.

Returning to the already used proportionality between these concentrations and the partial current densities, one obtains the latter as

$$i_A = \rho \bar{k}_A \bar{k}_B x_A / (M_A \bar{k}_B x_A + M_B \bar{k}_A x_B) \tag{5.27}$$

and

$$i_B = \rho \bar{k}_A \bar{k}_B x_B / (M_A \bar{k}_B x_A + M_B \bar{k}_A x_B). \tag{5.28}$$

* This holds as a kind of material balance equation, but only if there is negligible diffusion in the bulk.

An equation for the total current density similar to the sum of Eqs. (5.27) and (5.28) was derived by Mueller (1962).

The electrochemical rate constants pertain to the process of dissolution of the species *from the alloy*. The implications are discussed below. It is noteworthy that in some cases of rate-controlling charge transfer both partial currents and the total current density ($i = i_A + i_B$) follow the Tafel relation with potential, and also that in such cases rate constants for both components can be evaluated by measuring the total current density as a function of potential for different compositions of the alloy (Bockris *et al.*, 1969).

2. The Effect of Alloying on Rate Constants

As pointed out by Steigerwald and Greene (1962), three different cases can arise pertaining to eutectic mixtures, homogeneous solid solutions, and intermetallic compounds.

a. Eutectic Mixtures. These consist of fine crystallites of pure metals. The rate constants should thus be unaffected by alloying. Moreover, the above basic relations [Eqs. (5.22)–(5.28)] would appear to be inapplicable except for very fine dispersions of crystals. In coarsely crystalline mixtures, as pointed out by Stern (1958), the partial current density of a component at a given potential should be equal to that of a pure metal multiplied by the fraction of the surface area occupied by this metal. The latter can for a long time stay practically constant and in direct relation to the mole fraction of the metal in the alloy, so that the *apparent* steady state which is thus achieved may not obey the derived relations.

b. Homogeneous Solid Solutions. In general, the rate constants should differ from those for pure metals because the standard free energies of activation are affected by changes in the energies of the initial states for dissolution, first in alloying and then in partial dissolution of one component.

The changes in chemical potentials by alloying with respect to standard free energies of pure metals are determined by the heat of mixing. In many alloys this effect is so small that the alloys can be considered as ideal mixtures, and no change in rate constants should be expected. The second effect should be pronounced in the rate constant for the dissolution of the more-noble component. As the less-noble component is dissolved to attain the necessary relation between the surface concentrations [cf. Eq. (5.23)] the concentration of vacancies must be significantly increased, the average coordination number of the remaining atoms becomes smaller, and hence atoms are less strongly bound that in the corresponding pure metal. They are richer in energy, and if it is assumed that the activated state (partially charged and hydrated ions) does not depend on alloy composition, the increase in the heat content of atoms at the surface must be reflected in a decrease of the heat of activation and in the corresponding increase in the value of the rate constants (Bockris *et al.*, 1969) (but cf. the compensating effect on the rate constant of a change in reversible potential).

c. The Intermetallic Compounds. These should exhibit a behavior similar to that of solid solutions except that the solid-state differences (different and ordered crystal structure and/or stronger interatomic binding) may be greater than with solid solutions.

ACKNOWLEDGMENTS

Thanks are due to Miss Danka Jovanović for her help in reviewing the most recent literature and for technical aid in preparation of the manuscript. The authors are also indebted to Dr. A. Damjanović and to many other colleagues who have commented on parts of the manuscript.

REFERENCES

AMELINCKX, B. (1952). *Nature* **170**, 760.
ARKHAROV, V. (1936). *Tech. Phys. U.S.S.R.* **3**, 1072.
ARMSTRONG, R. D., FLEISCHMANN, M., and THIRSK, H. R. (1966). *J. Electroanal. Chem.* **11**, 208.
ATEN, A. H. W., and BOERLAGE, L. M. (1920). *Rec. Trav. Chim.* **39**, 720.
AVRAMI, M. (1939). *J. Chem. Phys.* **7**, 1103.
AVRAMI, M. (1940). *J. Chem. Phys.* **8**, 212.
AVRAMI, M. (1941). *J. Chem. Phys.* **9**, 177.
BAKISH, R., and ROBERTSON, W. D. (1955). *Acta Met.* **3**, 513.
BAKISH, R., and ROBERTSON, W. D. (1956). *Trans. AIME* **206**, 1277.
BANERJEE, B. C., and GOSWAMI, A. (1955). *J. Sci. Ind. Res. (India)* **14B**, 322.
BANERJEE, B. C., and GOSWAMI, A. (1957). *J. Sci. Ind. Res. (India)* **16B**, 144.
BANERJEE, B. C., and GOSWAMI, A. (1958). *J. Electrochem. Soc.* **106**, 20.
BANERJEE, B. C., and GOSWAMI, A. (1959). *J. Electrochem. Soc.* **107**, 590.
BANERJEE, T. (1952). *Symp. Electroplating Metal Finishing, India*, p. 29.
BARKER, G. C., FAIRCLOTH, R. L., and GARDNER, A. W. (1958). *Nature* **181**, 247.
BARKER, G. C. (1961). *Trans. Symp. Electrode Processes, Philadelphia, Pa., 1959*, p. 325. Wiley, New York.
BARNES, S. C. (1961). *Electrochim. Acta* **5**, 79.
BARNES, S. C., STOREY, G. G., and PICK, H. J. (1960). *Electrochim. Acta* **2**, 196.
BARTON, J. H., and BOCKRIS, J. O'M. (1962). *Proc. Roy. Soc.* **A268**, 485.
BATICLE, A. M. (1962). *Compt. Rend.* **254**, 668.
BAUER, H. H., and ELVING, P. (1958). *Anal. Chem.* **30**, 341.

BAUER, H. H., SMITH, D. L., and ELVING, P. (1960). *J. Am. Chem. Soc.* **82**, 2094.

BEACOM, S. E. (1959). *Plating* **46**, 814.

BECKER, R., and DÖRING, W. (1935). *Ann. Phys.* **24**, 719.

BEHR, B., DOJLIDO, J., and MALYSZKO, J. (1962). *Roczniki Chem.* **36**, 725.

BERZINS, T., and DELAHAY, P. (1955). *J. Am. Chem. Soc.* **77**, 6448.

BEWICK, A. (1961). Ph.D. Thesis, Univ. of Durham, Durham, England.

BOCKRIS, J.'O'M. (1954). *In* "Modern Aspects of Electrochemistry" (J. O'M. Bockris, ed.), Vol. I, Chapter 4. Butterworths, London.

BOCKRIS, J. O'M., and CONWAY, B. E. (1958). *J. Chem. Phys.* **28**, 707.

BOCKRIS, J. O'M., and DAMJANOVIĆ, A. (1964). *In* "Modern Aspects of Electrochemistry" (J. O'M. Bockris, ed.), Vol. 3, pp. 224–346. Butterworths, London.

BOCKRIS, J. O'M., and DRAŽIĆ, D. (1962). *Electrochim. Acta* **7**, 293.

BOCKRIS, J. O'M., and ENYO, M. (1962a). *J. Electrochem. Soc.* **109**, 48.

BOCKRIS, J. O'M., and ENYO, M. (1962b). *Trans. Faraday Soc.* **58**, 1187.

BOCKRIS, J. O'M., and KITA, H. (1962). *J. Electrochem. Soc.* **109**, 928.

BOCKRIS, J. O'M., and MATTHEWS, D. B. (1966). *Proc. Roy. Soc.* **A202**, 479.

BOCKRIS, J. O'M., and RAZUMNEY, G. A. (1967). "Fundamental Aspects of Electrocrystallisation." Plenum Press, New York.

BOCKRIS, J. O'M., and SWINKELS, D. M. (1964). *J. Electrochem. Soc.* **111**, 736, 743.

BOCKRIS, J. O'M., DRAŽIĆ, D. M., and DESPIĆ, A. R. (1961). *Electrochim. Acta* **4**, 325.

BOCKRIS, J. O'M., DEVANATHAN, M. A. V., and MUELLER, K. (1963). *Proc. Roy. Soc.* **A274**, 55.

BOCKRIS, J. O'M., DEVANATHAN, M. A. V., and REDDY, A. K. N. (1964). *Proc. Roy. Soc.* **A279**, 327.

BOCKRIS, J. O'M., RUBIN, B., and DESPIĆ, A. R. (1969). To be published.

BOZORTH, R. M. (1925). *Phys. Rev.* **26**, 390.

BRENNER, A. (1963). "Electrodeposition of Alloys. Principles and Practice." Academic Press, New York.

BUDEVSKI, E. (1966). *Electrochim. Metal.* **2**, 1.

BUDEVSKI, E., BOSTANOV, W., VITANOV, T., STOJNOV, Z., KOTZEVA, A., and KAISHEV, R. (1966a). *Phys. Status Solidi* **13**, 577.

BUDEVSKI, E., BOSTANOV, W., VITANOV, T., STOJNOV, Z., KOTZEVA, A., and KAISHEV, R. (1966b). *Electrochim. Acta* **11**, 1697.

BUNN, C. W., and EMMETT, H. (1949). *Discussions Faraday Soc.* **5**, 119.

BURTON, W. K., CABRERA, K. N., and FRANK, F. C. (1949). *Nature* **163**, 398.

BURTON, W. K., CABRERA, K. N., and FRANK, F. C. (1951). *Proc. Roy. Soc.* **243**, 299.

BUTLER, J. A. V. (1936). *Proc. Roy. Soc.* **A157**, 423.

CABRERA, K. N., and BURTON, W. K. (1949). *Discussions Faraday Soc.* **5**, 40.

CABRERA, K. N., and VERMILYEA, D. A. (1958). "Growth and Perfection of Crystals," p. 393. Wiley, New York.

CHERNOV, B. (1960). *Kristallografiya* **5**, 446.

CHRISTIANSEN, K. A., HOEG, H., MICHELSEN, K., NIELSEN, G. B., and NORD, H. (1961). *Acta Chem. Scand.* **15**, 300.

COCHRANE, W. (1936). *Proc. Phys. Soc. (London)* **48**, 723.

COHEN, J. B., and WEERTMAN, J. (1963). *Acta Met.* **11**, 996.

CONWAY, B. E. (1949). Ph.D. Thesis, Univ. of London, London.

CONWAY, B. E., and BOCKRIS, J. O'M. (1958). *Proc. Roy. Soc.* **A248**, 394.

CONWAY, B. E., and BOCKRIS, J. O'M. (1960). *Electrochim. Acta* **3**, 340.

DAMJANOVIĆ, A. (1965). *Plating* **52**, 1017.

DAMJANOVIĆ, A., and BOCKRIS, J. O'M. (1963). *J. Electrochem. Soc.* **110**, 1035.

DAMJANOVIĆ, A., PAUNOVIĆ, M., and BOCKRIS, J. O'M. (1965a). *J. Electroanal. Chem.* **9**, 93.

DAMJANOVIĆ, A., PAUNOVIĆ, M., and BOCKRIS, J. O'M. (1965b). *Electrochim. Acta* **10**, 111.

DAMJANOVIĆ, A., SETTY, T. H. V., and BOCKRIS, J. O'M. (1966). *J. Electrochem. Soc.* **113**, 129.

DE GROOT, S. R. (1952). "Thermodynamics of Irreversible Processes." North-Holland Publ., Amsterdam.

DELAHAY, P. (1953). *J. Am. Chem. Soc.* **75**, 1190.

DELAHAY, P. (1957). *J. Chim. Phys.* 369.

DELAHAY, P., and ARAMATA, A. (1962). *J. Phys. Chem.* **66**, 2208.

DELAHAY, P., and TRACHTENBERG, I. (1957). *J. Am. Chem. Soc.* **79**, 2355.

DELAHAY, P., and TRACHTENBERG, I. (1958). *J. Am. Chem. Soc.* **80**, 2094.

DESPIĆ, A. R. (1969). *Bull. Acad. Serbe Sci. Art, Cl. Sci. Math. Natur. Sci. Natur.* To be published.

DESPIĆ, A. R., and BOCKRIS, J. O'M. (1960). *J. Chem. Phys.* **32**, 389.

DESPIĆ, A. R., and JOVANOVIĆ, D. (1964). *CITCE Meeting, 15th, London*, p. 33. Macmillan (Pergamon), New York.

DESPIĆ, A. R., DRAŽIĆ, D. M., and ŠEPA, D. (1966). *Electrochim. Acta* **11**, 507.

DESPIĆ, A. R., DIGGLE, J., and BOCKRIS, J. O'M. (1968a). *J. Electrochem. Soc.* **115**, 507.

DESPIĆ, A. R., RAICHEFF, R. G., and BOCKRIS, J. O'M. (1968b). *J. Chem. Phys.* **49**, 926.

DESPIĆ, A. R., JOVANOVIĆ, D., and BINGULAC, S. (1969). *Electrochim. Acta.* To be published.

DIRKSE, T. P. (1962). *Z. Phys. Chem. (Frankfurt)* **33**, 387.

DUGDALE, I., FLEISCHMANN, M., and WYNNE-JONES, W. F. K. (1961). *Electrochim. Acta* **5**, 229.

ECONOMOU, N. A., and TRIVICH, D. (1961). *Electrochim. Acta* **3**, 292.

ECONOMOU, N. A., FISCHER, H., and TRIVICH, D. (1960). *Electrochim. Acta* **2**, 196.

EPELBOIN, I., and MOREL (1968). Thesis, Paris.

ERDEY-GRÚZ, T., and VOLMER, M. (1931a). *Z. Phys. Chem. (Leipzig)* A **150**, 203.

ERDEY-GRÚZ, T., and VOLMER, M. (1931b). *Z. Phys. Chem. (Leipzig)* A **157**, 165.

ERSHLER, B. N., and ROSENTAL, K. J. (1953). *Tr. Soveshch. po Elektrokhim. Akad. Nauk SSSR, Otd. Khim. Nauk, Moscow, 1950*, p. 446. Izd. AN SSSR, Moscow.

EVANS, D. J. (1952–1953). *Trans. Inst. Metal Finishing* **29**, 355.

FAUST, F. W., and JOHN, H. F. (1961). *J. Electrochem. Soc.* **108**, 109.

FAUST, F. W., and JOHN, H. F. (1963). *J. Electrochem. Soc.* **110**, 463.

FINCH, G. I., and SUN, C. H. (1936). *Trans. Faraday Soc.* **32**, 852.

FINCH, G. I., and WILLIAMS, A. G. (1937). *Trans. Faraday Soc.* **33**, 564.

FINCH, G. I., QUARRELL, A. G., and WILMAN, H. (1935). *Trans. Faraday Soc.* **31**, 1051.

FINCH, G. I., WILMAN, H., and YANG, L. (1947). *Discussions Faraday Soc.* **1**, 144.

FISCHER, H. (1954). "Elektrolytische Abscheidung und Elektrokristallisation von Metallen," p. 393. Springer, Berlin.

FISCHER, H. (1955). *Z. Elektrochem.* **59**, 612.

FISCHER, H. (1960). *Electrochim. Acta* **2**, 50.

FLEISCHMANN, M., and LILER, M. (1958). *Trans. Faraday Soc.* **54**, 1370.

FLEISCHMANN, M., and THIRSK, H. R. (1955). *Trans. Faraday Soc.* **51**, 71.

FLEISCHMANN, M., and THIRSK, H. R. (1960). *Electrochim. Acta* **2**, 22.

FLEISCHMANN, M., and THIRSK, H. R. (1963). *Advan. Electrochem. Electrochem. Eng.* **3**, 123.

FLORIANOVICH, G. H., SOKOLOVA, L. A., and KOLOTYRKIN, YA. M. (1969). *Electrochim. Acta* **12**, 879.

FOERSTER, F. (1931). "Elektrochemie Wässeriger Lösungen," p. 368. Barth, Leipzig.

FORTY, A., and FRANK, F. C. (1951). *Phil. Mag.* **42**, 670.

FORTY, A., and FRANK, F. C. (1952). *Phil. Mag.* **43**, 481.

FORTY, A., and FRANK F. C. (1953). *Proc. Roy. Soc.* **A217**, 263.

FOULKE, D. G., and KARDOS, O. (1956). *Tech. Proc. Am. Electroplaters' Soc.* **43**, 172.

FRANK, F. C. (1949a). *Discussions Faraday Soc.* **5**, 48.

FRANK, F. C. (1949b). *Discussions Faraday Soc.* **5**, 67.

FRANK, F. C. (1958). "Growth and Perfection of Crystals," p. 411. Wiley, New York.

FRANK, F. C., and VAN DER MERVE, J. H. (1949a). *Proc. Roy. Soc.* **A198**, 205, 216.

FRANK, F. C., and VAN DER MERVE, J. H. (1949b). *Proc. Roy. Soc.* **A200**, 125.

FRENKEL, J. (1946). *In* "Kinetic Theory of Liquids" (R. H. Doremus, B. W. Roberts, and D. Turnbull, eds.). Oxford Univ. Press (Clarendon), London and New York.

FRUMKIN, A. N. (1965a). *Elektrokhimiya* **1**, 394.

FRUMKIN, A. N. (1965b). *Elektrokhimiya* **1**, 1288.

FUKUDA, S. (1959). *Oyo Butsuri* **28**, 466.

GAVIOLI, G., and POPOFF, P. (1961). *Ric. Sci. Rend.* **1**, 193.

GERISCHER, H. (1953a). *Z. Elektrochem.* **57**, 605.

GERISCHER, H. (1953b). *Z. Phys. Chem. (Leipzig) A* **202**, 302.

GERISCHER, H. (1956). *Angew. Chem.* **68**, 20.

GERISCHER, H. (1958). *Z. Elektrochem.* **62**, 256.

GERISCHER, H. (1959). *Anal. Chem.* **31**, 33.

GERISCHER, H., and KRAUSE, M. (1957). *Z. Phys. Chem. (Frankfurt)* **10**, 264.

GERISCHER, H., and RICKERT, H. (1955). *Z. Metallk.* **46**, 681.

GIRON, I., and OGBURN, F. (1961). *J. Electrochem. Soc.* **108**, 842.

GLOCKER, R., and KAUPP, E. (1924). *Z. Physik* **24**, 121.

GORBUNOVA, K. M. (1938). *Dokl. Akad. Nauk SSSR* **20**, 467.

GORBUNOVA, K. M., and DANKOV, P. D. (1949). *Zh. Fiz. Khim.* **23**, 616.

GORBUNOVA, K. M., and POLUKAROV, YU. M. (1967). *Advan. Electrochem. Electrochem. Eng.* **5**, 249.

GORBUNOVA, K. M., and SUTYAGINA, A. A. (1958). *Zh. Fiz. Khim.* **32**, 785.

GORBUNOVA, K. M., and ZHUKOVA, A. I. (1949). *Zh. Fiz. Khim.* **23**, 605.

GORBUNOVA, K. M., POPOVA, O. S., SITYAGINA, A. A., and POLUKAROV, Y. M. (1957). Crystal growth. *Rept. Congr. Crystal Growth, 1st, March 1956, Moscow*, p. 58. Izd. Akad. Nauk SSSR, Moscow.

GRAF, L., and MORGENSTERN, W. (1955). *Z. Naturforsch.* **10a**, 345.

GRECHUKINA, T. N. (1956). *Izv. Kazansk. Filiala Akad. Nauk SSSR Ser. Khim. Nauk* **3**, 101.

GURNEY, R. W. (1931). *Proc. Roy. Soc.* **A134**, 137.

HAMILTON, D. R. (1963). *Electrochim. Acta* **8**, 731.

HAMILTON, D. R., and SEIDENSTICKER, R. G. (1960). *J. Appl. Phys.* **31**, 1165.

HEUSLER, K. E. (1958). *Z. Elektrochem.* **62**, 582.

HILLS, G. J., and IVES, D. J. G. (1951). *J. Chem. Soc.* p. 311.

HILSON, P. J. (1954). *Trans. Faraday Soc.* **50**, 385.

HIRATA, H., DOTO, H., and HARA, M. (1939). *Nippon Kinzoku Gakkaishi* **3**, 460.

HOAR, T. P., and HURLEN, T. (1958). *Proc. CITCE Meeting, 8th, Madrid, 1956*, p. 445. Butterworths, London.

HOAR, T. P., and WEST, J. M. (1958). *Nature* **181**, 835.

HORIUTI, J., and POLANYI, M. (1935). *Acta Physicochim. U.R.S.S.* **2**, 505.

HOWES, A. (1959). *Proc. Phys. Soc. (London)* **74**, 616.

HUME-ROTHERY, W., and RAYNOR, G. V. (1956). "The Structure of Metals and Alloys." Inst. of Metals, London.

HUNTINGTON, J. (1905). *Trans. Faraday Soc.* **1**, 324.

HURLEN, T. (1960). *Acta Chem. Scand.* **14**, 1533.

HURLEN, T. (1962a). *Acta Chem. Scand.* **16**, 1337.
HURLEN, T. (1962b). *Acta Chem. Scand.* **16**, 1353.
HUSH, N. S. (1961). *Trans. Faraday Soc.* **57**, 557.
IBL, N. (1961). *Chem. Ing. Tech.* **33**, 69.
IBL, N. (1963a). *Chem. Ing. Tech.* **35**, 353.
IBL, N. (1963b). *Galvanotech. Oberflaechenschutz* **4**, 265.
IMAI, H. (1958). *Sci. Hiroshima Univ. Ser. A* **22**, 291.
IMAI, H. (1959). *J. Electrochem. Soc. Japan (Japanese Ed.)* **27**, 55.
JONES, H. (1950). *Phil. Mag.* **41**, 663.
JOVANOVIĆ, D. (1969). Ph.D. Thesis, Univ. of Beograd, Beograd.
KABANOV, B., BURSTEIN, R., and FRUMKIN, A. N. (1947). *Discussions Faraday Soc.* **1**, 259.
KAISHEV, R. (1961). *Proc. Conf. Electrochem., 4th, Moscow*, p. 5. Consultant Bureau, New York.
KAISHEV, R., and BLIZNAKOV, G. (1949). *Compt. Rend. Acad. Bulgare Sci.* **1**, 123.
KAISHEV, R., BUDEVSKI, E., and MALINOVSKI, J. (1955). *Z. Phys. Chem. (Leipzig)* **204**, 348.
KAMBARA, T., and ISHII, T. (1961). *Rev. Polarog. (Kyoto)* **9**, 30.
KARDOS, O. (1956). *Tech. Proc. Am. Electroplaters' Soc.* **43**, 181.
KASPER, C. (1932). *J. Res. Natl. Bur. Std.* **9**, 353.
KASPER, C. (1940a). *Trans. Electrochem. Soc.* **77**, 353.
KASPER, C. (1940b). *Trans. Electrochem. Soc.* **77**, 365.
KASPER, C. (1940c). *Trans. Electrochem. Soc.* **78**, 131.
KIMBALL, G. E. (1940). *J. Chem. Phys.* **8**, 199.
KIRKALDY, J. S. (1959). *J. Physics* **37**, 739.
KITA, H., ENYO, M., and BOCKRIS, J. O'M. (1961). *Can. J. Chem.* **39**, 1670.
KOCHERGIN, S. M. (1953). *Zh. Tekhn. Fiz.* **23**, 995.
KOHLSCHÜTTER, V., and GOOD, A. (1927). *Z. Elektrochem.* **33**, 277.
KOHLSCHÜTTER, V., and TORRICHELLI, A. (1932). *Z. Elektrochem.* **38**, 213.
KOHLSCHÜTTER, V., and ÜBERSAX, F. (1924). *Z. Elektrochem.* **30**, 72.
KORYTA, J. (1962). *Electrochim. Acta* **6**, 67.
KOSSEL, W. (1927). *Nachr. Ges. Wiss. Göttingen* p. 135.
KOZLOVSKII, M. I., and LEMMLEIN, G. G. (1958). *Kristallografiya* **3**, 351.
KRUGLIKOV, S. S., KUDRYAVTSEV, N. T., VOROBYEVA, G. F., YARLIKOV, M. M., and ANTONOV, A. YA. (1963). *Dokl. Akad. Nauk SSSR* **149**, 911.
KRUGLIKOV, S. S., KUDRYAVTSEV, N. T., SOBOLYEV, R. P., ANTONOV, A. YA., and DRIBINSKY, A. V. (1964). *Trans. Inst. Metal Finishing* **42**, 129.
KRUGLIKOV, S. S., KUDRYAVTSEV, N. T., VOROBYEVA, G. F., and ANTONOV, A. YA. (1965). *Electrochim. Acta* **10**, 253.
KUMAR, D. M., and WILMAN, H., see WILMAN, H. (1957). *Acta Cryst.* **10**, 842.
LAINER, V. I. (1950). *Sb. Nauchn. Rabot Met. Moscow* p. 70.
LAINER, V. I., and YU, T.-J. (1963). *Zh. Prikl. Khim.* **36**, 121.
LAITINEN, H. A., TISCHER, R., and ROE, D. K. (1961). *Trans. Symp. Electrode Processes, Philadelphia, Pa., 1959*, p. 185. Wiley, New York.
LAW, J. T. (1952). Ph.D. Thesis, Univ. of London, London.
LAYTON, D. N. (1952). *J. Electrodepositors' Tech. Soc.* **28**, 239.
LOGAN, H. L. (1952). *J. Res. Natl. Bur. Std.* **48**, 99.
LOGAN, H. L. (1958). *J. Res. Natl. Bur. Std.* **61**, 503.
LOGAN, H. L. (1961). *J. Res. Natl. Bur. Std. C* **65**, 165.
LORENZ, W. (1953). *Naturwissenschaften* **40**, 778.
LORENZ, W. (1954a). *Z. Naturforsch.* **9a**, 716.

728   John O'M. Bockris and Aleksandar R. Despić

LORENZ, W. (1954b). *Z. Elektrochem.* **58**, 912.
LORENZ, W. (1958). *Z. Physik. Chem. (Leipzig)* **17**, 136.
LOSHKAREV, M. A., and KRYUKOVA, T. A. (1948). *Zh. Fiz. Khim.* **22**, 815.
LOSHKAREV, M. A., and KRYUKOVA, T. A. (1949). *Zh. Fiz. Khim.* **23**, 209.
LOSSEW, W. W. (1955). *Dokl. Akad. Nauk SSSR* **100**, 111.
LOSSEW, W. W. (1956a). *Dokl. Akad. Nauk SSSR* **107**, 432.
LOSSEW, W. W. (1956b). *Dokl. Akad. Nauk SSSR* **111**, 626.
LOSEV, V. V., MOLODOV, A. I., and GORODETZKI, V. V. (1967). *Electrochim. Acta* **12**, 475.
LOVREČEK, B., and MARINČIĆ, N. (1966). *Electrochim. Acta* **11**, 237.
LUSTMAN, B. (1943). *Trans. Electrochem. Soc.* **84**, 363.
MARCUS, R. A. (1957a). *J. Chem. Phys.* **26**, 867.
MARCUS, R. A. (1957b). *J. Chem. Phys.* **26**, 872.
MARTIROSYAN, A. P., and KRYUKOVA, T. A. (1953). *Zh. Fiz. Khim.* **27**, 851.
MATSUDA, H., and AYABE, Y. (1955). *Z. Elektrochem.* **59**, 494.
MATSUDA, H., and AYABE, Y. (1959). *Z. Elektrochem.* **63**, 1164.
MATSUDA, H., and AYABE, Y. (1962). *Z. Elektrochem.* **66**, 469.
MATSUNAGA, M. (1960). *Sci. Papers Inst. Phys. Chem. Res. (Tokyo)* **54**, 177.
MATTHEW, H. I., MUTUCUMARANA, T. DE S., and WILMAN, H. (1961). *Acta Cryst.* **14**, 636.
MATTSON, E., and BOCKRIS, J. O'M. (1959). *Trans. Faraday Soc.* **55**, 1586.
MEHL, W., and BOCKRIS, J. O'M. (1957). *J. Chem. Phys.* **27**, 817.
MEHL, W., and BOCKRIS, J. O'M. (1959). *Can. J. Chem.* **37**, 190.
MOELWYN-HUGHES, E. A. (1961). "Physical Chemistry." Pergamon Press, Oxford.
MORINAGA, K. (1955). *Nippon Kagaku Zasshi* **76**, 136.
MORINAGA, K. (1956). *Bull. Chem. Soc. Japan* **29**, 793.
MOTT, N. F. (1952). *Progr. Metal Phys.* **3**, 76.
MUELLER, W. A. (1962). *Corrosion* **18**, 73t.
MUELLER, W. A., and LORENZ, W. (1961). *Z. Phys. Chem. (Frankfurt)* **27**, 23.
OGBURN, F., PARETZKIN, B., and PEISER, H. S. (1964). *Acta Cryst.* **17**, 774.
OGBURN, F., BECHTOLDT, C., MORRIS, J. B., and DE KORANYI, A. (1965). *J. Electrochem. Soc.* **112**, 574.
OVENSTON, T. S. J., PARKER, C. A., and ROBINSON, A. E. (1957). *Trans. Electrochem. Soc.* **104**, 607.
PAMFILOV, A. V., LOPUSHANSKAYA, A. I., and IVCHER, T. S. (1961). *Ukr. Khim. Zh.* **27**, 598 [*Chem. Abstr.* **56**, 8461c (1962)].
PANGAROV, N. A. (1962). *Electrochim. Acta* **7**, 139.
PANGAROV, N. A. (1964). *Electrochim. Acta* **9**, 721.
PANGAROV, N. A., and DOBREV, D. (1962). *Compt. Rend. Acad. Bulgare Sci.* **15**, 519.
PANGAROV, N. A., and MICHAILOVA, W. (1963). *Dokl. Akad. Nauk SSSR* **153**, 119.
PANGAROV, N. A., and MICHAILOVA, W. (1964). *Bull. Inst. Chim. Phys. Acad. Bulgare Sci.* **4**, 111.
PANGAROV, N. A., and RASHKOV, ST. (1960a). *Bull. Inst. Phys. Chem.* **1**, 79.
PANGAROV, N. A., and RASHKOV, ST. (1960b). *Compt. Rend. Acad. Bulgare Sci.* **13**, 555.
PANGAROV, N. A., and VELINOV, V. (1966). *Electrochim. Acta* **11**, 1753.
PANGAROV, N. A., and VITKOVA, S. D. (1966a). *Electrochim. Acta* **11**, 1719.
PANGAROV, N. A., and VITKOVA, S. D. (1966b). *Electrochim. Acta* **11**, 1733.
PARSONS, R., and BOCKRIS, J. O'M. (1951). *Trans. Faraday Soc.* **47**, 914.
PASHLEY, D. W. (1956). *Advan. Phys.* **5**, 173.
PERSIANTSEVA, V. P., and TITOV, P. S. (1958). *Nauchn. Dokl. Vysshei Shkoly Khim. i Khim. Tekhnol.* **3**, 584.
PHILBERT, G. (1943). *J. Chim. Phys.* **40**, 157.

PICK, H. J. (1955). *Nature* **176**, 693.

PICK, H. J., STOREY, G. G., and VAUGHAN, T. B. (1960). *Electrochim. Acta* **2**, 165.

POLI, G., and BICELLI, L. P. (1959). *Met. Ital.* **51**, 399, 548.

POLUKAROV, YU. M., GORBUNOVA, K. M., and BONDAR, V. V. (1962a). *Zh. Fiz. Khim.* **36**, 1661.

POLUKAROV, YU. M., GORBUNOVA, K. M., and BONDAR, V. V. (1962b). *Zh. Fiz. Khim.* **36**, 1870.

PRICE, P. B., VERMILYEA, D. A., and WEBB, M. B. (1958). *Acta Met.* **6**, 524.

PRIGOGINE, I. (1955). *In* "Introduction of Thermodynamics of Irreversible Processes" (C. Charles, ed.), p. 75. Thomas, Springfield, Illinois.

RAICHEFF, R. G., and DESPIĆ, A. R. (1969). *Electrochim. Acta.* To be published.

RAICHEFF, R. G., DAMJANOVIĆ, A., and BOCKRIS, J. O'M. (1967). *J. Chem. Phys.* **47**, 2198.

RAMA-CHAR, T. L., and VAID, J. (1961a). *Plating* **48**, 871.

RAMA-CHAR, T. L., and VAID, J. (1961b). *Metal Finishing* **59**, 44.

RAMA-CHAR, T. L., and VAID, J. (1962). *Metalloberflaeche* **16**, 70.

RANDLES, J. E. B. (1947). *Discussions Faraday Soc.* **1**, 11.

RANDLES, J. E. B. (1961). *Trans. Symp. Electrode Processes, Philadelphia, Pa., 1959*, p. 209. Wiley, New York.

RANDLES, J. E. B., and SOMERTON, K. W. (1952). *Trans. Faraday Soc.* **48**, 951.

RAUB, E., and ENGEL, A. (1943). *Z. Elektrochem.* **49**, 89.

RAUB, E., and KRAUSE, D. (1944). *Z. Elektrochem.* **50**, 91.

RAUB, E., and WULLHORST, B. (1947). *Z. Metallforsch.* **2**, 41.

RAYNOR, G. V. (1949). *Progr. Metal Phys.* **1**, 1.

REDDY, A. K. N. (1963). *J. Electroanal. Chem.* **6**, 141.

REDDY, A. K. N., and WILMAN, H. (1959). *Trans. Inst. Metal Finishing* **36**, 97.

REDDY, T. B. (1966). *J. Electrochem. Soc.* **113**, 117.

RIUS, A., TORDESILLAS, I. M., and SACRISTAN, A. (1961). *Electrochim. Acta* **4**, 62.

ROGERS, G. T., and TAYLOR, K. J. (1963). *Electrochim. Acta* **8**, 887.

ROGERS, G. T., and TAYLOR, K. J. (1965). *Trans. Inst. Metal Finishing* **43**, 75.

ROITER, V. A., JUZA, V. A., and POLUYAN, E. S. (1939a). *Acta Physicochim. U.R.S.S.* **10**, 389 [*Chem. Abstr.* **33**, 6169[4] (1939)].

ROITER, V. A., POLUYAN, E. S., and JUZA, V. A. (1939b). *Acta Physicochim. U.R.S.S.* **10**, 845 [*Chem. Abstr.* **33**, 8123[9] (1939)].

SATO, R. (1959). *J. Electrochem. Soc.* **106**, 206.

SCORCHELETTI, C., STEPANOV, A., and KUBSHENKO, I. (1958). *Zh. Prikl. Khim.* **31**, 1823.

SEITER, H., and FISCHER, H. (1959). *Z. Elektrochem.* **63**, 249.

SEITER, H., FISCHER, H., and ALBERT, L. (1958). *Naturwissenschaften* **45**, 127.

SEITER, H., FISCHER, H., and ALBERT, L. (1960). *Electrochim. Acta* **2**, 97.

SEITZ, F. (1940). "Modern Theory of Solids," McGraw-Hill, New York.

SETTY, T. H. V. (1957). *J. Sci. Ind. Res. (India)* **16B**, 139.

SETTY, T. H. V., and WILMAN, H. (1955). *Trans. Faraday Soc.* **51**, 948.

SLUYTERS, J. H., and OOMEN, J. J. C. (1960). *Rec. Trav. Chim.* **79**, 1101.

SREE, V., and RAMA-CHAR, T. L. (1961a). *Metalloberflaeche* **15**, 301.

SREE, V., and RAMA-CHAR, T. L. (1961b). *Plating* **48**, 50.

SREE, V., PANIKAR, S. K., and RAMA-CHAR, T. L. (1961). *Bull. India Sect. Electrochem. Soc.* **10**, 8.

SROKA, R., and FISCHER, H. (1956). *Z. Elektrochem.* **60**, 109.

STABROVSKY, A. N. (1951). *Zh. Prikl. Khim.* **24**, 471.

STEIGERWALD, R. F., and GREENE, N. D. (1962). *J. Electrochem. Soc.* **109**, 1026.

STERN, M. (1958). *Corrosion* **14**, 329t.

STRANSKI, I. N. (1928). Z. Phys. Chem. (Leipzig) **136**, 259.

STRANSKI, I. N., and KAISHEV, R. (1934a). Z. Phys. Chem. (Leipzig) Ser. B **26**, 100.

STRANSKI, I. N., and KAISHEV, R. (1934b). Z. Phys. Chem. (Leipzig) Ser. B **26**, 114.

STRANSKI, I. N., and KAISHEV, R. (1934c). Z. Phys. Chem. (Leipzig) Ser. B **26**, 312.

STRANSKI, I. N., and KAISHEV, R. (1935a). Ann. Physik **23**, 330.

STRANSKI, I. N., and KAISHEV, R. (1935b). Physik. Z. **36**, 393.

STREINBERG, M. A. (1952). Nature **170**, 1119.

TAKAHASHI, N. (1952). Compt. Rend. **234**, 1619.

TAMAMUSHI, R., and TANAKA, N. (1959). Z. Phys. Chem. (Frankfurt) **21**, 89.

TAMAMUSHI, R., and TANAKA, N. (1963). Z. Phys. Chem. (Frankfurt) **39**, 117.

TAMAMUSHI, R., ISHIBASHI, K., and TANAKA, N. (1962). Z. Phys. Chem. (Frankfurt) **35**, 209.

TANAKA, N., and TAMAMUSHI, R. (1964). Electrochim. Acta **9**, 963.

THIRSK, H. R. (1939). Ph.D. Thesis, Univ. of London, London.

TURNER, D. R., and JOHNSON, G. R. (1962). J. Electrochem. Soc. **109**, 798.

VAGRAMYAN, A. T. (1959). Tr. 4-go Sovesch. po Elektrokhim., Moscow, 1956, p. 395. Izd. AN SSSR, Moscow.

VAGRAMYAN, A. T., and FATUEVA, T. A. (1959). Dokl. Akad. Nauk SSSR **128**, 773.

VAGRAMYAN, A. T., and FATUEVA, T. A. (1960). Dokl. Akad. Nauk SSSR **135**, 1413.

VAGRAMYAN, A. T., and GORBUNOVA, K. M. (1937). Acta Physicochim. U.R.S.S. **7**, 683.

VAGRAMYAN, A. T., and USACHEV, D. N. (1954). Dokl. Akad. Nauk SSSR **98**, 605.

VAGRAMYAN, A. T., USACHEV, D. N., and KLIMASENKO, N. L. (1961). Zh. Fiz. Khim. **35**, 647.

VAN CAKENBERGHE, J. (1951). Bull. Chim. Soc. Belges **60**, 3.

VAN DER MEULEN, P. A., and LINDSTROM, H. V. (1956). J. Electrochem. Soc. **103**, 390.

VAUGHAN, T. B. (1961). Electrochim. Acta **4**, 72.

VERMA, A. R. (1951). Phil. Mag. **42**, 1005.

VERMA, A. R. (1953). "Crystal Growth." Butterworths, London.

VERMILYEA, D. A. (1957). J. Chem. Phys. **27**, 814.

VERMILYEA, D. A. (1959). J. Electrochem. Soc. **106**, 66.

VERMILYEA, D. A. (1963). Advan. Electrochem. Electrochem. Eng. **3**, 211.

VETTER, K. J. (1952). Z. Naturforsch. **7a**, 328.

VETTER, K. J. (1953a). Z. Phys. Chem. (Leipzig) **202**, 1.

VETTER, K. J. (1953b). Z. Naturforsch. **8a**, 823.

VETTER, K. J. (1961). "Elektrochemische Kinetik," p. 605. Springer, Berlin.

VIELSTICH, W., and DELAHAY, P. (1957). J. Am. Chem. Soc. **79**, 1874.

VIELSTICH, W., and GERISCHER, H. (1955). Z. Phys. Chem. (Frankfurt) **4**, 10.

VOLMER, M. (1921). Physik. Z. **22**, 646.

VOLMER, M. (1922). Z. Phys. Chem. (Leipzig) A **102**, 267.

VOLMER, M. (1934). "Das Elektrolytische Krystallwachstum." Hermann, Paris.

WAGNER, C. (1951). J. Electrochem. Soc. **98**, 116.

WAGNER, C. (1961). Plating **48**, 997.

WASHI, V. (1957). Oyo Butsuri **26**, 147.

WATSON, S. A. (1960). Trans. Inst. Metal Finishing **37**, 144.

WATSON, S. A., and EDWARDS, J. (1957). Trans. Inst. Metal Finishing **34**, 167.

WILCOCK, A. (1956–1957). Trans. Inst. Metal Finishing **34**, 483.

WILMAN, H. (1955). Trans. Inst. Metal Finishing **32**, 281.

WOOD, W. A. (1935). Trans. Faraday Soc. **31**, 1248.

WRANGLÉN, G. (1955). Trans. Roy. Inst. Technol. Stockholm **94**, 1.

WRANGLÉN, G. (1960). Electrochim. Acta **2**, 130.

YANG, L., CHEIN, C., and HUDSON, R. G. (1959). J. Electrochem. Soc. **106**, 632.

Chapter 8

Fast Ionic Reactions

Edward M. Eyring

I. Introduction 731
II. Diffusion Controlled Reactions between Ions in Solutions 732
III. Diffusion Controlled Ionic Reactions: Relaxation Experiments 741
   A. Calculations of Specific Rates from Relaxation Times 742
   B. Measurement of Relaxation Times 745
IV. Somewhat Slower Fast Reactions 749
   A. Rupturing Intramolecular Hydrogen Bonds 750
   B. Formation of Complex Ions 752
   C. Metal Ion Hydrolysis and Polymerization 761
   D. Pseudo or Carbon Acid Dissociations 765
   References 769

I. Introduction

Since what is called "fast" today may well seem slow tomorrow, it is useful to explain at the beginning that we intend to restrict our attention to reactions between ions in liquid solutions having half reaction times ranging from tens of milliseconds down to about a nanosecond (10^{-9} sec). The reason that it is interesting to group chemical reactions in this manner is that within this time range we are able to follow the progress of many discrete reaction steps in complex overall reaction mechanisms.

Until the early 1950's, no one had directly determined half reaction times of ionic reactions in liquids under about a millisecond. Since there are a great many complex reactions in liquids with slow steps having half reaction times of minutes, hours, and even days, there was no shortage of interesting reactions to study, and indeed today the majority of reaction kineticists measure reaction rates that would not qualify as fast in the present context. The classic texts on reaction rates such as Glasstone *et al.*

(1941), Benson (1960), Frost and Pearson (1961), and Basolo and Pearson (1967) are, in fact, monuments to the tremendous fruitfulness of studying the mechanisms of slow reactions. However, for a really satisfying understanding of what is happening at the molecular level as, for example, the reaction of a hydrated cupric ion with ammonia in aqueous solution giving a beautiful blue complex ion, there is no substitute for the insights provided by Eigen, Porter, Connick, and a host of other fast reaction experimentalists in the last two decades. The vogue for fast reaction kinetics may fade in the years ahead, but the by then classical conclusions regarding mechanisms of discrete reaction steps will form a vitally important part of the preliminaries to whatever reaction kineticists and electrochemists then believe to be hot, new research topics.

We intend to consider here the kinetics of simple acid-base reactions such as neutralization and hydrolysis, polymerization of inorganic ions, and reactions between metal ions and various ligands to form complex ions. We will not treat homogeneous redox reactions, although elegant kinetic studies of such systems have been carried out with the relaxation techniques we will describe (e.g., Diebler (1960), Halpern et al. (1963), Hurwitz and Kustin (1966)). Neither will we treat enzyme catalyzed reactions that involve many rapid ionic reaction steps but whose complexity puts them beyond the reach of this abbreviated treatment.

A logical place to begin our discussion of fast reactions is the treatment of diffusion controlled reactions having very low activation energies of ~ 3.5 kcal/mole (Logan, 1967), attributable solely to the stepwise, jostled motion of reactant molecules or ions slipping by solvent molecules and toward one another in dilute solutions.

II. Diffusion Controlled Reactions between Ions in Solutions

The difficulties of extrapolating the kinetic theory of gases to collisions between reacting species in liquid solutions or of calculating partition functions for solvated reactants, not to mention the activated complex, required by an absolute rate theory treatment are very formidable. For practical results, we are presently forced to fall back on the phenomenological treatment of diffusion controlled reactions between molecules and ions or both in solutions evolved by von Smoluchowski (1915, 1917), Debye (1942), and Eigen (1954a). In this approach, measured reaction rates are correlated on the basis of our experimental knowledge of the macroscopic phenomena of viscosity, diffusion, and coulombic interactions between charged bodies. While this phenomenological approach is not a unique way of handling rates of diffusion controlled reactions (see Noyes (1961) for alternatives), it is probably the most widely used attack.

A structureless, isotropic solvent is postulated with the diffusion of reactants A and B toward one another described in terms of Fick's laws of diffusion. In Fick's empirical first law

$$J_A = -D_A \nabla C_A \qquad (2.1)$$

where J_A is the flux density or net flow of species A through unit area per unit time (typically in units of molecules $cm^{-2} \ sec^{-1}$), D_A is the diffusion coefficient characteristic of the species A at a specific temperature and in a specific medium (in units of $cm^2 \ sec^{-1}$), the operator

$$\nabla \equiv \frac{\partial}{\partial x} + \frac{\partial}{\partial y} + \frac{\partial}{\partial z},$$

and C_A is the concentration of the Ath species (in units of molecules cm^{-3}). Thus ∇C_A is the concentration gradient of species A. Molecules always flow in the direction of decreasing concentration gradient, so the negative sign in Eq. (2.1) is introduced to make the coefficient D positive. The diffusion coefficient is, in general, a function of concentration of the diffusing species, but in dilute solutions like those of interest to a fast reaction kineticist D can be considered to be constant.

Noyes (1961) suggested a helpful model from which we can get an intuitive feeling for the origin of the concentration gradients in a reaction mixture: We imagine two kinds of molecules A and B that may react with one another on colliding but do not exert long-range forces on one another. Furthermore, each kind of molecule is randomly distributed throughout the solvent as if the other kind were not even present. We intend to deduce the distributions of unreacted A and B molecules at some later time after chemical reaction has gotten underway but has not yet approached completion.

Let us visualize another solution identical in every way with this first solution except that A and B do not react. We further suppose that it is possible to tag A and B molecules in this hypothetical solution as "reacted" if they have experienced encounters that would have led to reaction in the solution we first considered. In this hypothetical second solution, the total number of A and B molecules is time independent as are also the distributions in space of A and B. We note too that there will be a flux of "unreacted" B molecules toward the remaining "unreacted" A molecules and an equal flux of "reacted" B molecules away from "reacted" A molecules.

In order that a net flux of molecules exist in such a solution there must also be a concentration gradient. In the neighborhood of any A molecule, the probable concentration of B molecules of both types, "reacted" and "unreacted," must be the same as near any other arbitrary site in the solution. However, near a "reacted" A molecule the probable concentration of "reacted" B molecules exceeds the average concentration of "reacted" B molecules throughout the solution, and near an "unreacted" A molecule the probable concentration of "unreacted" B molecules is less than the average throughout the solution. To grasp this latter statement consider the following: Sometimes an "unreacted" A molecule and an "unreacted" B molecule collide and a "reacted" A molecule and a "reacted" B molecule then separate. For all other interactions in the system ("unreacted" A encounters "reacted" B, for example), the status of the molecules does not change. "Reaction" means that "unreacted" A molecules approach "unreacted" B molecules more often than "unreacted" A molecules depart from "unreacted" B molecules. To exactly the same extent, "reacted" A molecules depart from "reacted" B molecules more often than "reacted" A molecules approach "reacted" B molecules. These excesses mean there is a net flux of "unreacted" B molecules toward "unreacted" A molecules and a net flux of "reacted" B molecules away from "reacted" A molecules. (Since the total concentrations of A and B molecules remain constant, there is no net flux of A and B molecules toward or away from each other.) If net fluxes of particular types of molecule do occur, these fluxes must be associated with concentration gradients; hence our earlier assertion that near an "unreacted" A molecule the probable concentration of "unreacted" B molecules is less than the average throughout the solution. By now denying the existence of "reacted" A and B molecules, we can transform our hypothetical system into something closely approximating the real one.

Cutting through the above thicket of verbiage, the important point is that the concentration gradient required to maintain a diffusion flux equal to the rate at which A molecules react is provided by the reaction's continuing depletion of B molecules near unreacted A molecules.

Returning now to Eq. (2.1), we should add that limiting equivalent conductivities λ_i or ion mobilities u_i are frequently tabulated rather than diffusion coefficients D_i. Thus we will find it helpful to recall that

$$D_i = (kT/|z_i|e_0)u_i \qquad (2.2)$$

where $k = 1.38 \times 10^{-23}$ J deg^{-1}, T is the absolute temperature, $|z_i|$ is the absolute value of the ionic charge, and $e_0 = 1.60 \times 10^{-19}$ C. With u_i expressed in cm^2 V^{-1} sec^{-1}, the diffusion coefficient D_i has the units

$cm^2 sec^{-1}$. The other useful relationship is that between limiting equivalent conductivities and ion mobilities

$$\lambda_i = Fu_i \tag{2.3}$$

where F is Faraday's constant, 96,500 C equiv^{-1}, and λ_i clearly has the units $cm^2 \Omega^{-1}$ equiv^{-1} when u_i is in $cm^2 V^{-1} sec^{-1}$. In Table I we have assembled values of the limiting equivalent conductivity for several cations and anions in five solvents.

In connection with Eq. (2.1), we should also append the comment that it is really valid only for a two-component system, and strictly applies to multicomponent systems only in the limit that interactions between fluxes of different species are negligible (Onsager, 1945).

In ionic solutions, the dominant coulombic interactions give rise to an additive perturbation term in Fick's first law. In general, if the flux of the Ath species is influenced by any type of external potential U, Fick's first law will take the form

$$J_A = -D_A[\nabla C_A + (C_A/kT)\, \nabla U]. \tag{2.4}$$

We next wish to obtain Fick's second law of diffusion by considering a nonsteady state in a one-dimensional system. Suppose that molecules diffuse in the positive χ direction through two equal 1 cm^2 areas at right angles to the χ axis and spaced a distance $\Delta\chi$ apart. The flow through the area element at χ is J_χ and that through the area element at $\chi + \Delta\chi$ is $J_{\chi+\Delta\chi}$. The parallelepiped between the two area elements has a volume $V = (1\ cm^2)\, \Delta\chi = \Delta\chi\ cm^3$.

During a time interval Δt, the number of moles moving in a positive χ direction and entering this volume through the area element at χ is $J_\chi\, dt$ while the number leaving this volume through the element at $\chi + \Delta\chi$ is $J_{\chi+\Delta\chi}\, dt$. The increase in the number of moles in the volume element is

$$\Delta n = J_\chi\, dt - J_{\chi+\Delta\chi}\, dt, \tag{2.5}$$

but since

$$J_{\chi+\Delta\chi} = J_\chi + (\partial J_\chi/\partial\chi)\, \Delta\chi \tag{2.6}$$

it follows that

$$\Delta n = -(\partial J_\chi/\partial\chi)\, \Delta\chi\, dt. \tag{2.7}$$

The corresponding increase in concentration within this volume element is

$$dc = \Delta n/\Delta\chi, \tag{2.8}$$

from which it follows that

$$dc = -(\partial J_\chi/\partial\chi)\, dt \tag{2.9}$$

TABLE I

LIMITING EQUIVALENT CONDUCTIVITIES, $\lambda°$ (cm^2 Ω^{-1} equiv^{-1}), OF IONS IN SEVERAL SOLVENTS[a]

Ion	Solvent: H$_2$O				D$_2$O	MeOH	EtOH	MeCN
	0°	10°	25°	45°	25°	25°	25°	25°
H$^+$	225		349.8	441.4				
Li$^+$	19.4	26.4	38.7	58.0		39.6	17.1	
Na$^+$	26.5	34.9	50.2	73.8	41.6	45.2	20.3	
K$^+$	40.7	53.1	73.6	103.6	61.4	52.4	23.6	
Cs$^+$	44	56.5	77.3	107.6	64.4	60.8	26.5	
Me$_4$N$^+$	24.1	30.9	44.4	65.0	36.6	68.7	30.0	94.2
Et$_4$N$^+$	16.4	21.9	32.2	48.0	26.4	60.5	29.5	84.6
Pr$_4$N$^+$	11.5	15.3	23.2	35.8	18.8	46.1		70.3
Bu$_4$N$^+$	9.6	12.6	19.3	30.4	15.6	38.9	19.2	61.4
Mg^{2+}	28.9		53.1					
La^{3+}	34.4		69.8					
OH$^-$	105		198.3					
F$^-$			55.3		44.8			
Cl$^-$	41.0	54.3	76.4	109.0	62.8	52.4	21.9	98.7
Br$^-$	42.6	56.2	78.2	110.7	64.7	56.5	24.0	100.7
I$^-$	41.4	55.4	77.0	108.8	63.8	62.8	26.1	102.7
SO$_4^{2-}$	41		80.0					
Solvent viscosities (centipoise)	1.787	1.306	0.8903	0.5963	1.096	0.5445	1.084	0.341

[a] From the compilations of Robinson and Stokes (1959) and Kay and Evans (1966).

and

$$\partial c/\partial t = -\partial J_x/\partial \chi. \tag{2.10}$$

At this point we note that for the one-dimensional case, Fick's first law, Eq. (2.1), reduces to

$$J_\chi = -D(\partial c/\partial \chi) \tag{2.11}$$

and we then obtain from Eqs. (2.10) and (2.11)

$$\frac{\partial c}{\partial t} = \frac{\partial}{\partial \chi}\left(D\frac{\partial c}{\partial \chi}\right) = D\frac{\partial^2 c}{\partial \chi^2} \tag{2.12}$$

where D is presumed independent of χ. This is Fick's second law of diffusion which can be rewritten for three dimensions with the added potential function U as

$$\frac{\partial C_A}{\partial t} = \nabla\left[D_A\left(\nabla C_A + \frac{C_A}{kT}\nabla U\right)\right]. \tag{2.13}$$

The maximum specific rate of reaction between the species A and B occurs when every collision between A and B leads to reaction. Since we are interested in relative rather than absolute positions of A and B, we may suppose that B is stationary and A undergoes random displacements at twice its true frequency. This can be called a "stationary reactive sink approximation" where the word sink implies reaction every time A collides with B. The concentration of A will vary from zero at a distance $r_A + r_B$ from B to the bulk equilibrium concentration $[A]$ at large distances from B. To minimize mathematical difficulties, we further assume spherical symmetry with the origin of coordinates at the sink B and the potential energy U a function of r only.

For practically interesting situations, we may assume a steady state, $\partial C_A/\partial t = 0$. It then follows from Eqs. (2.13) and (2.11) that

$$-J_A = \text{const} = D_A\left(\frac{\partial C_A}{\partial r} + \frac{C_A}{kT}\frac{\partial U}{\partial r}\right). \tag{2.14}$$

The rate of reaction divided by the number of B molecules we will denote by the quantity I_A. I_A is, in fact, the total flux of A molecules through a spherical surface of arbitrary radius r centered on the B sink, or

$$I_A = \int_0^{2\pi}\int_0^{\pi} J_A r^2 \sin\theta\, d\theta\, d\varphi = \text{const} \tag{2.15}$$

which from Eq. (2.14) is

$$I_A = 4\pi^2 D_A\left(\frac{\partial C_A}{\partial r} + \frac{C_A}{kT}\frac{\partial U}{\partial r}\right) = \text{const}. \tag{2.16}$$

The appropriate boundary conditions are $C_A = [A]$ and $U = 0$ at $r = \infty$ whereas at $r = r_A + r_B \equiv a$, $C_A = 0$ and $U = U(a)$. Multiplying both sides by $\exp(U/kT)$, we can integrate Eq. (2.16) thus:

$$I_A \int_a^\infty (\exp(U/kT)/r^2)\,dr = 4\pi D_A \int_0^{[A]} d[C_A \exp(U/kT)], \qquad (2.17)$$

whence multiplying and dividing by a we have

$$I_A = \frac{4\pi D_A a [A]}{a \int_a^\infty \exp(U/kT)\,dr/r^2}. \qquad (2.18)$$

Up to this point we have made the approximation that the B molecules are stationary whereas, in fact, they too diffuse. Therefore, we may take account of their thermal motion or, if you prefer, the Brownian motion of the coordinate system by replacing D_A in Eq. (2.18) by $D_A + D_B$. Since I_A is a rate of reaction for a single B sink, the total rate of chemical reaction R is obtained by multiplying Eq. (2.18) by $[B]$. We finally have for the maximum diffusion controlled specific rate

$$k_r = \frac{R}{[A][B]} = \frac{4\pi a(D_A + D_B)}{a \int_a^\infty \exp(U/kT)\,dr/r^2} \qquad \frac{\text{cc}}{\text{molecule-sec}}. \qquad (2.19)$$

If we choose to express k_r in units of $M^{-1}\,\text{sec}^{-1}$ that a solution kineticist would prefer, this becomes

$$k_r = \frac{4\pi Na(D_A + D_B)}{10^3 a \int_a^\infty \exp(U/kT)\,dr/r^2} \qquad \frac{\text{liter}}{\text{mole-sec}} \qquad (2.20)$$

where N is Avogadro's number. The steady-state approximation made in deriving Eq. (2.20) happily does not greatly limit its usefulness. The neglected transient contribution is responsible for less than a 1% error in k_r for half reaction times as long as 10^{-7} sec or longer (Noyes, 1961).

Since we are interested in rapid ionic reactions, the dominant potential is coulombic, i.e., $U = Z_A Z_B e_0^2 / \epsilon r$ where Z_A is the valence (a positive or negative integer) of the A reactant ion, e_0 is the charge on an electron (4.80×10^{-10} statcoulombs), and ϵ is the dielectric constant. When the integral in Eq. (2.20) is evaluated using this particular U, the resulting expression for the limiting specific rate is

$$k_r = \frac{4\pi N Z_A Z_B e_0^2 (D_A + D_B)}{10^3 \epsilon kT [\exp(Z_A Z_B e_0^2 / \epsilon kTa) - 1]} \qquad \frac{\text{liter}}{\text{mole-sec}}. \qquad (2.21)$$

The specific rates of a great many essentially diffusion controlled reactions have been directly measured, principally by relaxation techniques since their introduction by Eigen (1954b). A few representative values are shown in Table II along with limiting values calculated from Eq. (2.21) using a reaction distance $a = 7.5$ Å where we recall that a pair of reacting ions combine immediately at this distance. The factor by which entries in the two columns differ is frequently considered to be simply a steric factor. We should note, however, that there are minor inconsistencies in the derivation of Eq. (2.21) that might also account for these discrepancies, not to mention the sizable errors (as much as 10 to 20%) in the experimental values.

Bass and Greenhalgh (1966) objected to the "stationary reactive sink approximation" because it is only valid for the case where the reacting ions have equal concentrations. Watts (1966a,b) treated this and other difficulties and found that where there is no net reaction (as is true at equilibrium) Eq. (2.21) and the stationary sink approximation are valid. Since relaxation-method kinetic data are taken very near equilibrium, Eq. (2.21) is thus a satisfactory approximation to the diffusion controlled specific rate. Friedman (1966) has treated a hydrodynamic interaction between reacting solute species that, in principle, reduces the specific rate 15% below that calculated from Eq. (2.21). This result is also not vitally important because the experimental errors are usually of the same order of magnitude.

The form of Eq. (2.21) has an interesting consequence noted by Eigen (1954a). Where ion recombination is diffusion controlled, the rate constant for the reverse reaction must also contain the sum of the diffusion coefficients since the quotient of the two specific rates, the equilibrium constant, is independent of diffusion rates. Thus, the specific rate of the diffusive dissociation process $AB \rightarrow A^{Z_A} + B^{Z_B}$ turns out to be

$$k_d = \frac{3 Z_A Z_B e_0^2 (D_A + D_B)}{\epsilon k T a^3 [1 - \exp(- Z_A Z_B e_0^2 / \epsilon k T a)]}. \qquad (2.22)$$

The general form of the mechanism for an ion recombination-dissociation equilibrium is

$$A^+ + B^- \underset{k_{21}}{\overset{k_{12}}{\rightleftharpoons}} AB \underset{k_{32}}{\overset{k_{23}}{\rightleftharpoons}} C \qquad (2.23)$$

where the intermediate AB represents an ion pair. For a high dielectric constant solvent such as water in which the concentration of AB is small compared to those of A^+, B^-, and C, we have in the case of a diffusion controlled reaction the conditions $k_{21} \ll k_{23}$ and the overall forward rate constant $k = k_{12}$, whereas the overall reverse rate constant $k = k_{21} k_{32} / k_{23}$.

TABLE II

COMPARISON OF EXPERIMENTAL AND THEORETICAL RECOMBINATION RATE CONSTANTS IN WATER FOR DIFFUSION CONTROLLED REACTIONS.[a]

Reaction	k_R (theor.) (M^{-1} sec^{-1})	k_R (exp.) (M^{-1} sec^{-1})	T (°C)	$\sim\mu$ (M)	Ref.
$H^+ + OH^- \rightarrow H_2O$	1.3×10^{11}	$< 1.4 \times 10^{11}$ [c]	25	0	Eigen and DeMaeyer (1955)
$H^+ +$ phenol red dianion $\rightarrow$ monoanion	1.3×10^{11}	$> 7.2 \times 10^{10}$ [c]	15	2×10^{-4}	Ilgenfritz (1966)
$H^+ + F^- \rightarrow HF$	9.4×10^{10}	$< \sim 1.0 \times 10^{11}$ [c]	25	0	Eigen and Kustin (1960)
$H^+ + C_6H_5COO^- \rightarrow$ benzoic acid	8.9×10^{10}	$> 3.5 \times 10^{10}$ [c]	25	0	Eigen and Eyring (1962)
$H^+ + Me_3N \rightarrow Me_3NH^+$	5.3×10^{10} [b]	$> 2.5 \times 10^{10}$ [d]	25		Emerson et al. (1960)
$H^+ + UO_2OH^+ \rightarrow UO_2^{2+} + H_2O$	3.9×10^{10}	$> 1.7 \times 10^{10}$ [c]	25	4×10^{-4}	Cole et al. (1967)
$H^+ + AlOH^{2+} \rightarrow Al^{3+} + H_2O$	2.4×10^{10}	$> 4.4 \times 10^9$ [c]	25	10^{-3}	Holmes et al. (1968)
$H^+ + ThOH^{3+} \rightarrow Th^{4+} + H_2O$	4×10^{10}	$> \sim 7 \times 10^8$ [c]	25	10^{-3}	Eyring and Cole (1967)
$OH^- + C_3N_2H_5^+ \rightarrow$ imidazole $+ H_2O$	6.1×10^{10}	$> 2.5 \times 10^{10}$ [c]	25	0	Eigen et al. (1960)
$OH^- +$ adenine $\rightarrow C_5H_4N_5^- + H_2O$	2.9×10^{10} [b]	$> 1 \times 10^{10}$ [e]	25	0	Eigen et al. (1964)
$OH^- + HCO_3^- \rightarrow CO_3^{2-} + H_2O$	2.2×10^{11}	$> \sim 6 \times 10^9$ [e]	20	1.0	Eigen et al. (1964)
$OH^- + HPO_4^{2-} \rightarrow PO_4^{3-} + H_2O$	5.8×10^{10}	$> \sim 2 \times 10^9$ [e]	25	0.1	Eigen et al. (1964)
$OH^- + HATP^{3-} \rightarrow ATP^{4-} + H_2O$	6.8×10^{10}	$> 1.2 \times 10^9$ [f]	12	0.1	Eigen et al. (1964)

[a] Values of k_R (theor.) calculated from Eq. (2.21) using where necessary estimated values of diffusion coefficients and corrected to the experimental ionic strength μ with the relation log $k/k_0 = 1.02 Z_A Z_B \sqrt{\mu}$.

[b] Calculated from von Smoluchowski (1916): $k_R = 4\pi 10^{-3} Na(D_A + D_B)$ which is the limiting form of Eq. (2.21) for Z_A and/or $Z_B \rightarrow 0$.

[c] E-jump relaxation.

[d] NMR.

[e] Sound absorption.

[f] Temperature jump relaxation.

The important point to note here is that if we use Eq. (2.21) to calculate k_{12} in a low dielectric constant solvent such as 1,4-dioxane, we have not calculated the overall rate constant k.

It is also interesting to remark that Eq. (2.21) does not permit *a priori* calculations of k_{12} in a given solvent unless one is able to judiciously guess values for the parameters ϵ and a. As we noted above, Eigen and DeMaeyer (1958) deduced from relaxation-method kinetic measurements in water at room temperature that $a = 7.5$ Å. This distance would suggest that 2 or 3 water molecules separate the reacting ions and it is large enough to justify the use of the macroscopic dielectric constant for water which is 78.5 at 25°.

In summary, a chemist confronted with the need to know the rate and mechanism of an apparently rapid ionic reaction in water or some more obscure solvent could at the very least estimate an upper limit for the specific rates of forward and backward reactions from Eqs. (2.21) and (2.22) and his knowledge of the macroscopic properties of the solvent such as dielectric constant and viscosity.

Before considering a few well characterized general examples of non-diffusion controlled ionic reactions, let us digress and discuss how we would go about determining experimentally the specific rates of a few diffusion controlled reactions in some new solvent for which a value of the parameter a is still uncertain.

III. Diffusion Controlled Ionic Reactions: Relaxation Experiments

Nuclear magnetic and electron spin resonance (Strehlow, 1963a), polarography (Strehlow, 1963b), flash photolysis (Porter, 1968), pulse radiolysis (Ebert *et al.*, 1965), stopped and continuous flow techniques (Roughton and Chance, 1963), and several relaxation methods (temperature jump (Eigen and DeMaeyer, 1963), pressure jump (Hoffmann *et al.*, 1966), sound absorption (Stuehr and Yeager, 1965), and electric field jump (Eigen and De Maeyer, 1963) methods) all permit the experimentalist to follow the progress of solution reactions having half lives of less than a second. For the moment, our attention is fastened on diffusion controlled reactions between ions for which the expected half life is of the order of 10^{-6} to 10^{-7} sec for reactant concentrations of 10^{-5} to 10^{-4} M. Of the above techniques, the most suitable for following these diffusion controlled reactions would be sound absorption, the electric field jump (E-jump) relaxation method, and, in some cases, pulsed-laser flash photolytic (Witt, 1967) and pulse radiolysis techniques. We will arbitrarily limit our discussion to the applications of the electric field jump technique to the measurement of 0.1 to 10 μsec half reaction times.

A. CALCULATIONS OF SPECIFIC RATES FROM RELAXATION TIMES

The fundamental idea behind Eigen's relaxation methods is that a rapidly achieved chemical equilibrium can be displaced suddenly from equilibrium by a small jump in temperature T, pressure P, or electric field intensity E, and if the change in this external parameter is steep enough, it may be considered to be instantaneous compared to the experimental relaxation of the chemical system to the equilibrium concentrations characteristic of the new T, P, or E. The relaxation process can be recorded oscillographically as a function of time by spectrophotometric or conductometric detection of concentration changes in one or more reactants or products.

Consider for a moment the chemical equilibrium

$$A + B \underset{k_2}{\overset{k_1}{\rightleftharpoons}} D, \tag{3.1}$$

specific examples of which are $H^+ + OH^- \rightleftharpoons H_2O$ in pure water or $H^+ + AlOH^{2+} \rightleftharpoons Al^{3+} + H_2O$ in an aqueous aluminum chloride solution where the concentration of water can be taken to be essentially constant and can therefore be incorporated into k_2. If C_A denotes the instantaneous concentration of species A, etc., then the rate of production of D is given by

$$dC_D/dt = k_1 C_A C_B - k_2 C_D, \tag{3.2}$$

assuming that the forward and reverse reactions are first order in each species. Suppose now that following a sudden change in an external variable T, P, or E, the new equilibrium concentrations are $\bar{C}_A$, $\bar{C}_B$, and $\bar{C}_C$. These equilibrium concentrations are related to the instantaneous concentrations C_A, C_B, and C_C at some time $t > 0$ by the relations

$$C_A = \bar{C}_A + \delta C_A, \tag{3.3}$$

$$C_B = \bar{C}_B + \delta C_B, \tag{3.4}$$

$$C_D = \bar{C}_D + \delta C_D, \tag{3.5}$$

where δC_A, δC_B, and δC_D are the small ($\bar{C}_A \gg \delta C_A$, etc.), time dependent deviations of the concentrations from their final equilibrium values. Substituting $\bar{C}_A + \delta C_A$ for C_A, etc., in Eq. (3.2) we have

$$d(\bar{C}_D + \delta C_D)/dt = k_1(\bar{C}_A + \delta C_A)(\bar{C}_B + \delta C_B) - k_2(\bar{C}_D + \delta C_D) \tag{3.6}$$

which becomes

$$d\,\delta C_D/dt = k_1(\bar{C}_A\,\delta C_B + \bar{C}_B\,\delta C_A) - k_2\,\delta C_D, \qquad (3.7)$$

since

$$d\bar{C}_D/dt = 0 = k_1\bar{C}_A\bar{C}_B - k_2\bar{C}_D \qquad (3.8)$$

and for small perturbations $\delta C_A\,\delta C_B \approx 0$. From the conservation relation

$$\delta C_A = \delta C_B = -\delta C_D, \qquad (3.9)$$

it follows that Eq. (3.7) may be rewritten as

$$d\,\delta C_D/dt = -[k_1(\bar{C}_A + \bar{C}_B) + k_2]\,\delta C_D. \qquad (3.10)$$

It is evident that in order for the right-hand side of Eq. (3.10) to be dimensionally correct, the bracketed quantity must have the units of reciprocal time, i.e.,

$$1/\tau \equiv [k_1(C_A + C_B) + k_2] \qquad (3.11)$$

where τ is the so-called relaxation time or time constant for this chemical equilibration. Integration of the equation

$$d\,\delta C_D/dt = -\delta C_D/\tau \qquad (3.12)$$

with δC_D equal to a constant $\delta C_D(0)$ at $t = 0$ yields the solution

$$\delta C_D(t) = \delta C_D(0)\exp(t/\tau). \qquad (3.13)$$

Suppose, for example, spectrophotometric detection is used and the molecule D absorbs strongly at 550 mμ while A and B are both colorless. Then a monochromatic beam of this wavelength passing through the sample solution and incident on the green-light photosensitive cathode of a photomultiplier tube will give rise to an output voltage changing exponentially as a function of time following the initial sudden perturbation of the chemical equilibrium (by a jump in T, P, or E). The direction in which the equilibrium, Eq. (3.1), is displaced (toward or away from higher D concentration) is irrelevant. The time required for the initial voltage to change by $1/e = 1/2.718$ of its initial value is the required experimental relaxation time, τ.

Equation (3.11) has the very useful property of being an equation for a straight line. Thus if we plot τ^{-1} vs. $\bar{C}_A + \bar{C}_B$ for several different concentrations, the slope of the resulting straight line is the rate constant k_1 and the intercept on the τ^{-1} axis is k_2. Furthermore, the quotient $k_1/k_2 = \bar{C}_D/\bar{C}_A\bar{C}_B$ should equal the equilibrium constant, previously obtained by potentiometric titration or some other equilibrium technique, from which we calculated our values of $\bar{C}_A$ and $\bar{C}_B$ in the first place.

Czerlinski (1964a,b) has supplied an interesting answer to the following important question: What do we do with our τ^{-1} data if we do not have an equilibrium constant to begin with and, indeed, are not even sure that the equilibrium we are studying has the form of Eq. (3.1)? We certainly know the analytical concentrations of the species A, B, and D denoted by $C_A{}^0$, $C_B{}^0$, and $C_D{}^0$ that obtain at the instant of mixing up our sample system. Let us momentarily suppose that Eq. (3.1) does apply to our system and develop as an alternative to Eq. (3.11) an expression for τ^{-1} vs. some combination of $C_A{}^0$, $C_B{}^0$, and $C_D{}^0$ that we may compare with our kinetic data. From stoichiometry, we have

$$\bar{C}_A = C_A{}^0 - \bar{x}, \tag{3.14}$$

$$\bar{C}_B = C_B{}^0 - \bar{x}, \tag{3.15}$$

$$\bar{C}_D = C_D{}^0 + \bar{x}. \tag{3.16}$$

Judicious combinations of Eqs. (3.14)–(3.16) lead to

$$C_A{}^0 - \bar{C}_A = C_B{}^0 - \bar{C}_B, \tag{3.17}$$

$$C_A{}^0 - \bar{C}_A = -(C_D{}^0 - \bar{C}_D). \tag{3.18}$$

We will also find the equilibrium constant

$$K_{2,1} \equiv \frac{k_2}{k_1} = \frac{\bar{C}_A \bar{C}_B}{\bar{C}_D} \tag{3.19}$$

to be useful since in combination with Eqs. (3.17) and (3.18) it permits us to solve for any one of the equilibrium concentrations $\bar{C}_A$, $\bar{C}_B$, or $\bar{C}_D$ in terms of $K_{2,1}$, $C_A{}^0$, $C_B{}^0$, and $C_D{}^0$. Thus, for example, solving for $\bar{C}_A$, we obtain first

$$K_{2,1} = [\bar{C}_A(C_B{}^0 - C_A{}^0 + \bar{C}_A)]/(C_D{}^0 + C_A{}^0 - \bar{C}_A) \tag{3.20}$$

and then

$$\bar{C}_A{}^2 + (C_B{}^0 - C_A{}^0 + K_{2,1})\bar{C}_A - K_{2,1}(C_D{}^0 + C_A{}^0) = 0. \tag{3.21}$$

The root of interest is

$$\bar{C}_A = \{C_A{}^0 - C_B{}^0 - K_{2,1} + [(C_B{}^0 - C_A{}^0 + K_{2,1})^2 + 4K_{2,1}(C_D{}^0 + C_A{}^0)]^{1/2}\}/2. \tag{3.22}$$

The corresponding value of $\bar{C}_B$ is

$$\bar{C}_B = \{C_B{}^0 - C_A{}^0 - K_{2,1} + [(C_A{}^0 - C_B{}^0 + K_{2,1})^2 + 4K_{2,1}(C_D{}^0 + C_B{}^0)]^{1/2}\}/2. \tag{3.23}$$

It is expedient to consider initial conditions that markedly simplify these expressions for $\bar{C}_A$ and $\bar{C}_B$. A particularly satisfactory possibility is that for which $C_A{}^0 = C_B{}^0$. It follows from Eq. (3.17) that $\bar{C}_A = \bar{C}_B$ also when $C_A{}^0 = C_B{}^0$. Using these conditions and Eqs. (3.19) and (3.22)

$$1/\tau = k_2\{1 + [4(C_D{}^0 + C_A{}^0)/K_{2,1}]\}^{1/2}. \tag{3.24}$$

The experimentalist will actually find the square of Eq. (3.24)

$$\tau^{-2} = k_2{}^2 + 4k_1k_2(C_D{}^0 + C_A{}^0) \tag{3.25}$$

more useful since it presents a linear dependence on the analytical concentrations. If the data plotted in terms of Eq. (3.25) did not yield a straight line and only one chemical relaxation time was observed, it would be necessary to propose a new equilibrium such as $A + B \rightleftharpoons D + E$ or $2A \rightleftharpoons D$ and derive a new expression for τ in terms of the analytical concentrations analogous to Eq. (3.25) until such an equation is found that does permit a good fit of the experimental data.

B. MEASUREMENT OF RELAXATION TIMES

The rate constants for many different diffusion controlled reactions in water have been determined during the last decade by the sound absorption and dissociation field effect (electric field jump) relaxation techniques. In our own laboratory we have used the latter method to determine forward and backward rate constants for aqueous equilibria of the following type:

$$UO_2OH^+ + H^+ \overset{k_1}{\rightleftharpoons} UO_2^{2+} + H_2O, \tag{3.26}$$

$$AlOH^{2+} + H^+ \overset{k_2}{\rightleftharpoons} Al^{3+} + H_2O, \tag{3.27}$$

$$ThOH^{3+} + H^+ \overset{k_3}{\rightleftharpoons} Th^{4+} + H_2O. \tag{3.28}$$

As we see from Table II, $k_1 > k_2 > k_3$. The descending magnitude of these second-order rate constants is in excellent agreement with the predictions of Eq. (2.21) regarding the dependence of specific rate on ionic charge in diffusion controlled reactions and was semiquantitatively predictable from a generalization made by DeMaeyer and Kustin (1963).

In the electric field jump relaxation technique, a chemical equilibrium involving an incompletely dissociated electrolyte is perturbed by a suddenly applied dc field of $\sim 10^5$ V/cm. The relaxation of electrolyte concentrations from their equilibrium values at zero field to their equilibrium values in a high field gives rise to a rapid conductance change measurable with a Wheatstone bridge. The largest dissociation field effect, or second Wien effect (Wien and Schiele, 1931), would be anticipated for equilibria involving charge neutralization, but if the ionic species on the two sides of a chemical equilibrium differ markedly in equivalent conductance as in reactions (3.26)–(3.28) a relaxation should still be observable. In our apparatus (see Fig. 1), a rectangular high-

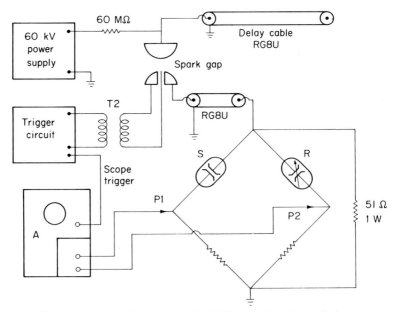

Fig. 1. Apparatus for the electric field jump relaxation technique.

voltage pulse of ~ 12 μsec duration is applied to a Wheatstone bridge. Two arms of the bridge consist of identical platinum-sealed-in-glass cells with electrode separations of 0.3 cm and a cell constant $a = 0.05$ cm^{-1}. Of these two cells, the sample cell S contains a dilute solution of, let us say, 10^{-4} M aqueous uranyl nitrate. The reference cell R contains a dilute solution of aqueous HCl of a concentration such that the two cells have the same electrical conductance measured on an impedance bridge operated at 10^3 Hz. A strong electrolyte solution in R is required to cancel a first Wien effect (Wien, 1927) as well as a slight resistive heating occurring in S. The other two arms of the bridge are conventional carbon resistors equal to one another but lower in resistance than S and R, i.e., $\sim 10^3$ Ω as opposed to $\sim 7 \times 10^3$ Ω. Rectangular high-voltage pulses are generated with approximately a kilometer of RG8U coaxial cable charged by a 60-kV low amperage power supply. Triggering of the pulse is accomplished with a commercial spark gap. Having balanced the high voltage probes P1 and P2 with low voltage rectangular pulses, the exponential process of going out of balance at high fields can be followed on a single sweep oscilloscope-differential preamplifier combination with a bandwidth of at least dc to 13 MHz. Typical oscilloscope traces showing the high voltage pulse applied to the bridge and a relaxation observed in an aqueous uranyl nitrate solution are found in Figs. 2 and

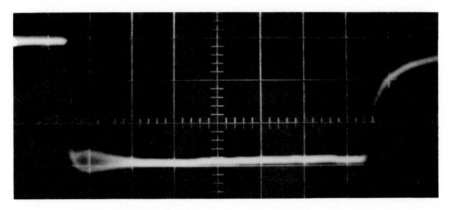

FIG. 2. Typical oscilloscope trace showing the high voltage pulse applied to the bridge. Major divisions on the vertical axis correspond to 5 kV. On the horizontal axis, they correspond to 1.0 μsec.

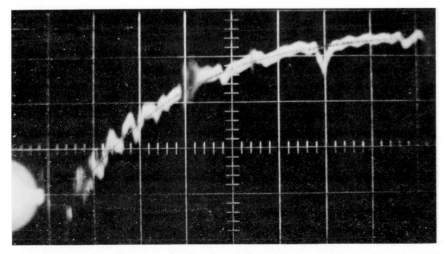

FIG. 3. Typical oscilloscope trace showing a relaxation observed in an aqueous uranyl nitrate solution. Major divisions on the vertical axis correspond to 20 V and on the horizontal axis to 1.0 μsec; $2.5 \times 10^{-5}\ M$ uranyl nitrate, pH $= 5.18$, 70°C.

3. This type of equipment has been described in considerable detail by Eigen and DeMaeyer (1963) and also by Rampton et al. (1967).

The physical sciences advance so rapidly that the reader may suspect that most of the really interesting experiments that are not prohibitively difficult have already been performed with an apparatus like that described above, whose origins go back to 1955. To dispel this concern, let us consider for a moment a few intriguing problems still left to be resolved by this or other kinetic techniques in the microsecond time range.

One of the guiding principles in fast reaction kinetics has been the supposition that the reaction

$$OH^- + H^+ \underset{\longleftarrow}{\overset{k_4}{\longrightarrow}} H_2O \tag{3.29}$$

is the fastest of the diffusion controlled reactions of ions in liquid water (Eigen and DeMaeyer, 1958). This proposition still appears to be correct. However, the rate constant $k_4 = 1.4 \times 10^{11} \, M^{-1} \, sec^{-1}$ at 25° determined by Eigen and DeMaeyer (1955) using the electric field jump relaxation method has been challenged by workers in two laboratories (Briere and Gaspard, 1967; Barker and Sammon, 1967), even though the above value of k_4 was confirmed once by Ertl and Gerischer (1962) using another relaxation technique, the microwave temperature jump. The principal objection raised to the Eigen and DeMaeyer experiment (1955) is the interelectrode distance of only 0.5 mm in the sample cell. Briere and Gaspard (1967) suggest that the relaxation Eigen and DeMaeyer observed may be attributable to ion migration over the interelectrode distance and not to the dissociation of H_2O modified by the applied intense electric field ($\sim 10^5$ V/cm). Since the value of k_4 reported by Eigen and DeMaeyer (1955) now serves as a bench mark in such diverse areas as radiation chemistry and enzyme kinetics, it would be interesting to have another value of k_4 determined by some other quite independent method.

An electric field jump apparatus with high voltage pulse lengths variable in duration from 0.1 to 100 μsec would be suitable for exploring a great many other interesting fast reactions in nonaqueous solvents where electrical resistance is high and hence the heating of the sample solution by long, high voltage pulses is not significant.

For instance, it should be possible to resolve a disagreement regarding the acid dissociation constant of benzoic acid in ethylene glycol. A value of p$Ka = 5.19$ at 20 to 22°C has been reported (Dulova and Lechkova, 1963) as well as p$Ka = 8.16$ at 30°C (Kundu and Das, 1964). By taking the quotient of the forward and backward rate constants for the reaction

$$\Phi—COOH \overset{k_5}{\underset{}{\rightleftharpoons}} \Phi—COO^- + H^+, \tag{3.30}$$

it would be possible to obtain an equilibrium constant that should clear up this discrepancy.

With an electric field jump relaxation-method apparatus like that described above, the instant that the high voltage pulse is applied to the Wheatstone bridge and throughout the pulse duration, the equilibrium constant

$$K = [\Phi—COO^-][H^+][\Phi—COOH]^{-1} \tag{3.31}$$

is slightly larger than it was in the absence of the field. Onsager (1934) developed a theoretical argument showing that in the successive reactions

$$\text{molecule} \overset{k_1}{\underset{k_{-1}}{\rightleftharpoons}} \text{ion pair} \overset{k_2}{\underset{k_{-2}}{\rightleftharpoons}} \text{ions,} \tag{3.32}$$

only the rate constant k_2 is dependent on electric field intensity, and in a field of 10^5 V/cm the linear dependence gives rise to an increase in k_2 seldom exceeding 5%. By extending the electric field intensity to 2×10^5 V/cm, it may be possible to measure the field dependence of both the rate and equilibrium constants and thus test the exactness of Onsager's theory.

The success of rate measurements on the equilibria of Eqs. (3.26)–(3.28) raises another interesting question regarding Onsager's theoretical treatment. His calculations were based on the displacement by an intense electric field of a charge neutralization equilibrium. In the case of the hydrolysis reactions (3.26)–(3.28), however, there is no charge neutralization although the differences in ion mobilities of the reactant and product species do give rise to a small, but still detectable, conductance change. Additional data on the electric field dependence of the amplitude of the conductance change for these noncharge-neutralization equilibria should suggest an extrapolation of Onsager's theory to this interesting specialized case.

IV. Somewhat Slower Fast Reactions

Now let us turn our attention to the kinetics and mechanisms of several general types of ionic reactions in solutions that are very rapid but still not nearly as rapid as the diffusion controlled reactions we have treated up to this point.

A. Rupturing Intramolecular Hydrogen Bonds

For instance, there are now many well-established cases of intramolecular hydrogen bonding in molecules or ions, which we denote by HA^{n-}, responsible for a retardation of the reaction

$$HA^{n-} + OH^- \overset{k_1}{\rightleftharpoons} A^{(n+1)-} + H_2O. \qquad (4.1)$$

Representative results are given in Table III. One example of HA^{n-} is

TABLE III

Specific Rates k_1 of Deprotonation of Intramolecularly Hydrogen Bonded Acids at 25° according to Eq. (4.1)[a]

2- and 2,2-Substituted malonic acid	Solvent H₂O			Solvent D₂O	
	pK_2	$K_1 K_2^{-1}$ [b] 10^4	k_1 ($10^7\ M^{-1}\ sec^{-1}$)	pK_2	k_1 ($10^7\ M^{-1}\ sec^{-1}$)
Isopropyl	5.57	0.06	340	6.07	~260
Methyl (1-methyl-butyl)	6.64	0.71	130	7.14	87
Diethyl	7.06	9.30	24	7.50	12
Ethylisoamyl	7.31	17	18	7.68	9.4
Ethylphenyl	7.10	32	17	7.45	8.5
Ethyl-n-butyl	7.25	13	16		
Di-n-butyl	7.36		14		
Di-n-heptyl	7.45		14		
Di-n-propyl	7.34	16	13		
Ethylisopropyl	8.07	150	5.0	8.42	3.2
Diisopropyl	8.58	330	4.6	8.93	3.1

[a] Miles *et al.* (1965, 1966).

[b] K_1 and K_2 are mixed acid ionization constants (Albert and Serjeant, 1962) for the loss of the first and second carboxylic acid protons, respectively.

shown in Fig. 4. In general, where the proton to be transferred is involved in a strong intramolecular hydrogen bond, there is a resulting disruption of the continuity of hydrogen bridging into the surrounding bulk solvent. This discontinuity of structure drastically reduces the speed with which the bound proton in HA^{n-} can be transferred by a Grotthuss type of conductivity (Moore, 1962) to the hydroxide ion located roughly two water molecules away in the bulk solvent. In cases of particularly strong intramolecular hydrogen bonding, as in rac-α,α'-di-t-butyl succinic acid monoanion ($k_1 = 2.3 \times 10^6\ M^{-1}\ sec^{-1}$; Haslam *et al.*, 1965b), the rate constant for proton removal is as much as four powers of ten slower than would be true for a diffusion controlled reaction.

Several comments of possible predictive value can be made regarding deprotonation of intramolecularly hydrogen bonded compounds, Eq. (4.1). Structural variations in the reactant HA^{n-} near the site of the intramolecular hydrogen bond, different basicities conferred upon X and Y by neighboring atoms, and conjugated ring systems that have the intramolecular hydrogen bond incorporated in them can separately or together produce variations in k_1 that dwarf differences in this specific

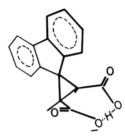

FIG. 4. An example of HA^{n-}.

rate stemming from the nature alone of the X and Y atoms in an X—$H \cdots Y$ intramolecular hydrogen bond. For example, for the O—$H \cdots O$ intramolecularly hydrogen bonded dicarboxylic acid mono-anions of tetramethylsuccinic acid and rac-α,α'-di-t-butyl succinic acid, we have $k_1 = 2.5 \times 10^8$ and $2.3 \times 10^6 \ M^{-1} \ sec^{-1}$, respectively (Haslam et al., 1965b). For the O—$H \cdots N$ intramolecularly hydrogen bonded azo dyes 6-methyl-4-(2-thiazolylazo)-resorcinol and 4-(2-pyridylazo)-resorcinol, we have $k_1 = 2.0 \times 10^8$ and $6.9 \times 10^5 \ M^{-1} \ sec^{-1}$, respectively (Inskeep et al., 1968). From the overlap in these ranges of k_1, it is evident that generalizations regarding relative hydrogen bond strengths, i.e., O—$H \cdots N$ stronger than O—$H \cdots O$ according to Freedman (1961), have little kinetic significance when applied to reaction (4.1). Similarly, a substantially lower value of k_1 for 3,3-diphenylcyclopropane-1,2-dicarboxylic acid ($4 \times 10^6 \ M^{-1} \ sec^{-1}$) than for N,N-dimethyl-o-aminobenzoic acid ($1 \times 10^8 \ M^{-1} \ sec^{-1}$) reported by Haslam et al. (1965a) contradicts the generalization (Eigen et al., 1964) that the most kinetically stable hydrogen bond configurations are 6-membered rings including conjugation. Values of the solvent deuterium oxide kinetic isotope effect k_1^H/k_1^D for various choices of X and Y in the X—$H \cdots Y$ intramolecular hydrogen bond do not vary systematically, so there is little hope of identifying such reactive bonds in biochemical systems from this type of kinetic data.

There are, however, a few generalizations regarding Eq. (4.1) that are successful. In the case of dicarboxylic acids, a large (pK_2-pK_1) in water is consistently associated with a comparatively small value of k_1. Also, the solvent deuterium oxide kinetic isotope effect k_1^H/k_1^D has been consistently greater than unity, indicating proton transfer, but has never been greater than about 2.

B. FORMATION OF COMPLEX IONS

Now let us consider another important group of frequently rapid reactions: metal complex formation reactions. The vast majority of rate studies of this type of reaction have been carried out in water, so we will be treating in general, the process of replacing one or more water molecules by a ligand in the first coordination sphere of a metal ion. Most of the experimental studies have been limited to alkali, alkaline earth, and divalent first row transition metals because of the considerable tendency of tri- and tetravalent metal ions to hydrolyze and polymerize. The attendant mechanistic complications by no means entirely preclude kinetic studies of these latter ions and considerable data is already available on a few important ones such as Fe(III), Al(III), and La(III).

Let us begin, however, by treating the case of divalent metals such as Mn^{++} and Mg^{++} which typically form complex ions with divalent anions in aqueous solution in just two stages. This mechanism, which is based primarily upon ultrasonic absorption relaxation spectra of 2–2 electrolytes, is schematically

$$M_{aq}^{++} + L_{aq}^{--} \underset{}{\overset{k_{12}}{\rightleftharpoons}} [M^{++}OH_2L^{--}]_{aq} \underset{}{\overset{k_{23}}{\rightleftharpoons}} ML_{aq} + H_2O. \qquad (4.2)$$

The step $1 \rightarrow 2$ is a diffusion controlled formation of an outer sphere complex or ion pair. (The outer or second coordination sphere is sometimes referred to as the contact solvation shell.) The subsequent step $2 \rightarrow 3$ is the rate limiting one in which the ligand replaces a first coordination sphere water molecule. Actually, the step $1 \rightarrow 2$ may consist of two or more steps, as evidenced by multiple relaxation times, but because the equilibrium between 1 and 2 is established so rapidly in comparison to the reaction $2 \rightarrow 3$, it is possible to treat the process $1 \rightarrow 2$ as a single step. For instance, three discrete relaxation times in aqueous $CoSO_4$ solutions have been observed by sound absorption techniques, the fastest two of which are attributed to the two-step nature of the process $1 \rightarrow 2$ (Kor, 1966).

Frequently, for a particular metal ion, the specific rate of ligand penetration, k_{23}, is the same within a power of ten for all ligands.

Furthermore, k_{23} is generally very nearly the same as k^*, the specific rate for water exchange in the reaction

$$M(H_2O)_6^{n+} + H_2O^* \overset{k^*}{\rightleftharpoons} M(H_2O)_5(H_2O^*)^{n+} + H_2O, \qquad (4.3)$$

determined usually by an NMR line broadening technique (Swift and Connick, 1962).

Since the outer-sphere complex formation constant $K_{12} = k_{12}/k_{21}$ varies little with the choice of L, the overall formation specific rate $k_F = K_{12}k_{23}$ is fairly independent of L. It then follows that the observed large variations in stability constant for a given metal ion with different ligands is attributable primarily to wide variations in the dissociation rate constant k_{32}. Representative values of k_{23} are shown in Table IV.

TABLE IV

EXPERIMENTAL VALUES OF THE SPECIFIC RATE k_{23} IN EQ. (4.2) FOR
SEVERAL LIGANDS REACTING WITH Ni^{++} AND Co^{++} AT $25°$

Incoming ligand	k_{23}, sec^{-1}		Ref.
	Ni(II)	Co(II)	
H_2O	$2.7 \times 10^{4\,a}$	$1.13 \times 10^{6\,a}$	Swift and Connick (1962)
Glycine	0.9×10^4	2.6×10^5	Hammes and Steinfeld (1962)
Diglycine	1.2×10^4	2.6×10^5	Hammes and Steinfeld (1962)
Imidazole	1.6×10^4	4.4×10^5	Hammes and Steinfeld (1962)
SO_4^{2-}	1.5×10^4	2×10^5	Eigen and Tamm (1962)
$HP_2O_7^{3-}$	1.2×10^4	5.3×10^5	Hammes and Morrell (1964)
$HP_3O_{10}^{4-}$	1.2×10^4	7.2×10^5	Hammes and Morrell (1964)
L-cysteine	1×10^4	3×10^5	Davies et al. (1968)

[a] These are actually values of k^* in Eq. (4.3).

Eigen and Wilkins (1965) have provided a particularly extensive compilation of k_{23} values for a variety of metal ions. In order to deduce values of k_{23} from measured values of k_F, it is necessary either to measure K_{12} or to estimate K_{12} in a manner outlined by Hammes and Steinfeld (1962) based on earlier theoretical treatments of the equilibrium between ion pairs and free ions (Fuoss, 1958). The appropriate equations are

$$K_{12} = (4\pi Na^3/3000) \exp[-U(a)/kT], \qquad (4.4)$$

$$U(a) = \frac{Z_1 Z_2 e^2}{a\epsilon} - \frac{Z_1 Z_2 e^2}{\epsilon(1 + \kappa a)}. \qquad (4.5)$$

$$\kappa^2 \equiv \frac{8\pi Ne^2}{1000\epsilon kT} \mu, \qquad (4.6)$$

where N is Avogadro's number, a is the distance of closest approach of the ion-pair partners, k is Boltzmann's constant, T is the absolute temperature, ϵ is the dielectric constant, Z_i is the charge on the ith ion, e is the charge on an electron, and μ is ionic strength. The usual choice for a would be 5–7.5 Å and for ϵ would be the dielectric constant of the bulk solvent. Since Eq. (4.4) is predicted from the Debye–Hückel theory of strong electrolytes, we cannot reasonably expect it to be valid for high ionic strengths or large valences of the entering ligands or both. Thus, for example, in treating the kinetics of cobalt(II) and nickel(II) complex formation with trivalent pyrophosphate and tetravalent tripolyphosphate, Hammes and Morrell (1964) deemed it more logical to deduce values of K_{12} from their measured values of k_F and the NMR values of k_{23}, i.e., $K_{12} = k_F/k_{23}$, since the constancy of k_{23} had been previously verified for a wide variety of ligands complexed with either nickel(II) or cobalt(II).

An interesting illustration of the use of the temperature jump relaxation method to determine k_{23}, k_{32}, and K_{12} in reaction (4.2) has been provided by Brintzinger and Hammes (1966). The particular complex they studied was that between nickel(II) and the dianion of methyl phosphate,

in water.

A schematic of a joule heating temperature jump apparatus utilizing spectrophotometric detection is shown in Fig. 5. The charge on a $\sim 0.1\ \mu F$ 30 kV capacitor C is permitted suddenly to pass through the sample cell S to ground by triggering the spark gap G. A resulting $\sim 10°$ rise in temperature of the ~ 1 ml sample solution volume between two metal electrodes to a final temperature of $25°$ will occur in $\sim 10\ \mu sec$ if the electrical resistance of the cell has been reduced to $100\ \Omega$ or less by added

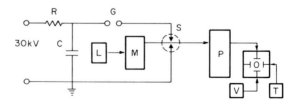

FIG. 5. Schematic of a temperature jump apparatus. R, resistor; C, capacitor; G, spark gap; S, sample cell; L, light source; M, monochromator; P, photomultiplier; O, oscilloscope; V, potentiometer; T, time base generator.

salt. In the nickel-methyl phosphate experiments, all the sample solutions were 0.1 M in NaCl. A constant ionic strength has the added advantage that ionic activity coefficients γ_i are constant for the set of experiments, and it is not necessary to include partial derivatives of the type $\partial \ln \gamma_i / \partial \ln c_i$ in expressions for τ as one must do in pressure jump relaxation-method studies (Hoffmann et al., 1966) that typically cover a wide range of ionic strengths.

Since the visible spectra of nickel(II), methyl phosphate, or the complex do not undergo marked changes with a $10°$ rise in temperature, it is convenient to add an acid-base indicator to the sample solutions. Brintzinger and Hammes made their solutions $2 \times 10^{-5} M$ in chlorophenol red, adjusted all solutions to pH 6.8, and then measured their relaxations at a wavelength of 573 mμ. Their relaxation data are shown in Table V.

TABLE V

EXPERIMENTAL RELAXATION TIMES AND CONCENTRATIONS IN A
TEMPERATURE JUMP RELAXATION-METHOD RATE STUDY OF THE
COMPLEXING OF NICKEL(II) BY METHYL PHOSPHATE[a]

τ (10^{-6} sec)	Total Ni(II) concentration ($10^{-2} M$)	Total methyl phosphate concentration ($10^{-2} M$)
85	10.0	10.0
95	4.00	4.00
100	1.60	1.60
120	0.64	0.64

[a] Brintzinger and Hammes (1966).

The coupled equilibria present in the sample solutions are the protolytic reaction of the chlorophenol red indicator

$$HIn^- \rightleftharpoons H^+ + In^{2-}, \tag{4.7}$$

the acid dissociation of the methyl phosphate

$$HL^- \rightleftharpoons H^+ + L^{--} \tag{4.8}$$

in this pH range, and the complexing reaction of primary interest

$$Ni(H_2O)_6^{++} + L^{--} \underset{k_{21}}{\overset{k_{12}}{\rightleftharpoons}} Ni(H_2O)_6L \underset{k_{32}}{\overset{k_{23}}{\rightleftharpoons}} Ni(H_2O)_{6-\chi}L + \chi H_2O, \tag{4.9}$$

where Brintzinger and Hammes have assumed that it is only the divalent anion of methyl phosphate that reacts with Ni^{++}. The equilibria of

Eqs. (4.7) and (4.8), as well as the formation of $Ni(H_2O)_6L$ are achieved very rapidly compared to the final equilibrium involving formation of $Ni(H_2O)_{6-x}L$. Using brackets to denote equilibrium concentrations, dropping ionic charges to simplify our notation, and denoting Ni^{++} by M, we may write the equilibrium constants for these three rapidly achieved equilibria as

$$K_I = \frac{[HIn]}{[H][In]} = 10^{6.2} \ M^{-1},$$

$$K_A = \frac{[HL]}{[H][L]} = 10^{6.2} \ M^{-1},$$

$$K_{12} = \frac{[ML_0]}{[M][L]}, \tag{4.10}$$

where the numerical values are for chlorophenol red and methyl phosphate, respectively.

In most temperature jump relaxation-method kinetic treatments of reaction (4.9) involving ligands other than methyl phosphate it is possible to make two simplifying assumptions: The concentration of the outer-sphere complex $Ni(H_2O)_6L$, denoted by $[ML_0]$, is negligibly small compared to the concentration of the inner-sphere complex $Ni(H_2O)_{6-x}L$, denoted by $[ML_i]$, and the rate of formation of the outer-sphere complex is very rapid compared to the loss of water molecules from the inner coordination sphere of the nickel(II) ion. Where both conditions apply, the appropriate expression for the relaxation time is Eq. (4.11) where α is a small factor accounting for effects

$$\tau^{-1} = K_{12}k_{23}\{[M]/(1 + \alpha) + [L]\} + k_{32} \tag{4.11}$$

of the rapid protolytic equilibria of the ligand and acid-base indicator (Hammes and Steinfeld, 1962). Equation (4.11) does not fit the data of Table V, and to achieve a fit it is necessary to drop one of the two foregoing assumptions. It is assumed instead that the concentration of ML_0 is significant compared to that of ML_i. On the other hand, the assumption that the formation of the outer-sphere complex or ion pair is rapid compared to the expulsion of inner-sphere water molecules is retained.

We find that we have the following six conservation relations, the first three of which follow from the equilibrium expressions (4.10):

$$K_{12}[M] \ \delta L + K_{12}[L] \ \delta M = \delta ML_0, \tag{4.12}$$

$$K_A[H] \ \delta L + K_A[L] \ \delta H = \delta HL, \tag{4.13}$$

$$K_I[H] \ \delta In + K_I[In] \ \delta H = -\delta In, \tag{4.14}$$

$$\delta L + \delta HL + \delta ML_o + \delta ML_i = 0, \tag{4.15}$$

$$\delta M + \delta ML_o + \delta ML_i = 0, \tag{4.16}$$

$$\delta H - \delta L - \delta In - \delta ML_i = 0. \tag{4.17}$$

Equations (4.15)–(4.17) are analogous to Eq. (3.9). Equation (4.14) follows from the additional conservation relation $\delta HIn = -\delta In$ as well as from the first of Eqs. (4.10). From Eqs. (4.12) to (4.14), we have

$$\delta L = (\delta ML_o - K_{12}[L]\,\delta M)/K_{12}[M], \tag{4.18}$$

$$\delta HL = K_A[H]\,\delta L + K_A[L]\,\delta H, \tag{4.19}$$

and

$$\delta In = -K_I[In]\,\delta H/(1 + K_I[H]), \tag{4.20}$$

respectively. Utilizing Eqs. (4.16), (4.18), and (4.19), we alter Eq. (4.15) to read

$$\frac{\{\delta ML_o + K_{12}[L](\delta ML_o + \delta ML_i)\}(1 + K_A[H])}{K_{12}[M]}$$
$$+ K_A[L]\,\delta H + \delta ML_o + \delta ML_i = 0. \tag{4.21}$$

Similarly, we may transform Eq. (4.17) with Eqs. (4.16), (4.18), and (4.20) to read

$$\delta H = \frac{\delta ML_i + (\{\delta ML_o + K_{12}[L](\delta ML_o + \delta ML_i)\}/K_{12}[M])}{1 + K_I[In]/\{1 + K_I[H]\}}. \tag{4.22}$$

The rate expression of interest, analogous to Eq. (3.7) above, is

$$d\,\delta ML_i/dt = k_{23}\,\delta ML_o - k_{32}\,\delta ML_i. \tag{4.23}$$

Clearly our objective is to find an expression for δML_o in terms of some function of δML_i, independent of δM, δH, δL, δHL, δIn, and δHIn, that we may use in reducing Eq. (4.23) to the form

$$d\,\delta ML_i/dt = \text{const} \cdot \delta ML_i \tag{4.24}$$

where const $= \tau^{-1}$. Equation (4.21) is almost the desired relationship between δML_o and δML_i. All we must do is replace δH in Eq. (4.21) by Eq. (4.22). After considerable manipulation of this new version of Eq. (4.21) but with the introduction of no additional relationships, we finally obtain

$$\delta ML_o = -\left[1 + \frac{1 + \dfrac{1}{K_A[H]} + \dfrac{[L]/[H] - K_{12}[L][M]/[H]}{1 + (K_I[In]/\{1 + K_I[H]\})}}{\dfrac{K_{12}[M]}{K_A[H]} + \dfrac{K_{12}[L]}{K_A[H]} + K_{12}[L] + \dfrac{K_{12}[L][M]/[H] + K_{12}[L]^2/[H]}{1 + (K_I[In]/\{1 + K_I[H]\})}}\right]^{-1} \delta ML_i \tag{4.25}$$

Now let us define

$$\alpha' \equiv \frac{\dfrac{[L]}{[H]}(1 - K_{12}[M])}{1 + \dfrac{K_{\mathrm{I}}[In]}{1 + K_{\mathrm{I}}[H]}} + \frac{1}{K_a[H]} \cong \frac{[L]/[H]}{1 + \dfrac{K_{\mathrm{I}}[In]}{1 + K_{\mathrm{I}}[H]}} + \frac{1}{K_a[H]} \quad (4.26)$$

where the approximation $1 \gg K_{12}[M]$ is presumed valid for the experimental conditions selected by Brintzinger and Hammes. Combining Eqs. (4.23), (4.25), and (4.26), we then obtain their expression for the relaxation time

$$\tau^{-1} = k_{32} + \frac{k_{23}}{1 + \dfrac{1 + \alpha'}{K_{12}[L] + \alpha' K_{12}([M] + [L])}}. \quad (4.27)$$

Because $\alpha' \gg 1$ under the conditions of their experiments, a further simplification of Eq. (4.27) was possible:

$$\tau^{-1} \cong k_{32} + \frac{k_{23}}{1 + \dfrac{1}{K_{12}([M] + [L])}}. \quad (4.28)$$

Quite typically in such a relaxation-method kinetic study, we are interested in manipulating our expression for τ in such a way as to make feasible a slope-intercept straight line plot of the experimental data. In this particular case the desired rearrangement of Eq. (4.28) is

$$\{1 + K([M] + [L])\}^{-1} = \tau k_{32}(K_{32} + 1) - K_{32} \quad (4.29)$$

where the overall formation constant

$$K = \frac{[ML_o] + [ML_i]}{[M][L]} = K_{12}(1 + K_{23}) = 81.3 \ M^{-1} \quad (4.30)$$

had been determined previously (Brintzinger, 1965). A plot of the data of Table V in terms of Eq. (4.29) yields an intercept $-K_{32}$ of -1, whence $K_{32} = 1$, and a slope $k_{32}(K_{32} + 1)$ of $\sim 1.4 \times 10^4$ from which it follows that $k_{23} = k_{32} = 7 \times 10^3 \ sec^{-1}$ with an experimental uncertainty of about $\pm 30\%$. We should add parenthetically that in relaxation-method rate measurements, the rate constants are seldom determined with a better than $\pm 10\%$ uncertainty. Finally, from $K_{23} = 1$ and Eq. (4.30), it follows that $K_{12} = 41 \ M^{-1}$.

The above derivation of τ is fairly typical of the approximate treatments encountered in relaxation-method rate studies. More elegant mathematical methods have been provided (Castellan, 1963; Kustin et al., 1965), but the novice may find more helpful a few other evaluations of τ

(Finholt, 1968; Swinehart, 1967; Eyring, 1964) for other chemical equilibria carried out in the same unsophisticated fashion as shown above.

The $k_{23} = 7 \times 10^3$ sec^{-1} found by Brintzinger and Hammes for the nickel(II)-methylphosphate complex is very like that found for other nickel(II) complexes, as well as the specific rate of first coordination-sphere water loss found by NMR spectroscopy (see Table IV). Thus, this system is just another good illustration of the principle that inner-sphere water expulsion is the rate-determining step in the formation of many metal complex ions. The nice feature of this particular rate study is the direct determination of the equilibrium constant K_{12}. Brintzinger and Hammes concluded from this work and other rate studies of nickel(II) complexed by trivalent and tetravalent phosphate anions that the ratio $[ML_i]/[ML_o]$ increases with increasing charge on the ligand.

The preceding discussion of metal complex formation requires some minor amendments. First of all, it is by no means universally true that expulsion of an inner-sphere water molecule is rate determining for complex ion formation. For example, the *trans* to *cis* isomerization of the diaquobisoxalatochromate (III) anion, $Cr(C_2O_4)_2(H_2O)_2{}^-$, proceeds much more rapidly (Aggett et al., 1968; Harris and Kelm, 1965) than the exchange of inner-sphere water molecules (Plane and Taube, 1952) though both types of reaction occur too slowly to qualify as fast in the present context. Also, the rate of expulsion of an inner-sphere water molecule is dependent on the nature of the other ligands in the first coordination sphere. Some ligands have a more pronounced loosening effect than others on first coordination-sphere waters (Hammes and Steinfeld, 1962; Margerum and Rosen, 1967). Davies et al. (1968) found, for example, that L-cysteine in the inner coordination sphere of Ni(II) and Co(II) causes a greater increase in the rate of loss of remaining water molecules than do the ligands glycylglycine and glycylsarcosine. In general, the binding of a —COO$^-$ functional group should reduce the positive charge density on the central metal ion, thus loosening the binding to the remaining inner-sphere water molecules.

Finally, in spite of the fact that a host of different ligands react with Ni(II) or with Co(II) at essentially constant rates, there are other bulkier ligands for which the rate-controlling process is not release of water but rather a sterically hindered formation of a metal chelate system. β-Alanine or β-aminobutyric acid reacting with Co(II) and L-carnosine reacting with Cu(II) are representative of this latter category (Pasternack and Kustin, 1968).

We can make one interesting extrapolation to organic chemistry of the preceding discussion of kinetics of metal complexing in aqueous solutions. Hogen–Esch and Smid (1966a,b) have identified spectroscopically contact and solvent-separated ion pairs in, for example, tetrahydrofuran (THF) solutions of Li^+ complexed by fluorenyl carbanion

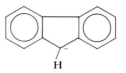

denoted hereafter by F^-. The equilibrium can be written in their notation as

$$Li^+ + F^- \underset{k_{12}}{\overset{}{\rightleftharpoons}} \underset{\substack{\text{solvent}\\\text{separated}\\\text{ion pair}}}{Li^+||F^-} \underset{k_{23}}{\overset{}{\rightleftharpoons}} \underset{\substack{\text{contact}\\\text{ion pair}}}{Li^+F^-} \underset{k_{34}}{\overset{}{\rightleftharpoons}} LiF \qquad (4.31)$$

which, of course, bears a striking resemblance to (4.2). The patron saint honored for first visualizing (4.31) for the benefit of the physical organic chemists is Winstein (Winstein et al., 1954). On the other hand, the sound absorption kineticists cite Eigen and Tamm (1962) for the quantitative treatment of (4.2) in the case of aqueous divalent metal sulfates.

Since Hogen–Esch and Smid (1966b) have supplied a complete set of equilibrium constants for (4.31) in THF, it is easy to predict relaxation times and rate constants for (4.31) based on the fact that rate constants for inner coordination-sphere solvent exchange in nonaqueous solvents such as methanol and dimethyl sulfoxide (Eigen and Wilkins, 1965; Thomas and Reynolds, 1966) are very like those found in water. Thus we assume k_{23} in (4.31) is like that for the reaction of $EDTA^{4-}$ with Li^+ in water at 25°, i.e., 5×10^7 sec^{-1} (Eigen and Maass, 1966). Neglecting the equilibrium between the states 3 and 4 and assuming the equilibrium between states 1 and 2 is achieved rapidly compared to that between states 2 and 3, we have the relaxation expression

$$\tau^{-1} = \cfrac{k_{23}}{\left[\cfrac{1}{K_{12}([F^-] + [Li^+])} + 1\right]} + k_{32}. \qquad (4.32)$$

The reciprocal of the bracketed denominator can be calculated from the equilibrium constants of Hogen–Esch and Smid (1966b) to be ~ 0.98, virtually independent of total fluorenyl lithium concentration. Thus, since $K_{23} = 2.9^{-1}$ and $k_{32} = k_{23} K_{23}^{-1} \cong 5 \times 10^7 \times 2.9 = 1.5 \times 10^8$ sec^{-1}, it follows that $\tau^{-1} \cong 2 \times 10^8$ sec^{-1} or $\tau \cong 5.3$ nsec. Such a relaxation may be measurable by sound absorption techniques, but an unequivocal assignment of a measured τ to (4.31) would be difficult. A recent ultrasonic study (Parker et al., 1966) of the helix-coil transition in aqueous poly-L-lysine illustrates the difficulty of assigning τ's measured by this technique. A more conclusive relaxation-method determination of τ for (4.31) utilizing the absorption of light by the contact ion pair at 349 mμ will probably await the development of subnanosecond rise time temperature or electric field jumps, possibly using a simultaneously Q-switched and mode locked laser (Heynau and Foster, 1968).

C. METAL ION HYDROLYSIS AND POLYMERIZATION

The hydrolysis reaction

$$Cr(H_2O)_6^{3+} \xrightarrow{k_1} Cr(H_2O)_5OH^{2+} + H^+ \tag{4.33}$$

bears at least a superficial resemblance to the expulsion of an inner-sphere water molecule

$$Cr(H_2O)_6^{3+} \xrightarrow{k_1^*} Cr(H_2O)_5^{3+} + H_2O \tag{4.34}$$

that we have considered above, yet the specific rates are very different. In the first case $k_1 = 1.4 \times 10^5$ sec^{-1} (Rich et al., 1969), whereas in the second $k_1^* = 3 \times 10^{-6}$ sec^{-1} (Hunt and Plane, 1954) at $\sim 25°$. Since the hydrolysis reaction involves the separation of repulsive charges as well as the migration of a proton that is very mobile in water, a marked difference between k_1 and k_1^* is to be expected. Monomeric hydrolyses of metal ions are generally quite fast as shown by the entries for UO_2^{2+}, Al^{3+}, and Th^{4+} in Table II.

These rapid hydrolyses are interesting for two reasons. First, in many cases they complicate the rate studies of complex ion formation which we have discussed previously. For instance, Miceli and Stuehr (1968) have found that the only reaction scheme that gives a quantitative fit of their relaxation data for sulfate complexing Al(III), Ga(III), and In(III) is

$$
\begin{array}{ccc}
Me^{3+} & + SO_4^{2-} \underset{k_B}{\overset{k_F}{\rightleftharpoons}} & MeSO_4^{+} \\
\Updownarrow & & \Updownarrow \\
MeOH^{2+} & + SO_4^{2-} \underset{k_B'}{\overset{k_F'}{\rightleftharpoons}} & HOMeSO_4 \\
+ & & + \\
H^+ & & H^+
\end{array}
\tag{4.35}
$$

where the two vertical equilibria are achieved rapidly compared to the horizontal equilibria. Another very rapid equilibrium, $H^+ + SO_4^{2-} \rightleftharpoons HSO_4^-$, is omitted from (4.35) for the sake of clarity. The equilibria of Eq. (4.35) are characterized by a single long relaxation time measurable by the temperature jump or pressure jump techniques. The derivation of an expression for τ has been given in some detail both by Cavasino (1968) and by Strehlow (1962). Miceli and Stuehr concluded that $k_F = 1.3 \times 10^3$, 2.8×10^4, and $4.8 \times 10^5 M^{-1} \sec^{-1}$ for Al(III), Ga(III), and In(III) whereas the corresponding values of k_F' were 5.8×10^5, 9.8×10^4, and $6.8 \times 10^6 M^{-1} \sec^{-1}$, respectively, for the monohydroxy ions. In other words, where metal ions have a high charge density and hydrolyze significantly, the monohydroxy product complexes much more rapidly than the unhydrolyzed metal ion.

Miceli and Stuehr (1968) also estimated values of $k_{23} = 1$, 30, and 500 $\sec^{-1}$, respectively, for Al(III), Ga(III), and In(III). Fiat and Connick (1968) had found with O^{17} NMR that the lifetimes of first coordination-sphere water molecules in fully hydrated Al(III) and Ga(III) are 7.5 and 5.5×10^{-4} sec, respectively.† These would correspond to $k^* \cong 1$ and $10^4 \sec^{-1}$, respectively. Clearly, the identity of k_{23} and k^* for Al(III) indicates an S_N1 mechanism for complexing of Al(III). On the other hand, in the case of Ga(III), the activation parameter data and the fact that $k^* \gg k_{23}$ are consistent with an S_N2 mechanism in which the coordination number increases in going from reactants to activated complex.

The other interesting problem to which the metal ion hydrolysis reactions of Table II are relevant is that of polymerization of inorganic hydroxy ions. The equilibrium constants

$$*K_1 = \frac{k_1}{k_{-1}} = \frac{[MeOH^{(n-1)+}][H^+]}{[Me^{n+}]}, \tag{4.36}$$

$$*K_2 = \frac{k_2}{k_{-2}} = \frac{[Me(OH)_2^{(n-2)+}][H^+]}{[MeOH^{(n-1)+}]}, \tag{4.37}$$

$$^+K_{22} = \frac{k_3}{k_{-3}} = \frac{[Me_2(OH)_2^{2(n-1)+}]}{[MeOH^{(n-1)+}]^2}, \tag{4.38}$$

$$*\beta_{22} = {}^+K_{22}(*K_1)^2 = \frac{[Me_2(OH)_2^{2(n-1)+}][H^+]^2}{[Me^{n+}]^2}, \tag{4.39}$$

as well as $*K_3$, $*K_4$, etc., corresponding to higher states of monomeric hydrolysis and values of $*\beta_{43}$, $*\beta_{5,3}$, etc., for more extensive polymerization have been reported for many metals (Sillen and Martell, 1964). Dimerizations corresponding to (4.38) have much longer half lives or relaxation times than do the protolytic reactions corresponding to (4.36) and (4.37) at comparable reactant concentrations. We would predict this

† Conversion of the NMR τ's to rate constants requires the following definitions: τ is the lifetime of a given H_2O molecule; $\tau/6$ is the lifetime of a given aquated ion, assuming a coordination number of six and $\tau/6 = k^{*-1}$ where k^* is the rate constant for exchange. Thus $k^* = 6/\tau = (6/7.5) \cong 1 \sec^{-1}$ in the case of Al(III).

relationship even if dimerization reactions were diffusion controlled because of coulombic repulsions between the dimerizing monomers as well as low mobilities of monomer compared to a proton. More explicitly, the rate constants at $\sim 25°$ for the overall dimerization reactions

$$2\,UO_2^{2+} + 2\,H_2O \rightarrow (UO_2)_2(OH)_2^{2+} + 2\,H^+, \qquad \mu = 0.5\,M, \qquad (4.40)$$

$$(VO_2)(OH)_2^- + (VO_2)(OH)_3^{2-} \rightarrow (VO_2)_2(OH)_5^{3-}, \qquad \mu = 0.5\,M, \qquad (4.41)$$

$$2\,HCrO_4^- \rightarrow Cr_2O_7^{2-} + H_2O, \qquad \mu = 0.1\,M, \qquad (4.42)$$

$$2\,ScOH^{2+} \rightarrow Sc_2(OH)_2^{4+}, \qquad \mu = 0.1\,M, \qquad (4.43)$$

are 116, 3.1×10^4, 1.8, and $1.3 \times 10^7\,M^{-1}\,sec^{-1}$, respectively (Whittaker *et al.*, 1965, 1966; Hurwitz and Atkinson, 1967; Swinehart and Castellan, 1964; Cole *et al.*, 1969). All four of these rate constants are markedly lower than those for the hydrolysis of UO_2^{2+}, Al^{3+}, Th^{4+}, and Cr^{3+} [see Table II and also Eq. (4.33)].

Overall rate constants for dimerization are not particularly satisfactory since they do not reveal the answer to the chemically interesting question: What is the nature of the rate-determining step? Glemser and Holtje (1966) tried to answer this question for the more complicated polymerization

$$7\,MoO_4^{2-} + 8\,H^+ \rightleftharpoons H_8(MoO_4)_7^{6-} \qquad (4.44)$$

in aqueous solutions of sodium molybdate, $Na_2MoO_4 \cdot 2\,H_2O$. From the slope of plots of $\log \tau^{-1} = n\,\log[H^+]$ and $\log \tau^{-1} = n\,\log[MoO_4^{2-}]$ they concluded that the rate-determining step is in this case

$$4\,MoO_4^{2-} + 6\,H^+ \rightleftharpoons Mo_4O_{10}(OH)_6^{2-} \qquad (4.45)$$

which, of course, is still not a single discrete step but only a slightly simpler overall reaction than (4.44). However, their work does foreshadow the identification of discrete relaxation steps in polymerization processes, assuming that the present $\sim 10\%$ precision in temperature and pressure jump relaxation times can be dramatically improved so that unequivocal values of the slope n can be obtained.

Eigen and Wilkins (1965) noted that the rate constant for the reaction

$$2\,FeOH^{2+} \xrightarrow{\ k_3\ } Fe_2(OH)_2^{4+} \qquad (4.46)$$

is $4.5 \times 10^2\,M^{-1}\,sec^{-1}$ at $25°$ (Wendt, 1962), which closely resembles the water exchange value for $FeOH^{2+}$ of $\sim 5 \times 10^2\,sec^{-1}$ (Seewald and Sutin, 1963). Thus the loss of inner coordination-sphere water from $MeOH^{(n-1)+}$ may be the rate-determining step in the dimerization and further polymerization of all such hydrolyzed metal ions. The specific rate k_3 is trivially easy to determine by the temperature jump method if

$^+K_{22}$ is known, but the water exchange rate for $Me(OH)^{(n-1)+}$ is still unknown for most cases and is substantially more difficult to determine.*

The relaxation kinetic methods can make an important contribution to the understanding of the equilibrium properties of aqueous solutions of hydrolyzed and polymerized metal ions. From the number of observed relaxation times, it is possible to deduce the number of chemical equilibria in the sample system, and from the magnitudes of the relaxation times one can further deduce the nature of the reactant species. Thus, for example, since in dilute, acidic, aqueous scandium(III) perchlorate only one relaxation is observed on a microsecond time scale (Cole et al., 1969), it is probable that only one protolytic reaction of the type $Me(OH)_m^{(n-m)+} \rightleftharpoons Me(OH)_{m+1}^{(n-m-1)+} + H^+$ occurs in this system.

Since we noted earlier in connection with Eqs. (3.26) through (3.28) that the charge of $MeOH^{n-1}$ seems to be the most important factor influencing the magnitude of the specific rate of the reaction $MeOH^{n-1} + H^+ \rightarrow Me^{n+} + H_2O$, one might well ask why k_{-1} for $CrOH^{2+}$ in Eq. (4.33) is only $\sim 7.8 \times 10^8 \ M^{-1} \ sec^{-1}$, whereas in the case of $AlOH^{2+}$ the corresponding specific rate is six times as large (Table II).

An explanation might involve either differences in dielectric constant near $CrOH^{2+}$ compared to $AlOH^{2+}$ or differences in the distances to which solvent water molecules are highly structured away from these two ions. Differences in both properties for these two ions would arise from differences in ionic radii (for Al^{3+}, $r = 0.50$ Å and for Cr^{3+}, $r = 0.69$ Å; Pauling, 1960) that in turn yield differences in electrostatic potential gradient near the surfaces of these isovalent ions.

Let us first attempt a rationale for the differences in k_{-1} based on dielectric constant. Noyes (1962) calculated an effective dielectric constant $\epsilon_{eff} = 1.583$ near Al^{3+} and $\epsilon_{eff} = 1.919$ near Cr^{3+}. Assuming that values of ϵ_{eff} for $AlOH^{2+}$ and $CrOH^{2+}$ are directly proportional to those for the corresponding M^{3+} ions, we conclude that the approaching proton would be repelled by $AlOH^{2+}$ over a greater distance than by $CrOH^{2+}$. Were this effect important, the measured k_{-1} for aqueous chromium(III) should have been larger than that for aluminum(III).

Now let us assume instead that a difference in structure-making of the solvent water by the MOH^{2+} ions is the dominant effect. The presumed higher charge density of $AlOH^{2+}$ than $CrOH^{2+}$ would give rise to a more highly ordered, ice-like structure of water molecules near $AlOH^{2+}$ than near $CrOH^{2+}$ (Kavanau, 1964). It is well known that protons have a markedly higher mobility in ice than in liquid water (Eigen and DeMaeyer, 1958). Thus the approach of H^+ to $AlOH^{2+}$ should be more

---

* Note added in proof: Wendt (1969) verified water loss is rate determining in the dimerization of $FeOH^{2+}$ and $VOOH^+$.

rapid than to $CrOH^{2+}$. This is consistent with the experimental k_{-1} values.

The experimental data therefore suggest that structure-making of solvent by ions of high charge density is more important for the kinetics of the reaction $H^+ + MOH^{2+} \rightarrow$ than is ϵ_{eff}. However, this argument overlooks a possibly important factor: Al^{3+} has an inert gas electronic configuration whereas incompletely filled d orbitals may play a role in the kinetics of aqueous chromium(III).

D. PSEUDO OR CARBON ACID DISSOCIATIONS

We will conclude our discussion of "somewhat slower" fast ionic reactions by treating the aqueous solution reactions of so-called pseudo acids, those in which a proton is removed directly or indirectly from a carbon atom. In the case of aqueous nitroparaffins such as nitroethane, $CH_3CH_2NO_2$, the loss of a proton from the tautomeric forms CH_3HCH—NO_2 and CH_3HC=NO_2H is slow enough for study by conventional kinetic techniques (Turnbull and Maron, 1943). On the other hand, acetylacetone (2,4-pentanedione) is a pseudo acid for which fairly rapid proton loss occurs through keto-enol tautomerization (Eigen et al., 1964).

Following Eigen et al. (1965), let us denote the keto form of a tautomeric compound in aqueous solution by KH, the enolic form by EH, and the enolate anion by E^-. The total acidity of a tautomeric compound in aqueous solution, in which KH is in equilibrium with EH, gives no information regarding the position of the tautomeric equilibrium. Since the product of dissociation E^- from either KH or EH is the same, the total dissociation constant K is given by a combination of those for the keto and enol forms:

$$\frac{1}{K} = \frac{1}{K_{KH}} + \frac{1}{K_{EH}} \qquad (4.47)$$

where, omitting charges to simplify our notation,

$$K = \frac{[H][E]}{[KH] + [EH]}, \qquad K_{KH} = \frac{[H][E]}{[KH]}, \qquad K_{EH} = \frac{[H][E]}{[EH]}. \qquad (4.48)$$

From these definitions, it follows that the ratio of the concentrations of enol to keto forms is

$$\frac{[EH]}{[KH]} = \frac{K_{KH}}{K_{EH}}. \qquad (4.49)$$

Thus, the keto form is the stronger acid if more of the compound exists in the enol form and vice versa.

If the tautomerization occurs so slowly that KH and EH can be prepared and studied separately, K_{KH} and K_{EH} can be directly determined by glass electrode pH measurements. Another approach is made possible by differing susceptibilities of the tautomeric forms to substitution reactions: In bromine titrations, the enol form reacts very rapidly with bromine whereas the keto form reacts only very slowly (Schwarzenbach and Felder, 1944). Infrared, ultraviolet, and NMR spectroscopic measurements also aid in the elucidation of these tautomeric systems. We will focus our attention, however, on the unique contribution made by relaxation techniques in those cases where one of the tautomeric forms is present only in very low concentration or the rate of proton transfer is faster than the titration reaction.

Since tautomerization in aqueous solution occurs via the enolate ion, the simplest mechanism to assume is that in which E^- recombines with the hydrated proton to yield either KH or EH:

$$KH \rightleftharpoons E^- + H^+ \rightleftharpoons EH. \qquad (4.50)$$

However, such a mechanism does not completely describe the system since account must be taken of base catalysis through enolation in forming the ketone. Thus, we have instead

$$
\begin{array}{ccc}
1 & & 2 \\
E^- + H^+ + E^- & \underset{k_{21}}{\overset{k_{12}}{\rightleftharpoons}} & EH + E^- \\
& k_{31} \Big\backslash \, k_{13} \qquad k_{32} \, \Big/\!\!\Big/ \, k_{23} & \\
& E^- + KH & \\
& 3 &
\end{array}
\qquad (4.51)
$$

Even this system, characterized by two relaxation times, does not fully explain results obtained with diacetylacetone, $CH_3COCH_2COCH_2COCH_3$ (Stuehr, 1967), but (4.51) will suffice for our present purposes. The shorter of the two relaxation times, τ_1, is associated with the diffusion controlled reaction between states 1 and 2. The longer relaxation time τ_2 is a function of the specific rates of the two slower steps between states 1 and 3, and 2 and 3, respectively, as well as the equilibrium parameters of the interaction between states 1 and 2. Assuming that τ_1 is in the microsecond time range and τ_2 is in the millisecond time range, it can be shown (Eigen et al., 1965; Stuehr, 1967) that

$$\tau_1^{-1} = k_{21} + k_{12}([H] + [E]), \qquad (4.52)$$

$$\tau_2^{-1} = k_{31} + k_{13}([H] + [E]) \frac{1}{1 + ([H] + [E])/K_{EH}}$$

$$+ k_{32}[E] + k_{23}[E] \frac{1}{1 + K_{EH}/([H] + [E])}. \qquad (4.53)$$

Unless one makes a felicitous choice of a tautomeric system that absorbs in the visible as, for example, methylindandione, the determination of τ_2 in a temperature jump experiment will require the addition of a rapidly equilibrating acid-base indicator such as methyl red. This additional equilibrium will introduce into Eq. (4.53) a small, readily derived correction factor (Eigen *et al.*, 1965) that we can ignore in our discussion.

Previous to Eigen's relaxation-method study of aqueous barbituric acid (Eigen *et al.*, 1965), there had been some interesting misconceptions regarding the acidic properties of this pseudo acid. It had been suggested on the basis of a superficial structural similarity of barbituric acid

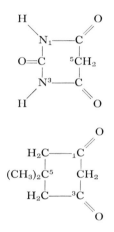

to dimedone

and to dihydroresorcinol that barbituric acid must also be nearly completely enolic (Schwarzenbach and Lutz, 1940). This was felt to be an adequate explanation for the comparatively high acidity of barbituric acid, $pK = 4.03$ at $25°$ (Biggs, 1956). The subsequent verification (Schwarzenbach and Felder, 1944) that dimedone is 95.3% monoenolic at $25°$ in 0.01 to 0.001 M aqueous solution with $pK = 5.25$ and $pK_{\mathrm{EH}} = 5.23$ appeared to fortify this view. The $—CH_2CO—$ group in barbituric acid was thought to enolize significantly in aqueous solution and that while tautomerism of the $—CONH—$ groups would make the enolate structure energetically more favorable, these latter groups produce no protons near pH 4 (Biggs, 1956; Arndt *et al.*, 1948; Fox and Shugar, 1952; Butler *et al.*, 1955; Albert and Phillips, 1956).

The relaxation time τ_1 expected in the microsecond time range for aqueous barbituric acid is not detectable with the electric field jump technique. The null result is sufficient proof that aqueous barbituric acid is almost entirely ketonic since dimedone and 5,5-diethylbarbituric acid (Veronal) both give readily detectable relaxations of the order of a microsecond by this same technique (Eigen *et al.*, 1965). Whether the recombination of a proton and the enolate ion of 5,5-diethylbarbituric acid occurs at an $R—O^-$ or an $R—N^-$ site cannot be determined since kinetic differences between oxygen and nitrogen acceptors are very small (Eigen, 1964).

Temperature jump experiments on aqueous barbituric acid with [E] $\gg$ [H], i.e., $C_0 \cong 10^{-2}\ M$ and pH = 5.3, yielded measurable relaxations in the millisecond time range. Equation (4.52) gives a good fit of Eigen's data plotted as τ_2^{-1} vs. [E] when the following rate constants are used: $k_{13} = 1.0 \times 10^5\ M^{-1}\ sec^{-1}$, $k_{31} = 10\ sec^{-1}$, $k_{23} = 7 \times 10^3\ M^{-1}\ sec^{-1}$, and $k_{32} = 91\ M^{-1}\ sec^{-1}$. The rate constant $k_{13} = 10^5\ M^{-1}\ sec^{-1}$, lying orders of magnitude below the diffusion controlled limit, indicates a relatively high activation energy barrier for reactions of CH acids, a finding that seems to be fairly general for CH acids (Bell, 1959). Evidently, hydrogen bridges form very slowly if at all in these cases, and ion recombination is further retarded when coupled with a charge displacement and structural change characteristic of tautomerization.

Finally, from the above rate constants it follows that for barbituric acid, $pK = 4.0$, $pK_{EH} = 2.1$, and $pK_{KH} = 4.0$, in striking contrast to the pK's for dimedone cited above that were confirmed in the same relaxation-method study (Eigen et al., 1965). The really important remark to make here is that just as in the case of metal ion hydrolysis considered earlier it was possible to deduce equilibrium constants for barbituric acid, previously inaccessible to the thermodynamicist, from relaxation kinetic data. It is precisely this capability of identifying new equilibria that is making relaxation techniques indispensable for the detailed understanding of enzyme catalyzed fast reactions (Eigen, 1968).

We have considered diffusion controlled ion recombinations, breaking of intramolecular hydrogen bonds, metal ion complexing, metal ion hydrolysis and polymerization, and reactions of pseudo acids primarily because it is now possible to generalize about these types of fast ionic reactions. For the sake of brevity we have omitted many pathologically complicated but nevertheless fairly well understood fast reactions that are the delight of the discriminating fast-reaction kineticist. Hopefully, the reader has felt rewarded but not particularly challenged by this introduction and will wish to acquaint himself with more complex fast reactions such as conformational changes in DNA (Crothers, 1964; Davison, 1966), polyamino acids (Hamori and Scheraga, 1967; Parker et al., 1968), and enzymes (French and Hammes, 1965).

ACKNOWLEDGMENT

This work was supported in part by Grant AM–06231 from the National Institute of Arthritis and Metabolic Diseases.

GENERAL REFERENCES

AMDUR, I., and HAMMES, G. G. (1966). "Chemical Kinetics: Principles and Selected Topics," Chapters 5 and 6. McGraw-Hill, New York.

CALDIN, E. F. (1964). "Fast Reactions in Solution." Wiley, New York.

CLAESSON, S., ed. (1967). "Fast Reactions and Primary Processes in Chemical Kinetics." Wiley (Interscience), New York.

CZERLINSKI, G. H. (1966). "Chemical Relaxation." Dekker, New York.

EIGEN, M., and WILKINS, R. G. (1965). In "Mechanisms of Inorganic Reactions" (Murmann, R. K., Fraser, R. T. M., and Bauman, J., eds.) pp. 55–80. Am. Chem. Soc., Washington, D.C.

FRIESS, S. L., LEWIS, E. S., and WEISSBERGER, A. (1963). "Technique of Organic Chemistry," Vol. VIII, Pt. II, pp. 703–1106. Wiley (Interscience), New York.

SPECIAL REFERENCES

AGGETT, J., MAWSTON, I., ODELL, A. L., and SMITH, B. E. (1968). *J. Chem. Soc. (A)*, p. 1413.

ALBERT, A., and PHILLIPS, J. N. (1956). *J. Chem. Soc.* p. 1294.

ALBERT, A., and SERJEANT, E. P. (1962). "Ionization Constants of Acids and Bases." Methuen, London.

ARNDT, F., LOEWE, L., and ERGENER, L. (1948). *Rev. Faculte' Sci. Univ. Istanbul Ser. A* **13**, 103.

BARKER, G. C., and SAMMON, D. C. (1967). *Nature* **213**, 65.

BASOLO, F., and PEARSON, R. G. (1967). "Mechanisms of Inorganic Reactions," 2nd ed. Wiley, New York.

BASS, L., and GREENHALGH, W. J. (1966). *Trans. Faraday Soc.* **62**, 715.

BELL, R. P. (1959). "The Proton in Chemistry," p. 161. Cornell Univ. Press, Ithaca, New York.

BENSON, S. W. (1960). "The Foundations of Chemical Kinetics." McGraw-Hill, New York.

BIGGS, A. I. (1956). *J. Chem. Soc.* p. 2485.

BRIERE, G., and GASPARD, F. (1967). *J. Chim. Phys.* **64**, 403.

BRINTZINGER, H. (1965). *Helv. Chim. Acta* **48**, 47.

BRINTZINGER, H., and HAMMES, G. G. (1966). *Inorg. Chem.* **5**, 1286.

BUTLER, T. C., RUTH, J. M., and TUCKER, G. F. (1955). *J. Am. Chem. Soc.* **77**, 1486.

CASTELLAN, G. W. (1963). *Ber. Bunsenges. Physik. Chem.* **67**, 898.

CAVASINO, F. P. (1968). *J. Phys. Chem.* **72**, 1378.

COLE, D. L., EYRING, E. M., RAMPTON, D. T., SILZARS, A., and JENSEN, R. P. (1967). *J. Phys. Chem.* **71**, 2771.

COLE, D. L., RICH, L. D., OWEN, J. D., and EYRING, E. M. (1969). *Inorg. Chem.* **8**, 682.

CROTHERS, D. M. (1964). *J. Mol. Biol.* **9**, 712.

CZERLINSKI, G. (1964a). *J. Theoret. Biol.* **7**, 435.

CZERLINSKI, G. (1964b). *J. Theoret. Biol.* **7**, 463.

DAVIES, G., KUSTIN, K., and PASTERNACK, R. F. (1968). *Trans. Faraday Soc.* **64**, 1006.

DAVISON, P. F. (1966). *J. Mol. Biol.* **22**, 97.

DEBYE, P. (1942). *Trans. Electrochem. Soc.* **82**, 265.

DeMAEYER, L., and KUSTIN, K. (1963). *Ann. Rev. Phys. Chem.* **14**, 5.

DIEBLER, H. (1960). *Z. Elektrochem.* **64**, 128.

DULOVA, V. I., and LECHKOVA, N. V. (1963). *Dokl. Akad. Nauk SSSR* **20**, 29.

EBERT, M., KEENE, J. P., SWALLOW, A. J., and BAXENDALE, J. H., eds. (1965). "Pulse Radiolysis." Academic Press, New York.

EIGEN, M. (1954a). *Z. Physik. Chem. (N. F.)* **1**, 176.

EIGEN, M. (1954b). *Discussions Faraday Soc.* **17**, 194.

EIGEN, M. (1964). *Angew. Chem. Intern. Ed. Engl.* **3**, 1.

EIGEN, M. (1968). *Quart. Rev. Biophys.* **1**, 3.

EIGEN, M., and DEMAEYER, L. (1955). *Z. Elektrochem.* **59**, 986.

EIGEN, M., and DEMAEYER, L. (1958). *Proc. Roy. Soc.* **A247**, 505.

EIGEN, M., and DEMAEYER, L. (1963). *In* "Technique of Organic Chemistry" (S. L. Friess, E. S. Lewis, and A. Weissberger, eds.), Vol. VIII, Pt. II, pp. 895–1054. Wiley (Interscience), New York.

EIGEN, M., and EYRING, E. M. (1962). *J. Am. Chem. Soc.* **84**, 3254.

EIGEN, M., and KUSTIN, K. (1960). *J. Am. Chem. Soc.* **82**, 5952.

EIGEN, M., and MAASS, G. (1966). *Z. Phys. Chem. (Frankfurt)* **49**, 163.

EIGEN, M., and TAMM, K. (1962). *Z. Elektrochem.* **66**, 93, 107.

EIGEN, M., and WILKINS, R. G. (1965). Mechanisms of inorganic reactions. *Advan. Chem. Ser.* **49**, 55–67.

EIGEN, M., HAMMES, G. G., and KUSTIN, K. (1960). *J. Am. Chem. Soc.* **82**, 3482.

EIGEN, M., KRUSE, W., MAASS, G., and DEMAEYER, L. (1964). *Progr. Reaction Kinetics* **2**, 287.

EIGEN, M., ILGENFRITZ, G., and KRUSE, W. (1965). *Chem. Ber.* **98**, 1623.

EMERSON, M. T., GRUNWALD, E., and KROMHOUT, R. A. (1960). *J. Chem. Phys.* **33**, 547.

ERTL, G., and GERISCHER, H. (1962). *Z. Elektrochem.* **66**, 560.

EYRING, E. M. (1964). *Surv. Progr. Chem.* **2**, 57–89.

EYRING, E. M., and COLE, D. L. (1967). *Nobel Symp. Fast Reactions and Primary Processes Chem. Kinetics, 5th, Stockholm, 1967* (S. Claesson, ed.), pp. 255–260. Wiley (Interscience), New York.

FIAT, D., and CONNICK, R. E. (1968). *J. Am. Chem. Soc.* **90**, 608.

FINHOLT, J. E. (1968). *J. Chem. Educ.* **45**, 394.

FOX, J. J., and SHUGAR, D. (1952). *Bull. Soc. Chim. Belges* **61**, 44.

FREEDMAN, H. H. (1961). *J. Am. Chem. Soc.* **83**, 2900.

FRENCH, T. C., and HAMMES, G. G. (1965). *J. Am. Chem. Soc.* **87**, 4669.

FRIEDMAN, H. L. (1966). *J. Phys. Chem.* **70**, 3931.

FROST, A. A., and PEARSON, R. G. (1961). "Kinetics and Mechanism," 2nd ed. Wiley, New York.

FUOSS, R. M. (1958). *J. Am. Chem. Soc.* **80**, 5059.

GLASSTONE, S., LAIDLER, K. J., and EYRING, H. (1941). "The Theory of Rate Processes." McGraw-Hill, New York.

GLEMSER, O., and HOLTJE, W. (1966). *Angew. Chem.* **78**, 756.

HALPERN, J., LEGARE, R. J., and LUMRY, R. (1963). *J. Am. Chem. Soc.* **85**, 680.

HAMMES, G. G., and MORRELL, M. L. (1964). *J. Am. Chem. Soc.* **86**, 1497.

HAMMES, G. G., and STEINFELD, J. I. (1962). *J. Am. Chem. Soc.* **84**, 4639.

HAMORI, E., and SCHERAGA, H. A. (1967). *J. Phys. Chem.* **71**, 4147.

HARRIS, G. M., and KELM, H. (1965). *In* M. Eigen and R. G. Wilkins, *Advan. Chem. Ser.* **49**, 55–67.

HASLAM, J. L., EYRING, E. M., EPSTEIN, W. W., CHRISTIANSEN, G. A., and MILES, M. H. (1965a). *J. Am. Chem. Soc.* **87**, 1.

HASLAM, J. L., EYRING, E. M., EPSTEIN, W. W., JENSEN, R. P., and JAGET, C. W. (1965b). *J. Am. Chem. Soc.* **87**, 4247.

HEYNAU, H. A., and FOSTER, M. C. (1968). *Laser Focus* **4**, August, 20.

HOFFMANN, H., STUEHR, J., and YEAGER, E. (1966). *Chem. Phys. Ionic Solutions, Intern. Symp. Electrochem. Soc., Toronto, 1964* (B. E. Conway and R. G. Barradas, eds.), pp. 255–279. Wiley, New York.

HOGEN-ESCH, T. E., and SMID, J. (1966a). *J. Am. Chem. Soc.* **88**, 307.

HOGEN-ESCH, T. E., and SMID, J. (1966b). *J. Am. Chem. Soc.* **88**, 318.

HOLMES, L. P., COLE, D. L., and EYRING, E. M. (1968). *J. Phys. Chem.* **72**, 301.

HUNT, J. P., and PLANE, R. A. (1954). *J. Am. Chem. Soc.* **76**, 5960.

HURWITZ, P. A., and ATKINSON, G. (1967). *J. Phys. Chem.* **71**, 4142.

HURWITZ, P. A., and KUSTIN, K. (1966). *Trans. Faraday Soc.* **62**, 427.

ILGENFRITZ, G. (1966). Ph.D. Thesis, Georg-August Univ., Goettingen.

INSKEEP, W. H., JONES, D. L., SILFVAST, W. T., and EYRING, E. M. (1968). *Proc. Natl. Acad. Sci. U.S.* **59**, 1027.

KAVANAU, J. L. (1964). "Water and Solute-Water Interactions," p. 54ff. Holden-Day, San Francisco, California.

KAY, R. L., and EVANS, D. F. (1966). *J. Phys. Chem.* **70**, 2325.

KOR, S. K. (1966). D.Sc. Thesis, Allahabad Univ., India.

KUNDU, K. K., and DAS, M. N. (1964). *J. Chem. Eng. Data* **9**, 82.

KUSTIN, K., SHEAR, D., and KLEITMAN, D. (1965). *J. Theoret. Biol.* **9**, 186.

LOGAN, S. R. (1967). *Trans. Faraday Soc.* **63**, 1712.

MARGERUM, D. W., and ROSEN, H. M. (1967). *J. Am. Chem. Soc.* **89**, 1088.

MICELI, J., and STUEHR, J. (1968). *J. Am. Chem. Soc.* **90**, 6967.

MILES, M. H., EYRING, E. M., EPSTEIN, W. W., and OSTLUND, R. E. (1965). *J. Phys. Chem.* **69**, 467.

MILES, M. H., EYRING, E. M., EPSTEIN, W. W., and ANDERSON, M. T. (1966). *J. Phys. Chem.* **70**, 3490.

MOORE, W. J. (1962). "Physical Chemistry," 3rd ed., p. 338. Prentice-Hall, Englewood Cliffs, New Jersey.

NOYES, R. M. (1961). *Progr. Reaction Kinetics* **1**, 129–160.

NOYES, R. M. (1962). *J. Am. Chem. Soc.* **84**, 513.

ONSAGER, L. (1934). *J. Chem. Phys.* **2**, 599.

ONSAGER, L. (1945). *Ann. N. Y. Acad. Sci.* **46**, 241.

PARKER, R. C., APPLEGATE, K., and SLUTSKY, L. J. (1966). *J. Phys. Chem.* **70**, 3018.

PARKER, R. C., SLUTSKY, L. J., and APPLEGATE, K. R. (1968). *J. Phys. Chem.* **72**, 3177.

PASTERNACK, R. F., and KUSTIN, K. (1968). *J. Am. Chem. Soc.* **90**, 2295.

PAULING, L. (1960). "The Nature of the Chemical Bond and the Structure of Molecules and Crystals," 3rd ed. Cornell Univ. Press, Ithaca, New York.

PLANE, R. A., and TAUBE, H. (1952). *J. Phys. Chem.* **56**, 33.

PORTER, G. (1968). *Science* **160**, 1299.

RAMPTON, D. T., HOLMES, L. P., COLE, D. L., JENSEN, R. P., and EYRING, E. M. (1967). *Rev. Sci. Instru.* **38**, 1637.

RICH, L. D., COLE, D. L., and EYRING, E. M. (1969). *J. Phys. Chem.* **73**, 713.

ROBINSON, R. A., and STOKES, R. H. (1959). "Electrolyte Solutions." Butterworths, London.

ROUGHTON, F. J. W., and CHANCE, B. (1963). *Tech. Org. Chem.* **8**, Pt. II, 703–792.

SCHWARZENBACH, G., and FELDER, E. (1944). *Helv. Chim. Acta* **27**, 1701.

SCHWARZENBACH, G., and LUTZ, K. (1940). *Helv. Chim. Acta* **23**, 1162.

SEEWALD, D., and SUTIN, N. (1963). *Inorg. Chem.* **2**, 643.

SILLEN, L. G., and MARTELL, A. E. (1964). "Stability Constants of Metal-Ion Complexes." Chem. Soc., London.

STREHLOW, H. (1962). *Z. Elektrochem.* **66**, 392.

STREHLOW, H. (1963a). *Tech. Org. Chem.* **8**, Pt. II, 865–893.

STREHLOW, H. (1963b). *Tech. Org. Chem.* **8**, 799–843.

STUEHR, J. (1967). *J. Am. Chem. Soc.* **89**, 2826.

STUEHR, J., and YEAGER, E. (1965). *Phys. Acoustics* **2A**, 351–462.

SWIFT, T. J., and CONNICK, R. E. (1962). *J. Chem. Phys.* **37**, 307.

SWINEHART, J. H. (1967). *J. Chem. Educ.* **44**, 524.

SWINEHART, J. H., and CASTELLAN, G. W. (1964). *Inorg. Chem.* **3**, 278.

THOMAS, S., and REYNOLDS, W. L. (1966). *J. Chem. Phys.* **44**, 3148.

TURNBULL, D., and MARON, S. H. (1943). *J. Am. Chem. Soc.* **65**, 212.

VON SMOLUCHOWSKI, M. (1915). *Ann. Physik* **48**, 1103.

VON SMOLUCHOWSKI, M. (1916). *Physik. Z.* **17**, 557, 585.

VON SMOLUCHOWSKI, M. (1917). *Z. Physik. Chem.* **92**, 129.

WATTS, A. M. (1966a). *Trans. Faraday Soc.* **62**, 2219.

WATTS, A. M. (1966b). *Trans. Faraday Soc.* **62**, 3189.

WENDT, H. (1962). *Z. Elektrochem.* **66**, 235.

WENDT, H. (1969). *Inorg. Chem.* **8**, 1527.

WHITTAKER, M. P., EYRING, E. M., and DIBBLE, E. (1965). *J. Phys. Chem.* **69**, 2319.

WHITTAKER, M. P., ASAY, J., and EYRING, E. M. (1966). *J. Phys. Chem.* **70**, 1005.

WIEN, M. (1927). *Ann. Physik* **83**, 327.

WIEN, M., and SCHIELE, J. (1931). *Z. Physik* **32**, 545.

WINSTEIN, S., CLIPPINGER, E., FAINBERG, A. H., and ROBINSON, G. C. (1954). *J. Am. Chem. Soc.* **76**, 2597.

WITT, H. T. (1967). *Nobel Symp. Fast Reactions and Primary Processes Chem. Kinetics, 5th, 1967, Stockholm* (S. Claesson, ed.), pp. 81–97. Wiley (Interscience), New York.

Chapter 9

Electrochemical Energy Conversion

M. EISENBERG

I. Introduction 774
II. Thermodynamics of Galvanic Cells and Related Systems 775
   A. Free Energy and Enthalpy in Galvanic Cells 775
   B. The Reversible Electrode Potential 779
   C. Effect of Temperature 780
   D. Effect of Pressure on the Cell EMF 780
   E. Types of Electrodes Occurring in Galvanic Systems 781
III. Electrode Kinetic Aspects of Energy Conversion 787
   A. The EMF of Closed Circuit Cells 787
   B. Charge-Transfer Polarization 790
   C. Crystallization Polarization 792
   D. Concentration Polarization in the Electrolyte 793
   E. Gas Side Concentration Polarization 794
   F. Energy Conversion Efficiency Considerations 795
IV. Theory of Porous Electrodes 797
   A. The Role of Porous Electrodes 797
   B. The Gas Diffusion Electrode 799
   C. Flooded Battery Electrodes 811
V. Energy Conversion in Fuel Cells and Related Dynamic Systems 824
   A. The Mechanism of the Hydrogen Electrode 825
   B. The Mechanism of the Oxygen Electrode 827
   C. Hydrocarbon Anodic Electrodes 833
   D. Liquid Diffusion Electrodes 835
   E. Fuel-Cell Systems 839
   F. Metal-Gas Hybrid Cells 840
VI. Regenerative Fuel-Cell Systems 842
VII. Batteries 846
   List of Symbols 853
   References 854

773

I. Introduction

The conversion of chemical energy into electrical energy is a phenomenon that has been known, at least, since the days of Galvani (1791), Volta (1800), and Davy (1802). The electrochemical means for generating electricity represented an early and convenient form of producing electric currents. Conventionally, a coupling of any two electrodes in a cell, which produced a positive emf with the result that current could flow through an external circuit, has been referred to as a galvanic cell.

The study of galvanic cells represents a major area of electrochemistry for two main reasons. First, galvanic cells have been used for the derivation of much of our most precise thermodynamic data; and secondly, the study of galvanic cells represented an essential part of the investigations in the development of practical battery and fuel-cell systems.

The fuel cell, which has been receiving renewed and extensive attention in recent years, is fundamentally no different than a galvanic cell, the basic difference being merely that the electrode reactants are continuously brought in from outside the cell and fed to the electrodes where they undergo, respectively, anodic oxidation and cathodic reduction; whereas, in galvanic cells, the active reacting materials are located on the electrodes within the cell, *a priori*. Thus, a fuel cell, by its operating mode, is analogous to an engine into which the fuel and oxidizer are continuously introduced. In this sense, the fuel cell could be referred to as an "electrochemical converter" or "galvanic engine." As will be discussed further in this chapter, hybrid systems are possible in which an electrode of a fuel-cell type and an electrode of a galvanic cell type are combined to provide an energy generating device. An example of such a device is the metal–air cell, such as the zinc–oxygen or zinc–air system. The continuously fed electrode of the fuel cell often employs a reactant which is either gaseous or liquid, i.e., in a form which could not readily be stored in appreciable quantities in a stationary, closed galvanic cell. Basically, the continuously fed fuel-cell electrode provides for the possibility of delivering energy for extended periods of time, much beyond those typical of a conventional battery.

In the last 20 years, and especially in the last decade, there has been considerable expansion in both fundamental and applied research in the field of electrochemical energy conversion. At the same time, problems arising in these areas have given rise to considerable fundamental studies in the area of electrochemical kinetics and especially in chemisorption on electrode surfaces, a study of considerable importance for understanding fuel cells. Another new area of investigations concerns the current distribution and mass transport in porous electrodes, both of the flooded type and of the gas diffusion type, which are important in energy producing batteries and fuel cells.

There has also been a considerable expansion of technology in many of these areas, especially in fuel cells and related devices, such as hybrid cells and regenerative systems. It is not the purpose of this chapter to concern itself with technology. Rather, it reviews the pertinent fundamentals of thermodynamics, electrode kinetics, and mass transfer theory to provide a better understanding of the electrode processes involved in electrochemical energy conversion and their effects on the performance of a complete cell. A general classification of the electrochemical energy conversion devices by systems or group types is provided, together with observations on their current research and development status.

II. Thermodynamics of Galvanic Cells and Related Systems

The thermodynamics of all galvanic cells, primary and secondary (i.e., rechargeable), and of related systems such as fuel cells and hybrid systems, is essentially the same. Classical thermodynamics of galvanic cells involves chemical cells as well as concentration cells with or without transference. While concentration cell effects may occur in practical battery or fuel-cell systems, they do not represent a type generally sought by design. The galvanic cells of interest are, therefore, chemical cells.

A. Free Energy and Enthalpy in Galvanic Cells

For a reversible chemical galvanic cell in which the overall reaction can be written as

$$aA + bB + \cdots \leftrightarrows cC + dD + \cdots, \tag{2.1}$$

it can be shown from elementary thermodynamic considerations that the chemical potential or partial molal free energy for each species, i, can be expressed as

$$\mu_i = \mu_i^0 + RT \ln f_i x_i \tag{2.2}$$

and the overall free energy (or Gibbs free energy)* change is

$$\Delta G = \Delta H - T \Delta S = \sum_i \nu_i \mu_i^0 = \sum_i \nu_i \mu_i^0 + RT \sum_i \nu_i \ln f_i x_i \tag{2.3}$$

* According to the recommendation of the CITCE Nomenclature Commission, G is called the "free enthalpy" which is the same as "Gibbs free energy" or "free energy."

where

ΔG is the free energy change of reaction,
ΔH is the enthalpy change of reaction,
ΔS is the entropy change of reaction,
μ_i is the chemical potential of species, i,
μ_i^0 is the chemical potential at the standard state; i.e., when the activity $f_i x_i = 1$,
f_i is the activity coefficient of species, i,
x_i is the mole fraction of species, i, and
ν_i is the stoichiometric number of species, i.

When all reactants and products are at their standard states, hence their activities are equal to one, Eq. (2.3) reduces to

$$\Delta G^\circ = \Delta H^\circ - T \Delta S^\circ = \sum \nu_i \mu_i \, . \tag{2.4}$$

The stoichiometric numbers, ν_i, are, of course, to be taken as negative for the reactants (A and B) and positive for the reaction products (C and D) in Eq. (2.1).

In galvanic cells operating at elevated temperatures in which either products or reactant may be gaseous (e.g., in fuel cells), deviations from ideal gas law behavior may be ignored when the pressures are reasonably low. It is possible, then, to substitute partial pressures, p_i, into Eq. (2.3), resulting in

$$\Delta G = \sum \nu_i \mu_i^0 + RT \sum \nu_i \ln p_i$$
$$= -RT \ln K_p + RT \sum \nu_i \ln p_i \tag{2.3a}$$

where

K_p is the equilibrium constant of the gas reaction.

From the first and second laws of thermodynamics, it can readily be shown that the maximum useful work obtainable for an isothermal discharge process of a galvanic cell at constant pressure (i.e., conditions readily met in practice) is equal to the change in the free energy of the system. Indeed, the free energy change

$$dG = dH - T \cdot dS \tag{2.5}$$

may be combined with the expression for the internal energy change

$$dU = d\underset{\sim}{Q} - dW = d\underset{\sim}{Q} - (p \, dV + dW_m) \tag{2.6}$$

where

dW_m is the maximum useful work, and
$p \, dV$ is the work of expansion.

It can be readily shown (Eisenberg, 1963a,b) that for constant pressure and temperature

$$dG = -dW_m \qquad (2.7)$$

i.e., that the change in free energy equals the useful work in the isothermal process. This basic relation shows that under isothermal conditions and constant pressure, a reversibly operating galvanic cell will *deliver* a maximum amount of useful work ($-dW_m$) which is equal to the change in the Gibbs free energy. This basic relation explains why an electrochemical device escapes the Carnot cycle limitations of any energy conversion process in which heat is an intermediate form of energy. This possibility of at least theoretical work outputs equal to the free energy change represents the basis for recurrent historic, and especially the recent resurgent interest and activity in the area of electrochemical fuel cells.

The free energy change for an isothermal and reversible cell is related to the cell emf, V_r, by the simple relation

$$\Delta G = -nFV_r \qquad (2.8)$$

where

ΔG is the free energy change in joules per mole,
F is the number of coulombs per equivalent $= 96,500$ (Faraday),
n is the number of electrons per reacting ion or molecule, and
V_r is the theoretical maximum cell voltage for the overall chemical reaction.

The open cell voltage (i.e., when no current is flowing) is the highest obtained experimentally. For a reversible cell, the open cell voltage is equal to the theoretical maximum voltage resulting from Eq. (2.8).

The Gibbs–Helmholtz equation for a reversible process:

$$\Delta G = \Delta H + T\left(\frac{\delta(\Delta G)}{\delta T}\right)_p, \qquad (2.9)$$

can be combined with Eq. (2.8) to yield the form

$$\Delta H = -nF[(V_r - T(\delta V_r/\delta T)_p)] \qquad (2.10)$$

which relates the enthalpy of the reaction to the cell emf and its temperature coefficient.

Since the theoretical emf of a galvanic cell (V_r) can be related to the enthalpy and entropy, the calculation of the effect of temperature changes or pressure changes on the value of the emf can be accomplished by the standard thermodynamic procedures which are used for the free energy or the equilibrium constant.

Considering the operating of an electrochemical fuel cell, it may be interesting to compare its theoretical energy output which is equal to the free energy change, ΔG, to the enthalpy (i.e., heat of reaction) for the overall process occurring in a cell, ΔH. In a thermal energy conversion cycle, ΔH represents, of course, the energy absorbed before the Carnot cycle efficiency is applied to it. Thus, the ideal thermal efficiency of an electrochemical fuel cell can be expressed as

$$\Delta G / \Delta H = nFV_r / \Delta H = 1 - (T \cdot \Delta S / \Delta H). \qquad (2.11)$$

Since for reactions useful for the generation of energy, ΔG and ΔH are always negative, the sign of the entropy change, ΔS, determines the ideal thermal efficiency as follows:

(a) When ΔS is positive, the ideal thermal efficiency η_T is greater than 100% and the cell delivers more electrical energy equivalent to the heat of reaction. Under isothermal conditions, such a cell absorbs an amount of heat equal to $T \cdot \Delta S$ from the surroundings (example: for the cell reaction $C + \frac{1}{2}O_2 \rightarrow CO$, $\eta_T = 124$ and 178% at 298 and $1000°K$, respectively).

(b) When ΔS is negative, then η_T is less than 100% and the cell liberates, together with the electrical energy, an amount of heat equal to $- T \cdot \Delta S$, which must be removed continuously to assure isothermal functioning of the cell (example: hydrogen–oxygen fuel cell reaction $H_2 + \frac{1}{2}O \rightarrow H_2O_{(g)}$, for which $\eta_T = 94$ and 78% at 298 and $1000°K$, respectively).

(c) When ΔS is zero, the efficiency η_T is 100% and there is no net heat transfer. The emf of the cell does not change with the temperature. (A close example is the electrochemical oxidation of methane: $CH_4 + 2O_2 \rightarrow CO_2 + 2H_2O$, for which η_T is 100% for a broad range of temperatures, and the cell emf, V_r, is 1.038 V.)

Thus, the principal attractiveness of galvanic cells for the generation of electrical energy lies in the fact that not only is a thermal Carnot cycle avoided, but also that due to the small magnitude of the entropy term $T \cdot \Delta S$ in comparison to the enthalpy change, ΔH, ideal thermal efficiencies close to 100%, and in a few cases over 100%, are at least theoretically possible. It was Ostwald (1894) who first clearly formulated the basic thermodynamic attractiveness of employing galvanic cells as a means of efficient conversion of chemical energy into electrical energy. The idea, applied even to solid carbonaceous fuels (see e.g., Schottky (1935)), recurs consistently over the last century and represents the main driving force behind the expanded research and development efforts in the recent decade.

B. The Reversible Electrode Potential

The thermodynamic treatment of galvanic cells may be concerned with the reactions of either complete cells or of individual electrodes. Since, on the other hand, in the field of electrochemical kinetics only the treatment of individual electrodes is rationally possible, it is preferable to consider the thermodynamics of single electrodes, cathodes or anodes, rather than of complete cells. The reversible potential of a single electrode, E_r, can be expressed by means of an expanded form of the Nernst equation. Thus, for a single electrode process $M \rightarrow M^{n+} + ne^-$, the oxidation potential can be written as

$$E_r + E_r^0 - (RT/nF) \ln(a_{M^{n+}}/a_M) \qquad (2.12)$$

or, more generally, the single electrode potential can be expressed as

$$E_r = E_r^0 - (RT/nF) \sum \ln a_i^{\nu_i} \qquad (2.12a)$$

where

n	represents the total number of electrons participating in the overall reaction,
a_i	is the activity of species, i,
ν_i	is the number of ions or molecules of species, i, for the reaction as written,
E_r	represents the reversible electrode potential, and
E_r^0	is the standard potential of the electrode when all substances, i, are at unit activity; i.e., in their standard states.

This is the generalized form of the well-known Nernst equation. For a complete galvanic cell, the open circuit emf, V_r, can be expressed in terms of thermodynamic quantities provided both electrodes are reversible:

$$V_r = E_{rA} - E_{rC}$$
$$= \left[E_{rA}^0 - (RT/nF) \sum_i \ln a_A^{\nu_A} \right] - \left[E_{rC}^0 - (RT/nF) \sum_i \ln a_C^{\nu_C} \right] \qquad (2.13)$$

where subscripts A and C refer to the anode and cathode, respectively.

It should be pointed out that the experimental verification of the above expression for the emf of a single cell is only possible if both electrodes are indeed at their thermodynamic reversible potential. In reality, open circuit conditions may or may not provide for a reversible electrode potential. Furthermore, in many experimental situations, especially in the case of fuel cells, simultaneous electrode reactions may proceed, and accordingly, mixed electrode potentials may prevail. Furthermore, should there be any liquid junction involved in the system, it must be accounted for through appropriate liquid junction potential calculations. Another practical problem which may arise in the employment of equations of the type (2.13) is that of the activity coefficient. Strictly speaking, individual activity coefficients for each ionic species, i, which is potential determining, should be used. On the other hand, the available methods for determining activity coefficients yield mean activity coefficients for the ions into which a solute may dissociate. While for very precise work on the emf of galvanic cells this may represent a serious problem, the use of mean activity coefficients yields fairly satisfactory results in most cases.

C. Effect of Temperature

By rearrangement of the Gibbs–Helmholtz equation, the effect of temperature on the reversible emf of a complete cell is expressed by

$$- nFV_r = \Delta H - T[\delta(nFV_r)/\delta T]_p. \qquad (2.14)$$

Since the term $[\delta(nFV_r)/\delta T]$ is the derivative of the free energy change for the overall reaction with respect to temperature at a constant pressure, it is therefore a measure of the entropy change, ΔS. Thus,

$$T[\delta(nFV_r)/\delta T]_p = T \cdot \Delta S = -q \qquad (2.15)$$

where q represents the heat quantity involved in the overall cell reaction.

The effect of a temperature change on the overall cell open circuit emf, V_r, can be calculated from experimental values of V_r and of its temperature coefficient. Typical temperature coefficients of the emf of galvanic cells around $298°K$ range from -6×10^{-4} to $+8 \times 10^{-4} V/°K$. This procedure involves, thus, the use of experimental values of the temperature coefficients with possible inaccuracies. In a more rigorous procedure, one can go back to the Gibbs free energy function of temperature and calculate the *standard* free energy change, $\Delta G°$, at the new desired temperature. From this, the *standard* reversible open cell emf, V_r^0, is then obtained (Eisenberg, 1963a).

D. Effect of Pressure on the Cell EMF

The effect of a pressure change on the emf of a complete cell at constant temperature is obtained by considering the effect on the partial molal free energies of each component, i. Thus, the total effect can be expressed by

$$\sum v_i \frac{\delta}{\delta p} \left(\frac{\delta G}{\delta x_i} \right) = \sum v_i \frac{\delta}{\delta x_i} \left(\frac{\delta G}{\delta p} \right). \qquad (2.16)$$

Since at a constant temperature $(\delta G/\delta p) = v$, it can be shown that the effect of a pressure change on the reversible total cell emf, V_r, is

$$-nF \frac{\delta}{\delta p}(V_r) = \sum v_i \left(\frac{\delta v}{\delta x_i}\right) = \Delta v \qquad (2.17)$$

where Δv represents the change of volume resulting from the reaction. Upon integration of Eq. (2.17), one can obtain the value of the reversible emf of the total cell at the higher pressure, p_2, when the value at the lower pressure, p_1, is known. Hence,

$$(V_r)_{p_2} = (V_r)_{p_i} - (1/nF) \int_{p_2}^{p_1} \Delta v \cdot dp. \qquad (2.18)$$

It is interesting to note from Eq. (2.18) that in a galvanic cell in which all products and reactants are in the form of solids or condensed phases in which the volume change resulting from the cell reaction is generally small, any changes in pressure would have a negligible effect on the total emf of the cell. On the other hand, in cells in which gaseous reactants or products occur, as is often the case in electrochemical fuel cells, the pressure correction expressed by Eq. (2.18) becomes significant indeed.

An alternative procedure which permits precise treatment of the pressure effects is to account for the effect of pressure on the potential of each individual electrode by employment of the generalized form of the Nernst equation (2.12a).

E. Types of Electrodes Occurring in Galvanic Systems

The first practical galvanic cells in rechargeable batteries generally involved electrodes of two types: the electrode of the first kind (metal–metal ion) and electrodes of the second kind (metal–insoluble metal oxide or salt–metal ion). With the recent rather great expansion and development of a great variety of other systems, a number of other basic types of electrodes are found to enter into electrochemical energy conversion systems. Rather than describe a great variety of types of electrodes, it is more useful to consider the basic classes or electrode types under which the large variety of electrodes can be grouped.

1. Electrode of the First Kind

This electrode consists of a metal in contact with an electrolyte solution containing ions of that metal with respect to which the electrode is reversible. The reversible reaction for such an electrode is generally written as shown in Section II,B, for which the oxidation potential is expressed by Eq. (2.12). The equilibrium of such an electrode is, in present kinetic concepts, assumed to be dynamic, i.e., to involve a balance of small but finite forward and reverse reaction rates at the open circuit.

Electrodes of the first kind with solid metals are involved in such typical processes as electrodeposition, anodic dissolution, electropolishing, etc. In some types of primary or secondary batteries, the anode often represents an electrode of the first kind. It is often referred to as a "soluble anode," e.g., zinc in alkaline AgO–Zn cells. When, instead of a solid metal, a liquid metal is employed with a suitable electrolyte containing ions of that metal, it also represents an electrode of the first kind. An example of the latter may be a fluid anode of an alkali metal in an elevated-temperature fused-electrolyte fuel cell.

2. Electrodes of the Second Kind

Physically, electrodes of the second kind are metal surfaces (either solid or liquid) covered with a layer of practically insoluble salts or oxides of the same metallic species. These electrodes are submerged in an electrolyte solution containing an anion common with the insoluble salt. Typical examples of electrodes of the second kind are the silver–silver chloride electrode in a hydrochloric acid solution, and the well-known calomel electrode consisting of a mercury electrode covered with a layer of mercurous chloride in contact with either a 1 normal or a saturated potassium chloride solution. An important difference from electrodes of the first type is that, whereas in the latter the passage of current is associated with the deposition or dissolution of metal ions, electrodes of the second kind can be viewed as having the flow of current associated with the transport of anions (although actually a metal ion discharge or dissolution is involved). For example, for the calomel electrode, which is widely used for pH measurements, the following two equilibria can be stated:

$$Hg_2Cl_2 \rightleftharpoons Hg_2^{2+} + 2\,Cl^-,$$
$$Hg_2^{2+} + 2e^- \rightleftharpoons 2\,Hg,$$

and the overall equilibrium is

$$Hg_2Cl_2 + 2e^- \rightleftharpoons 2\,Hg + 2\,Cl^-.$$

For the general reaction for an electrode of the second kind,

$$M + mA^{n-} \rightleftharpoons MA_m + (m \cdot n)e^-. \tag{2.19}$$

Using the solubility product of the salt, $K_s = a_M \cdot a_A{}^m$, the reversible emf can be written as:

$$E_r = E_r{}^0 - (RT/m \cdot nF) \ln(a_A{}^m/K_s). \qquad (2.20)$$

Combining K_s and $E_r{}^0$ into a new constant, Eq. (2.20) can be presented as

$$E_r = (E_r{}^0)' - (RT/m \cdot nF) \ln a_A{}^m. \qquad (2.21)$$

In aqueous battery systems, metal–metal oxides represent frequent typical examples of electrodes of the second kind used both as cathodes or as anodes. Examples of the former are alkaline $Ag/Ag_2O/Ag^+$ and alkaline $Hg/HgO/Hg^{2+}$; and of the latter, alkaline $Cd/CdO/Cd^{2+}$ and acid $Pb/PbO/Pb^{2+}$.

3. *Electrodes of the Third Kind*

Electrodes of this type are generally less common. They are employed when one desires a reversible electrode of metals which are very positive in the emf series of metals, i.e., readily decomposed by water. For instance, if it is necessary to prepare a reversible electrode of calcium in an aqueous electrolyte, one may proceed to build it according to the following scheme:

$$Zn|(ZnC_2O_4), (CaC_2O_4), Ca^{2+}.$$

The electrode consists of a metal covered with its insoluble salt and then covered with a second layer of insoluble salt with a common anion to the first layer and containing the desired cation (in this example calcium). This electrode is immersed in a solution containing the cation (Ca^{2+}). For the solubility products in the illustrated example,

$$K_1 = a_{Zn^{2+}} \cdot a_{C_2O_4^{2-}},$$
$$K_2 = a_{Ca^{2+}} \cdot a_{C_2O_4^{2-}},$$

and assuming that the activity of the $C_2O_4^{2-}$ ions is the same in saturated solutions of both layers, one can relate that

$$a_{Zn^{2+}} = (K_1/K_2)a_{Ca^{2+}}$$

and on substitution into the Nernst equation for a single zinc electrode obtain

$$E_r = E_r{}^0 - (RT/2F) \ln[(K_2/K_1)(1/a_{Ca^{2+}})]. \qquad (2.22)$$

Combining the constants, a convenient expression giving the effect of calcium-ion activity results:

$$E_r = (E_r{}^0)' + (RT/2F) \ln a_{Ca^{2+}}. \qquad (2.23)$$

4. *Amalgam or Alloy Electrodes*

Electrodes of this type represent a special group among the electrodes of the "first kind." The main difference is that the potential-determining species on the electrode side of the interface, e.g., the metal in the amalgam or alloy, is at an activity less than unity. Reflecting this for the general reaction,

$$(M)_{amalg} \rightleftharpoons (M^{n+})_{soln} + n \cdot e^-, \tag{2.24}$$

the emf is as follows:

$$(E_r)_{amalg} = (E_r^0)_{amalg} - (RT/nF) \ln(a_{M^{n+}}/a_M). \tag{2.25}$$

A typical example of such an electrode is sodium amalgam (in mercury) in contact with a sodium chloride solution. To date, amalgam or alloy electrodes have not yet acquired practical significance in fuel-cell technology. Numerous schemes have been proposed, however, involving the utilization of such electrodes. All of these are based on the fact that an exothermic process is involved in the decomposition of the amalgam, e.g., as is commonly done in the chlorine-caustic industry. Alkali metal amalgam electrodes offer the possibility of operation in aqueous electrolytes.

Several room temperature fuel-cell and hybrid systems have been considered in recent years, employing sodium amalgam electrodes. Such electrodes can be combined with a chlorine electrode or with an oxygen electrode with the resulting reaction products being sodium chloride and sodium hydroxide, respectively. Schemes of this type are sometimes combined with regenerative methods as discussed elsewhere in this chapter.

5. *Redox Electrodes*

Fundamentally, an oxidation-reduction (redox) electrode is one in which an unattackable metal is immersed in a solution containing both the oxidized and reduced species of either a simple or a complex ion. The metal serves as the electronic conductor and only the activities of the oxidized and reduced species affect the emf of the electrode. A redox electrode may be considered as a variant of the electrode of the first kind. For example, in a system consisting of metal (M) and its ion (M^{n+}), the former may be regarded as the reduced form and the latter as the oxidized form of the same species. However, the common conception of the redox electrode is as stated above. If several pairs of redox couples are present and reacting, a mixed redox electrode exists. For a single redox electrode involving two valence states of the cation, i.e., for

$$M^{n+} \rightleftharpoons M^{m+} + (m - n)e^-, \qquad (2.26)$$

the potential is given by

$$E_r = E_r^0 - \frac{RT}{(m - n)F} \ln \frac{a_{M^{m+}}}{a_{M^{n+}}}. \qquad (2.27)$$

Sometimes the redox process involves a change of the form of the reacting ion. For instance, for the permanganate reaction

$$Mn^{2+} + 4 H_2O \rightleftharpoons MnO_4^- + 8H^+ + 5e^-, \qquad (2.28)$$

the electrode potential is

$$E_r = E_r^0 - \frac{RT}{5F} \ln \frac{(a_{MnO_4^-})(a_{H^+})^8}{a_{Mn^{2+}}}. \qquad (2.29)$$

Other redox electrodes may involve solids or two gases; for instance, halogen electrodes can be regarded as redox electrodes. A number of redox electrodes have in recent years attracted the attention of those interested in the development of fuel cells; hence, the term "redox fuel cell."

While traditionally a redox electrode operates under conditions in which both the oxidized and reduced potentially determining species are in solution, it is not necessary that this be the case. As a matter of fact, many electrodes important in practical battery systems are redox electrodes *in the solid state*, i.e., where both forms are present in the solid state on the plate and remain essentially insoluble in the electrolyte. Examples are the electrodes $NiOOH/Ni(OH)_2$, MnO_2/Mn_2O_3, and PbO_2/PbO, occurring as cathodes in the nickel-cadmium, Leclanche, and lead-acid batteries, respectively.

Redox electrodes with both forms soluble in the electrolyte have in recent years been a subject of interest in certain accordingly named "redox fuel cells." One system proposed by Posner (1955) involved a stannous-stannic redox electrode (anode) in combination with a bromide-bromine redox electrode (cathode). To sustain a continuous operation, the systems of this type include regenerative oxidization and reduction steps usually performed in separate reactors outside the fuel cell.

6. *The Gas Diffusion Electrode*

Fundamentally, a gas diffusion electrode is one in which usually the reactant, or sometimes the product, is a gas. Three phases constitute the gas diffusion electrode; a solid phase of the electrode represented by the electron conductor with or without deliberately introduced catalytic materials, a liquid phase represented by the electrolyte, and a gas phase represented by the reactant or reaction product. The simplest version of a gas diffusion electrode is the standard hydrogen electrode, well known in electrochemical measurements. Electrodes of this type were originally introduced by Schmid (1923) more than forty years ago. As discussed elsewhere in this chapter, the correct understanding of the mechanism of the diffusion electrode, involving the film beyond the intrinsic meniscus of the electrolyte-gas interface, is relatively recent. In its simplest manifestation, a gas diffusion electrode encompasses an electrolyte and partly immersed electronic conductor which also provides the catalytic surface (a wire of platinum) and a reactive gas above the electrolyte space.

Since the reaction zone in the case of such a simple electrode is very limited, no appreciable currents can be passed and electrodes of this type would have no application in electrochemical energy conversion devices. For this reason, a significant version of the gas diffusion electrode involves a porous electrode matrix on one side of which is disposed the electrolyte and on the other, the reacting gaseous species. Hydrogen or hydrocarbon anode electrodes and oxygen or air cathodes in fuel-cell systems are typical examples of this type of a gas diffusion electrode.

The kinetic treatment of gas diffusion electrodes, especially its porous plate version, is complicated as far as current and potential distribution and mass transfer processes are concerned. Due to the considerable importance of the new investigations in this area, the subject will be treated in a separate section, further on. The thermodynamics of the gas diffusion electrode, on the other hand, can simply be handled on the basis of the generalized Nernst diffusion equation (2.12a) into which the proper activities for the gaseous components are entered. When the cell does not operate at very high pressures, the use of partial pressures is quite acceptable. When, for a complete galvanic cell, all reactants or products are gaseous, the chemical potential and the Gibbs free energy are expressed by Eq. (2.3a) from which the emf of the cell can be written as

$$V_r(p_i, T) = \frac{RT}{nF} \ln K_p - \frac{RT}{nF} \sum_i v_i \ln p_i$$

$$= V_r^{0\prime} - \frac{RT}{nF} \sum_i v_i \ln p_i. \tag{2.30}$$

For example, for the galvanic oxidation of carbon monoxide to carbon dioxide, Eq. (2.30) reads

$$V_r(p_i, T) = V_r^0 - \frac{RT}{2} \ln \frac{P_{CO} \cdot P_{O_2}^{1/2}}{P_{CO_2}}. \tag{2.30a}$$

7. *The Liquid Fuel Electrode*

For the sake of completeness of the discussion of the basic types of electrodes, one additional type which has become of interest in recent years as a result of investigations in the field of fuel cells should be mentioned. This electrode is based on the use of a liquid reactant which is generally dissolved in the electrolyte by one means or another. Examples of such liquid "fuel" electrodes are the anodic oxidation of alcohol or of hydrazine in an alkaline electrolyte, typically performed at a porous electrode matrix. An example is the anodic oxidation of hydrazine dissolved in the electrolyte to nitrogen and water in an alkaline supporting electrolyte according to

$$N_2H_{4(l)} + 4\,OH^- \rightarrow N_2 + 4\,H_2O + 4e^-. \tag{2.31}$$

The practical accomplishment of such electrodes may vary in approach. The reactant may be fed to the cell either directly or dissolved in the electrolyte through the backside of a porous anode which is facing a suitable cathode in the cell, or the circulating electrolyte may introduce the high concentration of the fuel between the cathode and the anode. Kinetic and mass balance considerations usually are effective in the selection of an optimum approach. From the point of view of the desirable electrolyte in variants with time of operation of a fuel cell, the reaction products should be readily removable from the system. When these reaction products are nitrogen and H_2O, as in the case of a hydrazine-oxygen fuel cell, their removal is usually simple. On the other hand, a reaction which produces CO_2 as a by-product obviously requires a CO_2 rejecting electrolyte, preferably an acid, a carbonate electrolyte, or a solid oxide electrolyte.

III. Electrode Kinetic Aspects of Energy Conversion

A. THE EMF OF CLOSED CIRCUIT CELLS

So far in the discussion of the open circuit potentials, it was assumed that the potential of each electrode in a galvanic cell is at its equilibrium value as obtained from the generalized Nernst equation. In reality, due to surface phenomena or specific chemisorption, electrodes may attain, at open circuit conditions, stable potentials of values different from the calculated equilibrium values. However, to simplify the discussion of electrode kinetic aspects pertinent to electrochemical energy conversion, it will be assumed that open circuit electrode potentials are equal to the equilibrium values. The kinetics of electrode processes, as well as other irreversible cell losses such as ohmic losses, will be discussed in this section to elucidate the nature of irreversibilities which affect performance of batteries and fuel cells. On the other hand, the effects of the geometric and hydrodynamic design of these devices upon the polarization characteristics of the cell will also be considered.

From the outset, it is useful to consider what happens when a load is applied to a galvanic cell. As shown in the schematic presentation of Fig. 1, a discharge or a charge load could be applied. An example of that would be a rechargeable or secondary battery. Considering at first the right-hand side of Fig. 1 (representing discharge conditions), each of the two electrodes shifts its potential, i.e., polarizes in opposite directions; that is, the cathode becomes less cathodic and the anode becomes less anodic. This in effect results in a decrease of the available cell voltage compared to the open circuit cell emf, $V_r^0 = E_C^0 - E_A^0$. This can be seen in Fig. 1,

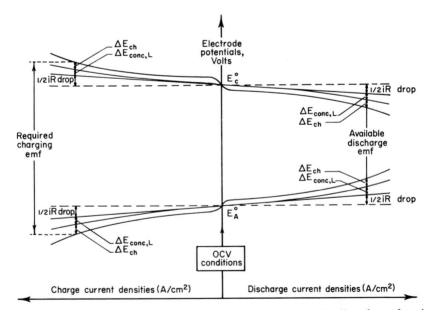

FIG. 1. Schematic presentation of potentials, polarizations, and cell emf as a function of current density on charge and discharge.

where, for the sake of a symmetrical presentation, the linear ohmic potential drop was equally distributed between cathode and anode. For simplicity, only two types of polarization, namely, concentration polarization, ΔE_{conc}, and charge-transfer polarization, ΔE_{ch}, at each electrode are shown in the diagram. In reality, however, as many as four components of polarization may occur at each of the electrodes. These components, discussed further on, are charge-transfer polarization (ΔE_{ch}), concentration polarization in the electrolyte ($\Delta E_{conc,L}$), gas side concentration polarization ($\Delta E_{conc,G}$) (applicable for the case of a gas diffusion electrode), and crystallization polarization (ΔE_{cryst}). The latter form is associated with a hindrance to the building of an ion into a crystal lattice on the electrode surface, or with the removal of an ion from such a lattice. This mode of polarization is involved in many battery systems, both primary and secondary, and occurs typically in the anodic dissolution process during the discharge of a metal anode, in the cathodic deposition during the charging process, and in corresponding processes at the cathode where the crystallites of a metal oxide or hydroxide may be involved. Thus, a broad statement for the total polarization of a given electrode may include the following terms*:

$$\Delta E_{T} = \Delta E_{ch} + \Delta E_{cryst} + \Delta E_{conc,L} + \Delta E_{conc,G}. \qquad (3.1)$$

Consequently, the voltage of a complete cell, V, at a load current, i, can be expressed as

$$V = (E_{r,A} - \Delta E_{T,A}) - (E_{r,C} - \Delta E_{T,C}) - i \sum R_i$$

$$= V_r - \Delta E_{T,A} + \Delta E_{T,C} - i \sum R_i \qquad (3.2)$$

where

$\Delta E_{T,A}$ is the sum total of all pertinent components of anodic polarization (a positive number),

$\Delta E_{T,C}$ is the sum total of all pertinent components of cathode polarization (a negative number),

* Still another mode of polarization, namely "reaction polarization" may be considered in certain cases. For instance, for redox electrodes where a slow *chemical equilibrium* might exist, affecting either the product or the reactant of an electrode reaction, i.e., before or after the charge transfer (see Vetter (1967), p. 107). This form of polarization has generally little significance in electrode processes typical of energy conversion devices. Since in Vetter's formulation, the sum total of the "diffusion polarization" and "reaction polarization" is equal to the concentration polarization, the concentration polarization, $\Delta E_{conc,L}$, as defined in this chapter is equivalent to his concept of diffusion polarization in the absence of slow chemical equilibria on the electrolyte side of the interface.

i is the total current through the cell, and
$\sum R_i$ is the sum of all internal ohmic cell resistances including electrolyte, any diaphragms, and resistances in the electrode bodies.

Thus, the various components of polarization at each of the electrodes and the ohmic potential drops inside the cell are responsible for the departure of the voltage of a closed cell from its thermodynamic open-circuit value, V_r. Since polarization and ohmic losses increase with current density, it is obvious from Eq. (3.2) why the overall operating cell voltage decreases with increasing current density. This increase of thermodynamic irreversibility of the working galvanic cell with increasing current, i.e., the reaction rate, is consistent with the departure from thermodynamic reversibility with increasing process rate in conventional chemical kinetics. In the remainder of the section, a brief discussion of the definitions and concepts of the individual modes of electrode polarization is given.

B. CHARGE-TRANSFER POLARIZATION

The charge-transfer polarization (or overvoltage), ΔE_{ch}, results from the hindrance of the charge-transfer reaction, specifically the hindrance of charged carriers across the electrical double layer. This mode of polarization has in the past been referred to (Agar and Bowden, 1939) as "activation polarization" and sometimes as "chemical polarization." However, modern developments in electrode kinetics require a distinction between the polarization associated with the actual charge-transfer process and that related to any slow crystal-transformation steps or to slow chemical equilibria. Thus, the older concept of activation polarization can be regarded as representing a sum of charge transfer, crystallization, and possibly reaction polarization modes. The present kinetic approach is preferable since the different rate-determining steps involve quite different processes and consequently different mathematical treatments. For this reason, lumping together of these modes of polarization would not be justified.

It is useful to express the charge-transfer polarization in terms of a dynamic model for a generalized redox electrode for which the total current density, I, passing through the double layer is the sum of the anodic and cathodic partial current densities, I_+ and I_-. Thus, the net current density passing through the electrode is from considerations of activation energies across the double layer,

$$I = I_+ + I_-. \tag{3.3}$$

For a simple charge-transfer reaction in the presence of an excess of inert electrolyte, Erdey-Grúz and Volmer (1930) derived the relation

$$I = k_+ \cdot c_r \cdot \exp\left(\frac{\alpha F}{RT} E\right) - k_- \cdot c_0 \cdot \exp - \left(\frac{(1 - \alpha)F}{RT} E\right) \qquad (3.4)$$

where

k_+ and k_- are the anodic and cathodic rate constants,

E is the electrode potential, and

α is the transfer coefficient.

This relation, originally obtained for the hydrogen electrode with $c_r = [H]$ and $c_0 = [H^+]$, was later generalized for redox electrodes by Eyring et al. (1939). When the exchange current density, i_0, which corresponds to the condition of dynamic equilibrium between a forward and reverse reaction is employed in Eq. (3.4), corresponding to the equilibrium potential E^0, and recognizing that the charge-transfer polarization, $\Delta E_{ch} = E - E^0$, the following basic relationship between the charge transfer polarization, ΔE_{ch}, and the current density is obtained:

$$I = i_0\left[\exp\left(\frac{\alpha F}{RT} \Delta E_{ch}\right) - \exp\left(-\frac{(1 - \alpha)F}{RT} \Delta E_{ch}\right)\right]. \qquad (3.5)$$

It is important to emphasize that the well-known relationship given in Eq. (3.5) is valid only as long as the concentrations of the oxidized and reduced species, c_0 and c_r, are independent of the current density, i.e., there are no mass transfer effects, a condition which is generally difficult to obtain or, strictly speaking, in Eq. (3.4), interfacial boundary concentration values obtained for c_0 and c_r by mass transfer calculations (Tobias et al., 1952), or by means of extrapolative experimental Schlieren techniques where applicable (Ibl et al., 1954).

Another important consideration is that only charge-transfer polarization is the form involved in Eq. (3.5) and other components of polarization should not be included. One often finds in the literature attempts to employ in this equation total polarization values obtained experimentally (i.e., ΔE_T) where only ΔE_{ch} should be used. Thus, Eq. (3.5) should not be employed when important contributions are made by other forms of polarization.

At appreciable current densities, when higher cathodic or anodic charge-transfer polarizations occur, $|\Delta E_{ch}| \gg RT/F$ and the first or the second term of Eq. (3.5) can be neglected, depending on the direction of the current. The well-known Tafel forms of the charge-transfer polarization thus result for the anodic and cathodic cases, respectively:

Anodic case:

$$\Delta E_{ch,A} = -\frac{RT}{\alpha F} \ln i_0 + \frac{RT}{\alpha F} \ln I \qquad (3.6)$$

or

$$\Delta E_{ch,A} = a + b \log I. \qquad (3.6a)$$

Cathodic case:

$$\Delta E_{ch,C} = \frac{RT}{(1 - \alpha)F} \ln i_0 + \frac{RT}{(1 - \alpha)F} \ln|I| \qquad (3.7)$$

or

$$\Delta E_{ch,C} = a + b \log|I|. \qquad (3.7a)$$

Thus, at not too small current densities, the linear relationship between the charge-transfer polarization and the logarithm of the current density as expressed by the Tafel-type equations (3.6a) and (3.7a) involve the "Tafel slope" b equal to $2.303RT/\alpha F$ or $2.303RT/(1 - \alpha)F$, which provides the basis for experimental determination of the transfer coefficient, α. By extending the Tafel lines according to Eqs. (3.6) and (3.7) to the equilibrium potential, i.e., $\Delta E_{ch} = 0$, the exchange current density, i_0, is obtained, as is commonly done in derivation of these electrode kinetic parameters from experimental data.

C. Crystallization Polarization

As indicated before, crystallization polarization, ΔE_{cryst}, is important in the electrode processes on solid electrodes in both primary and rechargeable batteries. This mode of polarization has not been properly appreciated in the past. However, advances in solid-state physics and chemistry in recent years have provided a basis for a theoretical treatment of crystallization polarization, a term introduced by Lorenz (1954). This mode of polarization is caused by the hindrance in the inclusion or release of metal atoms in the adsorption-like state called "ad-atoms" into and from the ordered lattice of solid metal electrodes.

For the sake of a definition of crystallization overvoltage, it must be assumed that charge-transfer polarization and concentration polarization are negligible, i.e., the processes to which they relate are in a state of near thermodynamic equilibrium. Hence, for the calculation of ΔE_{cryst}, one can apply the Nernst equation to the activities of the "ad-atoms," when in an equilibrium condition, and when a current flow occurs. Thus,

$$\Delta E_{cryst} = E - E^0 = (RT/nF) \ln(a_M(i)/a_M{}^0) \qquad (3.8)$$

where $a_M(i)$ and a_M^0 represent the metal "ad-atom" surface concentrations at current, i, and at equilibrium, respectively.

The treatment of Lorenz (1954) which was developed for electrocrystallization in the cathodic deposition and for anodic dissolution of metals can be extended to the crystallization of a galvanic cell in which ions are built into, or removed from, the ionic crystal of the cathode oxide or halide. For instance, in the charging process of a silver-silver oxide electrode, the concentration of the silver metal ad-atoms on the surface of an Ag_2O crystal can be analogously considered.

Modern theory of crystallization polarization, which is still in the process of development, concerns itself with the fundamentals of crystal growths and crystal dissolution in which the positions available for atoms or ions on the crystal surface are considered. An important aspect is the rate of surface diffusion and of charge transfer of the ad-atoms. The problem of diffusion which was solved for crystallization from the vapor phase of Burton *et al.* (1951) was considered by Lorenz (1954) for the case of electrodeposition or anodic dissolution. Detailed analytical expressions involving crystal lattice parameters such as the parallel growth step distance, have been obtained by Fleischmann and Thirsk (1960) and Damjanović and Bockris (1963). For a critical review of this work which is beyond the scope of the present chapter, the reader is referred to Vetter (1967, pp. 284–334).

D. Concentration Polarization in the Electrolyte

When a net electrochemical reaction occurs at a working electrode, certain changes in the concentration of the potential-determining species will occur in the immediate vicinity of the electrode surface. As a result of this concentration gradient, a shift in the electrode potential value occurs equal to the emf of a corresponding concentration cell without transference. Thus the concentration polarization in the electrolyte can be simply defined as

$$\Delta E_{conc,L} = [(RT)/(nF)] \ln(c_i/c_o) \qquad (3.9)$$

where c_i, c_o represent the interfacial and bulk concentrations of the reacting species. If more than one species is potential determining, for instance as is the case with a redox reaction, two gradients will usually form (a buildup of the concentration of one species and a decrease of the other) so that ratios of concentration of both species must be considered in the calculation of concentration polarization.

The maximum current density that can be passed (in the case of a consumptive electrode process) depends, of course, upon the rate of mass transfer. This value, the limiting current density, I_L, is obtained when the concentration difference is the maximum, i.e., $\Delta c = c_0$, the bulk concentration. It is simple to show from Eq. (3.9) that concentration polarization within the electrolyte of a cell is

$$\Delta E_{\text{conc},L} = (RT/nF) \ln[(I_L - 1)/I_L] \tag{3.10}$$

where I_L is the limiting current density for the consumptive process and I is the applied current density. Concentration polarization also occurs, of course, at the opposite electrode at which a species is generated and the concentration builds up. For this case, its value is

$$\Delta E_{\text{conc},L} = (RT/nF) \ln[(I_L' + I)/I_L]. \tag{3.11}$$

Except for the case of surface blocking by the generated species, the concentration polarization in this case is smaller and reaches at the limiting conditions $(I = I')$ a maximum value of $(RT/nF) \ln 2$. Concentration polarization within the electrolyte of a cell can be very significant when either low bulk concentrations are employed, or when large concentration gradients result from poor mass transfer within the fuel cell.

The limiting current density for the electrode process can be related to the mass-transfer coefficient, k_L, and to the diffusion boundary layer thickness, δ, as follows:

$$I_L(1 - t_i)/nF = k_L \cdot c_0 = (D/\delta)c_0 \tag{3.12}$$

where

    c_0 is the bulk concentration of the potential-determining species, and

    t_i is its transference number.

The mass transfer coefficient, k_L, represents the combined effect of diffusion and all modes of convection prevailing in the cell.

For a more detailed discussion of the effects of hydrodynamics and mass transfer on the rate processes of electrochemical cells, the reader is referred to Tobias et al. (1952), Ibl (1959), Levich (1962), and Eisenberg (1962a,b).

E. Gas Side Concentration Polarization

Gas side concentration polarization arises in the operation of a working electrode when either the reactant or a product are in gaseous states, and a change in the partial pressure of the potential-determining gaseous species at the reaction zone occurs in respect to its partial pressure in the bulk of the gaseous phase. A typical example is the operation of an air electrode in a fuel cell or a metal-air cell. When the reaction proceeds at a finite rate, i.e., current density, the partial pressure of oxygen near the reaction zone within the pores of the electrode is usually diminished as compared to its partial pressure in the bulk of the air feed stream to the cell.

This change, according to the thermodynamic equilibrium equation, must result in a shift of the emf of the electrode by an amount corresponding to the value of a gas concentration cell without transference. This shift can be termed as "gas side concentration polarization" (Eisenberg, 1962a,b), and can be shown to be equal to

$$(\Delta E_{conc})_G = (RT/nF) \ln[(p_A)_1/(p_A)_2]. \tag{3.13}$$

In the operation of some types of fuel cells, more than one gaseous concentration gradient must be considered within the body of the electrode. For instance, at the anode in a carbon monoxide high-temperature fuel cell, there is a countercurrent diffusion of the reactant, CO, and the by-product, CO_2. Thus, the equation for gas side concentration polarization for this electrode involves the ratios of partial pressures of both of these species in the bulk of the anodic gas stream and near the reaction zone. By use of Maxwell's diffusion equations and gas mass-transfer theory, it has been shown (Eisenberg, 1962a,b) how the partial pressures of the potential-determining species can be calculated for a given current density and porosity of the electrode.

F. ENERGY CONVERSION EFFICIENCY CONSIDERATIONS

When efficiencies of a galvanic cell (a battery or a fuel cell) are discussed, it is important to distinguish between the following types of efficiency concepts, each related to a different aspect.

(1) On the basis of thermodynamic considerations, the ideal thermal efficiency, η_T, of a galvanic cell has already been expressed [see Eq. (2.11)] as follows:

$$\eta_T = 1 - \frac{T \cdot \Delta S}{\Delta S} = \frac{nFV_r}{\Delta H}. \tag{2.11a}$$

Thus, the *ideal thermal efficiency* can simply be calculated from thermodynamic data, provided the initial and final states of all reactants and products are known.

(2) For an operating cell, it is possible to express a *voltage efficiency*, η_v, as a ratio of the cell voltage, V, for a given current load, to the reversible open cell emf, V_r. Thus,

$$\eta_v = V/V_r \tag{3.14}$$

where, for the evaluation of V_r, actual activities of the potential-determining species and not their standard states are used, as indicated by Eq. (2.13).

(3) The energy output of a fuel-cell system can also be lowered if the reactants are simultaneously consumed through a nonelectrochemical side reaction or are lost because of physical or mechanical circumstances. In cases like these, the desired electrochemical (faradaic) utilization of the reactants can be less than 100% of the theoretical coulombic capacity and one can define the *faradaic efficiency* as

$$\eta_F = \frac{\text{Coulombs obtained}}{(\text{Total moles consumed}) \times nF}. \tag{3.15}$$

The *overall efficiency* of an operating fuel cell, η_0, may thus be expressed as the product of the individual efficiency terms:

$$\eta_0 = \eta_T \eta_v \eta_F = (-nFV/\Delta H)\eta_F = (-nF/\Delta H)(V \cdot \eta_F). \tag{3.16}$$

Thus optimization of the overall operating efficiency of a given fuel-cell system involves the maximum of the product, $V \cdot \eta_F$. In most cases, the faradaic efficiency will be important only with respect to one capacity or rate-limiting electrode, e.g., with respect to the anodic fuel electrode in a cell using air as the cathodic oxidant. Thus, in such cases, η_F refers to the anode fuel and Eq. (3.16) is satisfactory for overall energy efficiency calculations.

When, in the general case, the faradaic efficiency at both the cathode and anode ($\eta_{F,A}$ and $\eta_{F,C}$, respectively) are of interest, then recognizing that the closed-circuit cell potential can be written as

$$V = E_C + (-E_A) - 1 \cdot h/k, \tag{3.17}$$

the general form of the overall energy conversion efficiency can be obtained:

$$\eta_0 = (-nF/\Delta H)[E_C \cdot \eta_{F,C} + (-E_A)\eta_{F,A} - Ih/k] \tag{3.18}$$

where

E_C, E_A	are the cathode and anode potentials at closed circuit (V),
I	is the apparent current density (A/cm²),
h	is the electrolyte thickness (cm), and
k	is the electrolyte specific conductivity (mhos/cm).

In solid matrix type fuel cells and in batteries with separators, the *effective* value of the h/k parameter for the combined electrolyte-diaphragm system should be employed.

IV. Theory of Porous Electrodes

A. THE ROLE OF POROUS ELECTRODES

In the operation of practical electrochemical energy conversion devices, the use of porous electrodes becomes necessary to increase the reaction area per unit volume (or the effective apparent current density per unit area of cell cross section). In the case of gaseous reactants, as in fuel cells, the porous electrode is also essential to provide for a stable separation between gas and electrolyte. For an electrode in a galvanic cell at which both products and reactants are either solid (e.g., as in traditional batteries) or at which a liquid reactant is soluble in the electrolyte (e.g., as in a hydrazine or alcohol fuel cell), the ideal electrode is a porous, high surface area, fully flooded plate.

The two types of electrodes are illustrated in Fig. 2. The gas side pores in the gas diffusion electrode are preferably made larger and the liquid side pores somewhat smaller, to enhance its capability for maintaining a stable triphase equilibrium (electrode-gas-electrolyte). The advantage of an effective pore diameter gradient for the achievement of this objective can be appreciated from the relation between the pore radius, r, and the differential pressure across the electrode, ΔP (Eisenberg, 1963a,b):

$$r = (2\sigma \cdot \cos \theta)/\Delta P. \qquad (4.1)$$

The differential pressure, ΔP, is in the balanced condition, equal to the capillary pressure. Thus, the smaller the pore diameter, the larger the differential pressure that can be supported, i.e., the larger the possible pressure of the gaseous reactant supplied to the cell and the higher the operating current densities. If a gradient of pore diameters in the direction normal to the surface of the electrode can be obtained by employing powders of different particle size in the construction of the electrode, as illustrated in Fig. 2a, then the visible gas phase boundary can be fixed approximately where the two porosity zones in the electrode structure meet. This was recognized by Bacon (1954), who employed a nickel electrode with a fine structure (2–5 μ) towards the electrolyte solution and a coarse pore structure (20–30 μ) towards the gas side in the construction of an efficient medium temperature, high pressure hydrogen-oxygen fuel cell.

With reference to Fig. 2a and in light of the findings discussed in the following subsection, it must be understood that only the visible boundary corresponding to the intrinsic meniscus is located at the boundary of the two porosities. The thin electrolyte of the order of 1 μ continues to extend all the way along the pore walls (i.e., over the surface of the coarser structural particles) towards the gas phase. As will be shown, it is in this zone where most of the reaction occurs and where most of the current is carried.

A porosity gradient is not, of course, the only method available for the successful maintenance of a stable triphase equilibrium in a porous gas diffusion electrode. As can be seen from Eq. (4.1), a hydrophobic treatment of the electrode (e.g., by applying a thin film of waxes or teflon to the surface of the porous electrode) would also provide for an increase in the stable differential pressure, ΔP, for a given effective pore radius.

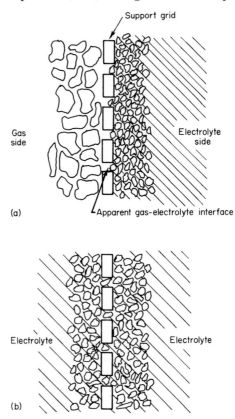

FIG. 2. Schematic presentation of two types of porous electrodes. (a) Porous gas diffusion electrode. (b) Flooded electrode.

The flooded electrode, on the other hand, is mechanistically simpler. Here, the ideal condition is the complete penetration of the electrolyte through the entire porosity of the electrode. A high volume percentage porosity is desirable and the pores should be preferably free of constrictions and tortuosities. As illustrated in Fig. 2b, this electrode again is constructed of particles which should have sufficiently good contacts to provide for electronic conductivity to and from each other and to the grid. The ideal flooded electrode should not have completely closed pockets from which there would be no passage through the bulk electrolyte. Since the electrode reaction takes place almost exclusively in the pores, the external surface of an electrode is relatively small compared to the combined wall surfaces of the pore. For a flooded electrode as employed in batteries, the reactant may be supplied right within the solid matrix (e.g., a porous metal oxide cathode or dissolving metal anode). A flooded electrode as employed in a fuel cell or a hybrid system may consist of a solid matrix structure with the reactant dissolved in the electrolyte and circulated either in parallel with the outer electrode face or by being forced through the porous structure of the electrode [e.g., a liquid hydrazine anode on a porous nickel matrix electrode (see Section V,D)].

The reaction in a porous electrode is distributed over the walls of the pores and the rate at any point is dependent upon the conditions of potential, current density, and concentration of pertinent species which prevail at that point. These local conditions, in turn, are governed by the mass transfer of the pertinent species and the transport of current to and from a given point. Thus, knowledge of the distribution of the electrode reaction is necessary to complete the characterization of the electrode behavior.

The mathematical models most useful in the discussion of the two types of porous electrodes are covered in the following subsections.

B. The Gas Diffusion Electrode

1. Mechanism of Mass Transfer and Current Distribution

The porous gas diffusion electrode represents a theoretically most complex problem inasmuch as a variety of simultaneous and consecutive rate processes occurs, complicated by the complex structure of the electrode plate itself. Only in recent years have these problems begun receiving proper attention by a number of investigators, as will be discussed further on. Early efforts would generally seek an idealized model in which the porous electrode plate was assumed to consist of cylindrical, uniform diameter pores octagonal to the face of the electrode and whose pore radii would follow a simple distribution function around one most frequent radius [see e.g., Justi *et al.* (1960)]. In such simplified models, other restrictive assumptions have often been made such as the requirement of a uniform electrolyte concentration or potential in the porous structure. Such an assumption can certainly not be supported for the case of real electrodes in which the pores are small and the reaction rates reasonably high. Thus, any kinetic relationships and polarization values calculated with models of this type cannot be considered useful. The first efforts towards a more realistic model of a porous electrode are due to Euler and Nonnenmacher (1960).

For the case of the gas diffusion electrode, it is interesting to consider the basic mass-transfer mechanism at the gas-electrode-electrolyte interface on a simple model before considering the complexities of a porous electrode. Such a simplified model in the form of a solid platinum wire, partially immersed into an acid electrolyte in a hydrogen gas atmosphere, was first considered by Will (1963). Variations of the total current passing through the platinum wire electrode for the anodic oxidation of hydrogen as a function of the length of the wire extending *above* the electrolyte level for a constant applied voltage have been found. Thus, a thin electrolyte film must exist above the intrinsic (observable) meniscus and most of the charge transfer reaction, in this case, hydrogen oxidation to H^+, occurs in a narrow region adjacent to the upper edge of the meniscus. Will also showed that surface migration of hydrogen along the platinum wire was not a significant mode of mass transfer. The diffusion rate of hydrogen through the liquid electrolyte film above the intrinsic meniscus was determined to be the rate controlling step. In other words, hydrogen first dissolves in the electrolyte of the thin film, then diffuses across the film to the electrode surface where it is adsorbed prior to the charge-transfer reaction. In a further inquiry into this mechanism, Bennion and Tobias (1966a) employed a vertical, cylindrical electrode, sectioned into two portions separated by a thin insulator for the study of oxygen reduction. This permitted the measurement of the current distribution to the two sections, especially of the portion of the total current carried by the upper section which was exposed only to the thin film above the intrinsic meniscus. This portion of the current was measured as a function of the distance of the insulating gap between the two sections above the bulk electrolyte level between the cylindrical electrode and the inner wall of a concentric glass cylinder (see Fig. 3). Silver and nickel electrodes were employed. Figure 4 shows a typical set of results obtained by Bennion and Tobias (1966a) for the percentage of total current carried by the top section of the electrode as a function of these distances from the bulk electrolyte level (Δ_L). Despite certain experimental difficulties concerning distortion of the 1 μ thin film by the much wider insulating separation, the studies confirmed the existence of this very thin film of electrolyte above the intrinsic meniscus (of the order of magnitude of wavelengths of visible light) through which a major portion of the current passes. Appreciable fractions of the total current are carried by portions of the electrode as much as 20 mm above the intrinsic meniscus. It was found

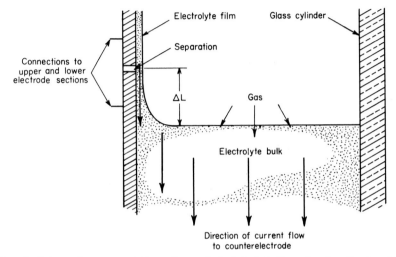

FIG. 3. Schematic representation of gas electrolyte interface showing the insulating gap.

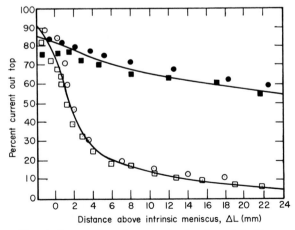

FIG. 4. Penetration of current in the electrolyte film.

that the thicker the film and the smaller the applied current, the further the current distributes over the electrode surface covered by the film. Furthermore, as illustrated in Fig. 4, for a given applied current the site of reaction is concentrated nearer the intrinsic meniscus on silver electrodes than on nickel electrodes.

Assuming a constant film thickness, Bennion and Tobias (1966b) constructed a mathematical, one-dimensional model which describes cathodic reduction of oxygen in a potassium hydroxide electrolyte. For the overall electrode reaction of

$$O_2 + 2\,H_2O + 4e^- = 4\,OH^-,$$

oxygen is assumed to reach the electrode by diffusion in the x direction through the upper film beyond the intrinsic meniscus, as shown in Fig. 5.

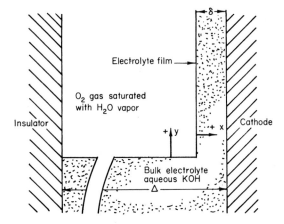

FIG. 5. Mathematical model of the gas-electrolyte-electrode interface.

In the bulk electrolyte region, oxygen is assumed to move only in the y direction. The transfer current density, J, which is proportional to the rate of reaction at the electrode, is taken as positive for the cathodic reduction of oxygen. In the thin film in which oxygen migrates towards the electrode in the x direction, OH^- ions move down the film (y direction) while water reaches the reaction side either by diffusion in the y direction countercurrent to the OH^- ions, or by condensation from the gas phase onto the film surface and subsequent cocurrent diffusion with the oxygen. From the set of equations for the fluxes of all ionic species and of water, and the electroneutrality principle, mass transport equations yield an expression for the current density corresponding to the oxygen diffusion rate:

$$J = \frac{FD_1(p/p^0)c_1{}^0 - c_1}{\delta} \qquad (4.2)$$

where

D_1 is the diffusion coefficient of oxygen at the concentration of KOH existing at the point under consideration in the film (cm^2/sec),

c_1 is the oxygen concentration ($moles/cm^3$),

$c_1{}^0$ is saturation concentration of oxygen at 1 atm ($mole/cm^3$-atm),

p is partial pressure of oxygen (atm), and

p^0 is oxygen reference pressure (1 atm).

For the bulk region of the electrolyte, the steady-state oxygen transport is expressed by the relation

$$D_1 \frac{\delta^2 c_1}{\delta_y{}^2} = \frac{J}{4F\Delta} \qquad (y \leq 0) \qquad (4.3)$$

in which Δ represents the width of the annular bulk electrolyte region below the meniscus (cm).

Applying this mathematical structure to the case of an oxygen electrode and employing the reaction mechanism summarized by Vetter (1967, pp. 632–642), Bennion and Tobias (1966b) also obtained a relation between the transfer current density, J, and the polarization. The principle of charge conservation also leads to the following relations between the transfer current density, J, and the current density in the solution, I_s, which is assumed to vary only in the y direction:

$$\begin{aligned} J &= -S(dI_s/dy) & \text{(at } y \geq 0), \\ J &= -\Delta(dI_s/dy) & \text{(at } y \leq 0). \end{aligned} \qquad (4.4)$$

The nonlinear first- and second-order differential equations were transformed into finite differences from equations and solved numerically on a digital computer. Measured or literature values for the physical properties of the system were employed. The agreement between the theoretical predictions for current distribution and the previously discussed experimental results for the percentage of the current out of the top electrode section as a function of distance, ΔL, was very satisfactory. A typical set of curves obtained for the current density distribution as a function of total peripheral current and as a function of position beyond the intrinsic meniscus is reproduced in Fig. 6. The experimental results indicated a strong dependence of the total applied current on the thickness of the film, δ, beyond the intrinsic meniscus. A range of film thickness from 0.33 to 2.25 μ has been calculated in reasonable agreement with optical interferrometric measurements. For the case of the oxygen electrode in a KOH electrolyte, the numerical results indicated that the activity of oxygen throughout the entire length of the film is very close to unity. However, below the meniscus in the bulk of the electrolyte, this activity drops rapidly. Thus, the dissolution rate of oxygen in the electrolyte of

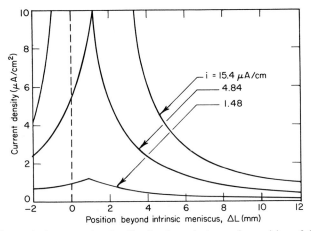

Fig. 6. Theoretical current density distribution relative to the position of the meniscus. Low current range.

the film and its transport across to the film-solid electrode interface are not controlling in the case of an alkaline oxygen electrode in which appreciable extended thin films can develop, i.e., where the flooding of the electrode by the electrolyte is effectively prevented. If oxygen transport is not rate limiting, then the charge transfer polarization and the ohmic resistance drop in the thin electrolyte film beyond the intrinsic meniscus may become rate determining. This may account for the differences observed with nickel and silver oxygen electrodes. Differences in conclusions between Bennion (1964) for the oxygen electrode and Will (1963) for the hydrogen electrode on platinum (where the transport of hydrogen gas to the electrode surface through the film and the ohmic drop were found to be rate controlling) is simply due to the fact that the charge-transfer polarization for hydrogen oxidation on the platinum is significantly lower than the corresponding polarization for the oxygen reduction on bright silver or nickel; hence mass transfer control becomes *relatively* significant. Again, on a surface more catalytically active for the reduction of oxygen, the transport rate of oxygen gas through the film to the electrode surface would also probably be found to be rate limiting in some balance with the limitation imposed by the ohmic drop in this thin film of solution.

Hence, the gas transport through the thin film above the intrinsic meniscus may or may not be rate limiting, depending on the relative catalytic activity of the electrode surface and its effect on the charge-transfer polarization. The hydrogen and oxygen gas diffusion electrodes play a major role on electrochemical energy conversion devices for which these findings have some practical consequences. These can be summarized as follows:

(1) The fact that for a gas diffusion electrode little or no reaction occurs below the intrinsic meniscus, i.e., in the bulk electrolyte region, and that appreciable current densities are required in the thin film up to 20 mm beyond that meniscus, leads to the suggestion that an ideal porous electrode plate is one in which the pores are open rather than dead-ended with minimum tortuosity and in which the maintenance of a stable equilibrium between the electrolyte liquid phase and the gas phase provides for *well-developed thin films* beyond the intrinsic meniscus to facilitate the gas transport to the electrode surface, i.e., solid pore walls.

(2) On electrocatalytically active electrodes, the hydrogen diffusion rate through the thin film and the ohmic drop may be rate controlling. Thus, high porosity electrodes capable of maintenance of a stable tri-phase equilibrium, i.e., having a large portion of unflooded micropores and the employment of high conductivity electrolytes are indicated.

(3) The high solubility rate of the reactant gas in the electrolyte and its effective transport through the thin film are important for the mass transfer kinetics. In the case of the oxygen electrode in alkaline solutions, gas transport through the film should not be rate limiting for a stable, unflooded porous electrode which is catalytically active.

(4) The high concentrations of KOH calculated for the upper portions of the thin film of the alkaline oxygen gas diffusion electrode (Ksenzhek, 1962a) underscore the desirability of supplying water as vapor in the oxygen gas phase as water is continuously consumed at the oxygen electrode.

(5) The experimentally well-known fact that the cathodic oxygen electrode polarization is generally high suggests directing the attention to the high charge-transfer polarization and associated electrocatalyst problems.

The studies on these simple models show, thus, that the mechanism for the gas diffusion electrode consists of the following steps: dissolution of the gas in the thin electrolyte film beyond the intrinsic meniscus, its diffusion across the film towards the appropriate catalytic sites of the solid electrode, chemisorption at the active sites, followed by charge-transfer reaction. This is the general scheme regarding the details of the chemisorption and charge-transfer steps. If these two steps occur rapidly, e.g., on a good catalytic surface, then the mass transport of the hydrogen through the thin liquid film may become rate limiting.

For a discussion of the nonmass transport aspects of the mechanisms of the hydrogen and oxygen electrodes, see Sections V,A and V,B, respectively.

2. A Porous Gas Electrode with Two Scales of Pore Structure

Unlike the sintered metal matrix electrodes, in which the pores are all much of the same size, or nearly so, even when dual layer electrodes (e.g., Bacon (1954) -type) are used, most active fuel-cell electrodes are of a carbon structure in which two systems of pores, considerably different in size, are superimposed upon one another. A system of large or *macropores* (with a diameter range of 3 to 50 μ) is filled with the reactant gas. From these *macropores*, a system of smaller or *micropores* branches out (with a diameter range of only 0.01 to 0.1 μ). These are flooded with electrolyte due to strong capillary forces. This micro-macro structure exists throughout the body of the electrolyte. There are no distinct separate layers. The reactant gas enters the macropores from the gas phase behind the electrode plate and dissolves in the electrolyte at the entry openings to the micropores. The gas then diffuses through the electrolyte solution to the reaction sites on the solid walls of the micropores. The ionic species which participate in the reaction and carry the current are transported by diffusion and migration through the micropore system between the reaction site and the bulk electrolyte on the liquid side of the electrode plate. A model of such an electrode was proposed by Grens (1966) as shown in the simplified presentation in Fig. 7. While an actual porous electrode consists of a most complex arrangement of unconnected or interconnected pores in a conducting solid matrix, it is possible to adopt this one-dimensional model, in which the pore systems are considered to be separate, so long as the characteristic dimensions of both types of pores are small compared with the distances over which significant changes in concentrations and potential occur. The dissolved gas concentration at the entry to the micropores can be assumed to be saturated with the gas reactant, but this condition does not hold throughout the micropore system. To consider the concentration gradient of the reactant gas at the entry to the micropores, Grens (1966) proposed to treat the initial segment of the micropores as "*linking pores*" joining the macropore and micropore phases. In these linking pores, the dissolved gas concentration changes sharply with the distance away from the macropores, i.e., distance y in Fig. 7. This one-dimensional electrode model was then analyzed with the assumptions of a Volmer-type polarization expression,

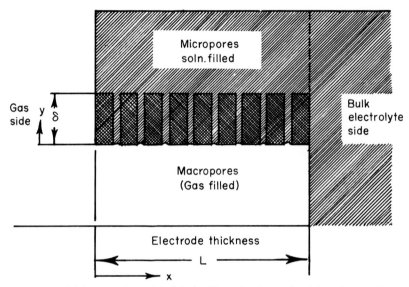

FIG. 7. Model for gas electrode with double scale of porosity (after Grens, 1966).

of a uniform gas composition in the macropore structure (a fairly reasonable assumption as long as reaction rates are moderate), of an isopotential electrode matrix, of the absence of hydrodynamic flows in the macropores, of constant values of transport properties, and of an isothermal operation. From the fundamental mass flux equations which account for the diffusion of the dissolved gas and for the diffusion and migration of the two ionic species in the electrolyte, and by applying the principles of electroneutrality and conservation of the dissolved gas in the micropores, the following system of equations was obtained:

$$\frac{d^2 c_g}{dx^2} + \frac{\omega}{q D_g}(U_g - U_R) = 0, \qquad (4.5)$$

$$U_R = \frac{a}{nF}\left(1 - \frac{A\delta}{\omega}\right) i^s, \qquad (4.6)$$

$$\frac{d^2 \bar{E}}{dx^2} + \frac{\omega z_2 \nu}{q D_1 c_1 0 (z_2 - z_1)}(U_G - U_g + U_R) = 0, \qquad (4.7)$$

where

i^s	is the transfer current density (A/cm²),
Φ	is $F(E - E_r)/RT$,
$\bar{E}$	is exp $F(E - E_r)/RT$,
ω	is the micropore tortuosity,
Q, q	are the macro- and microporosity, respectively,
A, a	are the macro- and micropore specific surface areas (cm²/cm³),
D_g	is the diffusion coefficient for dissolved gas in the micropores (cm²/sec),
c_g	is the concentration of the dissolved gas (g-moles/cm³),
U_g	is the source term for dissolved gas entering micropores (from the linking pores) at $y = \delta$ (g-moles/cm³/sec),
U_R	is the sink term for gas consumption for micropore walls (g-moles/cm³/sec),
U_G	is the sink term for gas dissolving into linking pores (g-moles/cm³/sec), and
$(U_G - U_g)$	is the gas reacting in linking pores per unit volume of total electrode (g-moles/cm³/sec).

Equations (4.5)–(4.7), together with the boundary conditions for the dissolved gas concentration,

$$\text{at } x = 0: \qquad c_g = c_g{}^0, \qquad \frac{d\Phi}{dx} = \frac{d\bar{E}}{dx} = 0,$$

$$\text{at } x = L: \qquad c_g = 0, \qquad \Phi = \Phi^0 \rightarrow \bar{E} = 1, \tag{4.8}$$

represent the micropore behavior.

It should be pointed out that the sink term, U_G, in the macropores and the source term, U_g, for dissolved gas entering the micropores are both functions of $\bar{E}(\Phi)$, i.e., of the extent of the electrode polarization. The concentration, c_g, must be determined by consideration of the processes occurring in the "linking pores." Grens then analyzed the transport relationships in the "linking pores" which are here considered only in the y direction normal to the local macropore (instead of the x direction). Equations analogous to (4.5) and (4.7) are then obtained and the boundary conditions at $y = 0$ and $y = \delta$ (see Fig. 7) are fixed for the concentration of the reactant gas dissolved in the electrolyte. Through a change to dimensionless variables, a system of nonlinear differential equations in terms of dimensionless group parameters was obtained.

This system required numerical techniques for solution, and a simultaneous first-order iteration process was employed in computations carried out with a digital computer for the plausible case of an oxygen electrode operating on a silver matrix. The parameters for such an oxygen cathode in a 6.9 N KOH electrolyte at 25°C were taken as follows:

Electrons transferred in rate-determining step, m (for Volmer equation)	1
Transfer coefficient, α	$\frac{1}{2}$
Exchange current density, i_0	2×10^{-6} A/cm^2
Diffusion coefficient of OH$^-$ ion, D_1	2×10^{-5} cm^2/sec
Diffusion coefficient of dissolved O$_2$, D_g	0.6×10^{-5} cm^2/sec
Oxygen solubility for 1 atm pressure, $c_g{}^0$	7×10^{-8} g-moles/cm^3

The structural parameters of the silver matrix electrode selected in the example were as follows:

Plate thickness, L	0.1 cm
Macroporosity, Q	30% (with 2 μ pores)
Microporosity, q	30% (with 0.1 μ pores)
Macro region specific surface area, A	600 cm^2/cm^3
Micro region specific surface area, a	1.2×10^5 cm^2/cm^3
Tortuosity factor, ω	2
Linking pore length, δ	0.1 μ (i.e., assumed equal to 1 micropore diameter)

Since the relationship between the apparent current density and the total electrode polarization is of particular interest, it was computed for the above set of values which represent a reasonably typical example. It should be noted here that in the procedure employed, the Volmer-type expression which applies to the charge-transfer polarization, was combined with the mass flux equations so that the total effect of charge transfer and of mass transfer of all reaction-participating species has been determined. The resulting polarization $(E - E_r)$ corresponds, therefore, to the total polarization, ΔE_T, discussed in Section III.

The results of Grens for the dual macro-microporous oxygen electrode for the above parameters are shown in Fig. 8, as curve A. The calculated current density was found to be sensitive to the assumed length, δ, of the linking pores. Since the scales in Fig. 8 were chosen to correspond to a typical Tafel plot (ΔE vs. log apparent c.d.), it will be noted that the total polarization is *not* linear with the logarithm of the current density. Rather, ΔE_T is nearly linear directly with the current density. This relationship results from the reduced dissolved gas concentrations in the micropore structure at higher electrode currents, which in part offset the almost exponential increases of the reaction rate with increasing electrode polarization.

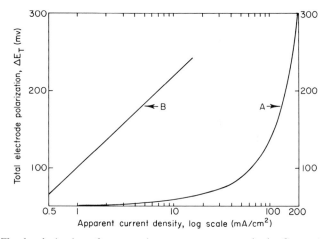

Fig. 8. Total polarization of macro-microporous oxygen cathode. Curve A—Electrode matrix with macroporosity $Q = 30\%$. Curve B—Electrode matrix with macroporosity $Q = 0.01\%$.

Curve B in Fig. 8 shows the effect of a much reduced macroporosity ($Q = 0.01\%$ compared to $Q = 30\%$). The sharp reduction in the macroporosity substantially increases the mass-transfer resistance for the gaseous reactant, resulting in a sharp increase in the total polarization for a given current density. Under such conditions, the activity of the electrode is much diminished and the curve becomes almost linear with log c.d.; in other words, closer approximating a Tafel relationship one would expect to find on a smooth, nonporous electrode surface.

Thus, the macro-microporous gas diffusion electrode model shows that a Tafel relationship will be approximated only for electrodes with very small macroporosities. For electrodes with reasonable macro-porosity (20 to 60% in real working systems), the total polarization is more nearly directly linear with the apparent current density. This model offers the possibility of extension towards investigations of the effects of porosity distribution in gas diffusion electrodes with a broad range of pore sizes.

C. FLOODED BATTERY ELECTRODES

In the analysis of a porous flooded electrode, its characterization in the form of an idealized pore geometry is not very useful due to the complete randomness of the configurations and of the electrode matrix. Simplified models have been repeatedly proposed (e.g., Winsel (1962), Austin (1963)). The analysis of both steady state and transient behavior of flooded isothermal porous electrodes represents a difficult and, at the same time, challenging task in which certain assumptions must be made to afford a theoretical grasp. It is important, however, to avoid assumptions and simplifications which cannot be defended from the point of view of basic realities encountered in a real working porous electrode. Since idealized geometrical pore structures can certainly not be defended, it is better not to engage into the detailed configuration of the porous body, but to treat the entire electrode as a homogeneous microscopic region of electrolyte with reacting species and with a distributed source of current. This distributed source of current represents the reaction which occurs at electrode-electrolyte interfaces throughout the body of the flooded electrode. One important and most essential assumption is to accept a *one-dimensional model* in which the gradients perpendicular to the overall direction of the current flow are ignored and the variables are functions only of one space dimension which is perpendicular to the face of the electrode. Thus, the distribution of current, potential, and concentration of individual species into the depth of the electrode from its outer face is investigated. This model requires that the electrode be macroscopically uniform and that the characteristic dimensions of the structure of the matrix be small in comparison to the distances over which there is significant variation in current or in concentrations. Due to the generally small pore dimensions in most electrodes of interest in batteries, these conditions commonly prevail.

A one-dimensional model was analyzed by Ksenzhek and Stender (1956) in which the electrolyte throughout the porosity was of a uniform concentration and the matrix offered no electrical resistance. The model resulted in an expression for the overall electrode polarization as a function of time for short periods. The model was also extended to the inner surface of a cylindrical electrode and equations were obtained for the local overpotential for an applied alternating current of a sinusoidal wave. This model was tested experimentally using small nickel tubes with inserted capillary probes (Ksenzhek and Stender, 1957). Ksenzhek (1957) later obtained the activation energy for the overall reaction in a porous electrode at steady state conditions. Ksenzhek (1962a) later extended his work to the case of infinitely thick porous electrodes filled with electrolyte of such a high conductivity that the local electrode potential remains constant throughout the system. The analysis of the distribution of current in the electrode is then reduced to that of the distribution of a heterogeneous chemical reaction in a porous material, thus permitting the employment of equations from this field of hetero-geneous kinetics. Employing a Tafel-type expression for the polarization, Ksenzhek obtained a polarization curve for the electrode which had a slope equal to twice the Tafel slope of an equivalent solid plate electrode surface, but with lower total polarization values over the range of most practical current densities. In a following paper, Ksenzhek (1962b) considered the polarization of thin porous electrodes (extending his earlier results for infinitely thick electrodes) with the assumption of a uniform electrolyte concentration within the porous structure.

A clear derivation of the one-dimensional model of a flooded electrode of uniform electrolyte concentration was presented by Euler and Non-nenmacher (1960) as a case applicable to the carbon-manganese dioxide electrode of the Leclanche battery. In this model, uniform conductivities were assumed for both the conducting electrode matrix and for the electrolyte. The effect of changes of concentration within the electrolyte was ignored. Analytical solutions were derived for a model in which a linear polarization was represented by an equivalent resistance to the charge transfer. Several examples have been calculated with this model for thin layers of a carbon-manganese dioxide electrode. In an experi-mental study designed to characterize the pore dimensions and true surface area of manganese dioxide electrodes, Euler (1961) measured the effect of capacitance of electrodes for various frequencies of applied alternating current. Since under these conditions no changes in the concentration of species in the electrolyte can be expected, this experi-ment was in agreement with the assumptions of the derivation.

An analysis of the steady state operation of flooded porous electrodes of finite thickness, where both reactants and products are neutral non-ionic substances, was carried out by Gurevich and Bagotsky (1964). Using the Volmer equation and allowing for variations in reactant and product concentration, relationships between the polarization and the limiting current density were obtained for conditions of low and high polarization.

Most previous models assumed for simplicity that the electrode matrix itself has a high conductivity and, therefore, its resistance can be ignored. In an extension of the work of Euler and Nonnenmacher (1960), Newman and Tobias (1962) considered a one-dimensional model with a resistive matrix. A Tafel-type polarization and a uniform electrolyte concentration were assumed and solutions obtained for the case of a binary electrolyte.

Thus, studies on flooded porous electrodes have generally assumed that the electrolyte remains (in the electrode depth) of a uniform and a constant (with time) conductivity. Many analyses are based on the assumption of a uniform electrolyte composition, a condition which, structurally speaking, exists only when no current is drawn. In a detailed analytical treatment in which the restrictions of constant electrolyte conductivity on concentrations were no longer made, Grens and Tobias (1964) developed a procedure which is amenable to any type of local polarization equation. The analysis provides for a description of the electrode behavior both in the steady state and in transient operation.

The theory of the porous electrode is essential to a modern understanding of electrochemical energy conversion. At the same time, the flooded porous electrode represents the practical electrode occurring in all battery systems and in certain types of fuel cells. Since the formulation of the problem provides an opportunity to discuss transport phenomena in a porous electrode with emphasis on mass transport, the analysis approach of Grens and Tobias (1964) is given here in some detail.

At the solid-electrolyte interface of the porous electrode, a reaction of the general type

$$\sum_i \nu_i M_i^{z_i} \rightarrow -ne^-$$ (4.9)

is occurring. The reactants and products of Eq. (4.9) may be ionic or uncharged species in the electrolyte or in the solid electrode matrix. The description of the processes occurring within the pores of the electrode can thus be made in terms of the kinetics of the electrode reactions occurring at the solid-electrolyte interface at the pore walls. The transfer current density at the pore wall is obviously a function of both the potential and the concentration of all species. Such an expression, e.g. Volmer-type equation, is applicable to reactions of the first order and includes both forward and reverse reaction times, as shown below:

$$\frac{i^s}{i_0} = \frac{c_r}{c_r^{\,0}} \exp\left[\frac{\alpha nF}{RT}(E - E_0)\right] - \frac{c_p}{c_p^{\,0}} \exp\left[\frac{(\alpha - 1)nF}{RT}(E - E_0)\right]$$ (4.10)

where

i^{s} is the transfer current density at the pore wall,
i_0 is the exchange current density, and
r and p   refer to reactants and products, respectively.

The other symbols have their usual meanings. The conditions prevailing at any position in the electrolyte filling a pore are described by the potential in the electrolyte, E, and the concentrations of all pertinent species, c_i. Since the entire electrode matrix can be assumed to exist at some constant potential, it is convenient to specify the potential in the electrolyte, E, with reference to that matrix potential. It is customary to represent the mass-transfer flux for any species, i, in the electrolyte by the expression,

$$N_i = D_i \nabla C_i - Z_i e \cdot u_i e_i \cdot \nabla E + V \cdot c_i \qquad (4.11)$$

where

e is the electron charge,
u_i is the ionic mobility, and
V is the bulk electrolyte velocity,

in which the three terms represent respectively the diffusion under a concentration gradient, migration under a potential gradient, and the convection due to bulk movement of the electrolyte.

Since the electric current density in the electrolyte, I, is carried by all the charged species, it is related to the sum of the fluxes:

$$I = F \sum_i z_i N_i. \qquad (4.12)$$

In electrolytic mass transport, the electroneutrality principle can be assumed throughout the regions except in the double layer. Thus for each ionic species, one can write

$$\sum_i z_i c_i = 0. \qquad (4.13)$$

The principle of conservation for each species, i, requires that the continuity equation be applied as follows:

$$\frac{dc_i}{dt} = -\nabla \cdot N_i + S_i \qquad (4.14)$$

where the source term, S_i, is related to the transfer current density at the electrolyte pore wall interface as follows:

$$S_i = (A \cdot \nu_i / nF) i^{\mathrm{s}} \qquad (4.15)$$

where

A is the specific matrix surface per unit volume (cm^2/cm^3), and
ν_i is the stoichiometric coefficient.

In the operation of an electrode, the electrical time constant associated with the capacitive effect in charging the double layer is very short in comparison to the mass-transfer time constant. Thus, ignoring capacitive effects, one can write an equation for conservation of charge as follows:

$$-\nabla\cdot I - A\cdot i^s = 0 \tag{4.16}$$

in which the transfer current density, i^s, is related to the divergence of the current density, I, in the electrolyte. Consequently, the source of any species can be obtained by combining Eq. (4.15) and (4.16):

$$S_i = (\nu_i/nF)\nabla\cdot I. \tag{4.17}$$

Upon substitution of Eqs. (4.10) and (4.17) into the continuity equations of type (4.14) for each species, a system of conservation equations results as follows for each species, i:

$$dc_i/dt = \nabla\cdot(D_i\nabla c_i) + z_i e\nabla\cdot(u_i c_i\nabla E) + V\cdot\nabla c_i + (\nu_i/nF)\nabla\cdot I. \tag{4.18}$$

Hence, the transport processes occurring in the electrolyte within the pores of the electrode under isothermal conditions are completely described by this system of equations, together with the electroneutrality condition [Eq. (4.13)], provided, of course, that a suitable polarization expression is employed to characterize the divergence of the current, $\nabla\cdot I$.

Before these basic equations can be applied to the case of a porous flooded electrode, certain assumptions and restrictions are necessary. The assumptions made by Grens and Tobias (1964) were as follows:

(1) A one-dimensional model is applicable with transport phenomena considered in direction x perpendicular to the face of the electrode.

(2) No bulk flow of electrolyte is assumed in the pores, i.e., velocity $V = 0$.

(3) The matrix is assumed to exist at a single potential due to its high conductivity in comparison to that of the electrolyte.

(4) All transport properties, e.g., mobilities and diffusion coefficient, are assumed to be constant and independent of concentration.

(5) The morphology of the electrode matrix is assumed to remain unchanged while the reaction proceeds; i.e., parameter a, the specific matrix surface per unit volume of the porous electrode, remains unchanged.

(6) The effect of transport phenomena in the electrolyte just outside the porous structure is assumed to be accountable by an equivalent diffusion boundary layer of thickness δ.

(7) The system is assumed to be isothermal.

The schematic of such a one-dimensional electrode model is given in Fig. 9. For this one-dimensional model, Eq. (4.18) can be reduced to the

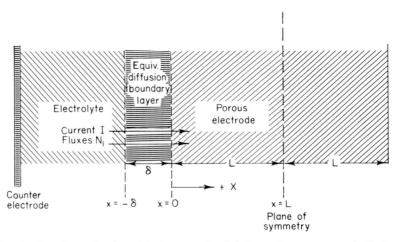

FIG. 9. One-dimensional model of porous flooded electrode as macroscopically homogenous region.

following expression for each species, i.

$$\frac{dc_i}{dt} = D_i \frac{dc_i}{dx} + z_i e u_i \frac{\delta}{\delta x}\left(c_i \frac{\delta E}{\delta x}\right) + \frac{v_i}{nF}\frac{\delta I}{\delta x}. \tag{4.19}$$

Here, E, is the potential in the solution with reference to the constant potential of the matrix which was assumed to be isopotential and set to zero as a reference. The following boundary conditions can now be formulated:

(1) Before any electrode reaction has taken place, the electrolyte is unaltered in the pores, i.e., is the same as in the bulk outside the porous electrode structure; hence, initially,

$$c_i = c_i{}^0 \qquad (t = 0).$$

(2) In the bulk electrolyte, the concentrations remain unchanged; hence,

$$c_i = c_i{}^0 \qquad (x = -\delta).$$

(3) At the face of the electrode and at the face of the diffusion boundary layer, the current density in the electrolyte is the same as the total applied current density; hence,

$$I = I_{ap} \qquad (x = 0, x = -\delta),$$

where I_{ap}, the applied current density, is the same as the one normally referred to as apparent current density. At the plane of symmetry, i.e., at $x = L$, symmetry conditions must be met, i.e., there is no net flux of current or concentration of potential gradients; hence,

$$I = 0,$$
$$\frac{dc_i}{dx} = \frac{dE}{\cdot \, dx} = 0 \qquad (x = L).$$

Using the simple Nernst–Einstein relation for the ionic mobilities:

$$u_i = D_i/kT, \tag{4.20}$$

Eq. (4.20) becomes

$$\frac{dc_i}{dt} = D_i \frac{\delta c_i}{\delta x} + z_i D_i \frac{F}{RT} \left(\frac{\delta c_i}{\delta x} \frac{\delta E}{\delta x} + c_i \frac{\delta^2 E}{\delta x^2} \right) + \frac{v_i}{nF} \frac{\delta I}{\delta x}. \tag{4.21}$$

Similar treatment applied to Eq. (4.12) yields the expression

$$I = -F \overline{\sum_i} z_i D_i \left(\frac{\delta c}{\delta x} + z_i c_i \frac{\delta E}{\delta x} \right). \tag{4.22}$$

In addition, a relation between the current density and the electrode potential and the concentrations of all species, i, is needed of the general form

$$(\delta I/\delta x) = -A \cdot f(E, c_i) \tag{4.23}$$

where A represents the specific matrix surface per unit volume and the porous electrode, f, represents a suitable function of the potential-determining species, i. The Volmer or Erdey–Grúz type relations may serve this purpose. Equations (4.22)–(4.23) now provide the characterization of the electrode model and are ready for mathematical solution.

In their solution, Grens and Tobias (1964) employed the following transformations:

$$\begin{aligned} x &= x/L; & c_i &= c_i/c_k{}^0; \\ \tau &= D_k t/L^2; & \Phi &= F(E - E_r)/RT, \end{aligned} \tag{4.24}$$

where

E_r is the equilibrium potential of the electrode at the bulk electrolyte composition, and

k represents a nonreacting species present in large concentrations in the bulk electrolyte (e.g., supporting electrolyte species which thus can serve as a reference concentration).

This permitted the rearrangement of the differential equations into forms involving the following dimensionless parameters:

$$I^i = (I/I_{ap}); \qquad \pi_i = (D_i/D_k); \qquad \gamma_i = (c_i^0/c_k^0);$$
$$\beta = \frac{I_{ap} \cdot L}{nFD_k c_k^0}; \qquad \Delta = (\delta/L). \tag{4.25}$$

An interesting parameter resulting from these transformations is

$$\xi = A \cdot L^2 \cdot i_0/nFD_k c_k^0 \tag{4.26}$$

which characterizes the exchange current density within the context of the physical properties of the porous electrodes (properties A and L) and the transport properties of the bulk electrolyte (c_k^0 and D_k, the bulk concentration and diffusion coefficient of a nonreacting reference species present in large concentrations). The dimensionless parameters given in Eqs. (4.24) and (4.25) establish *criteria of similarity* for comparing porous electrodes. These dimensionless parameters give insight into the relationships between the independent variables of the system and provide the basis for comparison between operating porous flooded electrodes.

With these parameters and dimensionless variables, the system of differential equations can now be presented as

$$\frac{I}{\pi_i}\frac{\delta c_i}{\delta t} = \frac{\delta^2 c_i}{\delta x^2} + z_i\left(c_i\frac{\delta^2 \Phi}{\delta x^2} + \frac{\delta c_i}{\delta x}\frac{\delta \Phi}{\delta x}\right) + \frac{v_i}{\pi_i}\beta\frac{\delta I'}{dx}, \quad i = I, \ldots, N, \tag{4.27}$$

$$\sum_i z_i c_i = 0, \tag{4.28}$$

$$\frac{\delta I'}{dx} = -Af(\Phi, c_i). \tag{4.29}$$

The function, f, which represents the polarization expression, now also employs dimensionless variables. When the Volmer-type polarization expression is introduced into Eq. (4.29), the following form is obtained:

$$\frac{\delta I^i}{\delta x} = \frac{\xi}{\beta}\left\{\frac{c_r}{\gamma_r}\exp[\alpha n\Phi] - \frac{c_p}{\gamma_p}\exp[(\alpha - 1)n\Phi]\right\} \tag{4.30}$$

in which parameter, ξ, as defined by Eq. (4.26), includes the exchange current density, i_0, and the physical properties of the porous electrode, as well as the bulk transport properties of an unchanging reference species in the bulk electrolyte.

Thus, the treatment of the one-dimensional model of the porous flooded electrode with unchanging physical characteristics can be reduced to a system of N parabolic partial differential equations for each of the N species present in the electrolyte, as shown in Eq. (4.27). The condition of electroneutrality [Eq. (4.21)] can be employed to eliminate any one of the dimensional concentrations, c_i. For this reason, the set of equations of type (4.27) involves $(N - 1)$ concentrations and the dimensionless parameter, Φ, in the solution of this problem; using the Volmer expression for the polarization relation (Eq. 4.30), Grens and Tobias (1964) employed a numerical analysis, since the differential equations are nonlinear. This involved a numerical procedure including the use of Crank–Nicholson symmetric finite difference forms in terms of independent variables x and τ. A digital computer program was employed which required input information for charge numbers, stoichiometric coefficients, diffusion coefficient ratios, and dimensionless bulk concentrations of the species present in the electrolyte.

The computational program was applied for the analysis of an *idealized cadmium anode* as an example of an electrode of the second kind (metal-"insoluble" metal oxide-electrolyte) as one finds in a nickel-cadmium battery. It was assumed that the geometry of the electrode matrix (internal and external) does not undergo any changes during the half-cell discharge reaction:

$$Cd + 2OH^- = Cd(OH)_2 + 2e^-. \qquad (4.31)$$

Designating the ionic species of the electrolyte, OH^- and K^+, with indices 1 and 2, respectively, the $5N$ KOH solution usually employed in such a battery leads to the parameters:

Valence numbers:

$$z_1 = -1, \qquad z_2 = +1$$

Stoichiometric numbers:

$$\nu_1 = +2, \qquad \nu_2 = 0.$$

The following system parameters and electrode properties were assumed for the computation:

Concentrations: $c_1^0 = c_2^0 = 5.10^{-3}$ moles/cm^3
Transfer coefficient: $\alpha = 0.5$

Effective electrode thickness: $L = 0.1$ cm
Electrode porosity: 50%
Specific surface area of electrode: $A = 10^4$ cm²/cm³

Grens and Tobias examined the current distribution for a cadmium anode over a range of apparent current densities from 0.01 to 1.9 A/cm² (corresponding to a range of parameter β values of 0.1 to 19). Figure 10

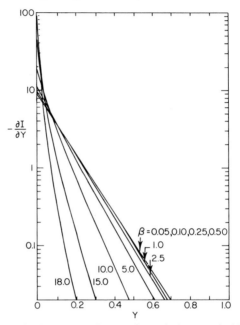

FIG. 10. Current distributions at steady state for cadmium anode (5N KOH), $\xi = 5.0$, $\Delta = 0.1$.

shows the distribution of the current density in the electrolyte of the pores within the matrix, i.e., $\delta I/\delta x$, as a function of the depths of penetration, x, for various levels of apparent current density, I_{ap}, which is expressed through the parameter, β, for a fixed equivalent transfer layer thickness, Δ. The latter is assumed to be 0.1 and parameter ξ assumed to be equal to 50. It is interesting to note that as β (i.e., apparent current density) is reduced below $\beta = 0.5$, the distribution is no longer sensitive to the current drain, i.e., the β value. In other words, at low current densities the distribution pattern or ratio remains constant. At higher current density levels, the distribution is significantly affected by the magnitude of apparent current density and the penetration into the depths of the electrode decreases rapidly as the apparent current density increases.

Parameter ξ has, as would be expected from its definition Eq. (4.26), a profound effect on the distribution. This can be seen best by inspection of Fig. 11 in which the distribution is plotted against the depths of penetration for three different values of ξ at a reasonably low apparent current density corresponding to $\beta = 1.0$. For values of ξ of less than 5 at low current densities, the current distribution within the porosity of the electrode approaches uniformity.

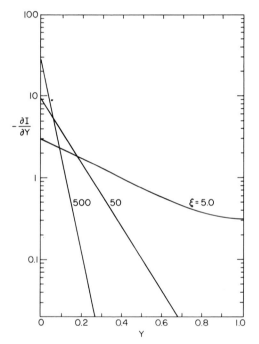

FIG. 11. Current distributions at steady state for cadmium anode (5N KOH), $\Delta = 0.1$, $\beta = 1.0$.

The transfer layer thickness (expressed through the dimensionless parameter, Δ) should, of course, have an influence on the electrode polarization. The polarization of the electrode expressed through parameter, Φ, at the electrode-electrolyte interface, i.e., at position $X = 0$, is of interest from the point of view of comparison to experimentally determined values. The computer results obtained for the electrode polarization parameter, Φ_0 at $X = 0$ (exclusive of the resistive iR potential drop across the transport layer) is illustrated in Fig. 12 as a function of the apparent current density (parameter β) for sets of values of parameters ξ and Δ. It should be noted that both scales in Fig. 12 are logarithmic. Thus, a *linear relationship* exists between the polarization Φ_0, and the current density directly (as expressed by β), rather than the logarithm of the current density. This brings out the important distinction between the polarization characteristic of a porous electrode and that of a planar solid electrode. While a solid electrode leads, for a Volmer–Tafel relation, to the well-known *logarithmic* relation between

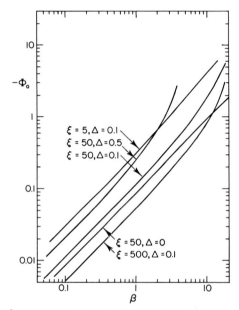

FIG. 12. Electrode overpotential at steady state for cadmium anode (5N KOH).

the polarization and the current density, for the porous electrode, a *linear relation* results. Tobias and Grens find for this linear relation, under the conditions assumed in their computations, a range of slopes from -0.087 to -0.303. The linear relation fails above β values of 5.

From a consideration of the distribution of the transfer current in the cadmium anode, it can be seen that the portion of the electrode depth contributing significantly to the current is rather small. In other words, the penetration of the reaction is limited. It is convenient to define the depth of penetration as that depth of the electrode from its frontal face in which 90% of the electrode reaction occurs, X_{90} (which corresponds to the value of X, at which the current in the electrolyte, I', has fallen to 10% of its value at the electrode face where the normalized current density $I' = 1.0$). It is useful to view the penetration as a function of the apparent current density, I_{ap} [as expressed through parameter β; see

Eq. (4.25)]. This is shown in Fig. 13 for various parameters ξ and Δ. The effect of boundary layer thickness (expressed through the dimensionless parameter, Δ) is illustrated for the constant parameter $\xi = 50$. For a given system in geometry, an increasing boundary layer thickness may rapidly reduce the penetration of the reaction; in other words, the portion of the electrode that participates. The depths of penetration

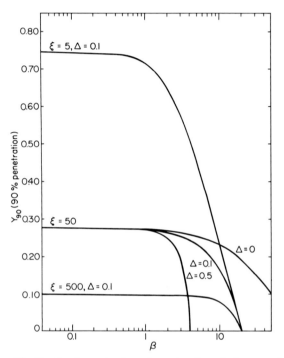

FIG. 13. Depth of penetration of reaction in cadmium anode.

reach asymptotic limiting values for decreasing values of apparent current density (decreasing β). This is in line with the previously pointed out characteristic of deeper distribution of the current density as the total level of the apparent current density decreases. It is interesting to note that for a given value of apparent current density (or β), the depths of penetration drop rapidly as ξ increases. With reference to Eq. (4.26), it can be seen that the reaction penetration or extent of utilization of a porous flooded electrode drops rapidly when

(1) the specific surface area of the matrix (A, cm^2/cm^3) is lower; and
(2) the thickness of the electrode is increased with all other parameters constant (note ξ proportionality to L^2).

It should be noted that the penetration becomes zero when the value of β, corresponding to the limiting superficial current density for the transfer layer thickness, Δ, is reached, as would be expected on the basis of mass-transfer considerations.

Another most interesting result from this model is that for very reversible electrode reactions, i.e., high values of the exchange current density i_0 [see Eq. (4.26)], the penetration is limited to a narrow portion nearest its frontal face (see bottom curve in Fig. 13).

While the Grens–Tobias model represents a generalized approach to the one-dimensional flooded porous electrode and provides new insights into this complex problem, it should strictly be applied only to those cases where the electrode matrix does not, as indicated previously, undergo changes during the reaction process. Such an electrode might be a redox electrode on an inert porous flooded electrode matrix. In the case of the $Cd/Cd(OH)_2$ "insoluble" electrode, changes in the matrix geometry occur due to the difference in the molar density of the reaction product. For this reason, the model is even less applicable to those cases in which the reaction product is soluble, as for the case for an alkaline zinc anode whose discharge

$$Zn + 4\,OH^- = Zn(OH)_2^= + 2e^- \qquad (4.32)$$

profoundly changes the matrix structure and also necessitates the consideration of another mass-transfer problem, namely, the transport of the zincate ions from the interior of the electrode matrix towards the free external electrolyte.

Where there are changes in the electrode matrix by anodic dissolution (or electrocrystallization upon charging) or by the formation of a new "insoluble" solid face, these affect not only A in parameter, ξ, but also result in changes in the local polarization caused by the passage of current with time. Such an analysis would require consideration of the effect of solid state transformations on current density distribution and overall polarization as the reaction proceeds with time.

V. Energy Conversion in Fuel Cells and Related Dynamic Systems

As pointed out earlier in this chapter, a fuel cell represents a dynamic form of a galvanic cell in which reactants and oxidants are continuously supplied to the cell. Since the mechanical transport of these reactants is facilitated when they are in a liquid or gaseous form, gas diffusion and liquid diffusion electrodes are prevalent in these devices, rather than solid reactants. On the other hand, solid reactants in the form of metal anodes are becoming of increasing interest in a related class of dynamic galvanic devices, namely *metal-gas cells.* These represent hybrids between fuel cells and batteries, in which the metal electrode is resident *in situ,* as in a conventional battery and the cathode, usually in the form of a gas diffusion electrode, is continuously supplied, as in a fuel cell.

In this section, rather than discuss the performance of complete fuel-cell assemblies, the reaction mechanisms of individual half-cells will be treated.

A. THE MECHANISM OF THE HYDROGEN ELECTRODE

The hydrogen electrode represents a subject of enormous interest to electrochemists since the early days of electrode kinetics (Tafel, 1905). The literature on the hydrogen electrode, especially for the platinum group of metals, is vast. It is not the purpose of this chapter to delve comprehensively into this area. However, the large number of studies in recent years pertinent to fuel cells have contributed significantly towards a better understanding of the hydrogen electrode in both acid and alkaline media. The hydrogen-oxygen fuel cell is at present by far the best-known and relatively best-developed fuel cell system.

Practical electrodes in fuel cells may involve the utilization of a metal (e.g., sintered nickel) or carbon-conducting and supporting matrix in which an appropriate electrocatalyst is disposed, together with possible conducting powders, binders, or nonwetting agents. To simplify the understanding of the mechanism of the hydrogen electrode, it is therefore useful to concentrate simply on the electrocatalyst as an electrode surface. A survey of the massive literature on fuel-cell research and development indicates the general preponderance of the platinum metals as the most effective electrocatalysts. While important efforts are under-way to develop more economical electrocatalysts, it is platinum which so far retains the major significance in this area. For this reason, the discussion of the mechanism of the hydrogen oxidation mechanism in this section will simply be reduced to a discussion for platinum.

The overall anodic oxidation process for hydrogen

$$H_2 + 2 OH^- \rightarrow 2 H_2O + 2e^- \qquad \text{(in alkaline media)} \qquad (5.1)$$

or

$$H_2 \rightarrow 2 H^+ + 2e^- \qquad \text{(in acid media)} \qquad (5.2)$$

can be divided into the following steps:

(1) mass transfer of the dissolved hydrogen through the electrolyte film to the electrode-electrolyte interface (see Section IV),

(2) adsorption of molecular hydrogen at the surface of the electrode,

(3) dissociative process (Tafel or Heyrovsky reaction),

(4) charge-transfer process, and

(5) proton solvation (in the case of acid electrolyte) and product (H_3O^+ or H_2O) mass transport away from the electrode.*

Process steps (2) and (3) have been the subject of a great many studies by various investigators (e.g., Frumkin (1932, 1933, 1937, 1963)). Depending on the course of step (2), we may encounter either the Tafel–Volmer mechanism (Tafel, 1905; Erdey–Grúz and Volmer, 1930):

$$(H_2)_{ad} \rightarrow 2H_{ad} \qquad (\text{Tafel } R \times n)$$

followed by Volmer charge transfer $R \times n$:

$$H_{ad} + OH^- \rightarrow H_2O + e^- \qquad (\text{in alkaline soln}), \qquad (5.3)$$

$$H_{ad} \rightarrow H^+ + e^- \qquad (\text{in acid soln}) \qquad (5.4)$$

or the Heyrovsky–Volmer mechanism (Heyrovsky, 1927a, b):

$$(H_2)_{ad} + OH^- \rightarrow H_{ad} + H_2O + e^- \qquad (\text{in alkaline soln}), \qquad (5.5)$$

$$(H_2)_{ad} \rightarrow H_{ad} + H^+ + e^- \qquad (\text{in acid soln}). \qquad (5.6)$$

The H_{ad} is oxidized anodically according to the Volmer reaction steps given above.

The performance of the mechanism depends on the conditions of the anodic oxidation and on the chemical and catalytic effects of the electrode. It is possible to divide the electrocatalytic metals into two groups in respect to their anodic hydrogen oxidation (Vielstich, 1961). In the first group of metals (to which Hg and Ag belong), hydrogen is not chemisorbed and the anodic oxidation is negligible. The second group (which includes the platinum metals) is marked by considerable chemisorption and relative ease of the chemisorption of hydrogen. Measurements on platinum, indium, rhodium, and palladium [(Breiter *et al.*, 1955) and (Aikasjan and Fedorowa, 1952)] indicate that the rates of chemisorption and hydrogen oxidation are determined by the transfer process.

---

* In an alkaline system where H_2O is the product, the transport "away" from the electrode, i.e., from the reaction zone, can be either into the liquid (electrolyte) phase or into the gas phase by counter-diffusion into the hydrogen feed stream in the backside of the electrode. In an acid system, the water product may be produced primarily at the cathode and may therefore be evaporated into the oxygen stream.

B. The Mechanism of the Oxygen Electrode

The oxygen electrode, especially in alkaline media, and the related air electrode have played an early role in electrochemical energy conversion devices. Grove (1839) employed an oxygen electrode in the first experimental hydrogen-oxygen fuel cell ever built. Most electrochemical fuel-cell systems today, and the metal-air cells (see Section V,F) employ the oxygen or the air electrode. Even in the field of long established carbon-air cell batteries, the so-called "air-depolarized electrode" was employed. Despite this early practical employment and widespread use of the oxygen electrode, elucidation of its mechanism is rather recent and to date still not complete.

Except for certain regenerative fuel-cell systems (see Section VI) in which the anodic evolution is involved, the major interest in the oxygen electrode in an energy-conversion device centers on the cathodic reduction process of the molecular oxygen which is in solution in the electrolyte. The formation of hydrogen peroxide, H_2O_2, as an intermediate product in the reduction of oxygen on a dropping mercury electrode was first observed by Heyrovsky (1927a,b). Two equal polarographic consecutive waves have been obtained; the first one corresponding to the reaction

$$O_2 + 2\,H^+ + 2e^- \rightarrow H_2O_2 \qquad (5.7)$$

and the second one corresponding to the reduction to water according to

$$H_2O_2 + 2\,H^+ + 2e^- \rightarrow 2\,H_2O. \qquad (5.8)$$

In 1941, Laitinen (see Laitinen and Kolthoff (1941)) found hydrogen peroxide in the solution following its cathodic reduction on a platinum microelectrode. In 1943, Berl (1943), who was concerned with the mechanism in a galvanic cell employing carbon, found hydrogen peroxide as an intermediate in the reaction on carbon electrodes. He proposed that the actual electrochemical reduction is only to the peroxide stage according to

$$O_2 + 2\,H^+ + 2e^- = H_2O_2 \qquad \text{(in acids).} \qquad (5.9)$$

Berl's work was confirmed and extended to alkaline media by Weiss and Jaffe (1948):

$$O_2 + H_2 + 2e^- = HO_2^- + OH^- \qquad \text{(in alkalis).} \qquad (5.10)$$

The extensive studies by Krasilshchikov (1947, 1949, 1952, 1953, 1954) included Ag, Pt, Au, Ni, and other metals as electrodes. For silver and gold electrodes, the following mechanism was proposed:

$$O_2 + e^- = O_2^- \tag{a}$$
$$O_2^- + H_2O = HO_2 + OH^- \tag{b}$$
$$HO_2 + e^- = HO_2^- \tag{c}$$
$$HO_2^- + H_2O = H_2O_2 + OH^- \tag{d}$$
$$H_2O_2 + e^- = OH + OH^- \tag{e}$$
$$OH + e^- = OH^-. \tag{f}$$

With polarographic techniques, Bagotsky and Yablokova (1953) confirmed the same mechanism for both acids and bases for a dropping mercury electrode. Using a variation of the Levich rotating disc electrode, namely, the disc-ring electrode, at which the ring section is capable of sensing unstable reaction intermediates, Frumkin and co-workers (1959) found evidence of the production of H_2O_2 at low overvoltages followed by its reduction at higher overvoltages on a gold amalgam electrode.

Direct experimental evidence of the peroxide mechanism was obtained by Davies *et al.* (1959) using ^{18}O isotope enriched oxygen for reduction on carbon electrodes. In the H_2O_2, the same relative amounts of $H^{18}O-^{18}OH$, $H^{16}O-^{16}OH$ were found as in the oxygen stream originally used. Hence, it is clear that the O—O bond is not broken during the reduction to the peroxide state.

The more recent work by Sawyer and Interrante (1961), who employed chronopotentiometry for the reduction of oxygen on platinum, palladium and nickel, supports the peroxide mechanism for these metals.

For the overall reduction of oxygen to peroxide

$$O_2 + 2\,H^+ + 2e^- \rightarrow H_2O_2,$$

the very complete studies by Krasilshchikov (1947, 1949, 1952–1954) resulted in the following relation for the electrode potential on silver as a function of oxygen concentration pH and current density:

$$E = A + \frac{RT}{0.5F}\ln[O_2] - \frac{RT}{0.5F}\ln|I|$$

(in acid and neutral solns),   (5.11)

$$E = A' + \frac{RT}{1.5F}\ln[O_2] + \frac{RT}{1.5F}\ln[H^+] - \frac{RT}{1.5F}\ln|I|$$

(in alkaline solns).   (5.12)

Equation (5.11) was confirmed for acid solutions on bright platinum (Winkelmann, 1956), and for acid and neutral solutions on mercury electrodes (Bagotskii and Yablokova, 1953).

If the transfer coefficient, α, can be assumed to be 0.5, the following more useful equation forms can be written:

$$I = -k_- \cdot [O_2] \cdot \exp - \left(\frac{(1 - \alpha)FE}{RT} \right)$$

(in acid and neutral solns), (5.11a)

$$I = -k_-' \cdot [O_2] \cdot [H^+] \cdot \exp - \left(\frac{(2 - \alpha)FE}{RT} \right)$$

(in alkaline solns). (5.11b)

From these, the electrochemical reaction orders are

$$Z_{0,O_2} = +1, \qquad Z_{0,H^+} = 0 \qquad \text{(acid)},$$
$$Z_{0,O_2} = +1, \qquad Z_{0,H^+} = +1 \qquad \text{(alkaline)}$$

These reaction orders are in agreement with the basic mechanism reported by steps (a)–(d) discussed above.

Once hydrogen peroxide is produced, three possibilities exist as to its further reaction:

(1) It may undergo further electrochemical reduction;
(2) It may undergo chemical catalytic decomposition at the surface of the electrode to oxygen and water;
(3) It may diffuse to the bulk of the electrolyte, buildup to a steady-state concentration, and then decompose by homogeneous redox reaction in the bulk.

Which course of action will take place depends a great deal on the nature of the electrode surface and the electrolyte used. In some cases, a combination of processes (1), (2), or (3) may take place. Experimental results obtained in strong alkaline solutions, especially on silver (Krasilshchikov, 1947, 1949, 1952–1954; Sawyer and Seo, 1962) indicate that the hydrogen peroxide is reduced electrochemically to hydroxide, probably in accordance with steps (a), (e), and (f) of the previously discussed mechanism, before any significant amount of peroxide can diffuse away from the electrode-electrolyte interface. The accumulation of hydrogen peroxide in the electrolyte bulk of a galvanic cell such as a fuel cell would, of course, be undesirable on simple considerations of the equilibrium potentials. Thus, in an alkaline solution, the reduction potential of oxygen to the hydroxide state is

$$O_2 + 2\,H_2O + 4e^- \rightarrow 4\,OH^-, \qquad -E^0 = 0.401 \text{ V} \qquad \text{(red. pot.).} \quad (5.13)$$

On the other hand, for the reduction to the peroxide stage in an alkaline electrolyte,

$$O_2 + H_2O + 2e^- \rightarrow HO_2^- + OH^-, \qquad -E^0 = 0.076 \text{ V} \qquad \text{(red. pot.),} \quad (5.14)$$

a substantially lower emf would result. Now, considering the second possibility discussed above, namely the catalytic decomposition of the peroxide at the surface of the electrode to oxygen and water, it can be shown that this also would result in larger emf values, simply due to the changes in activity. It can be readily calculated from the Nernst equation (for the reaction as written)

$$-E = -0.076 - \frac{RT}{2F} \ln \frac{(^aHO_2^-)(^aOH^-)}{(^pO_2)(^aH_2O)} \qquad (5.15)$$

that (ignoring changes in activity coefficients), the reduction potential of the oxygen electrode at 25°C is increased by $+0.290$ V if the steady state concentration of the peroxide ion can be held to a value of 10^{-10} mole/liter. Indeed, by use of silver or certain oxide catalysts on the cathode, the peroxide ion can be rapidly decomposed (into oxygen and hydroxyl ion) so that such low, or even lower, values of peroxide concentration can be obtained. Thus, in practice, a good decomposition catalyst may, by the decrease of the steady state activity of the peroxide ion, result in reaching emf values reasonably close to a reduction potential of 0.401 V. Many electrode materials, such as silver oxide, nickel oxide, platinum oxide, etc., are known to be very effective catalysts for hydrogen peroxide decomposition in alkaline media. This, therefore, leaves the question of the second part of the mechanism (i.e., the peroxide to hydroxide steps) open to discussion as to whether an electrochemical mechanism (steps (e) and (f), above) or a purely catalytic decomposition is the prevailing mechanism when electrodes of this type are employed. If it can be assumed that all of the hydrogen peroxide is rapidly reduced, i.e., that H_2O_2 is in equilibrium with OH^-, then the reaction can be represented by an overall reduction of oxygen to hydroxide ions. Thus, the deviations from the potential given in Eq. (5.14) are due to the fact that—as a result of catalytic decomposition—the steady state H_2O_2 activity is much less than unity. The catalytic decomposition of the peroxide yields oxygen which then is again available for the reaction. The net result is, therefore, that the faradaic yield is four equivalents per mole of oxygen in the same way as it would be if the reaction, expressed by Eq. (5.13), would be potential-determining. The mechanism of the oxygen reduction has been determined by measurements of the hydroxyl ion and peroxide ion concentrations (Berl, 1943), as well as through investigations of the partial pressure of oxygen during the reaction (Kordesch and Martinola, 1953).

It is important to recognize the dual role of an effective catalytic decomposition of the intermediate peroxide ion. First, this decomposition allows a positive shift in the oxygen reduction potential from -0.076 V towards the $+0.401$ V value of the four-electron mechanism, Eq. (5.13), and thus an increase in cell emf. Secondly, the faradaic yield is increased from 2 towards 4 electron-equivalents per mole of consumed oxygen. The experimentally determined open-circuit potential of an alkaline oxygen electrode can be used as a measure of the peroxide ion decomposition capabilities of the electrode and its catalysts. This is an interesting case in which one finds how the kinetics of a reaction may affect the thermodynamic aspects of an electrode process.

In acid solutions, the reduction of oxygen is far more difficult than in alkaline media due to the fact that even the first step, oxygen to peroxide, proceeds with a high degree of irreversibility. Furthermore, the peroxide ion is very stable in acids. This is one of the major electrochemical problems in fuel cell technology since the desirability of an acid electrolyte in hydrocarbon fuel cells is obvious from the point of view of removal of the carbon dioxide by-product.

If the electrochemical reduction of peroxide to hydroxide is considered, the previously discussed overall mechanism of the oxygen electrode can now be restated by including the effect of the electrode surface. It is obvious that an adsorption step for the oxygen on the metal surface must occur once the oxygen dissolved in the electrolyte has arrived at the electrolyte surface. If the adsorption site is denoted by M, the reaction mechanism can be expressed as a sequence of the following seven steps:

(1) $$O_2 + M = MO_2$$
(2) $$MO_2 + e^- = MO_2^-$$
(3) $$MO_2^- + H_2O = MHO_2 + OH^-$$
(4) $$MHO_2 + e^- = MHO_2^-$$
(5) $$MHO_2^- + H_2O = MH_2O_2 + OH^-$$
(6) $$MH_2O_2 + e^- = MOH + OH^-$$
(7) $$MOH + e^- = M + OH^-.$$

Of these steps, (1) or (2) is rate controlling (Krasilshchikov, 1947, 1949, 1952–1954). If the other steps proceed rapidly, the products and the reactants of those steps can be related by equilibrium constants. If the charge-transfer step (2) is rate controlling, the following expression for a charge-transfer current can be obtained (Bennion, 1964). For reaction (2), one can write

$$J = [k_2]MO_2 \exp[(\alpha_2 - 1)FE/RT] - k_{-2}[MO_2^-]\exp[\alpha_2 FE/RT].$$
(5.16)

For reaction (1), the equilibrium can be expressed

$$K_1 = [MO_2]/[O_2][M].$$
(5.17)

Combining reaction steps (3)–(7) as one equilibrium reaction in which the reduction proceeds from MO_2^- to $4\,OH^-$, an equilibrium constant can be expressed in terms of the activities and the electrode potential, E, as follows:

$$K_{3-4} = \frac{[OH^-]^4[M]}{[MO_{2^-}][H_2O]^2} \exp\left(\frac{3FE}{RT}\right).$$
(5.18)

Combining these equations to eliminate activities of the intermediate species, the rate equation in terms of a charge-transfer current is obtained for the case of control by the charge-transfer step (2):

$$J = k_c [O_2] \exp\left[\frac{(\alpha_2 - 1)FE}{RT}\right] - k_a \frac{[OH^-]^4}{[H_2O]^2} \exp\frac{(\alpha_2 + 3)FE}{RT}.$$
(5.19)

Through similar procedures, it can be shown that if step (4) (which leads to the creation of the peroxide ion HO_2^-) is rate controlling, the charge-transfer current is

$$J = k_c \frac{[O_2][H_2O]}{[OH^-]} \exp\left[\frac{(\alpha_4 - 2)FE}{RT}\right] - k_a \frac{[OH^-]^3}{[H_2O]} \exp\left[\frac{(\alpha_4 + 2)FE}{RT}\right].$$
(5.20)

If the potential, E, is sufficiently negative, i.e., an appreciable cathodic polarization exists, Eqs. (5.19) and (5.20) reduce to the simpler expressions obtained by Krasilshchikov (1961) for acid solutions and alkaline solutions, respectively.

In the case of silver which is an important electrocatalytic surface for the reduction of oxygen in an alkaline solution, step (4) was found to be rate controlling (Krasilshchikov, 1961). The same is probably true about gold, mercury, and carbon electrodes. The exchange current densities reported in the literature for the cathodic reduction of oxygen vary widely from 10^{-7} to 10^{-4} A/cm^2. Values as low as 10^{-20} have been reported, but these are obviously for very poor electrocatalytic surfaces.

Many electrochemical energy conversion systems are based, of course, on the use of air as the cathodic reactant, instead of pure oxygen. The basic electrode kinetic mechanism of the reaction is the same as discussed above. The only additional consideration is the gas side concentration polarization associated with the boundary-layer effects when oxygen is being consumed and must transfer through stagnant nitrogen within the pores of the electrode. When air is used as a depolarizer in the case of an alkaline electrolyte, the prior removal of CO_2 is an important consideration to avoid a build-up of carbonates in the electrolyte and thus assure its invariance. This precaution is not necessary in acid or neutral electrolytes.

C. Hydrocarbon Anodic Electrodes

On elementary thermodynamic considerations of energetics, the hydrocarbons represent a most attractive group of anodic fuels. This realization has in recent years given rise to much intensified applied and basic research in the area of direct hydrocarbon-oxygen fuel cells (see e.g., Baker (1965)). Studies in the 1950's have been almost exclusively devoted to molten carbonate electrolytes, following historical precedent in this area (see e.g., Broers (1958)). Complete butane-air fuel cells operating at reasonably high ideal thermal efficiencies have been demonstrated (Baker and Eisenberg, 1964). The technical and material difficulties of high temperature operation with fused salt electrolytes have given impetus to efforts to find more favorable conditions for the direct anodic oxidation of hydrocarbons. Grubb (1962) reported direct oxidation of methane, ethane, propane, and other lower paraffins on platinum black catalyst in a concentrated phosphoric acid electrolyte at 65°C. Galvanostatic and gas chromatographic studies have been made by Niedrach (1964) on C_1 through C_4 hydrocarbons, both saturated and unsaturated, on platinum black, Teflon-bonded fuel-cell anodes in a variety of aqueous electrolytes. In dilute sulfuric electrolyte, all low molecular weight aliphatic hydrocarbons show some adsorption on platinum black at room temperature. The surface coverage is low for methane, intermediate for saturated hydrocarbons, and high for the unsaturated. The rates of adsorption are relatively high in the acidic electrolyte. However, in alkaline and carbonate electrolytes, adsorption of saturated hydrocarbons is very slow, while the adsorption rate of unsaturated hydrocarbons remains high.

During equilibration on open circuit many surface reactions, including dehydrogenation, hydrogenation, cracking, and polymerization occur. These are preceded by dissociative chemisorption with formation of some hydrogen on the surface. This dissociating hydrogen oxidizes readily. However, the carbonaceous surface species do so with high overvoltages. An increase in temperature was found, as expected, to lower that overvoltage.

The mechanism of the anodic oxidation of hydrocarbons is naturally more involved and difficult to elucidate. Through the use of a multipulse potentiodynamic transient technique, Gilman (1965) studied the oxidation of ethane on platinum wire microelectrodes and found that the primary adsorption step corresponds to the formation of an ethyl radical on the surface as follows:

$$C_2H_{6(g)} + S \rightarrow S\text{---}C_2H_5 + H^+ + e. \tag{a}$$

This occurs at potentials at which reasonable current densities are drawn from the electrode. Another equivalent process is one in which there is no charge transfer and the hydrogen also becomes surface bound:

$$C_2H_{6(g)} + 2\,S \rightarrow S\text{---}C_2H_5 + S\text{---}H. \tag{b}$$

The adsorbed hydrogen is readily oxidized by the Volmer reaction:

$$S\text{---}H \rightarrow S + H^+ + e. \tag{c}$$

Gilman suggested two reaction paths for the adsorbed ethyl radical:

(1) The first results in the formation of four surface bonds without the rupture of the C—C linkage. Simultaneously, a 3-electron charge-transfer process takes place:

$$S\text{---}C_2H_5 + 3\,S \rightarrow S_4\text{---}C_2H_2 + 3\,H^+ + 3e. \tag{d}$$

(2) The second reaction path of the C_2H_5 radical involves, according to Niedrach (1965), the breakage of the C—C bond, combined with further charge-transfer reaction for the freed hydrogen, as shown by the following general equation:

$$S\text{---}C_2H_5 + S \rightarrow 2\,S\text{---}CH_a + (5 - 2a)H^+ + (5 - 2a)e. \tag{e}$$

The above is in line with observations of C_1 species formation during ethane or propane adsorption on the electrode, even at room temperature (Niedrach, 1964).

The C_1 species appears to react readily with water to form a partially oxygenated compound according to the general scheme

$$S\text{---}CH_a + bH_2O + cS \rightarrow S_{(c+1)}\text{---}CH_dO_b + (a + 2b - d)H^+ + (a + 2b - d)e. \tag{f}$$

The specific identity of the oxygenated species has not been established. Reactions (e) and (f), above, correspond to cracking and reforming reactions, respectively. Therefore, catalysts which promote these reactions are potentially interesting electrocatalysts.

It appears that the electrolytic process for anodic oxidation of hydrocarbons involves the cracking of high molecular weight hydrocarbons to form C_1 radicals and C_1 oxygenated species before the final oxidation to carbon dioxide.

D. Liquid Diffusion Electrodes

As discussed towards the end of Section II, when the reactant is soluble in the electrolyte of the fuel cell it can be supplied continuously to the cell either by diffusion through the backside of a porous electrode matrix or by being carried to the electrode by means of the electrolyte circulation. In the latter case, the supply stream of electrolyte enriched in the reactant ("fuel") can be introduced in front of the working electrode, i.e., between the cathode and the anode, or in the case of a porous electrode plate, through the backside of the electrode. Combinations of such feeding methods may, of course, be employed. When a porous electrode is used (instead of a solid metal sheet), a liquid diffusion electrode prevails analogous to the flooded porous electrode discussed in Section IV. Depending on the method of reactant supply (e.g., frontal, backward, or combination), the boundary conditions may have to be restated for the Grens–Tobias (1964) equations. The problem can then be solved by numerical techniques, as discussed in Section IV,C.

Hydrazine has been demonstrated to represent a very promising anodic reactant in fuel cells employing alkaline electrolytes (Eisenberg, 1963a,b). A hydrazine anode is particularly attractive in comparison to a hydrogen electrode, simply because hydrazine remains very active and the problem of electrocatalysis is much simplified. In addition, hydrazine with a liquidity range close to that of water represents a fuel which is readily storable, unlike hydrogen for which gaseous or cryogenic storage involves considerable weight penalties in terms of storage requirements. In a typical realization of a hydrazine electrode, 0.5 to 2.0 molar solution of N_2H_4 may be employed in a 6 molar potassium hydroxide electrolyte. The electrode matrix may be a porous nickel plaque, catalyzed with platinum black or with suitable alloys of the platinum metal group. Under these conditions, the reaction proceeds quantitatively to nitrogen and water in accordance with

$$N_2H_{4aq.} + 4\,OH^-_{aq.} = N_{2(g)} + 4\,H_2O\,(1) + 4e^-, \qquad E_A^0 = -1.16\ \text{V}. \quad (5.21)$$

This "liquid anode" is readily compatible with the oxygen electrode for which the alkaline electrolyte is also most favorable. Hence, in combination with an oxygen cathode,

$$O_{2(g)} + 2\,H_2O + 4e^- = 4\,OH^-, \qquad E_c{}^0 = 0.401 \text{ V}, \qquad (5.13a)$$

a hydrazine-oxygen fuel cell with a theoretical emf of 1.56 V is formed. When the peroxide mechanism is predominant at the oxygen cathode, the corresponding theoretical cell emf is only 1.12 V. It is interesting to note that the theoretical energy density for a hydrazine-oxygen system is 1185 Wh/lb of combined reactants, compared to 1620 for the hydrogen-oxygen system.

For a porous nickel electrode, catalyzed with a platinum metal, the anodic oxidation of hydrazine involves a relatively low charge-transfer polarization. The process is essentially controlled by the mass-transfer polarization which in turn is a strong function of system geometry, electrode porosity, reactant concentration, feeding rates, and the resulting system hydrodynamics. Since hydrazine to some extent will also spontaneously decompose on the surface of the electrodes at open circuit conditions, i.e., when no electrochemical reaction is employed, one must consider the faradaic utilization efficiency under operating conditions. If this decomposition rate can be expressed in the form, $N_D = k_D \cdot A \cdot c$, where k_D represents the rate constant, and A and c the apparent area and concentration respectively, then the faradaic efficiency of the anode can be expressed as

$$\eta_F = \frac{I/nF}{i/nF + k_D \cdot c}.$$

Thus, improvement in faradaic efficiency should result from an increase in current density, all other variables being constant. Indeed, as shown from the upper curve of Fig. 14, faradaic efficiencies rise rapidly with increasing current density, and current densities as high as 200 mA/cm^2 may reach 94%. Figure 14 also shows how reasonably small the total net anodic polarization is even at appreciable current densities, e.g., 0.16 V at 200 mA/cm^2 of apparent electrode surface area.

Polarization studies for the anodic oxidation of hydrazine on nickel and on platinized nickel electrodes have been carried out by Eisenberg (1962a,b). For the platinum black electrode, a Tafel slope $b = 0.120$ was obtained (see Fig. 15) which points to a slow discharge step as controlling. On the basis of these data as well as open circuit potential measurements, the hydrazine electrode mechanism was analyzed as

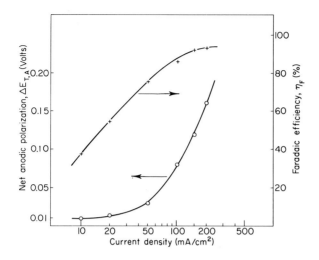

FIG. 14. Polarization and Faradaic efficiency for a hydrazine anode.

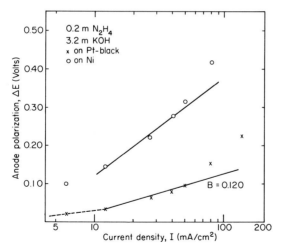

FIG. 15. Hydrazine anode polarization.

including an active hydrogen-type electrode in the following series of steps:

(1) $\qquad\qquad$ $N_2H_4 \xrightarrow{\text{Pt}} 4\,H(+N_2)$

(2a) $\qquad\qquad$ $4\,H \rightleftharpoons 2\,H_2$ $\qquad$ (2b) $\quad 4\,H \rightleftharpoons 4\,H_{ads}$

$\qquad\qquad\qquad\qquad\qquad$ alternatives

(3) $\qquad\qquad$ $2\,H_2 \rightleftharpoons 4\,H$

(4) $\qquad\quad$ $OH^- + H_{ads} \underset{\overleftarrow{}}{\overset{i_{-4}}{\rightleftharpoons}} H_2O + e^-$ $\qquad$ (discharge step).

The value of the Tafel slope can be explained if a slow discharge step (4) is assumed to be controlling, according to the Volmer mechanism for an alkaline solution. For an assumed transfer coefficient of $\alpha = 0.5$, the experimental results of Fig. 15 yield an exchange current density, $i_0 = 7.4 \times 10^{-3}$ A/cm^2. This comparatively high value of the exchange current density indicates a fairly reversible nature of the hydrazine electrode on black platinum.

At current densities above 70 mA/cm^2 concentration, polarization enters the picture more significantly and deviations from the Tafel slope occur, as would be expected.

When uncatalyzed nickel electrodes were used, substantially higher values of polarization were experienced (see Fig. 15), indicating that steps (1) and (2b) may be rate limiting. On the other hand, active catalysts such as platinum black, make the production of atomic hydrogen (step 1) and the adsorption-desorption step (2b) occur readily so that only the charge-transfer step (4) becomes relatively rate controlling.

Alcohols, such as methanol, also represent attractive liquid fuels in fuel cells. Since carbon dioxide is one of the products of the anodic reaction, the electrolyte, in order to remain invariant, must reject the product. This eliminates alkaline and neutral electrolytes from consideration and focuses attention on acids, such as HF, and on carbonate electrolytes. In such media the overall anodic reaction is

$$CH_3OH + 3\,CO_3^= \rightarrow 4\,CO_2 + 2\,H_2O + 6e^-.$$

Oxidation to formates represents one of the side reactions which occur to some extent. In acid media, the polarization on platinum is higher for the anodic oxidation of methanol. Platinum-ruthenium alloy catalysts prepared by borohydride precipitation techniques have been reported to be more effective than platinum alone (Heath, 1964).

Cesium carbonate electrolytes have recently been found to offer, at an operational temperature of approximately 120°C, conductivity levels which make a methanol-air fuel cell relatively attractive (Cairns and

Bartosik, 1964). Ideal thermal efficiencies of 42% can be obtained at operating cell emf's of 0.55 V at an apparent current density of 20 mA/cm², with Teflon-bonded platinum electrodes.

E. Fuel-Cell Systems

While this chapter is principally concerned with the mechanisms of the individual electrode reactions which are part of the overall energy conversion process in a galvanic cell, a few basic requirements for the successful coupling of cathodic and anodic electrodes into a balanced operating fuel cell should be studied:

(1) The net overall cell reaction must be such as to provide for electrolyte invariance in continuing operation.

(2) Consequently, the fuel-cell system design must provide for the removal of by-products at the rate at which they form.

(3) The design of electrodes and cells must prevent the reaching of an anodic reactant to a cathode and vice versa in order to avoid not only losses of reactants, but also interference with the individual electrode processes.

(4) When electrolyte circulation is employed, the hydrodynamic design of the system and the circulation rate must be optimized from a mass transfer point of view for optimum system performance.

Once the basic physicochemical and electrochemical parameters for both electrode reactants are available, the design of a fuel-cell system can be rationalized in the form of an optimization procedure designed to achieve one or more goals, such as maximizing power density. An example of the latter (Eisenberg, 1962a,b) is illustrated in Fig. 16 which shows single cell power densities per unit cross-sectional area as a function of the current density and linear circulation velocity of the electrolyte between the parallel electrodes in the cell.

The dynamic design of a fuel-cell system profoundly influences the mass-transfer polarization characteristics of the system which expresses itself as concentration polarization, $\Delta E_{conc,L}$, on the electrolyte side; and concentration polarization, $\Delta E_{conc,G}$, in a gas phase. Since these forms of polarization may represent the cause of major power losses in a practical fuel-cell system operating at appreciable current densities, the electrochemical engineering aspects of mass transfer both in the electrolyte and in the gas phases of the system deserve considerable attention.

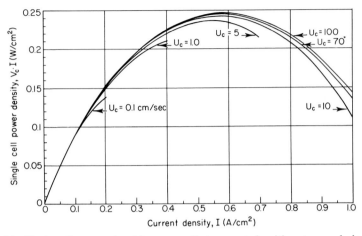

FIG. 16. Single cell power densities vs. applied current densities at several electrolyte velocities. (H_2—O_2) fuel cell.

F. METAL-GAS HYBRID CELLS

As pointed out in the beginning of this section, metal-air cells represent a hybrid between a battery and a fuel cell, by combining *in situ* a metal anode with a dynamically supplied gas cathodic electrode. As far as the cathodic reactant is concerned, the halogen gases, nitrogen tetroxide, and

TABLE I

THEORETICAL ENERGY DENSITIES OF METAL-OXYGEN GALVANIC SYSTEMS[a]

Metal	Cell Reaction Product	Cell emf V_R^0	Energy Density Wh/lb Metal
Be	$Be(OH)_2(C)$	3.02	8142
Al	$H_2AlO_3^-(Ag)$	2.75	3718
Mg	$Mg(OH)_2(C)$	3.09	3090
Ca	$Ca(OH)_2(C)$	3.42	2075
Mn	$Mn(OH)_2(C)$	1.96	868
Zn	$Zn(OH)_2(C)$	1.65	614
Fe	$Fe(OH)_2(C)$	1.28	556
Co	$Co(OH)_2(C)$	1.13	466
Ni	$Ni(OH)_2(C)$	1.12	464
Cd	$Cd(OH)_2(C)$	1.21	262

[a] Based on Fleischer (1968).

oxygen may be considered. For both historical and technical reasons, only oxygen cathodes have reached a measure of development. In the form of low rate zinc-"air depolarized" cells, the alkaline zinc-oxygen system has been used for a long time.

The attraction of metal-oxygen galvanic systems lies in the high theoretical energy densities available from a number of combinations, as shown in Table I for a selected number of metals, regardless of the practicality of the system.

The achievement of a practical system is often limited by physico-chemical and electrochemical considerations such as thermodynamic stability towards the electrolyte, kinetic behavior of the metal, passivation phenomena, etc. The past five years have witnessed a rising interest in metal-air cells (Fleischer, 1968), and particularly in more efficient forms of the zinc-oxygen system.

The reactions in the zinc-oxygen cell can be presented.

$$\text{Anode:} \quad Zn + 4\,OH^- \rightarrow ZnO_2^= + 2\,H_2O + 2e^-,$$
$$E_r = 1.216 \text{ V} \qquad (5.22)$$

$$\text{Cathode:} \quad \tfrac{1}{2}O_2 + H_2O + 2e^- \rightarrow 2\,OH^-, \qquad E_r = -0.401 \text{ V} \quad (5.23)$$
$$\overline{Zn + \tfrac{1}{2}O_2 + 2\,OH^- \rightarrow ZnO_2^= + H_2O,}$$
$$V_r = 1.617 \qquad (5.24)$$

Of interest is the mechanism of zinc oxidation in the alkaline electrolyte. It has been found that the hydroxyl ion activity determines whether the reaction product is $Zn(OH)_4^=$ or $Zn(OH)_2$ (Kober and West, 1967). In low OH^- activity solutions, the solid phase orthorhombic E-$Zn(OH)_2$ is the anodic oxidation product. Farr and Hampson (1967) suggested the following mechanism for the zinc anode on the basis of double-impulse transient studies:

$$Zn + OH^- \xrightarrow[\text{slow}]{\text{very}} ZnOH_{ad} \qquad (a)$$

$$ZnOH_{ad} \xrightarrow{\text{slow}} Zn(OH)_{ad} + e^- \qquad (b)$$

$$Zn(OH)_{ad} + OH^- \xrightarrow{\text{fast}} Zn(OH)_2 + e^- \qquad (c)$$

$$Zn(OH)_2 + 2\,OH^- \xrightarrow{\text{fast}} Zn(OH)_4^=. \qquad (d)$$

The effect of the hydroxyl ion can be anticipated in going from step (c) to step (d).

The passivation of the zinc anode in the alkaline electrolyte is of considerable importance not only in metal-air cells, but also in alkaline batteries. Landsberg (1957) has correlated experimental results of anodic dissolution of zinc in sodium hydroxide with the expression in the form

$$I - I_0 = Bt^{-1/2} \qquad (5.25)$$

where t is the time to passivation, and B and i_0 are constants, the latter increasing with the hydroxyl ion concentration in the electrolyte.

The functional dependence on time in this equation indicates a diffusional process. Eisenberg *et al.* (1961) analyzed the passivation problem from a mass-transfer point of view and concluded that in the presence of natural convection in the cell (depending on the relative position of the anode within the cell) different relationships would apply. Indeed, only for passivation experiments at high current densities with a duration of less than 20 sec (i.e., a time too short for convective flow patterns to establish) was a linear relationship with $t^{-1/2}$ established in the following form:

$$I - 0.0636 = 0.846 \cdot t^{-1/2} \qquad (5.26)$$

where the I value has a range of 0.25 to 0.60 A/cm^2 and t has a range of 2.5–19 sec. However, when the process proceeds at lower current densities and free convection boundary layers develop, e.g., at a vertical electrode or horizontal electrode facing downward in a gravitational field, considerably larger limiting current densities are possible before anode passivation ensues. These phenomena studied on smooth zinc anodes become naturally more complicated in a porous zinc anode. Here, the role of natural convection is minimized and the major mass transport occurs by diffusion.

VI. Regenerative Fuel-Cell Systems

The electrochemical fuel cell is sometimes referred to as a chemical fuel cell to emphasize the direct conversion of chemical into electrical energy. It is possible, however, to combine the advantages of galvanic energy conversion with other sources of energy in closed loop regenerative systems of one type or another. The following types of regenerative fuel-cell systems will be briefly discussed:

(a) thermally regenerative systems,
(b) electrically regenerative systems, and
(c) photochemically regenerative systems (including nuclear radiation sources).

In a thermal regenerative system, the fuel cell operating at a temperature, T_1, is combined with a regenerator operating at a higher temperature, T_2, as shown in Fig. 17. The reaction product of the fuel cell, AC, is decomposed into anodic and cathodic reactants A and C, respectively, in the regenerator. These are then fed into the fuel cell which provides its electrical energy output to the load. The thermally regenerative fuel

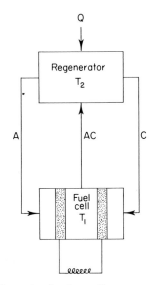

FIG. 17. Schematic of a thermally regenerative fuel cell.

cell can be analyzed in terms of a Carnot cycle (Eisenberg, 1963a) and the useful work of the cycle related to the thermodynamic functions of the cycle by the relations:

For $\Delta C_p < 0$:

$$W = (-\Delta S_I)(T_2 - T_1)$$
$$= -\Delta G_I - \Delta G_{III} - \left\{ \int_{T_1}^{T_2} \Delta C_p \cdot dT - T_2 \int_{T_1}^{T_2} \Delta C_p \cdot d \ln T \right\}, \quad (6.1)$$

For $C_p > 0$:

$$W = \Delta S_{III}(T_2 - T_1)$$
$$= -\Delta G_I - \Delta G_{III} - \left\{ \int_{T_1}^{T_2} \Delta C_p \cdot dT - T_1 \int_{T_1}^{T_2} \Delta C_p \cdot d \ln T \right\}. \quad (6.2)$$

T_1 and T_2 are the temperatures of the fuel cell and regenerator respectively.

Subscript I   refers to the reaction A + C = AC occuring in the fuel cell.

Subscript III   refers to the dissociation reaction AC = A + C occurring in regenerator temperature T_2.

The difference between an ordinary Carnot cycle and the one represented by a thermal regenerative fuel cell is seen in the bracketed expressions in the above equations. These expressions represent the energy dissipated due to the irreversibilities of the heat and cooling steps. These quantities are always positive and correspond to the lowering of the efficiency of the cycle as compared to that of an ideal Carnot cycle. In the selection of chemical systems for consideration in a thermally regenerative fuel cell, it is desirable that the free energy change in the fuel cell (ΔG_I) shall be as negative as possible. ΔG_{III} in the regenerator and ΔC_p should be as close to zero as feasible. In addition, practical considerations suggest that preference should be given to chemical systems which would yield a reasonably large open circuit emf. A minimum value of 0.5 V appears desirable.

Table II lists some systems of potential interest for thermally regenerative fuel cells.

TABLE II

THERMODYNAMIC DATA FOR SOME CHEMICAL SYSTEMS IN THERMALLY REGENERATIVE FUEL CELLS

Fuel-cell reaction	V_r^0 (V at 298°K)	Ideal thermal efficiency (cell alone) (η_T, %, at 298°K)	V_r^0 (V at 700°K)	Ideal thermal efficiency (cell alone) (η_T, %, at 700°K)
$Li + \frac{1}{2} H_2 = LiH$	0.731	78	0.435	63
$Ti + Cl_2 = TiCl_2$	2.083	84	1.965	80
$Co + Cl_2 = CoCl_2$	1.46	86.8	—	—
$Cd + I_2 = CdI_2$	1.04	100	0.838	63
$Ga + \frac{3}{2} I_2 = GaI_3$	0.839	80	0.554	54

The thermal efficiencies in Table II pertain only to the cell itself and not to the overall regenerative cycle. In practice, the problem of dissociation kinetics and separation of reactants A and C may offer difficulties in the development of a system. The lithium hydride cell has been studied for some time (e.g., Werner et al. (1959)).

The use of heat from nuclear reactors in combination with thermally regenerative fuel cells has been proposed occasionally. Clearly, systems of this type are thermally regenerative since nuclear radiation is not involved. The use of nuclear heat sources is currently contemplated in combination with bimetallic galvanic cells in which a molten alkali metal anode operates against a cathode constituted by its alloy with another metal, usually of the Group V-B or VI-B. A lithium–tellurium thermal cell of the following scheme was investigated by Shimotake *et al.* (1967):

Li	LiCl– LiI –LiF	Li in
anode	49 at % 59% 2%	Li–Te alloy cathode
	electrolyte	

A cell of this type can be used as a simple galvanic concentration cell. When the lithium activity in the cathode is low, appreciable current densities can be achieved at the elevated temperatures (typical at 400–500°C). For a complete thermal regenerative system, a portion of the lithium-tellurium alloy liquid cathode might be circulated through a regenerator (heat source) which would remove the lithium and return it to the supply stream to the liquid anode.

As already indicated, in some circumstances it may be desirable to employ electrical energy to decompose the reaction product of a fuel-cell reaction into reactants to be recycled. Thus, in this concept, an electrolysis is combined with the galvanic reaction. Such an approach may be considered when it is desirable to *store* electrical energy. Thus, an electrically or electrolytically regenerative fuel cell simply would function in place of a storage battery since it would permit the storage of dc energy. Such a system can be accomplished by use of a separate electrolyzer; for instance, in the case of a hydrogen-oxygen fuel cell, the use of water electrolyzers (Lee, 1960). Alternatively, the electrolysis may be performed in the fuel cell itself. Considering the fact that energy storage efficiency in conventional batteries is rather high, electrically regenerative fuel cells may not be advantageous except in certain special missions where large quantities of energy would have to be stored in any one cycle.

The photochemical regeneration approach may principally involve any useful radiation in the electromagnetic spectrum, regardless of source. A study relating to the utilization of the visible and near ultraviolet light was carried out by Eisenberg and Silverman (1961). In the light absorbing compartment of the regenerator unit, an organic dye such as proflavine (D) in solution is photochemically bleached by ascorbic acid, H_2A.

$$hv + D + H_2A \rightarrow DH_2 + A. \tag{6.3}$$

The solution containing the reduced dye, DH_2, is electrochemically oxidized by a suitable redox couple such as stannic-stannous fluoride contained in the dark side (opposite half-cell) of the regenerator unit. Thus,

$$DH_2 \rightarrow D + 2\,H^+ + 2e, \tag{6.4a}$$

$$Sn^{+4} + 2e \rightarrow Sn^{+2}. \tag{6.4b}$$

Hence, the overall process in the regenerative unit is the oxidation of ascorbic acid and the reduction of stannic ion,

$$H_2A + Sn^{+4} \rightarrow A + 2\,H^+ + Sn^{+2} \tag{6.5}$$

with the passage of current through an external circuit. In the dark "fuel cell," the reaction is reversed and the stannous ion reduces the oxidized ascorbic acid electrochemically.

$$A + 2\,H^+ + 2e \rightarrow H_2A, \tag{6.6a}$$

$$Sn^{+2} \rightarrow Sn^{+4} + 2e. \tag{6.6b}$$

Power may be drawn from either cell or both. To preserve the mass balances for the reactions it is, of course, necessary that currents drawn from each cell be equal to each other. Thus, the anodic reaction involves the electrochemical utilization of organic *leuco* dyes which have been obtained *in situ* through the photoreduction process. In practice, systems of this type are limited to low power density, basically due to the low quantum yields for the power producing chemical species. For the cited work, the quantum yield was found to be 2.3% for an absorption peak of 440 μ. The low photochemical efficiency limits, thus, performance of photoregenerative fuel cells based on such organic dyes.

VII. Batteries

In this section, batteries are defined as galvanic cells, in which usually solid cathodic and anodic reactants are present *in situ* on the plates within the cell enclosure. Attention is given to comparative energetics of systems and to some common aspects, rather than to the detailed characteristics of individual batteries.

Depending on the manner of use, batteries can be classified as *primary* (serving one useful discharge) or *secondary* (rechargeable for repeated use). Sometimes a given galvanic system may be employed either as a primary or as a secondary battery. However, the criteria for a stable secondary battery, capable of an extensive cycle life, are much more critical. The requirements include the following:

(a) The active cathode oxide or salt must have an exceedingly low solubility in the electrolyte in order to minimize the concentration of noble metal ions in the electrolyte, which upon diffusion (across separators) towards the anode would be reduced to the noble metal form (cementation), resulting in the creation of local action cells conducive to a rapid deterioration of the anode.

(b) The electrocrystallization characteristics of each of the electrodes, both for the discharge and charge reaction, should not involve major changes in molar density to provide the basis for the structural integrity of the porous electrode and the preservation of desirable microporosity characteristics in going from the state of charge to the state of discharge and reverse.

(c) The anodic product of the discharge reaction (usually a hydroxide or a halide salt) should be insoluble in the electrolyte to assure the availability of the salt and to reduce the mass-transfer problem within the pores of the plate when recharge to the metallic state is undertaken.*

(d) All electrode active species in both states of oxidation must be stable towards the electrolyte to avoid undesirable side reactions or losses through dissolution.

(e) The anode oxidation product must not be passivating at reasonably high levels of current density, i.e., it must not form dense nonconducting films, since the diffusion and migration of anode metal ions from the side of the metal through the lattice of the oxide or salt layer constitutes the basic transport mechanism in the discharge of these electrodes of the second kind. (For a discussion of such a mechanism in the alkaline cadmium electrode, see Milner and Thomas (1967).)

Since not too many electrochemical systems are capable of meeting the above criteria to a high degree, it is not surprising that the number of stable secondary batteries available is considerably less than that of primary batteries.

* The considerable mass-transfer polarization in the case of a soluble anode leads to a rapid reaching of limiting conditions conducive to electrodeposition of the anode metal in dendritic or powdery nonadherent forms, as best exemplified by the soluble zinc anode in alkaline electrolytes.

In considering battery systems, it may be useful to compare their *theoretical energy densities* (e.g., watt-hours per pound). This energy density takes into account the weight of active materials only (ignoring the weight of electrode supports, electrolyte, structures, etc.), and for simplicity, the computation is usually carried out using equilibrium potentials (frequently even without corrections for deviations from unit activity). This method of comparison is, therefore, only useful from the point of view of relative energetics of the system. The relative merits of individual potentially useful cathode or anode materials can also be viewed separately if one compares the weight requirement for a given coulombic yield and the individual electrode potentials. Such an example is given in Tables III and IV (Uhler *et al.*, 1963).

The potentials used in these tables are approximate standard potentials in aqueous acids (E_a) and in bases (E_b), respectively. The weight requirements for a 500 Ah coulombic yield is listed in the first column in both tables.

The energy density, E.D., in watt-hours per pound can be simply calculated for various couples using the data of Tables III and IV in the relation:

$$\text{E.D.} = 500 \frac{[E_A - E_C]}{[\text{lb}_a + \text{lb}_c]} \quad (\text{Wh/lb}). \quad (6.7)$$

TABLE III

THEORETICAL PROPERTIES OF SOME CATHODE MATERIALS

Cathode material	lb$_c$ lb/500 Ah	Theor. cathode pot., E_C E_a	E_b
Oxygen	0.33	−1.23	−0.401
m-Dinitrobenzene	0.59	−0.87	
Sulfur	0.66	−0.141	0.48
Fluorine	0.78	−2.87	2.87
$Cu^{2+} \rightarrow Cu^0$	1.3	−0.34	
$CuF_2 \rightarrow Cu^0$	2.09	−0.6	
$CuCl_2 \rightarrow Cu^0$	2.77	−0.07	
Chlorine	1.46		−1.37
Cupric oxide	1.65		0.224
Manganese dioxide (2e)	1.77	−1.28	
$Pd^{2+} \rightarrow Pd^0$	2.20	−0.83	
$Ag^{2+} \rightarrow Ag^0$	2.23	−0.79	
Silver II oxide	2.54		−0.57
Mercuric oxide	4.55		+0.098
Lead dioxide	4.90	+1.685	

When electrode materials of low equivalent weight and high emf are employed, large theoretical energy densities result. At the same time, however, the emf of such a battery may normally exceed the decomposition potential of water (range 1.7–2.1 V), thus making it necessary to resort to nonaqueous solvent systems.

TABLE IV

THEORETICAL PROPERTIES OF SOME ANODE MATERIALS

Anode material	lb_a lb/500 Ah	Theor. anode pot., E_A E_a	E_b
Hydrogen	0.0415	0.0	0.83
Hydrocarbons	0.085–0.1		
Borohydrides	0.11–0.12		
Beryllium	0.185	1.7	2.3
Lithium	0.285	3.0	3.0
Aluminum	0.37	1.67	2.35
Titanium	2.49	0.95	
Magnesium	0.50	2.34	2.67
Chromium	0.71	0.71	1.2
Iron	1.14	0.44	0.87
Manganese	1.14	1.05	1.47
Zinc	1.34	0.76	1.22
Cadmium	2.31	0.402	0.815
Lead	4.26	0.126	0.578

With rising technological demands upon battery systems, there has been a recent surge of interest in nonaqueous high energy batteries. There are three types of nonaqueous electrolytes which may be considered:

(a) fused salt systems,
(b) salt solutions in inorganic solvents, e.g., NH_3 or SO_2, and
(c) salt solutions in organic solvents with a liquidity range in the vicinity of room temperature.

Of these three groups, the last has been receiving an increasing measure of attention, since it offers the possibility not only of achieving higher energy densities, through the use of active electrode couples, but also because it offers a broader range of operational temperatures (around room temperature) than available from aqueous electrolyte systems. Using the data of Tables III and IV or similar data, the theoretical energy densities for potential nonaqueous batteries shown in Table V have been calculated for cells of type $Li/LiCl\|MX_n|M$.

The emf values in Table V are standard potentials in aqueous solutions. They are not, therefore, strictly applicable to the case of non-aqueous solvents which might be employed. Since the electrolyte systems used with lithium anodes would be either fused halide salts or salt solutions in *aprotic* solvents, H^+ and OH^- ions are no longer available for the current transport through the electrolyte and, consequently, oxide or hydroxide electrodes cannot be employed if stoichiometric compatability and electrolyte invariance are to be assured. Consequently, metal-metal halide electrodes of the second kind, in which the metal salt is essentially insoluble in the electrolyte, must be employed.

TABLE V

THEORETICAL GRAVIMETRIC ENERGY DENSITIES
(E.D.) FOR NONAQUEOUS CELLS OF TYPE
$Li | LiCl \| MX_n | M$

Cathode material $MX_n \rightarrow M$	E.D. Wh/lb	Theoretical cell voltage
CoF_3	970	3.63
CuF_2	746	3.54
$CuCl_2$	503	3.06
$CuBr_2$	292	2.8
NiF_2	620	2.9
$NiCl_2$	435	2.5
CrF_3	692	2.46
$CrCl_3$	492	2.41
$CrCl_2$	378	2.13
$FeCl_3$	562	2.8
$FeCl_2$	392	2.4
AgF	375	4.1
$AgCl$	228	2.9

A discussion of the individual systems of Table V in respect to their ability to meet both stability and compatability requirements, either for primary batteries or for secondary batteries, would be beyond the scope of this chapter. One important over-riding consideration which is analogous to the problem of water decomposition in an aqueous battery is that the solvent must not be decomposed at the open circuit or closed circuit emf at which the cell would operate. With emf values in excess of 3 V, many solvents may have to be ruled out. Another important general consideration is that the solvent itself should not have readily oxidizable or reducible groups, thus avoiding its eventual destruction in the battery.

Despite the rather recent interest in the area of high energy-density organic electrolyte batteries capable of performing in the vicinity of room temperature, some systems such as AgCl-Li and $CuCl_2$-Li are reaching the practical stage of development (see e.g., Wilburn *et al.* (1967)). Little research has been devoted so far to the problem of secondary organic nonaqueous batteries. As discussed in the beginning of this section, the compatability and requirements criteria (a)–(e) are much more critical for secondary batteries. On the basis of preliminary results (Eisenberg, 1967), the Li-$CuCl_2$, Li-CuF_2, and Li-AgCl systems shown in the right-hand side of Table VI appear to offer some promise.

Five aqueous secondary batteries are also given in Table VI for comparison. As can be seen, the theoretical energy densities of the copper halide nonaqueous systems exceed the energy density of the silver-zinc system (208 Wh/lb, the highest available in an aqueous battery) by *factors* of 2.5–3.5. Since the potential increase in *practical operational* energy density may also be expected to reach approximately the same factors, it is not surprising that systems of this type have been receiving considerable attention. The operational energy densities given for the three nonaqueous systems in Table VI are based on estimates or extrapolations.

Energy density per unit weight (or volume) is not, of course, the only important criterion in the selection of battery power sources. Often requirements of shelf life, in the case of primary batteries, cycle life in the case of secondary batteries, rate capabilities, and of operational temperatures, may make systems of lower energy density preferable. In the final analysis, the usefulness of a battery system may be decided not on thermodynamic considerations of energetics, but on the physicochemical and electrochemical characteristics which are decisive for its stability and range of capabilities.

Historically, batteries represent probably the oldest electrochemical devices known to man. Since the days of Volta and Daniel (early nineteenth century) reasonably useful primary cells have been available. The lead-acid battery, a rather remarkably stable and reversible system, has been in use over a hundred years. Perhaps this early availability accounts for the fact that conventional batteries have been taken for granted for too long and little research effort has been devoted to this area over the years. Only in the last twenty years has this field begun benefiting from advances in the areas of electrode kinetics (see e.g., Vetter (1967), p. 698, etc.), x-ray crystallography (e.g. Falk (1960), Salkind and Bruins, 1962), and the recognition that many long-established electrode active materials are nonstoichiometric semiconductor compounds through which both electrons and metal ions are transported

M. Eisenberg

TABLE VI

SECONDARY BATTERIES

Types	Aqueous Batteries					Nonaqueous Batteries		
	Lead acid	Nickel-iron	Nickel-cadmium	Silver-zinc	Silver-cadmium	Li/Copper chloride	Li/Copper fluoride	Li/Silver chloride
Energy density:								
a. Theoretical on active matl. (Wh/lb)	115	138	107	208	120	503	746	230
b. Operational (Wh/lb)	5–15	7–10	12–15	25–50	16–25	(25–175)	(25–200)	(25–90)
(Wh/in^3)	4–1.0	0.7–1.0	0.7–1.0	1.7–3.5	1.4–2.9	—	—	—
Operating voltage per cell—mid-point	1.7–1.95	1.2	1.1–1.23	1.3–1.5	1.0–1.10	2.3–2.8	2.3–2.8	2.3–2.8
Cycles	200–500	2000+	300–3000	10–200	300–1500	—	—	50–200 (exper)
Operating temperature range (°F)	−40 to +140	−10 to +115	−40 to +140	0 to +140	−10 to +140	−40 to +165	—	—

(see e.g., for manganese dioxide, Vetter (1963)). These new approaches to the solid state half-cell reactions, on the one hand, and the progress in the analysis of transport processes and current distribution in porous flooded electrodes (see Section IV,C), on the other hand, should hopefully, lead towards a better understanding of the complex processes occurring in the porous solid electrodes of primary and secondary batteries.

LIST OF SYMBOLS

a_i	Activity of species i
A	Specific surface of electrode matrix ($cm^2/cm^3 = cm^{-1}$)
c_i	Concentration of species, i, at a given point in the electrolyte (g-mole/cm^3)
$c_i{}^0$	Concentration of species, i, in the electrolyte bulk (g-mole/cm^3)
c_0	Ionic concentration in bulk of solution (moles/cm^3)
D	Individual ionic diffusion coefficient (cm^2/sec)
D_i	Diffusion coefficient of species, i (cm^2/sec)
e	Electron charge (1.60×10^{-11} coulomb)
E	Potential of the polarized electrode or potential in the electrolyte in respect to a point of reference (V)
$E_r{}^0$	Standard reversible oxidation potential of the electrode when all substances, i, are at unit activity, i.e., in their standard states (V)
E_r	The reversible (equilibrium) electrode potential (at prevailing activities A for anode, C for cathode) (V)
E_A, E_C	Anode or cathode total potential vs. some reference electrode (at current) (V)
ΔE_T	Total polarization of a given electrode (V)
ΔE_{ch}	Charge-transfer polarization (V)
ΔE_{conc}	Concentration polarization of a given electrode (on electrolyte side) (V)
$(\Delta E_{conc})_G$	Concentration polarization due to gradient on the gas side (V)
F	Faraday equivalent (96,500 coulomb/equiv.)
ΔG	Free energy of reaction (cal/mole)
G^*	Activation free energy (cal/mole)
f_i	Activity coefficient for species, i ($a_i = f_i \cdot x_i$)
h	Electrolyte thickness (cm)
H	Enthalpy (cal/mole)
i	Total current (A)
I	Current density in the electrolyte (A/cm^2)
i_0	Exchange current density for electrode reaction (A/cm^2)
i'	I/I_{ap} dimensionless current density in electrolyte
I_{ap}	Apparent current density per unit apparent surface area of electrode (i.e., current density at entrance to the pore, A/cm^2)
I_s	Current density in the solution
i^s	Transfer current density at the pore wall (A/cm^2)
J	Transfer current density
k	Boltzmann constant (1.38×10^{-16} erg/$°K$)
k_L	Mass transfer coefficient (cm/sec)
k_D	Decomposition rate coefficient (cm/sec)

L	Half thickness of porous electrode plate (cm)
N_i	Fluxes of species, i (g-mole/cm²-sec)
n	Number of Faradays of charge transfer per g-mole of reactant (+4 cathodic reaction)
N	Total number of operating pores per unit apparent electrode area
P	Volume percent porosity
P_c	Capillary pressure (atm)
P_i	Partial pressure of gas component, i (atm)
Q	Heat evolved (cal)
R	Gas constant (1.987 cal/mole °K)
R	Ohmic resistance of a single cell (Ω)
r	Pore radius (effective) (cm)
S	Entropy (cal/mole °K)
S_i	Source rate for species, i (g-mole/cm²-sec)
T	Temperature (°K)
t	Time (sec)
t_i	Transference number of ion, i
u_i	Ionic mobility of species, i (cm/sec-dyn)
V	Bulk electrolyte velocity (cm/sec)
V_r^0	Standard reversible cell emf (V)
V_r, V	Reversible (open) and closed cell voltage, respectively (V)
v	Volume (cm³)
x	Distance into porous electrode (cm)
x_i	Mole fraction for species, i
X	Distance into porous electrode (dimensionless) $= x/L$
z_i	Charge number of species, i
α	Transfer coefficient in the polarization equation
β	$I_{ap} \cdot L/nFD_k C_k^0$
γ_i	C_i^0/C_k^0
δ	Equivalent transfer layer thickness next to outer face of electrode (cm)
Δ	Equivalent transfer layer thickness (dimensionless)
θ	Dispersion angle of pore structure
ν_i	Stoichiometric coefficient of species, i
ξ	$AL^2 i_0/nFD_k C_k^0$
π_i	D_i/D_k
τ	Dimensionless time, $D_k t/L^2$
Φ	Dimensionless potential, $F(E - E_r)/RT$
ω	Tortuosity factor of pore structure

SPECIAL REFERENCES

AGAR, J. N., and BOWDEN, F. P. (1939). *Proc. Roy. Soc.* **A169**, 206.

AIKASJAN, E. A., and FEDOROWA, A. I. (1952). *Dokl. Akad. Nauk. SSSR* **86**, 1137.

AUSTIN, L. G. (1963). "Fuel Cells" (G. J. Young, ed.), Vol. II, p. 95. Reinhold, New York.

BACON, A. (1954). *Beama J.* **61**, 6.

BAGOTSKY, V. S., and YABLOKOVA, I. E. (1953). *Zh. Fiz. Khim.* **27**, 1663.

BAKER, B. S., ed. (1965). "Hydrocarbon Fuel Cell Technology." Academic Press, New York.

BAKER, B. S., and EISENBERG, M. (1964). *Electrochem. Technol.* **2**, 258.
BENNION, D. N. (1964). Phenomena at gas electrode-electrolyte interface. Thesis, Univ. of California, Berkeley, California.
BENNION, D. N., and TOBIAS, C. W. (1966a). *J. Electrochem. Soc.* **113**, 589.
BENNION, D. N., and TOBIAS, C. W. (1966b). *J. Electrochem. Soc.* **113**, 593.
BERL, W. G. (1943). *Trans. Electrochem. Soc.* **83**, 253.
BREITER, M. W., KNORR, C. A., and VOLKL, W. (1955). *Z. Elektrochem.* **59**, 681.
BROERS, H. J. (1958). High temperature galvanic fuel cells. Ph.D. Thesis, Univ. of Amsterdam, Amsterdam.
BURTON, W. K., CABRERA, K. N., and FRANK, F. C. (1951). *Phil. Trans. Roy. Soc. Ser. A* **243**, 299.
CAIRNS, E. J., and BARTOSIK, D. C. (1964). *J. Electrochem. Soc.* **111**, 1205.
COLLINS, D. H. (1963). "Batteries." Macmillan, New York and Pergamon Press, Oxford.
DAMJANOVIĆ, A., and BOCKRIS, J. O'M. (1963). *J. Electrochem. Soc.* **110**, 1035.
DAVIES, M. O., CLARK, M., YEAGER, E., and HOVORKA, F. (1959). *J. Electrochem. Soc.* **106**, 56.
DAVY, H. (1802). *Nicholson's J. Nat. Phil.* **144**.
EISENBERG, M. (1962a). Some design and scale-up considerations for electrochemical fuel cells. *Advan. Electrochem. Electrochem. Eng.* **2**, 235.
EISENBERG, M. (1962b). Mechanism and kinetics of the hydrazine electrode. *Electrochem. Soc. Meeting, Boston, October 1962, Abstract.*
EISENBERG, M. (1963a). Thermodynamics of electrochemical fuel cells. "Fuel Cells" (W. Mitchell, ed.), p. 24. Academic Press, New York.
EISENBERG, M. (1963b). A hydrazine fuel cell for aerospace applications. *Proc. Ann. Power Sources Conf.* **17**, 97.
EISENBERG, M. (1967). High energy non-aqueous battery systems. *Proc. Symp. Power Systems for Electric Vehicles, Columbia Univ., New York, April 1967,* p. 209. U.S. Dept. of Health, Educ. and Welfare, Washington, D.C.
EISENBERG, M., and SILVERMAN, H. P. (1961). *Electrochim. Acta* **5**, 1.
EISENBERG, M., BAUMAN, H. F., and BRETTNER, D. M. (1961). *J. Electrochem. Soc.* **108**, 909.
ERDEY-GRÚZ, T., and VOLMER, M. (1930). *Z. Physik. Chem. (Leipzig)* **A150**, 203.
EULER, J. (1961). *Electrochim. Acta* **4**, 27.
EULER, J., and NONNENMACHER, W. (1960). *Electrochim. Acta* **2**, 268.
EYRING, H., GLASSTONE, S., and LAIDLER, K. J. (1939). *J. Chem. Phys.* **7**, 1053.
FALK, S. U. (1960). *J. Electrochem. Soc.* **107**, 661.
FARR, J. P., and HAMPSON, N. A. (1967). *J. Electroanal. Chem.* **13**, 433.
FLEISCHER, A. (1968). Survey and analysis on metal-air cells. Tech. Rept. AFAPL-TR-68-6. Air Force Systems Command, Wright Patterson Air Force Base, Ohio.
FLEISCHMANN, M., and THIRSK, H. R. (1960). *Electrochim. Acta* **2**, 22.
FRUMKIN, A. N. (1932). *Z. Physik. Chem. (Leipzig)* **A160**, 116.
FRUMKIN, A. N. (1933). *Z. Physik. Chem. (Leipzig)* **A164**, 121.
FRUMKIN, A. N. (1937). *Acta Physiochim. U.R.S.S.* **7**, 475.
FRUMKIN, A. N. (1963). *Advan. Electrochem. Electrochem. Eng.* **3**, 287–389.
FRUMKIN, A. N., NEKRASOV, L. N., LEVICH, B., and IVANOV, J. (1959). *J. Electroanal. Chem.* **1**, 84.
GILMAN, S. (1965). The study of hydrocarbon surface processes by the multipulse potentiodynamic (MPP) method. *In* "Hydrocarbon Fuel Cell Technology" (B. S. Baker, ed.), p. 349. Academic Press, New York.
GRENS, E. A. (1966). *Ind. Eng. Chem. Fundamentals* **5**, 542.

GRENS, E. A., and TOBIAS, C. W. (1964). *Ber. Bunsenges. Physik. Chem.* **68**, 236.

GROVE, W. R. (1839). *Phil. Mag.* **14**, 127.

GRUBB, W. T. (1962). *Proc. Ann. Power Sources Conf.* **16**, 31.

GUREVICH, I. G., and BAGOTSKY, V. S. (1964). *Electrochim. Acta* **9**, 1151.

HEATH, C. E. (1964). *Proc. Ann. Power Sources Conf.* **18**, 33.

HEYROVSKY, J. (1927a). *J. Cas. Cesk. LeKarn.* **7**, 242.

HEYROVSKY, J. (1927b). *Rec. Trav. Chim.* **46**, 582.

IBL, N. (1959). *Helv. Chim. Acta* **42**, 117.

IBL, N., BARRADA, Y., and TRUMPLER, G. (1954). *Helv. Chim. Acta* **37**, 583.

JUSTI, E., PILKHUN, M., SCHEIBE, W., and WINSEL, A. (1960). "Hochbelastbare Wasserstoff-Diffusions-Electroden." Akad. Wiss., Mainz.

KOBER, F. P., and WEST, H. (1967). Anodic oxidation of zinc in alkaline solution. *Electrochem. Soc. Meeting, Chicago, October 1967*, Ext. Abstr. 28.

KORDESCH, K., and MARTINOLA, F. (1953). *Monatsh. Chem.* **84**, 39.

KRASILSHCHIKOV, A. I. (1947). *Zh. Fiz. Khim.* **21**, 849, 855.

KRASILSHCHIKOV, A. I. (1949). *Zh. Fiz. Khim.* **23**, 332.

KRASILSHCHIKOV, A. I. (1952). *Zh. Fiz. Khim.* **26**, 216.

KRASILSHCHIKOV, A. I. (1953). *Zh. Fiz. Khim.* **27**, 389.

KRASILSHCHIKOV, A. I. (1954). *Zh. Fiz. Khim.* **28**, 1286.

KRASILSHCHIKOV, A. I. (1961). *Soviet Electrochem. Proc. Conf. Electrochem. (English Transl.), 4th, Moscow, 1956*, **II**. Consultants Bureau, New York.

KSENZHEK, O. S. (1957). *Ukr. Khim. Zh.* **23**, 443.

KSENZHEK, O. S. (1962a). *Zh. Fiz. Khim.* **36**, 121.

KSENZHEK, O. S. (1962b). *Zh. Fiz. Khim.* **36**, 331.

KSENZHEK, O. S., and STENDER, V. V. (1956). *Dokl. Akad. Nauk. USSR* **107**, 280.

KSENZHEK, O. S., and STENDER, V. V. (1957). *Zh. Fiz. Khim.* **31**, 117.

LAITINEN, H. A., and KOLTHOFF, I. M. (1941). *J. Phys. Chem.* **45**, 1061.

LANDSBERG, R. (1957). *Z. Physik. Chem. (Leipzig)* **206**, 291.

LEE, J. M. (1960). Research on a 500-Watt solar regenerative H_2—O_2 fuel cell. 2nd Semi-Ann. Rept., 1960, Contract DA36-039-SC-85259. U.S. Army Signal R & D Lab. Fort Monmouth, New Jersey.

LEVICH, V. G. (1962). "Physicochemical Hydrodynamics." Prentice-Hall, Englewood Cliffs, New Jersey.

LORENZ, W. (1954). *Z. Naturforsch.* **9a**, 716.

MILNER, P. C., and THOMAS, U. B. (1967). The nickel-cadmium cell. *Advan. Electrochem. Electrochem. Eng.* **5**, 1.

NEWMAN, J. S., and TOBIAS, C. W. (1962). *J. Electrochem. Soc.* **109**, 1183.

NIEDRACH, L. W. (1964). *J. Electrochem. Soc.* **111**, 1309.

NIEDRACH, L. W. (1965). "Hydrocarbon Fuel Cell Technology" (B. S. Baker, ed.). Academic Press, New York.

OSTWALD, W. (1894). *Z. Elektrochem.* **1**, 122.

POSNER, A. M. (1955). *Fuel* **34**, 330.

SALKIND, A. J., and BRUINS, P. F. (1962). *J. Electrochem. Soc.* **109**, 356.

SAWYER, D. T., and INTERRANTE, L. V. (1961). *J. Electroanal. Chem.* **2**, 310.

SAWYER, D. T., and SEO, E. T. (1962). *J. Electroanal. Chem.* **3**, 410.

SCHMID, A. (1923). "Die Diffusionselektrode." Stuttgart.

SCHOTTKY, W. (1935). *Wiss. Veroffentl. Siemens-Werken* **14**, 1.

SHIMOTAKE, H., ROGERS, G. L., and CAIRNS, E. J. (1967). Performance characteristics of a lithium-tellurium cell. *Electrochem. Soc. Meeting, Chicago, Illinois*, Ext. Abstr. 18.

TAFEL, J. (1905). *Z. Physik Chem. (Leipzig)* **50**, 641.

Tobias, C. W., Eisenberg, M., and Wilke, C. R. (1952). *J. Electrochem. Soc.* **99**, 359c.
Uhler, E. F., Stockdale, G., Ritterman, P., and Lozier, G. S. (1963). Investigations of new cathode-anode couples for secondary batteries using molten salt electrolytes. Rept. No. ASD-TDR-63-115, Contract No. AF33(657)-7758. Wright-Patterson Air Force Base, Ohio.
Vetter, K. J. (1963). *Z. Electrochem. Soc.* **110**, 597.
Vetter, K. J. (1967). "Electrochemical Kinetics," English ed. Academic Press, New York.
Vielstitch, W. (1961). *Chem. Ing. Tech.* **33**, 75.
Volta, A. (1800). *Phil. Trans. Royal Soc.* **90**, 403.
Weiss, R. S., and Jaffe, S. S. (1948). *Trans. Electrochem. Soc.* **93**, 128.
Werner, R. C., Shearer, R. E., and Ciarlarello, T. A. (1959). *Proc. Ann. Power Sources Conf.* **13**, 122.
Wilburn, N. T., Almerini, A. L., and Bradley, C. L. (1967). Comparative high rate performance of reserve batteries with organic or magnesium perchlorate electrolyte. ECOM-2880. U.S. Army Electron. Command, Fort Monmouth, New Jersey.
Will, F. G. (1963). *J. Electrochem. Soc.* **110**, 145, 152.
Winkelmann, D. (1956). *Z. Elektrochem.* **60**, 731.
Winsel, A. (1962). *Z. Elektrochem.* **66**, 287.

Chapter 10

Fused-Salt Electrochemistry

GEORGE E. BLOMGREN

I. The Nature of Fused Salts 859
   A. Introduction 859
   B. The Nature of Pure Fused Salts 860
   C. The Nature of Fused-Salt Solutions 866
II. EMF Measurements of Fused Salts 874
   A. Introduction 874
   B. Standard States 875
   C. Reversible Electrodes 876
   D. Cells with Liquid Junction 878
   E. Formation Cells 880
   F. Results of EMF Studies 883
III. Electrical Transport Processes 887
   A. Electrical Conductivity and Transference Numbers 887
   B. The Electrical Double Layer 892
IV. Electrochemical Kinetics 893
   A. Introduction 893
   B. Steady-State and Intermediate-Time Methods 894
   C. Impedance and Pulse Techniques 897
   D. Summary 899
   References 899

I. The Nature of Fused Salts

A. INTRODUCTION

The field of fused-salt electrochemistry has taken on increasing importance in recent years because of newly developing applications for fused salts as electrolytes for electrochemical processes and galvanic cell systems, and also as heat-transfer media for nuclear reactors and other systems in which the electrochemical corrosion properties are extremely important. Fused salts are also of considerable theoretical interest in electrochemistry. They represent the ultimate in ionic concentration, and advances in the understanding of their behavior should help in understanding other concentrated solutions. Some electrochemical reactions of fused salts are among the fastest reactions known, and study of this kind of system represents a great challenge to theoretical electrochemistry.

It seems advisable to discuss some of the physical properties of fused salts and the present state of understanding of the nature of pure fused salts and solutions of fused salts before proceeding to electrochemical properties. Such a discussion follows in this section. Section II of this chapter discusses equilibrium properties of fused salts as they relate to electrochemistry, in particular the emf of electrodes in fused-salt electrolytes. Section III deals with electrical transport properties of fused salts and what they can tell us about the nature of these systems. Finally, Section IV discusses the subject of electrode processes for fused-salt electrolytes and the electrochemical methods which have been used.

B. The Nature of Pure Fused Salts

Fused salts are generally thought of as completely dissociated ionic liquids. This gives a consistent picture, e.g., for the alkali halides, for which the lattice energy of crystals on the one hand are well accounted for on the basis of this assumption (Mayer and Mayer, 1940), while the bonding energy of gaseous molecules (alkali halide monomers and dimers) are also accounted for (Bauer and Porter, 1964). Certain other salts, such as silver and thallium halides and the entire class of glass-forming oxides and many sulfides, undoubtedly have a greater or lesser contribution of covalent type of bonding. While this complication makes any theoretical discussion difficult, these partially covalent materials are often of interest, and thus many experimental studies have been carried out upon them. Returning to the nature of the simpler, purely ionic salts, it is first necessary to consider the potential energy of interaction between ions before proceeding to a description of the fused state. For this the work of Born, Mayer, and others (for references see Mayer and Mayer, 1940) on the lattice energy of the solid state can be examined. Here it is found that by far the largest contributions to the lattice energy arise from the Coulomb force of interaction, the potential of which decreases as $1/r$ and thus has long-range effects, and the repulsive force due to overlap of electron clouds as the ions approach each other, and which therefore acts only at short range. Several forms have been assumed for the repulsive potential, including an r^{-n} dependence, where n goes from 6 to 9 depending on the material, as well as an exponential dependence on the distance of approach. Other terms which contribute to the lattice energy include ion-induced dipole interactions with an r^{-4} dependence, induced-dipole–induced-dipole interactions (van der Waals interaction) with an r^{-6} dependence, dipole–quadrupole and higher terms which are quite small, and the quantum-mechanical zero-point energy from the lattice vibrational modes. The effect of the charge–dipole interactions is to introduce many-bodied forces into the problem, but these can be taken into account, to a good approximation, by the introduction of a dielectric constant in the Coulomb and r^{-4} terms (see Stillinger (1964) for a complete discussion of the potential energy of interaction). The result of these considerations is that the potential energy can be expressed as a pairwise additive function of the interion separations with a large long-range part which is attractive or repulsive depending on whether the two ions are of unlike or like charge, and several short-range parts the most important of which is the repulsive part. The qualitative effect of this potential energy on the distribution of ions in a fused salt is to force a certain amount of short-range charge ordering on the ions. This is due to the fact that since only the cation–anion Coulomb term is attractive, the cations will have mostly anions for first neighbors, and *vice versa*. This ordering will not be preserved to long range, however, because such an ordered system could not exhibit the observed fluidity of fused salts. This is also precisely the kind of behavior which is shown by radial distribution curves obtained from x-ray and neutron-diffraction measurements on fused salts (Levy and Danford, 1964). Here several peaks are seen which can clearly be interpreted as arising from short-range order, with the first peak due primarily to cation–anion neighbors (it occurs at a slightly shorter distance and shows fewer first neighbors than in the solid) and the second peak due mainly to anion–anion and cation–cation neighbors (slightly longer distance and fewer second neighbors than in the solid). Even though more peaks are observed, it is difficult to ascribe definite charge relations to them.

These general expectations of theory seem to be verified by experiment, but attempts to put a rigorous theory on a quantitative basis have not been pursued very far. The general lines along which such a theory can be developed have been outlined by Stillinger *et al.* (1960), but the theory has not been carried to a quantitative level. Some rather drastic physical approximations have been introduced, such as considering the fused salt as a fluid of equal-sized, uncharged particles in order to calculate the mean distance of closest approach (Stillinger, 1961) and to relate the properties of surface tension, compressibility, and expansivity to each other (Mayer, 1964).

Another approach which has been taken is to use a more intuitive theory in order to calculate results which are helpful in investigating the relationship of fused salts to other liquids and interpreting the physical properties in terms of the parameters of the theory. One such theory which has been applied to fused salts is the significant structure theory of Eyring *et al.* (1958). Two versions of the theory involving slightly different assumptions have been developed by Blomgren (1960) and Carlson *et al.* (1960). The results of the calculations indicate that this theory can be successfully applied to fused salts if adequate account is taken of the Coulomb force and the resultant increase in the lattice energy over nonionic liquids. Good agreement with experiment is obtained for such properties as the entropy and temperature of fusion, the molar volume of the liquid as a function of temperature, and temperature and entropy of vaporization (boiling point). Values of the critical properties are also predicted. These await experimental verification, however, as no critical-point measurements have as yet been made. While the theory gives good agreement with experiment, the derivation does not proceed in an orderly way from statistical mechanics, but rather rests on a number of *ad hoc* assumptions concerning the form of the partition function, which are difficult to test for validity.

Two other theories of pure fused salts have been advanced which relate macroscopic properties of liquids through statistical-mechanical models of the liquid state. The first theory of this type is the hole theory, which presumes a statistical distribution of small holes within the fluid. The properties of these holes are calculated with the aid of hydrodynamic equations to determine their kinetic properties, and the macroscopic surface tension to calculate the reversible work of hole formation. The results of calculations applied to fused salts, as reviewed by Bloom and Bockris (1964), show that the average sized hole is of about ionic size. The theory has been used to calculate the volume change on fusion, isothermal compressibility, and thermal expansivity of fused salts, with fairly good agreement with experiment. The theory is limited, however, in two senses. Because it does not start with a partition function, there is no general way of proceeding directly to calculate all thermodynamic quantities (such as free energy, enthalpy, entropy, equation of state, etc.), but must rely on secondary considerations and assumptions. In addition, the application of macroscopic equations of hydrodynamics and surface tension to ionic-sized holes casts a doubtful light on the theoretical basis.

The second theory which relates macroscopic quantities is the theory of corresponding states. The application of this theory to pure fused salts, originally made by Reiss *et al.* (1961), has been reviewed by Luks and Davis (1967). This theory relies on a simplified potential function with a single parameter such that the pair potential between like ions is given solely by the Coulomb potential

$$
\begin{aligned}
U_{\alpha\alpha}(r) &= (Z_\alpha e)^2/\kappa r, & r &> 0, \\
U_{\beta\beta}(r) &= (Z_\beta e)^2/\kappa r, & r &> 0,
\end{aligned}
\tag{1.1}
$$

where Z_α and Z_β are the charges on ions of species α and β, and κ is a microscopic dielectric constant, and the potential is therefore purely repulsive. This is justified on the basis that like ions rarely approach each other closer than the second-neighbor distance of the crystal, as is revealed by the x-ray diffraction results referred to earlier. As a result, the distance is always beyond the effective range of the non-Coulombic interactions. The pair potential between unlike ions is then given by

$$
\begin{aligned}
U_{\alpha\beta}(r) &= \infty, & r &\le \lambda, \\
&= Z_\alpha Z_\beta e^2/\kappa r, & r &> \lambda,
\end{aligned}
\tag{1.2}
$$

where λ is identified with the sum of the ionic radii and is a hard core cutoff parameter to the attractive Coulomb potential. For symmetrical salts the pair potential energy can be written in terms of the dimensionless variable

$$
r^* = r/\lambda
\tag{1.3}
$$

as

$$
U_{\alpha\alpha}(r) = U_{\beta\beta}(r) = Ze^2/\kappa\lambda r^* = [(Ze)^2/\kappa\lambda]U^*(r^*),
\tag{1.4}
$$

and

$$
\begin{aligned}
U_{\alpha\beta}(r^*) &= \infty, & r^* &\le 1, \\
&= -[(Ze)^2/\kappa\lambda]U^*(r^*), & r^* &> 1,
\end{aligned}
\tag{1.5}
$$

so that the total intermolecular potential U becomes

$$
U(r_1, r_2, \ldots, r_{2N}) = (Z^2e^2/\kappa\lambda)\, U^*(r_1{}^*, r_2{}^*, \ldots, r_{2N}^*).
\tag{1.6}
$$

The partition function Z is

$$
\begin{aligned}
Z &= \int_v \cdots \int_v \exp(-U/kT)\, dv_1 \cdots dv_{2N} \\
&= \lambda^{6N} \int_{v/\lambda^3} \cdots \int_{v/\lambda^3} \exp\left(-\frac{Z^2e^2}{\kappa\lambda kT}\, U^*\right) dv_1{}^* \cdots dv_{2N}^* \\
&= \lambda^{6N} I(\tau, \theta, N),
\end{aligned}
\tag{1.7}
$$

where τ is a reduced temperature and θ a reduced volume defined, respectively, by

$$\tau = \lambda\kappa kT/Z^2e^2 \quad \text{and} \quad \theta = V/\lambda^3. \quad (1.8)$$

Furthermore, a reduced pressure may be defined by

$$\pi = \pi(\tau, \theta) = (\kappa\lambda^4/Z^2e^2)P. \quad (1.9)$$

The reduced quantities are dimensionless under these definitions. The reduced temperatures for the alkali halides and some alkaline earth oxides were compared at the melting point, which was assumed to be a corresponding state for the different materials. The reduced melting temperatures as calculated by Luks and Davis (1967), with the dielectric constant taken as unity, are given in Table I. Also given in the table are the reduced volumes which have been calculated for the alkali halides. It is readily seen from the table that the temperatures can be broken up into three groups: the alkaline earth oxides, the lithium halides, and the rest of the alkali halides. The agreement within these groups is quite good, although there is a definite trend toward higher reduced temperatures with increasing interionic distances. The reduced volumes, on the other hand, show a downward trend with increasing interionic distance, and here the lithium halides are not separated from the rest of the alkali halides. These differences were ascribed by Reiss *et al.* (1961) in part to variations in dielectric constant, but they must also be due to differences among the solids. For example, several of the cesium halides melt from a body-centered-cubic structure rather than from the face-centered structure from which the other alkali halides melt, and one would not expect these cases to have corresponding states. This points up the difficulty in using these simple considerations, e.g., to determine details of the potential-energy function. It is surprising that the theory nevertheless works as well as it does. Further expressions have been derived relating the pressure and surface tension to the melting parameters, and the results of these calculations are also quite consistent (Reiss *et al.*, 1961).

The status of the theory of pure fused salts may be summarized by stating that the intuitive theories present models which have had considerable success in providing agreement with experiment, but have not given much information on the role of the Coulomb potential or the other terms of the potential energy. More general considerations have led to some confidence in the belief that the important factors in the potential are the Coulomb term and the repulsive term, and the other

terms do not, in most cases, influence the behavior of fused salts significantly. However, a more exacting theoretical treatment still awaits
development.

On the experimental side, one of the primary problems of working
with fused salts is the preparation and purification of the materials. Most

TABLE I

REDUCED TEMPERATURE AND VOLUME AT MELTING POINT FOR SYMMETRICAL SALTS

Salt	Melting point T_m (°K)	Interionic distance λ (cm) $\times 10^8$	Reduced temp. $\tau_m \times 10^2$	Melting volume $V_m \left(\dfrac{cc}{mole}\right)$	Reduced volume $\theta_m \times 10^{-24}$
MgO	3073	2.10	0.96	—	—
CaO	2873	2.40	1.03	—	—
SrO	2733	2.54	1.04	—	—
BaO	2198	2.75	0.90	—	—
NaF	1265	2.31	1.75	21.55	1.75
NaCl	1074	2.81	1.81	37.74	1.70
NaBr	1023	2.98	1.82	44.08	1.67
NaI	933	3.23	1.80	54.74	1.64
KF	1129	2.67	1.81	30.36	1.60
KCl	1045	3.14	1.96	48.80	1.62
KBr	1013	3.29	2.00	56.03	1.58
KI	958	3.53	2.03	67.76	1.51
RbF	1048	2.82	1.77	—	—
RbCl	988	3.29	1.95	53.70	1.51
RbBr	953	3.43	1.96	61.06	1.51
RbI	913	3.66	2.00	73.29	1.49
CsF	955	3.01	1.72	41.63	1.53
CsCl	918	3.47	1.90	59.85	1.43
CsBr	909	3.62	1.97	67.94	1.43
CsI	834	3.83	2.04	81.73	1.45
LiF	1121	2.01	1.35	14.40	1.78
LiCl	887	2.57	1.36	28.30	1.66
LiBr	823	2.75	1.36	34.12	1.64
LiI	718	3.02	1.32	43.05	1.56

salts are not available commercially in high purity, and for some of the
techniques of purification the reader is referred to the handbook by
Janz (1967). These problems become especially important in kinetic
measurements of electrochemical reactions.

C. The Nature of Fused-Salt Solutions

1. Simple Solutions

Considerable progress has been made in the development of theoretical understanding of the nature of simple fused salt solutions in which complex ion formation does not occur. The properties of theoretical interest are the free energy, the enthalpy, the entropy, and the volume of mixing. The volume of mixing is determined from density measurements of the mixture and of the pure salts. The thermodynamic properties may be determined by thermal methods, including phase-diagram studies, or emf methods. The reader is referred to the work of Kleppa (1965) for a review of thermal methods and the chapter by Førland (1964) for a review of phase-diagram methods. Emf measurements will be considered in the next section.

The simplest model of uni-univalent fused-salt solutions is one advanced by Temkin (1945) in which it is assumed that the different types of anions are randomly distributed among the anions, and cations among the cations, but that no mutual mixing occurs. This leads to a model of an ideal fused-salt mixture for which the mixing quantities $H^{(M)} = E^{(M)} = V^{(M)} = 0$, and the entropy and free energy of mixing, e.g., for a mixture of n_A moles of cation A, n_B moles of cation B, n_C moles of anion C, and n_D moles of anion D, are given by the random-mixing formula

$$\frac{G_{id}^{(M)}}{RT} = \frac{A_{id}^{(m)}}{RT} = \frac{-S_{id}^{(M)}}{R}$$

$$= n_A \ln\left[\frac{n_A}{n_A + n_B}\right] + n_B \ln\left[\frac{n_B}{n_A + n_B}\right]$$

$$+ n_C \ln\left[\frac{n_C}{n_C + n_D}\right] + n_D \ln\left[\frac{n_D}{n_C + n_D}\right]. \qquad (1.10)$$

This is a general formula for fluid or solid mixtures and only assumes that the cation volume and anion volume are available only to cations and anions, respectively. It is possible to make the model more restrictive with the assumption of a lattice structure for the melts and thus take into account divalent materials with the introduction of lattice vacancies. For example, for a divalent cation halide mixed with a univalent cation halide, a cation vacancy can be included in the lattice for each divalent cation introduced. The manner in which the vacancy is introduced into the mixing formulas depends on whether or not the vacancy is assumed to be always associated with the divalent cation or is randomly mixed with the cations. For specific formulas and the solutions for which they apply the reader is referred to the review by Førland (1964).

A second type of mixing theory is based upon a rigid-lattice model for the fused salt. That is, the ions are assumed to be fixed on the sites of a quasicrystalline lattice in both the pure salts and in mixtures, and the thermodynamic mixing properties result from differences in the interaction energies among the different kinds of ions. Generally, a sodium chloride lattice structure is assumed for the mixture, with a lattice parameter calculated from a molar volume given by an arithmetic or geometric mean (weighted by the mole fractions) of the molar volumes of the pure salts. Here only the form of the variation of the lattice parameter is important, since the volume does not appear explicitly in the final results for free energies and other properties. The theory has been carried out to various degrees of complexity from this point. The next step in the development of the lattice theory is to consider the microscopic interaction energy between the various kinds of pairs of ions (restricted to first and second neighbor pairs in the theories developed so far), find the average number of each kind of pair, and calculate the total energy of the mixture and thus the free energy of the mixture from the phase integral. The free energy of mixing is then given by the formula

$$A^{(M)} = A_{12} - X_1 A_1 - X_2 A_2, \tag{1.11}$$

where A_{12} is the free energy of the mixture, A_1 and A_2 are the free energies of the pure components, and X_1 and X_2 are the mole fractions of the components. The excess free energy of mixing is given by

$$A^{(E)} = A^{(M)} - A_{id}^{(M)}, \tag{1.12}$$

where $A_{id}^{(M)}$ is given by the Temkin formula, Eq. (1.10), for the ideal solution. The restriction of the theory to second-nearest neighbors implies that the long-range Coulomb forces will not contribute to the mixing properties. Within the limitations of the model this can only be strictly true when both pure solutions and the mixture all have the same molar volume, but it is not too bad an approximation when the molar volumes are within about 10% of each other. The mixing properties which are measured by emf techniques are the excess partial molar free energy and the excess partial molar entropy if the temperature dependence of the emf is measured. For the excess partial molar free energy of one component in the random-mixing approximation for binary solutions, e.g., two salts AC and BC with a common anion, the result obtained by Hildebrand and Salstrom (1933) is

$$\bar{A}_{AC}^{(E)} = RT \ln \gamma_{AC} = X_B{}^2 \omega, \qquad \omega = \omega_{AB} - \tfrac{1}{2}(\omega_{AA} + \omega_{BB}), \tag{1.13}$$

where γ_{AC} is the activity coefficient of the first salt, X_B is the ion fraction of the second salt, and ω_{AB}, ω_{AA}, and ω_{BB} refer to the second-nearest-neighbor interactions of the different kinds of ions, and ω is determined from a fit of the data. When $\omega = 0$ the excess partial molar free energy goes to zero, in agreement with the Temkin ideal solution. It has been found, however, that a single temperature-independent interaction energy is not sufficient to fit the data for many systems even if more elaborate statistical methods are used to determine the average number of pairs of various kinds of ions (Blomgren, 1962).

For reciprocal salt solutions involving two kinds of cations and two kinds of anions several levels of approximation have been advanced. The random-mixing approximation has been developed by Flood et al. (1954) and gives for the excess partial molar free energy for cations A and B and anions C and D for the salt AD in the mixture

$$\bar{A}_{AD}^{(E)} = X_B X_C \omega = RT \ln \gamma_{AD}, \qquad \omega = \omega_{AD} + \omega_{BC} - \omega_{AC} - \omega_{BD},$$

(1.14)

where X_B and X_C are the ion fractions of cation B and anion C, respectively. This is a nearest-neighbor approximation only. Flood et al. (1954) then take the microscopic energy difference ω to be equal to the free-energy change for the metathetical reaction AC + BD = AD + BC. Blander and Braunstein (1960) carried this analysis a step further by replacing the random-mixing approximation by the quasichemical approximation of Guggenheim (1952). This leads to an expression for the excess partial molar free energy,

$$\bar{A}_{AD}^{(E)} = Z_1 RT \ln \frac{(1 - \bar{X}_1/n_A)}{X_D},$$

(1.15)

where Z_1 is the number of first neighbors and $\bar{X}_1$ is obtained from a solution of the equation

$$\bar{X}_1(N_B - N_C + \bar{X}_1)/(N_A - \bar{X}_1)(N_C - \bar{X}_1) = \exp(-\omega_1/Z_1 RT).$$ (1.16)

This is by no means the extent of the problem, however, since the entire excess free energy for binary solutions arises from next-nearest-neighbor interactions in the lattice model, and therefore the effect of these interactions should be included in the general theory for reciprocal salt solutions. This was done by Blomgren (1962), and for the excess partial molar free energy leads to

$$\bar{A}_{AD}^{(E)} = Z_1 RT \ln\left[\frac{(1 - \bar{X}_1/n_A)}{X_D}\right] + Z_2 RT \ln\left[\frac{\beta_2 + X_A - X_B}{X_A(\beta_2 + 1)}\right]$$
$$+ Z_2 RT \ln\left[\frac{\beta_3 + X_D - X_C}{X_D(\beta_3 + 1)}\right],$$

(1.17)

where the first term is the same as that given by Blander and Braunstein (1960), Z_2 is the number of second-nearest neighbors, and β_2 and β_3 are given by

$$\beta_2 = \{1 + 4X_AX_B[\exp(\omega_2/Z_2kT) - 1]\}^{1/2}$$
$$\beta_3 = \{1 + 4X_CX_D[\exp(\omega_3/Z_2kT) - 1]\}^{1/2}. \tag{1.18}$$

A convenient thermodynamic method of breaking down the free energy for the reciprocal salt solution into terms relating to the binary mixtures and other terms was given by Kleppa (1965). The equation for the excess free energy is

$$A^{(E)} = X_AX_C \Delta A^{\circ}_{13} + X_BA^{(E)}_{14} + X_AA^{(E)}_{23} + X_DA^{(E)}_{34} + X_CA^{(E)}_{12} + A^{(E)}_{1234}, \tag{1.19}$$

where $A^{(E)}_{14}$, $A^{(E)}_{23}$, $A^{(E)}_{34}$, and $A^{(E)}_{12}$ are the excess free energies for the binary mixtures of salts 1(B–C), 2(A–C), 3(A–D), and 4(B–D), ΔA°_{13} is the free-energy change for the metathetical reaction $AC + BD = AD + BC$ and $A^{(E)}_{1234}$ is a free-energy term which applies only to the ternary or reciprocal salt solutions. It is possible to relate in a simple way the complete lattice theory of Blomgren (1962) to the thermodynamic analysis of Kleppa (1965) by taking the zeroth-order approximation for the second-nearest-neighbor interactions and expanding in a power series the quasichemical approximation for the first-neighbor interaction in the manner of Førland (1964), leading to an equation for the excess free energy

$$A^{(E)} = X_AX_C\omega_1 + X_AX_B\omega_2 + X_CX_D\omega_3 - \tfrac{1}{2}X_AX_BX_CX_D(\omega_1{}^2/Z_1kT). \tag{1.20}$$

If the first term is identified with the free energy of the metathetical reaction, ΔA°_{13}, in the manner of Flood et al. (1954), the second term with the excess free energies of the binary mixtures from $X_AX_B\omega_2 = X_DA^{(E)}_{34} + X_CA^{(E)}_{12}$, the third term with the excess free energies from $X_CX_D\omega_3 = X_BA^{(E)}_{14} + X_AA^{(E)}_{23}$, and the fourth term with $A^{(E)}_{1234}$, the lattice theory can be fully related to the thermodynamic equation. The degree to which Eq. (1.19) can be expected to hold will be discussed in the next section. At least two serious problems exist with the lattice-theory approach developed up to this point. One is that the excess volume is specified to be nearly zero and that the excess partial molar entropy is taken to be due solely to deviations from random mixing and predicted to be always negative. Experimental results do not uniformly verify these assumptions (Blomgren, 1962). A further problem is that no attempt is made to relate the interaction energy parameters, ω, to known intermolecular forces of the system, so the theory is semi-empirical in nature and relies on obtaining a best fit of experimental data by adjusting the parameters. A first attempt to deal with these problems was made by Blander (1961), who treated a one-dimensional chain of atoms with a Coulomb attraction and hard-sphere repulsion, and a single different atom in order to determine the sign of the energy change. This approach says nothing about the volume behavior, since it cannot be easily extended to three dimensions. A more elaborate attempt to explain the excess energy of mixing on the basis of intermolecular potential functions with a lattice model was made by Blomgren (1962). The method was to take the observed volume of the mixture and allow the ions of the lattice to relax to positions of minimum potential energy. The same sign as the experimentally determined energy of solution was obtained for a dilute solution calculation when both repulsive interactions and Coulomb interactions were brought into the calculation. The quantitative comparison was about 50% in error, but the importance of these interactions in determining the energy of mixing was clearly shown.

Perhaps the most elaborate theoretical development has been of the conformal solution theory. The application of this theory to molten salts was initiated by Reiss et al. (1962) and is in part based on the potential function considerations of the theory of corresponding states. The central idea of the theory is that if the pair potential between uni-univalent ions of opposite charge can be written as

$$
\begin{aligned}
u(r) &= \infty, && r \le \lambda \\
&= -q^2/\kappa r, && r > \lambda,
\end{aligned}
\tag{1.21}
$$

where q is the electronic charge, κ an effective dielectric constant, and λ is taken to be the sum of the ionic radii, the pair potential can be written in the more general form

$$
u(r) = (1/\lambda)f(r/\lambda),
\tag{1.22}
$$

and the free energy for a mixture and also the pure salts can be expanded in power series in the parameters λ_i for the different cation–anion combinations. This is the same pair potential as that of the theory of corresponding states for pure fused salts discussed earlier. The pair potential for like ions is taken as

$$
u(r) = q^2/\kappa r,
\tag{1.23}
$$

in which there is no length parameter, since it is presumed that there is seldom any like-ion contact. Through the process of power series expansion the excess free energy of mixing becomes

$$A^{(E)} = X_1 X_2 \left(\frac{kT}{2}\right)\left(\frac{\lambda_1 - \lambda_2}{\lambda}\right)^2 N^3 \{\epsilon - \omega + N(\omega - \alpha^2)\}, \quad (1.24)$$

where λ is the length parameter of a reference substance, which can be taken to be the same for all symmetrical salts; ϵ, ω, and α are complicated integrals which have not been evaluated but depend only on T, V, and λ and can also be taken to be the same value for all symmetrical salts (see Luks and Davis (1967) for derivations and discussion). The important points are that the concentration enters only as the product $X_1 X_2$ (as in the lattice theory) and the only difference among salts comes in through the term $(\lambda_1 - \lambda_2)^2$. A similar form applies to the heat of mixing, and by the introduction of other non-Coulombic forces, e.g., van der Waals attraction, Blander (1962) and Davis and Rice (1964) have included another term independent of the length parameter which when applied to the experimental heats of mixing (Hersh and Kleppa, 1965), leads to the equation

$$\Delta H_M = X_1 X_2 \left[U_0 - 340\left(\frac{\lambda_1 - \lambda_2}{\lambda_1 \lambda_2}\right)^2\right], \quad (1.25)$$

where U_0 is a calculable quantity related to the van der Waal's constants of the pair interactions, and the number 340 is obtained from a fit of experimental data. The subtleties of the theory are extensive, and the interested reader is referred to the work of Luks and Davis (1967) for a thorough discussion.

The theory of conformal solutions has been extended to reciprocal salt solutions by Blander and Yosim (1963). The result of the theory gives for the excess free energy

$$A^{(E)} = X_A X_C \Delta A_{13}^\circ + X_C A_{14}^{(E)} + X_D A_{23}^{(E)} + X_A A_{34}^{(E)} + X_B A_{12}^{(E)}$$
$$+ X_A X_B X_C X_D \left(\frac{1}{\lambda_1} + \frac{1}{\lambda_4} - \frac{1}{\lambda_2} - \frac{1}{\lambda_3}\right)^2 \lambda^2 P \quad (1.26)$$

where the notation is the same as Eq. (1.19) and P is again related to a set of complicated integrals involving T, V, and λ for a reference salt. Thus an equation of the same form as the thermodynamic result, Eq. (1.19), is obtained, and the individual binary terms are calculable from the conformal solution results on binary solutions. Thus only the last term involves a quantity unique to the ternary solution, and its evaluation will be discussed in the next section.

A few points about the conformal solution theory of fused salts should be made. One is that, in contrast to its application to non-Coulombic fluids, the theoretical development has not yet shown the relation of the complicated integrals to the thermodynamic properties of a reference fluid. Therefore it is difficult to obtain information on other excess properties such as entropy and volume. The attempts at integral evaluation made by Luks and Davis (1967) would seem to be a much more difficult approach to obtaining the desired quantities, although since the theory goes to second order in the length parameters, it would undoubtedly be necessary to invoke some model to evaluate these high-order terms. Also, the effects of ignoring like-ion repulsion have not been evaluated, even though this is the only reason the theoretical analysis can be carried out. The theory in its simpler forms, however, has led to considerable insight into the nature of fused-salt solutions and has demonstrated the major effect of ion size on the properties of fused-salt solutions.

2. Complex Solutions

The study of complex solutions containing polyvalent cations of alkaline earths, transition metals, and rare earths and various anion types has evoked considerable controversy concerning the postulation of complex ions, association in the liquid state, etc. The controversy has arisen primarily because of the inadequate theoretical foundation for interpretation of the data. For example, the existence of a maximum or minimum in equivalent conductivity vs. composition plots for binary mixtures has often been cited as evidence for complex ion formation, but VanArtsdalen and Yaffe (1955) found a minimum for the simple LiCl–KCl system. Therefore it is necessary to use considerable caution before ascribing unexpected effects to the existence of complex ions when bulk thermodynamic or electrokinetic measurements are being performed.

The more recent applications of spectroscopic techniques to fused-salt solutions has in many cases put the existence of complex ions and network formation on a much firmer foundation. Visible spectroscopy has found wide application in the studies of transition-metal ions, in which the absorption bands found are due primarily to transitions of electrons in the d shells. The peak positions, shapes, and intensities are sensitive to the surroundings of the transition-metal ion, and by analysis based on ligand field theory the symmetry of the surrounding anions can be deduced. Thus it was possible for Gruen and McBeth (1959) to show that Ni^{++} in an alkali chloride melt has tetragonal four-coordination at low temperature and tetrahedral four-coordination at high temperature. The reader can benefit from the excellent reviews by Gruen (1964) and Smith (1964).

The use of Raman and infrared spectroscopy have also led to many interesting results in establishing the existence of complex ions. The experiments here are not limited to the transition-metal ions and many kinds of systems have been studied (for a review see James (1964)). Experimental techniques are difficult, however, since window materials normally used in IR spectroscopy and containers used in Raman cells are frequently attacked by fused salts. To overcome this, a special cell has been designed to study the Raman effect which does not require either the exciting light or the scattered light to pass through windows in contact with the fused salt (Bues, 1955). A general method for the study of infrared spectra has been designed by Wilmshurst and Senderoff (1958) and uses a reflection technique and analysis of the reflected beam intensity by the Kramers–Kronig relations to obtain the absorption coefficient of the sample. A problem of interpretation arises with the data, however, since the lattice frequencies of the crystalline salt, which absorb in the infrared, persist into the liquid state. This indicates that larger clusters of ions (in a noncomplexing situation) can also give rise to IR bands. Thus one must be very careful before assigning bands to complex ions until investigating possible latticelike mode assignments. Another method of IR study involves a platinum mesh screen on which the molten salt is suspended and the transmission is observed (Greenberg and Hallgren, 1960).

Other spectral methods such as electron spin resonance, nuclear magnetic resonance, and nuclear quadrupole resonance have not been widely used, most probably because of the difficulties associated with the experimental techniques for these high-temperature and high-conductivity liquids.

The situation for complex solutions of fused salts is that in some cases spectroscopic methods have been used to define the nature of the species present, and these solutions can be considered to be well understood. In many other cases, however, we have limited information on macroscopic properties only. As the techniques of purification and temperature control progress, more precise data on macroscopic properties should become available, and as these results are combined with spectral studies considerable progress in the understanding of complex solutions should evolve.

II. EMF Measurements of Fused Salts

A. INTRODUCTION

The emf of galvanic cells has a twofold importance in electrochemistry. First, it is important in that it leads directly to the change in the Gibbs free energy for the cell reaction giving rise to the potential through the relation

$$\Delta G = -nFE, \tag{2.1}$$

where n is the number of electrons involved in the half-cell reaction, F is Faraday's constant, and E is the measured emf of the cell. Also, if the temperature dependence of the emf is studied, the entropy change of the cell reaction follows from

$$\Delta S = nF \, dE/dT \tag{2.2}$$

and the enthalpy change from

$$\Delta H = \Delta G + T \, \Delta S \tag{2.3}$$

Secondly, the emf is the equilibrium point of potential measurement, and all polarizations measured in the study of electrochemical kinetics must be expressed with reference to this potential.

The following types of galvanic cell have been studied in fused salts: Concentration cells with transference, Jacobi–Daniell-type cells, formation or chemical cells in both pure fused salts and in solution, and concentration cells of the metal alloy type. Cells of the first two types are subject to liquid junction potentials which are in general unknown, but can be approximated in some cases or can be assumed to be negligibly small in other cases. In cases where the junction potential is unknown but relatively constant, the electrode can still be used as a reference electrode for kinetic studies even though the potential is unknown. Alloy concentration cells can be used for thermodynamic analysis of the alloys, e.g., amalgam cells, but also have recently been studied as possible energy sources (Foster, 1967). In fact, the use of fused salts as electrolytes in galvanic cells for energy sources has been of increasing interest as requirements for high-rate batteries have become more widespread. The high conductivity of fused salts leads to consideration of them for such devises as fuel cells utilizing hydrocarbon and oxygen fuels. Since most of the work on these systems has been of an applied nature, the reader is referred to the work of Baker (1965) and Kronenberg (1965) for further information on this interesting topic.

A related type of study which has received considerable attention, especially in the Soviet Union, is the study of decomposition potentials. This technique gives an approximation to the free energy of formation for the reacting ions in the medium by means of an extrapolation to zero current and equating the potential so obtained (the decomposition potential) to the formation potential and thus the free energy. For further details see Delimarskii and Markov (1961).

Measurement of emf of fused-salt galvanic cells has not as yet been able to approach the accuracy of the best measurements of aqueous solutions because of the experimental difficulties brought about by the high temperature. Even though results are often quoted to tenths of millivolts, the agreement among different experimenters and with thermo-chemical data is seldom better than several millivolts on even the simplest systems (Dijkhuis *et al.*, 1968).

B. STANDARD STATES

As implied by the theoretical equations of Section I, the most useful concentration units for fused-salt solutions are mole fractions and ion fractions, although occasionally equivalent fractions are also used. Since there is no indifferent or uncharged medium present with fused salts, the units of molarity, molality, and normality commonly used with aqueous and other solutions are not very useful in a description of thermo-dynamic properties (at least in comparison with present theories), although molarity units will be used in obtaining the equivalent con-ductance in a later section. If an exact theory of fused-salt solutions existed, it undoubtedly would be desirable to convert to molarity units. However, since most studies to date have not made any attempt to control the volume, and solutions have been made at their equilibrium pressures, the mole or ion fraction description has proved adequate.

The choice of a standard state will depend on the property being measured and will be taken as a matter of convenience and adequacy of the description. Some commonly used definitions of activities and activity coefficients for the salt $B_r D_s$ are (following the notation of Blander (1964))

$$\bar{G}_2 = \mu_2 = \mu_2{}^\circ + RT \ln a_2 = \mu_2{}^* + RT \ln a_2{}^* = \mu_2{}^\square + RT \ln a_2{}^\square,$$
(2.4)

$$\gamma_2 = a_2/X_B{}^r X_D{}^s = \gamma_B{}^r \gamma_D{}^s,$$
(2.5)

$$\gamma_2{}^* = a_2{}^*/X_B{}^r X_D{}^s = \gamma_B{}^{*r}\gamma_D{}^{*s},$$
(2.6)

$$\gamma_2{}^\square = a_2{}^\square/X_B{}^r X_D{}^s = \gamma_B{}^{\square r}\gamma_D{}^{\square s},$$
(2.7)

where $\mu_2{}^\circ$ is the chemical potential of the pure liquid salt at the temperature T (it may be necessary to refer to the supercooled liquid if the melting point of the component is above the temperature of the solution being studied) and the activity coefficient goes to unity for the pure liquid. The term $\mu_2{}^*$ is the chemical potential of the standard state chosen such that the activity coefficient $\gamma_2{}^*$ goes to unity as the solute concentrations go to zero, and $\mu_2{}^\square$ is the chemical potential of the pure solid. The most convenient standard state for intercomparison of solutions and comparison to theory is the pure liquid at the temperature studied. This eliminates complications due to the fusion point of high-melting solids, although experimentally one must always be wary that the liquidus on the phase diagram has not been crossed and that solid may be present. The single ion activity coefficients $\gamma_B{}^r$, etc., have even less meaning than for aqueous solutions, since they are not experimentally determinable and no theory comparable to the Debye–Hückel theory exists for fused salts, but if they are used, they must be used in the combinations shown.

C. REVERSIBLE ELECTRODES

The first requirement for the measurement of the emf of a galvanic cell is that the two electrodes must each be reversible to one ionic constituent of the fused salt or solution. Implicit in the definition of reversibility is that no chemical or electrochemical (corrosion) reaction occurs with the electrode other than the one sought. This fact has led to great difficulty with fused salts, since, e.g., alkali metals and most alkaline earth metals tend to dissolve in their fused salts, thus setting up a nonequilibrium condition. Also, since displacement reactions tend to proceed easily, a more-active metal cannot generally be studied in a solution with the ion of a less-active metal. A further limitation is the low melting point of many metals compared to their salts, which leads to great difficulty in handling the electrode if the vapor pressure is too high. For these reasons relatively few reversible electrodes have been studied, but a considerable amount of information has been collected with these electrodes.

The kinds of electrodes and experimental details concerning them have been reviewed by Laity (1961), Albyshev et al. (1965), and Dijkhuis et al. (1968). Among the metal electrodes which have been studied are Ag, Zn, Pb, Cd, and alloys of Ce and Mg. Silver electrodes have presented the least difficulty, since the melting point of the metal is higher than those of its salts, and it is insoluble in its salts and their solutions; thus it has been used more frequently. The other pure metals (Zn, Pb, and Cd) are all liquids at the melting temperatures of many of their salts and also have sufficient solubility in their salts to require special care in making the measurements. The Ce alloys were required because it was found by Senderoff *et al.* (1960) that the pure metal was sufficiently soluble to cause a continuous drift in the emf, and these workers found that Ce–Sn alloys behaved reversibly. The requirements given for such an alloy electrode were: (1) the diluent metal should be very noble compared to the alloyed cerium, so as not to contribute to the emf, (2) the alloy should consist of two phases over a considerable range of composition to maintain a constant cerium activity, (3) the activity of cerium in the alloy should be low enough so as not to reduce Ce^{3+} or other metal ions in the solution, and (4) the alloy must behave reversibly as a Ce/Ce^{3+} electrode. Both Ce–Bi and Mg–Bi alloys were investigated by Neil *et al.* (1965) and found to be reversible. The success of these electrodes indicates that there are undoubtedly many other possible alloys which could be studied, while the possibilities with pure metals are sharply limited by the problems discussed above.

Another type of electrode which has been widely used is the anion gas electrode such as Cl_2/C, Br_2/C, I_2/C, and NO_2–O_2/Pt. The chlorine electrode has proved very valuable, but requires special treatment to obtain reversibility. Senderoff and Mellors (1958) have described its preparation, and their data on the $Ag|AgCl|Cl_2|C$ cell has served as a standard for many subsequent studies. The other halogen electrodes and the NO_2–O_2 electrode also require special handling, and the reader is referred to the reviews mentioned above for details.

Finally, an electrode which uses a glass separator without liquid junction as a cation indicator electrode has recently been developed. Cells involving these electrodes differ from ordinary concentration cells, since the glass tends to conduct monovalent cations only. Thus with a salt of a monovalent cation on one side of the glass and a solution of the same cation and a divalent cation on the other side the measured emf corresponds to the partial molar free energy of solution of the mono-valent cation salt, and the junction potential is zero. Various versions of this type of cell have been developed by Førland and Ostvold (1966), Dijkhuis and Ketelaar (1966), Notz and Keenan (1966), and Sternberg and Herdlicka (1966). This development, like the use of alloy electrodes, opens up many possibilities for new studies. One limitation of the electrode is the low conductivity of the glass, which necessitates the use of very-high-impedance null detectors. The sensitivity and accuracy of the electrode could possibly be improved by using in place of glass a material like β-alumina, which has recently been shown to be an excellent cation-only conductor by Yao and Kummer (1967).

D. Cells with Liquid Junction

The two principal types of cells with liquid junctions are concentration cells and Jacobi–Daniell cells. An example of a concentration cell is

$$\text{Ag}\begin{vmatrix} \text{AgNO}_3 \ (X_1') \\ \text{AgCl} \ (X_2') \end{vmatrix}\begin{vmatrix} \text{AgNO}_3 \ (X_1'') \\ \text{AgCl} \ (X_2'') \end{vmatrix}\text{Ag}, \qquad (2.8)$$

where the AgNO_3 and AgCl are at different concentrations in the left and right parts of the cell. The measured potential of the cell arises from two sources, the difference in chemical potential of the silver ion in the two media (i.e., the differing concentrations) and the diffusion potential due to the concentration gradients in the neighborhood of the interface (or junction) between the electrolytes. The theory of the diffusion potential has been worked out in detail, and the reader is referred to the work of Wagner (1966) and Klemm (1964) for derivations. The exact expression for the diffusion potential between two electrode compartments I and II connected by a liquid junction is

$$\Delta\phi_{\text{diff}} = -\int_{\text{I}}^{\text{II}} (1/\kappa) \sum_i c_i b_{ij} \, d\mu_i, \qquad (2.9)$$

where c_i and μ_i are the concentration and chemical potential of ion i, respectively, b_{ij} is the mobility of ion i relative to ion j—the reference ion in a mobility or transference number experiment—and κ, the specific conductance, is given by

$$\kappa = \sum_i c_i e_i b_{ij}, \qquad (2.10)$$

where e_i is the charge on ion i. The cell potential is then given by the sum of the potential difference at the electrodes (i.e., minus the change in the Gibbs free energy for the electrode process per equivalent divided by Faraday's constant) and the diffusion potential. The same equations expressed in ion fractions and transference numbers can be found in the work of Wagner (1966). Now, for the cell Eq. (2.8), if A, C, and D represent Ag^+, Cl^-, and NO_3^-, respectively, and if dilute solutions of AgCl in AgNO_3 are used, the cell potential is given approximately by

$$\Delta\phi = \frac{RT}{F} \frac{b_{\text{CD}}}{b_{\text{AD}}} \frac{C_{\text{C}}^{(\text{I})} - C_{\text{C}}^{(\text{II})}}{C_{\text{A}}}. \qquad (2.11)$$

Klemm (1964) shows that for mole ratios $C_{\text{C}}^{(\text{II})}/C_{\text{A}} = 0$ and $C_{\text{C}}^{(\text{I})}/C_{\text{A}} = 0.34$ and the measured mobility ratio $b_{\text{CD}}/b_{\text{AD}}$ at 250°C (Monse, 1957) the calculated cell potential is -1.1 mV. On the other hand, measurements of the cell by Schwarz (1941) for the same conditions have given the result $\Delta\phi = +2.0$ mV. Thus even in this simple case the experimental difficulties of emf measurements and of maintaining a well-defined boundary do not permit a correct prediction of either sign or magnitude of the potential. This illustrates the problem of fused salts with a liquid junction. However, in some cases this type of cell has proved useful; e.g., in the case of the cell

$$\text{Ag}\left|\begin{matrix}\text{AgNO}_3\ (X_1')\\\text{NaNO}_3\ (X_2')\end{matrix}\right|\begin{matrix}\text{AgNO}_3\ (X_1'')\\\text{NaNO}_3\ (X_2'')\end{matrix}\right|\text{Ag} \qquad (2.12)$$

studied by Laity (1957). In this system the diffusion potential (Klemm, 1964) is given, after simplification, by

$$\Delta\phi_{\text{diff}} = -\frac{1}{F}\int_{\text{I}}^{\text{II}} \frac{c_A b_{AB}\, d\mu_{AC}}{c_A b_{AC} c_B b_{CB}}, \qquad (2.13)$$

where A, B, and C are Ag^+, Na^+, and NO_3^-, respectively. Here transference number experiments by Duke *et al.* (1957) have shown that $b_{AB} = 0$ within experimental error, i.e., that Ag^+ and Na^+ have the same mobility over a wide concentration range. This fact means that the diffusion potential vanishes and the cell potential is given by

$$E = (RT/F)[\ln(X_{AC}^{(\text{II})}/X_{AC}^{(\text{I})}) + \ln(\gamma_{AC}^{(\text{II})}/\gamma_{AC}^{(\text{I})})]. \qquad (2.14)$$

The results of this work will be discussed in Section II,F. Extensive work in other concentration cells has been carried out, but mostly in the dilute region, where, for cases in which the free-energy change is large compared to the expected diffusion potential, results of reasonable percentage accuracy have been obtained. Notable among these measurements have been the emf studies of reciprocal salt solutions in which a binary system is in the left cell compartment and the reciprocal salt system is in the right cell compartment; e.g., in the cell

$$\text{Ag}\left|\begin{matrix}\text{AgNO}_3\\[6pt]\text{NaNO}_3\end{matrix}\right|\begin{matrix}\text{AgNO}_3\\\text{NaNO}_3\\\text{NaCl}\end{matrix}\right|\text{Ag.} \qquad (2.15)$$

Blander (1964) summarizes the data on systems which have been studied in this way. In nearly every case no correction is made for the liquid junction potential. The studies are generally made in solutions dilute in both the silver salt and the halide salt, and it is generally argued that the diffusion potential will be small. However, its magnitude relative to the measured potential has not yet been thoroughly studied, e.g., by transference number experiments.

The other important kind of cell with liquid junction is the Jacobi–Daniell cell and work on this type of cell has been thoroughly reviewed by Delimarskii and Markov (1961). The emf of this type of cell is generated by the oxidation of a metal to a salt at the anode and the reduction of the salt of another more-noble metal to the metal at the cathode. The simplest kind of this type of cell is the following:

$$A|AC|BC|B, \qquad\qquad (2.16)$$

in which A is the anode metal and BC the cathode salt. The junction has been made variously by capillary tubes joining the compartments, with glass and fritted disk partitions, and simple superposition of the less-dense upon the more-dense salt. The cell emf for Eq. (2.16) is given by the difference in the formation potential of the salts AC and BC at the temperature of the study plus the diffusion potential. The diffusion potential is seldom small and is sometimes very large, of the order of several hundred millivolts. Furthermore, the required sharp boundaries are very difficult to establish, so that the predictability of the observed diffusion potential is poor. For this reason the Jacobi–Daniell type of cell has seldom been used to establish precise thermodynamic properties.

E. Formation Cells

Most of the results of interest for thermodynamic properties and comparison with theories have been obtained on formation cells. The simplest type of formation cell is

$$A|AC|C, \qquad\qquad (2.17)$$

in which A is a metal electrode and C is a nonmetal electrode, generally a halogen-gas–carbon electrode. As mentioned in Section II,A, even these simplest of formation cells do not lead to consistent results or agreement with thermal data to better than a few millivolts. The entropy is even more difficult to determine precisely, because of the relatively large uncertainty in the emf. It is also observed that the higher the temperature which is studied, the greater is the divergence from thermal data. Table II, which is taken from Dijkhuis et al. (1968), shows the situation for the cell

$$Ag|AgCl|Cl_2, \qquad\qquad (2.18)$$

which has been widely studied and has perhaps the best characterized and most reversible of all fused-salt anodes and cathodes. Even here the agreement among investigators is of the order of several millivolts, and the disparity with the thermal data increases with increasing temperature to about 15 mV at 900°C. This is undoubtedly due in part to lack of reversibility of the chlorine electrode at the higher temperatures. In fact, even at 500°C an elaborate pretreatment of the carbon electrode is required before reversibility is established (see Dijkhuis *et al.* (1968)).

TABLE II

VALUES OF E^0 FOR THE CELL Ag|AgCl|Cl$_2$ AT VARIOUS TEMPERATURES[a]

E^0 (mV)					
at 500°C	600°C	700°C	800°C	900°C	Ref.
895.0	868.4	842.1	815.8	789.5	Senderoff and Mellors (1958)
898.3	869.0	841.1	814.7	789.8	Panish *et al.* (1958)
893.9	864.5	838.5	816.0	796.5	Leonardi and Brenet (1965)
890.9	863.4	835.9	808.3	780.7	Murgulescu and Sternberg (1957)
899.7	871.1	842.5	814.0	785.4	Salstrom (1934)
898.8	865.8	832.7	799.7	766.6	Stern (1956)
896	870	848	826	805	Thermal data

[a] From Dijkhuis *et al.* (1968).

Possible irreversibility of the silver electrode as well cannot be overlooked, however (Førland and Dijkhuis, 1961). Other simple formation cells which have been widely studied include Cd|CdCl$_2$|Cl$_2$; Pb|PbCl$_2$|Cl$_2$; Zn|ZnCl$_2$|Cl$_2$; Ag|AgBr|Br$_2$; and Pb|PbBr$_2$|Br$_2$ (see Dijkhuis *et al.* (1968)). Problems similar to the above have been found with all of these cells, with the additional problems of finite metal solubility in the salt with cells involving cadmium and lead, both of which are liquids at the temperatures of their molten salts. Other formation cells with only a single fused-salt electrolyte have involved alloy anodes to get around the solubility problem as discussed in Section II,C.

More complex formation cells have involved binary mixtures of fused salts with a common anion or a common cation. In this type of cell the emf is related to the partial molar free energy of the salt to which the electrodes are reversible through the relation

$$\bar{G}_{AC} = -nFE, \tag{2.19}$$

e.g., for the cell

$$A|AC,BC|C. \tag{2.20}$$

It is of course necessary to choose an anode metal A more noble than the metal from the salt BC so that a displacement reaction does not occur. In principle, one could obtain the partial molar free energy of the other component through the use of the Gibbs–Duhem relation and the methods of Lewis and Randall (1923). The accuracy of the measurements made heretofore is not sufficient, however, to accomplish this with any degree of certainty, and most workers simply give the data for the partial molar free energy of the component studied directly. Dijkhuis *et al.* (1968), however, have given a simple way of expressing the chemical potential of the other component and thus the free energy of mixing. If the excess free energy for the cell Eq. (2.20) can be expressed as a simple power series, as has been shown to be the case from thermal data, in the form

$$G^{(E)} = X_A X_B (a + b X_B + c X_B{}^2), \tag{2.21}$$

where a, b, and c may be temperature-dependent but are composition-independent constants, the expressions for the excess partial molar free energy of the components AC and BC are

$$\bar{G}_{AC}^{(E)} = [a - b + 2(b - c)X_B + 3cX_B{}^2]X_B{}^2 \tag{2.22}$$

and

$$\bar{G}_{BC}^{(E)} = (a + 2bX_B + 3cX_B{}^2)X_A{}^2. \tag{2.23}$$

The excess partial molar free energy can be related to the experimental emf by use of the Temkin relation

$$\bar{G}_{AC}^{(id)} = RT \ln X_A \tag{2.24}$$

and the standard state defined as the pure liquid. This leads to

$$\bar{G}_{AC}^{(E)} = -n(E - E_{AC}^0)F - RT \ln X_A, \tag{2.25}$$

where E_{AC}^0 is the emf of the formation cell for pure AC and may be obtained from thermal data. The excess partial molar entropy is similarly related to the experimental emf by

$$\bar{S}_{AC}^{(E)} = n\left(\frac{dE}{dT} - \frac{dE_{AC}^0}{dT}\right)F + R \ln X_A. \tag{2.26}$$

Thus from an experimental emf study the excess partial molar free energy and entropy can be obtained with the aid of Eqs. (2.25) and (2.26) and, by fitting these data to Eq. (2.22), the excess partial molar free energy of the other component (BC) and the excess free energy may be obtained from the constants a, b, and c and Eqs. (2.23) and (2.21), respectively.

Dijkhuis *et al.* (1968) recommended a similar procedure to obtain the excess entropy, but the precision of excess partial molar entropies obtained from Eq. (2.26) and the experimental emf temperature coefficients is so poor that it does not at present seem to be a worthwhile approach.

A more-active metal can be studied by the use of an alkali-metal–glass-indicator electrode. For example, in the cells of the type studied by Ostvold (1966),

$$Cl_2|NaCl|Glass|NaCl–MCl_2|Cl_2, \qquad (2.27)$$

where M was Mg, Ca, Sr, and Ba the measured emf is directly related to the partial molar free energy of mixing of the more-active metal, i.e., sodium chloride. In a case like this the sodium indicator electrode can be used to determine the NaCl partial molar free energy, and a formation cell with a magnesium anode, e.g., could be used to determine the partial molar free energy of $MgCl_2$. This sort of experiment could well be used to better determine the correctness of the analysis given by Dijkhuis *et al.* (1968).

F. RESULTS OF EMF STUDIES

One use of results of emf studies is to attempt to establish an electrochemical series based on single electrode potentials similar to that for aqueous solutions. Delimarskii and Markov (1961) have discussed this problem in detail. From their discussion it is apparent that while it is possible to discuss such a series with reference, e.g., to a sodium indicator electrode as a potential zero, with a given anion or a given fused-salt electrolyte no generalization emerges which allows a single electrochemical series to be established. Of course, the broad trends of the aqueous series are maintained, e.g., the alkali metals are more active than the transition metals, but metals closer in potential frequently change places in the series depending on the electrolyte and occasionally on the temperature. This implies that in studying electrochemical properties of fused-salt solutions, one should be careful to identify the reacting metal ion and not rely on the aqueous or other general electrochemical series.

As discussed in Section II,E, emf measurements provide basic thermodynamic data which can be used to test theories of fused salts. In particular, the excess partial molar free energies are useful quantities to compare with theoretical predictions. A number of simple solutions of uni-univalent salts have been found to be in reasonable agreement with the predictions of regular solution theory, Eq. (1.13), as well as to the theory of conformal solutions, Eq. (1.24), at least as far as the concentration dependence at a given temperature is concerned. That is, the excess partial molar free energy of one component divided by the square of the mole fraction of the other component is a constant. The solutions which follow this concentration dependence to within the experimental uncertainty are shown in Table III along with the interaction parameter

TABLE III

INTERACTION PARAMETERS FROM REGULAR
SOLUTION THEORY FOR BINARY MIXTURES

Mixture	ω (cal/mole)	Temp. (°C)
(Li–Ag)Cl	~ 2100	800
(Na–Ag)Cl	~ 800	800
(K–Ag)Cl	~ -1500	650
(Li–Ag)Br	1880	550
(Na–Ag)Br	1050	600
(K–Ag)Br	-1480	600
(Rb–Ag)Br	-2580	550
(K–Ag)I	~ -2000	600
(Na–Ag)NO$_3$	840	300–350
Ag(Cl–Br)	~ 2000	600
Ag(Cl–I)	450	600
Ag(Br–I)	1150	600
K(Cl–Br)	700	800
Na(Cl–Br)	400	800

ω from regular solution theory. The values in Table III have been adapted from the reviews of Blander (1964) and Dijkhuis *et al.* (1968) and references to the original data can be found in those sources. No attempts have yet been made to fit these emf data into the framework of the conformal solution theory to see how the values depend on ionic radii and the non-Coulombic terms, although it is apparent that increasing alkali-metal cation size makes the interaction parameter more negative for the common anion series, in qualitative agreement with conformal solution theory. The point should be made that while the experimental results give the excess partial molar Gibbs free energy and the theory gives the excess partial molar Helmholtz free energy, the difference should be small, since the vapor pressures for most salts at the temperatures of study are very low. It is also of interest that no significant effects of nonrandom mixing, which would require application of the quasichemical or some other approximation for the lattice theory, have been observed in binary solutions. It is possible, however, that any such effects would be masked by the uncertainties in the data, and the more precise thermal studies of Kleppa (1965) have shown that substantial asymmetries in the concentration dependence of the heat of mixing exist which could arise from nonrandom mixing as well as other sources.

The temperature dependence of the emf has received some attention, and for two systems with quite accurate data at different temperatures the interaction parameter ω has been shown by Blomgren (1962) to be nearly linearly dependent on the temperature. These results are shown in Table IV. There is no theoretical basis for understanding this temperature dependence; the lattice theory treats ω as an experimentally-determined parameter and the conformal solution theory treats the temperature dependence as that of a reference substance, although this

TABLE IV

TEMPERATURE DEPENDENCE OF THE INTERACTION PARAMETER ω [a]

| System | ω (cal/mole) | | | |
	at 600°C	700°C	800°C	900°C
(Ag–Li)Cl	2240	2140	2060	1940
(Ag–Na)Cl	—	1120	940	750

[a] From Blomgren (1962).

aspect has not as yet been discussed in detail. Since the excess partial molar free energy seems to fit the regular solution theory, the analysis of Dijkhuis et al. (1968) discussed earlier is simplified and Eq. (2.23) becomes

$$\overline{G}_{BC}^{(E)} = \omega X_A^2 \tag{2.28}$$

and Eq. (2.21) becomes

$$G^{(E)} = \omega X_A X_B. \tag{2.29}$$

Thus by combining derived values for the excess free energy with measured excess enthalpies the excess entropies for these solutions are obtained. Table V shows the results for a number of solutions at the 50 mole % concentration. It is evident from the table that all but one of the values of the excess entropies are small and negative, and for the (Ag–Na)Cl system, if the temperature corrected value of ω is used, the excess entropy is nearly zero. Even though the entropy values are small, the high temperature makes them quite significant in the $TS^{(E)}$ term for the free energy. The negative value of the excess entropy is in the direction one would expect for nonrandom mixing. No detailed attempts have yet been made to explain the magnitudes of the excess entropy, however.

TABLE V

EXCESS ENTHALPIES, FREE ENERGIES, AND ENTROPIES FOR 50 MOLE % BINARY MIXTURES

Mixture	T (°K)	$H_{0.5}^{(E)}$ [a]	$G_{0.5}^{(E)}$ [b]	$TS_{0.5}^{(E)}$	$S_{0.5}^{(E)}$
(Ag–Li)Cl	933	490	525 (545)[c]	−35 (−55)	−0.04 (−0.06)
(Ag–Li)Br	850	368	470	−102	−0.12
(Ag–Na)Cl	933	310	200 (290)	+110 (+20)	+0.1 (+0.02)
(Ag–Na)Br	850	210	263	−53	−0.06
(Ag–Na)NO$_3$	623	150	210	−60	−0.1
(Ag–K)Cl	933	−550	−400	−150	−0.2
(Ag–K)Br	850	−536	−370	−166	−0.20
(Ag–Rb)Br	850	−922	−645	−277	−0.33

[a] From Hersh et al. (1965).
[b] From Blander (1964).
[c] Values in parentheses from Blomgren (1962).

The situation for binary solutions in which one or more of the ions is multivalent is considerably more complicated. In some cases the excess partial molar free energy behaves like a regular solution, but in general, even for simple alkaline-earth–alkali-metal salt solutions, there are very large deviations from regular solution behavior. In some cases these deviations can be explained reasonably by the formation of a complex ion, e.g., in the $CeCl_3$–alkali-chloride solutions studied by Senderoff et al. (1960). The best situation for explaining the deviations by complex ion formation are cases in which molecular spectroscopic methods have established the existence of the complex ions and the emf results can be explained in terms of them. Such a case is offered by the $NiCl_2$–alkali-chloride solutions studied by Hamby and Scott (1968). Visible spectroscopy has established the presence of $NiCl_4^{--}$ ions and the emf results are shown to be compatible with their presence. It is much to be desired that more combined spectroscopic–electrochemical studies of this type will be carried out. In other solutions involving salts such as $CdCl_2$, $ZnCl_2$, and $PbCl_2$ in which a number of complex species are suspected to be present from spectroscopic studies, it would be expected that an exact thermodynamic analysis applied to emf results would be quite complicated, and little progress has been made in this direction. The major difficulty in treating the solutions thermodynamically is that the activity coefficients of the complex ions are unknown and are not expected to be constant with concentration changes, so that the methods commonly used for aqueous solutions do not apply.

Another type of solution which has been studied by the emf method is the reciprocal salt solution. Most of these studies have been made in concentration cells with no liquid junction correction (Blander, 1964), so that the results are somewhat questionable even though dilute solutions have been used. A comparison of the suitability of various theories to explain the thermodynamic properties of reciprocal salt solutions has been made by Blander and Topol (1965). They have concluded that the lattice theory of Flood *et al.* (1954) is insufficient to explain the results, and that the conformal solution theory of Blander and Yosim (1963) is adequate in one case (LiF–KCl), but not in another (NaNO$_3$–AgCl), but that a complete lattice theory including binary and ternary terms can explain the data on both solutions. More tests of experimental data on reciprocal salt solutions would be desirable to verify these conclusions.

III. Electrical Transport Processes

A. Electrical Conductivity and Transference Numbers

The electrical transport properties of fused salts to be discussed in this section are the electrical conductivity and transference numbers. The electrical double layer will be discussed in Section III,B. A great number of measurements of conductivity of fused salts have been made, going back to the nineteenth century, but relatively few reliable transference experiments have been carried out. Many references to the literature in this field have been given by Blomgren and VanArtsdalen (1960), Klemm (1964), Sundheim (1964), and Delimarskii and Markov (1961).

The experimental techniques for measuring fused-salt conductivities have been well worked out. The greatest problem faced in measuring conductivity, aside from the materials and temperature problems common to all fused-salt studies, is due to the fact that the specific conductivity is so high for the typical fused salt (of the order of 1–10 ohm^{-1} cm^{-1}) that the cells must be specially designed to obtain cell constants of the order of a few hundred cm^{-1} in order to use conventional bridges. To this end, the conducting path generally makes use of capillary tubes. Sundheim (1964) and Janz (1967) have reviewed the types of cells employed and give experimental details. Along with the measurement of conductivity, it is necessary to measure densities of the salts in order to obtain the equivalent conductivity from the formula

$$\Lambda = \kappa V, \tag{3.1}$$

where κ is the specific conductivity and V is the equivalent volume (in cc/equiv.) obtained from a measurement of the density ρ and the formula

$$V = \left(\sum_i X_i E_i\right)/\rho, \tag{3.2}$$

in which X_i is the mole fraction of the ith salt of equivalent weight E_i. These formulas can be used both for pure salts and solutions. Klemm (1964) gives extensive tables of data on pure salts and solutions.

It was recognized very early that conductivity measurements provide a method of distinguishing between fused salts and molecular liquids. For example, Biltz (1924) observed that most molten materials have specific conductances higher than 10^{-2} ohm^{-1} cm^{-1} or lower than 10^{-4} ohm^{-1} cm^{-1}, thus forming a natural division between salts and molecular liquids. It should also be noted that some largely-undissociated liquids which are semiconductors in the solid state, such as some sulfides, selenides, etc., retain electronic conductivity in the liquid and thus have anomalously high conductivities.

Beyond this qualitative generalization, however, it has been found difficult to develop a theory of conductivity of fused salts. Many workers (see VanArtsdalen and Yaffe (1955) for references) have treated the conductivity of pure fused salts as a rate process by use of the equation

$$\Lambda_i = A_i \exp(-\Delta H^{\ddagger}/RT), \tag{3.3}$$

where A_i is a constant for the system and $\Delta H^{\ddagger}$ is the activation enthalpy for the conduction process. If a sufficient temperature range is used, it is usually found that $\Delta H^{\ddagger}$ is temperature dependent, and no explanation for this behavior has been found. Sundheim (1964) has pointed out that the rate-process approach tacitly assumes that the temperature dependence is measured with constant volume, whereas the experiments are conducted at constant pressure. The effect of volume changes with temperature should thus be brought into the theoretical analysis, and this is likely to prove difficult.

Other approaches have related the conductivity to the measurement of other transport properties. For example, Frenkel (1946) proposed the relation

$$\Lambda^m \eta = \text{const}, \tag{3.4}$$

where η is the viscosity and m is to be determined from the temperature dependence of a given salt. When $m = 1$ this becomes Walden's rule, which has been applied to dilute solutions in aqueous and nonaqueous solvents. Even though Eq. (3.4) has been found to be fairly widely applicable, it does not lead to much understanding of the conduction process, as pointed out by Yaffe and VanArtsdalen (1956). Another proposed relation among transport properties is the Nernst–Einstein equation relating self-diffusion coefficients of the individual ions in a pure salt to the conductivity:

$$(Z_1 D_1 + Z_2 D_2)/RT = \Lambda/F^2, \tag{3.5}$$

where Z_i and D_i are the charge and self-diffusion coefficient of the ion i, respectively, and the gas constant R is expressed in joules. Equation (3.5) has been found not to hold for most fused salts for which self-diffusion coefficients have been measured by tracer techniques (see Sundheim (1964) for data and references). This is not too surprising, since the equation was developed for dilute aqueous solutions. The reason it does not hold is that the assumption is tacitly made that the ions move independently of one another as shown by the development of irreversible thermodynamics (see below). The independent motion is a reasonable assumption for dilute solutions and it also permits the determination of individual ion conductivities in dilute solutions. It could not be expected to hold for highly concentrated fused salts.

The conductivity of solutions of fused salts is even less susceptible of analysis. Almost no solutions show additivity of the component conductivities, and both positive and negative deviations from additivity have been observed for both simple binary solutions and for those expected to interact through complex formation, etc. (see Klemm (1964) for data). Therefore conductivity measurement is not a reliable method for detecting the presence of chemical interaction among ions in a fused-salt solution, although it has often been used in this way.

The measurement of transference numbers in fused salts has aroused a considerable amount of controversy, particularly for pure fused salts. The problems involved are both experimental and theoretical and are caused by the fact that there is no intervening medium, like water in aqueous solutions, which can permit concentration changes to occur and provide a reference system relative to which the ion motion is measured. In a typical transference experiment for a pure fused salt a plug with fine pores is inserted between the electrode compartments in order to prevent hydrodynamic backflow which would otherwise occur and negate the effect of different ion velocities. These plugs have, in general, a very high resistance compared to the rest of the cell, so that the main effect occurs within the plug. Many ingenious methods to determine the effect of different ion velocities have been devised, and the reader is referred to the review of Sundheim (1964) for details. If none of the ion momentum is exchanged with the cell (and if fluid flow is totally prevented), the transference number is easily predicted from momentum-conservation considerations, and (Sundheim, 1957) is given by

$$t_A = M_C/(M_A + M_C),\qquad (3.6)$$

where M_A and M_C are the cation and anion molecular weights, respectively. Experimental results have often been found not to be in accord with this simple result, a fact which implies that momentum is exchanged unequally for the different ions within the plug. A model based on separate lamellar ionic flows with the plug exerting a different viscous drag on each flow has been developed, but not carried sufficiently far to obtain numerical results (see Sundheim (1964)). At the present stage of knowledge a transference number experiment with a pure fused salt does not provide fundamental information on relative ionic velocities, but rather is related to unknown interactions of the salt with the porous plug.

Transference numbers in fused-salt solutions can be unambiguously determined by standard methods. Hittorf-type cells may be used if precautions are taken to prevent convective mixing, and this may be accomplished by a diaphragm, a narrow tube, or a vertical upward temperature gradient between the electrode compartments. Moving-boundary cells can be used if the different mixtures are arranged in a vertical tube with the most-dense fluid at the bottom, and if the current is passed in such a direction as to cause a self-sharpening boundary to form. The use of emf cells with liquid junction to determine transference numbers is not a sufficiently sensitive technique, as discussed earlier. For more details the reader is referred to the work of Klemm (1964).

Transport properties of fused salts can be put in a comprehensive and rigorous framework with the aid of the thermodynamics of irreversible processes. This theory develops in a systematic way the interrelationships among transport properties when the fluxes of the various species are linearly related to the forces present in the system. The theoretical background is well treated by de Groot (1951) and the application to fused salts is given in Klemm (1964) and Sundheim (1964), so only a brief review will be given here. The notation of Laity (1959) will be followed. For fused salts with N ionic species the phenomenological equations relating generalized forces to fluxes are given by

$$K_i = -(\nabla\mu_i \pm Z_i F \nabla\phi) = \sum_{k=1}^{N} r_{ik} X_k (v_i - v_k) \qquad (i = 1, 2, \dots N),$$

(3.7)

where K_i, μ_i, Z_i, and v_i are the force, chemical potential, numerical value of the charge, and the average velocity of the ith species, respectively, X_k is the mole fraction of the kth species, and r_{ik} is a "friction coefficient" which is to be determined from experiment. The theory states that the friction coefficients obey the Onsager relations:

$$r_{ik} = r_{ki}.$$

(3.8)

The use of the Gibbs–Duhem equation

$$\sum_{i=1}^{N} X_i K_i = 0$$

(3.9)

allows the N equations given by Eq. (3.7) to be reduced to $N - 1$ equations. All linear transport properties (viscosity and many chemical reactions are nonlinear) follow from the solutions of Eqs. (3.7)–(3.9) for the particular case. The results are summarized below for conductivities, transference numbers, and diffusion coefficients:

a. Pure Fused Salts (Two Ionic Species)

$$\Lambda/F^2 = [(Z_+ + Z_-)/r_{+-}],$$

(3.10)

$$D_{++}/RT = (Z_+ + Z_-)/(Z_+ r_{+-} + Z_- r_{++}),$$

(3.11)

$$D_{--}/RT = (Z_+ + Z_-)/(Z_- r_{+-} + Z_+ r_{--}).$$

(3.12)

b. Binary Solutions (Two Cationic Species 1 and 2, and One Anionic Species 3)

$$\frac{\Lambda}{F^2} = \frac{(Z_1 Z_2 + X_{13} Z_2 Z_3 + X_{23} Z_1 Z_3)(Z_3 r_{12} + X_{13} Z_1 r_{23} + X_{23} Z_2 r_{13})}{Z_2 Z_3 X_{13} r_{12} r_{13} + Z_1 Z_3 X_{23} r_{12} r_{23} + Z_1 Z_2 r_{13} r_{23}},$$

(3.13)

$$t_2 = \frac{X_{23}(Z_3 r_{12} + Z_2 r_{13})}{Z_3 r_{12} + Z_1 X_{13} r_{23} + Z_2 X_{23} r_{13}},$$

(3.14)

$$\frac{Z_3 D'_{12}}{RT} = \frac{Z_1 Z_2 + X_{13} Z_2 Z_3 + X_{23} Z_1 Z_3}{(Z_1 X_{23} + Z_2 X_{13})(Z_3 r_{12} + Z_2 X_{23} r_{13} + Z_1 X_{13} r_{23})},$$

(3.15)

$$\frac{D_{11}}{RT} = \frac{Z_1 Z_2 + X_{13} Z_2 Z_3 + X_{23} Z_1 Z_3}{Z_2 Z_3 X_{13} r_{11} + Z_1 Z_3 X_{23} r_{12} + Z_1 Z_2 r_{13}},$$

(3.16)

$$\frac{D_{22}}{RT} = \frac{Z_1 Z_2 + X_{13} Z_2 Z_3 + X_{23} Z_1 Z_3}{Z_2 Z_3 X_{13} r_{12} + Z_1 Z_3 X_{23} r_{22} + Z_1 Z_2 r_{23}}, \tag{3.17}$$

$$\frac{D_{33}}{RT} = \frac{Z_1 Z_2 + X_{13} Z_2 Z_3 + X_{23} Z_1 Z_3}{Z_2 Z_3 X_{13} r_{13} + Z_1 Z_3 X_{23} r_{23} + Z_1 Z_2 r_{33}}. \tag{3.18}$$

The X_{ij} are the mole fractions of the salt ij in the mixture, D_{ii} are self-diffusion coefficients, which must be determined from isotopic tracer experiments, D'_{12} is an interdiffusion coefficient, and t_2 is the transference number of the second cation relative to an anion transference number of zero. It is clear that there are as many friction coefficients as there are experiments which can be performed, so that no new information is introduced by this analysis. Some interesting results do follow, however. It is clear from Eqs. (3.10)–(3.12), e.g., that the conductivity of a pure fused salt will not, in general, be proportional to the sum of the self-diffusion coefficients, as in Eq. (3.5), unless the friction coefficients r_{++} and r_{--} are equal to each other or are negligibly small. Laity (1960) has found that in most systems r_{++} is much smaller than r_{+-}, but r_{--} is of the same order of magnitude as r_{+-}, thus indicating that anion–anion interactions are primarily responsible for deviations from the Nernst–Einstein equation. Because complete transport data have not been measured on any one system, few generalizations have resulted from Eqs. (3.13)–(3.18), but Laity (1960) has found that, except in dilute solutions, the friction coefficients are not independent of concentration. If they had been found to be independent of concentration, considerable simplification would have resulted. In summary, more complete transport measurements are necessary before this interesting approach can show its full potential.

B. THE ELECTRICAL DOUBLE LAYER

At first thought it might appear that the presence of an electrical double layer must be primarily due to specific adsorption and that diffuse double layers would not be supported in such concentrated ionic media as fused salts. Recent double-layer-capacity experiments, however, on fused nitrates and halides of the alkali metals have shown the opposite to be the case (Hills and Power, 1968; Ukshe and Bukun, 1961). In these measurements the double-layer capacity as a function of potential was found to lie on a parabola symmetrical about the point of zero charge. The minimum capacities appeared to be around 20 $\mu F/cm^2$. The absence of specific adsorption is reflected in the symmetry of the parabola. If the ionic distribution functions were known, an alternating excess and deficit of one charge species which would likely be rapidly attenuated with distance from the electrode would fully account for the observed double-layer behavior. However, in the absence of a detailed theory (comparable with the Debye–Hückel theory of aqueous solutions) of ionic distributions for fused salts no quantitative statements can be made. Since concentrated aqueous solutions behave in an opposite way, the importance of solvent adsorption, in that case in the inner Helmholtz plane, is clear. The experimental situation in fused salts is often clouded by the presence of Faradaic reactions. Since Faradaic processes in fused salts are generally fast, it is often difficult to resolve the double-layer charging and the reaction. In some systems adsorption also appears to occur, but in a way which is highly potential-dependent. Relatively few studies have been carried out in this theoretically and experimentally important area, but see the work by Hills and Power (1968) for experimental techniques and further references.

Many earlier studies had shown a pronounced frequency dependence of the double-layer capacity, but the most recent work discussed above would indicate that this frequency dispersion is likely due to Faradaic processes from melt impurities and spurious effects from experimental problems such as melt creepage.

IV. Electrochemical Kinetics

A. Introduction

A number of reviews of the techniques and results of electrochemical kinetic studies in fused-salt media have appeared. These include the works by Laitinen and Osteryoung (1964), Liu et al. (1964), Graves et al. (1966), Delimarskii and Markov (1961), Inman (1968), and Graves and Inman (1968). The conclusions of all of these authors is that kinetics of fused-salt reactions seem to be of two types: (1) very slow reactions involving gas–anion electrode couples, and (2) very fast reactions involving metal–metal-ion couples, with very few reactions falling between these extremes. One reason for this phenomenon is that the passivating films so commonly found in aqueous electrochemistry are seldom stable in high-temperature melts, so that nearly all metal-dissolution reactions are fast. Also, the sorts of catalysts used for gas-electrode reactions in aqueous solutions seem to lose their activity in the high-temperature systems, possibly because of impurity poisoning, or annealing of active sites and crystal growth (loss of surface area) at the high temperature. Because of this, steady-state methods and scanning and pulse methods with intermediate time regimes have served only for obtaining kinetic information on gas electrodes and determining mass transport properties along with some mechanism speculations for metal electrodes. These studies will be taken up first. The attempts to measure kinetic properties of the very fast metal–metal-ion electrode reactions seem to be starting to yield to advanced pulse techniques, but have in the past been very difficult and controversial. These measurements will be considered in the final section. The many fine studies on fused salts with analytical polarography have been reviewed by Laitinen and Osteryoung (1964) and Delimarskii and Markov (1961), and will not be considered further in this chapter.

The two greatest problems which beset kinetic studies in fused salts are (1) the presence of electroactive impurities and the difficulty of their removal, and (2) the presence of parasitic chemical or electrochemical (corrosion) reactions in the melts with all of the materials in the systems, often including the electrodes themselves and newly-deposited materials on the electrodes. Purification procedures are steadily improving, and with considerable effort quite pure melts can be obtained (see Janz (1967)). Progress toward the solution to the second problem is greatly assisted by the discovery and application of new inert materials to fused systems, but remains a subject of active inquiry and stands in need of further advance. The other problem of electrode–melt interaction is part of the kinetic process and must be taken into account and corrected for in the kinetic measurements themselves.

B. Steady-State and Intermediate-Time Methods

Most of the electrode processes studied by dc steady-state methods involve the reduction of anions to form gaseous products. The simplest example is the reduction of halide ions to form halogen gas. It appears in this case that the reaction is fast, unlike oxyanion electrodes, but is severely complicated by attack of the halide melt on the graphite electrode (Shams El Din, 1961). The evolution of oxygen from OH^-, NO_3^-, CO_3^-, O^{--}, AlO_2^-, and SO_4^{--} has been studied in various melts (see Graves *et al.* (1966) for references), and no single mechanism has been found to be operative for the oxygen-evolution process. In fact, for sulfate and carbonate a change in the mechanism occurs for high-temperature or high-current density, with the Tafel slope changing from $RT/3F$ to a larger value, which has not yet been definitely established (the high values of the current lead to a questionable steady-state situation). The following proposed mechanism is consistent with the low-current Tafel slope of $RT/3F$ (Janz and Saegusa, 1961):

$$O^{--} + M \rightarrow MO + 2e, \tag{4.1a}$$

$$O^{--} + MO \rightarrow M + O_2 + 2e, \tag{4.1b}$$

where the reaction Eq. (4.1b) is the rate-limiting step, the O^{--} ion is used symbolically in place of the oxyanion studied, and M is a surface atom of the solid. Oxygen evolution from nitrates also shows a change in Tafel slope with increasing temperature, from $2RT/F$ below 250°C to RT/F above 280°C, although the reaction mechanisms have not yet been established. For oxide-ion and aluminate-ion discharge it appears that the two-electron charge-transfer step Eq. (4.1a) is the rate-limiting step. Many hydrogen-evolution studies have also been carried out in various fused salts. In particular, liquids KHF_2, $KHSO_4$, and HCl in LiCl–KCl have all been used for hydrogen-evolution studies with dc methods. Various workers do not agree on results, however, and the subject is open to further inquiry (see Graves $et\ al.$ (1966) for references).

The intermediate-time methods of polarography, linear-sweep volt-ammetry, and chronopotentiometry are being increasingly used for fused-salt solutions. Most of the polarographic studies have been carried out with solid electrodes, although some low-temperature melts have been studied with dropping mercury electrodes and some studies have used other liquid-metal electrodes. While polarography is discussed theoretically as a steady-state method, a frequent observation with recording polarographs is the presence of a peak rather than a plateau, which is an indication of nonsteady-state conditions. For polarograms showing true steady-state conditions the current–voltage relation should obey either the Kolthoff–Lingane equation

$$E = \text{const} + (2.3RT/nF)\log(i_1 - i), \tag{4.2}$$

where i_1 is the polarographic limiting current when a metal ion is re-duced to metal at unit activity, or the Heyrovsky–Ilkovic equation

$$E = \text{const} + (2.3RT/nF)\log[(i_1 - i)/i], \tag{4.3}$$

if the reduced metal ion is free to diffuse, e.g., into a liquid metal drop. It has been found, however, that the behavior is not predictable, espe-cially at solid electrodes, and sometimes is intermediate between these two cases. The general equation for alloy formation at a solid electrode advanced by Delimarskii and Gorodskii (1961) is

$$E = \text{const} + \frac{2.3RT}{nF}\log\left[\frac{k_s(i_1 - i)}{a_m - k_m(i_m' - i_m)}\right], \tag{4.4}$$

where a_m is the maximum activity of the metal, k_s relates surface ionic activity to current density, k_m relates surface metal activity to the intermetallic diffusion current i_m, the maximum value of which is i_m'. Equation (4.4) has been found useful, since it reduces to Eq. (4.2) when $i_m = i_m'$ and to Eq. (4.3) when $i_1 \ll i_m'$ and also covers the intermediate cases. The actual physical situation, however, is very difficult to ascertain. The result of these uncertainties is that such quantities as diffusion coefficients of electroactive species obtainable in general from i_1 are uncertain, and conclusions concerning mechanisms and overall reaction schemes are also uncertain. The use of polarography as an analytical tool is not affected, however, as long as the behavior of the polarogram with concentration changes is established. The reader is referred to Laitinen and Osteryoung (1964) and Graves et al. (1966) for more details.

The use of chronopotentiometry in fused salts was pioneered by Laitinen and his co-workers, and the technique is reviewed by Laitinen and Osteryoung (1964). For fast electrode reactions under linear diffusion control, to which the Sand equation can be applied, such as metal-deposition reactions, the method gives reliable values of the diffusion coefficient of the electroactive species if sufficient care is taken in the experiment. One of the reasons for good results with fused salts is that reactions like metal deposition are so rapid that, with adjustment of experimental conditions, transition times in the millisecond to 0.1-sec range can easily be achieved and problems of double-layer charging with shorter times and convection with longer times can be obviated. Diffusion coefficients for many metal ions in a variety of fused-salt solutions have been determined in this way. Another use of the technique has been to develop conclusions concerning the qualitative aspects of the reaction such as diffusion control, solid or soluble products, presence of consecutive reactions (electrochemical, chemical, or both), etc., by applying the chronopotentiometric criteria to the experimental results. Such a use of the technique has been applied by Senderoff and Mellors (1966) to the deposition of refractory, hard metals from fluoride baths. These authors find that the reduction of Ta(V) ions occurs in two steps with a fast, reversible three-electron transfer preceding a final, slow, irreversible two-electron transfer by the equations

$$\begin{aligned} \text{TaF}_7^{2-} + 3e &\rightarrow \text{TaF}_2(s) + 5\text{F}^- \\ \text{TaF}_2(s) + 2e &\rightarrow \text{Ta}^\circ + 2\text{F}^-. \end{aligned} \tag{4.5}$$

The work represents a good merging of physical and electrochemical techniques, as the existence of TaF_7^{2-} as the sole tantalum species in the equilibrium melt has been established by Fordyce and Baum (1966) by an infrared reflection study. While studies of this type do not produce the kinetic parameters for the reaction, they do provide valuable mechanistic information.

The increasingly popular technique of linear-sweep voltammetry has also been applied to fused salts. Anders and Plambeck (1968) studied the electrolysis of fused lithium perchlorate and found the overall reactions to be

$$
\begin{aligned}
&\text{Li}^+ + \text{e} \rightarrow \text{Li}^\circ \\
&\text{ClO}_4^- \rightarrow \tfrac{1}{2}\,\text{Cl}_2 + 2\,\text{O}_2 + \text{e},
\end{aligned}
\qquad (4.6)
$$

at the cathode and anode. The anode process was complicated by an additional step forming chlorate ion, which induced a chemical reaction in the melt. This technique will undoubtedly find wide use in the future for qualitative reaction studies.

C. IMPEDANCE AND PULSE TECHNIQUES

Since the rates of so many fused-salt reactions are known to be very fast, there are only two types of measurement which could be expected to give rate data on these electrochemical processes. The first is the impedance technique, which uses an applied ac field and an ac bridge to measure the impedance as a function of frequency. The second is to apply a very-rapid-rise step-function pulse of constant voltage or constant current and to observe the current or voltage as a function of time.

The impedance technique requires a separate measurement of the solution resistance and double-layer capacitance in the absence of the electroactive ion. Then an impedance analysis is carried out based on an equivalent circuit having the solution resistance in series with a parallel network which has the double-layer capacitance in one arm and the polarization resistance (R_r) and pseudocapacity (C_r) for the electrode reaction in the other arm. The circuit analysis gives R_r and C_r in terms of the electrode area A, the circular frequency ω, the diffusion coefficient of the electroactive species D, and the specific rate constant for the electrode reaction k through the equations

$$
R_r = \frac{RT}{n^2 F^2 AC}\left[\left(\frac{1}{2D\omega}\right)^{1/2} + \frac{1}{k}\right], \qquad (4.7)
$$

$$
C_r = \frac{n^2 F^2 AC}{RT}\left(\frac{2D}{\omega}\right)^{1/2}, \qquad (4.8)
$$

$$
R_r - \frac{1}{\omega C_r} = \frac{RT}{n^2 F^2 AC}\frac{1}{k}, \qquad (4.9)
$$

where C is the concentration of the electroactive species. The method has been applied to a number of systems (see Laitinen and Osteryoung (1964) for a review), but the results have been confused by impurities and experimental problems leading to frequency dependence of the separately-measured double-layer capacity and solution resistance. Even in the absence of these experimental problems there is serious doubt about the validity of the circuit analysis for these cases of fast, but not infinitely fast, electrode reactions, to the effect that the double-layer capacity may not be constant or even separable in the presence of the Faradaic processes (see Graves and Inman (1968)). For this reason it seems likely that pulse techniques will prove to be of greater importance in establishing the true rates of the fast fused-salt electrode reactions.

Laitinen *et al.* (1960) have pioneered the application of pulse methods for fused salts and have examined both voltage-step and double-current-pulse techniques. In the voltage-step method a constant voltage pulse is imposed across the whole cell, and the current decay following the pulse is studied. A blank, in the absence of the electroactive species, is taken to give the circuit resistance and the double-layer capacity, and the effects of these quantities are subtracted out in the analysis. The analysis which was used assumes a separation of double-layer current and Faradaic current. These workers also studied a number of the same reactions with a double-current-pulse technique. In this method a large current pulse with a short time constant is superimposed on a lower current pulse and the voltage transient of the second pulse is studied. The purpose of the first pulse is to charge the double layer prior to the application of the second pulse, and its height and length are adjusted in such a way that the potential–time plot of the second pulse has a horizontal tangent at the instant of termination of the first pulse. The properties of the transient are then taken to be characteristic of the Faradaic process, and the kinetic parameters are determined from the appropriate analysis. It was found, however, that the dependence of the apparent exchange current on the length of the charging pulse was greater than predicted by the theory and that calculated diffusion coefficients were much too low.

An explanation for the problems associated with the above techniques for fused salts has recently been advanced by Graves and Inman (1968). These authors believe that many of the problems are associated with the inseparability of double-layer charging and Faradaic processes for these fast reactions. By an interesting analysis they have presented criteria for distinguishing those reaction which can be studied by the classical methods of analysis. They find that the Pb(II)|Pb couple in LiCl–KCl eutectic fulfills these criteria, and thus this system can be studied with a conventional single-current-pulse technique. The results of this work are compared in Table VI with results obtained by Laitinen *et al.* (1960) on this and other systems. The comparison for the lead system studied by the two groups shows that additional work needs to be done. While the comparison for cadmium and zinc shows reasonable agreement for the two methods (voltage-step and double-current-pulse), further work should establish the independence of double-layer charging by the method of Graves and Inman (1968).

TABLE VI

Comparison of Kinetic Parameters from Various Pulse Techniques in LiCl–KCl Eutectic at 450°C

Electrode	Method [a]	Transfer coefficient, α	Rate constant k^0 (cm/sec)	Ref.
Cd(II)/Cd	VS	0.13 (± 0.05)	0.4	Laitinen *et al.* (1960)
	DCP	0.13 (± 0.05)	0.4	Laitinen *et al.* (1960)
Zn(II)/Zn	VS	0.10 (± 0.05)	0.4	Laitinen *et al.* (1960)
	DCP	0.16 (± 0.05)	0.3	Laitinen *et al.* (1960)
Pb(II)/Pb	DCP	0.38 (± 0.06)	0.01	Laitinen *et al.* (1960)
	CP	0.22 ($\pm 20\%$)	0.07	Graves and Inman (1968)

[a] VS: Voltage step. DCP: Double current pulse. CP: Current pulse.

D. Summary

While it is clear that many basic problems remain for the study of electrochemical kinetics in fused salts, the field has advanced rapidly in recent years and techniques of great promise are under development. Future work should produce results of interest to the entire field of fast reactions because of the unusually high rates of the reactions in fused salts. It is hoped that many workers will be stimulated to enter this difficult, but exciting field.

Special References

Albyshev, A. F., Lantratov, M. F., and Morachevskii, A. G., ed. (1965). "Reference Electrodes for Fused Salts." Sigma Press, Washington, D.C.
Anders, U., and Plambeck, J. A. (1968). *J. Electrochem. Soc.* 115, 598.
Baker, B. S., ed. (1965). "Hydrocarbon Fuel Cell Technology." Academic Press, New York.

BAUER, S. H., and PORTER, R. F. (1964). In "Molten Salt Chemistry" (M. Blander, ed.), pp. 607–680. Wiley (Interscience), New York.

BILTZ, W. (1924). Z. Anorg. Allgem. Chem. 133, 312.

BLANDER, M. (1961). J. Chem. Phys. 34, 697.

BLANDER, M. (1962). J. Chem. Phys. 36, 1092.

BLANDER, M., ed. (1964). "Molten Salt Chemistry," pp. 127–237. Wiley (Interscience), New York.

BLANDER, M., and BRAUNSTEIN, J. (1960). Ann. N. Y. Acad. Sci. 79, 838, Art. 11.

BLANDER, M., and TOPOL, L. E. (1965). Electrochim. Acta 10, 1161.

BLANDER, M., and YOSIM, S. T. (1963). J. Chem. Phys. 39, 2610.

BLOMGREN, G. E. (1960). Ann. N. Y. Acad. Sci. 79, 781.

BLOMGREN, G. E. (1962). J. Phys. Chem. 66, 1500.

BLOMGREN, G. E., and VANARTSDALEN, E. R. (1960). Ann. Rev. Phys. Chem. 11, 273–306.

BLOOM, H., and BOCKRIS, J. O'M. (1964). In "Fused Salts" (B. R. Sundheim, ed.), pp. 1–64. McGraw-Hill, New York.

BUES, W. (1955). Z. Anorg. Allgem. Chem. 279, 104.

CARLSON, C. M., EYRING, H., and REE, T. (1960). Proc. Natl. Acad. Sci. U.S. 46, 333.

DAVIS, H. T., and RICE, S. A. (1964). J. Chem. Phys. 41, 14.

DE GROOT, S. R. (1951). "Thermodynamics of Irreversible Processes." North-Holland Publ., Amsterdam.

DELIMARSKII, IU. K., and GORODSKII, O. V. (1961). Zh. Fiz. Khim. 35, 687.

DELIMARSKII, IU. K., and MARKOV, B. F. (1961). "Electrochemistry of Fused Salts." Sigma Press, Washington, D.C.

DIJKHUIS, C. G. M., and KETELAAR, J. A. A. (1966). Electrochim. Acta. 11, 1607.

DIJKHUIS, C. G. M., DIJKHUIS, R., and JANZ, G. J. (1968). Chem. Rev. 68, 253.

DUKE, F. R., LAITY, R. W., and OWENS, B. (1957). J. Electrochem. Soc. 104, 299.

EYRING, H., REE, T., and HIRAI, N. (1958). Proc. Natl. Acad. Sci. U.S. 44, 683.

FLOOD, H., FØRLAND, T., and GRJOTHEIM, K. (1954). Z. Anorg. Allgem. Chem. 276, 289.

FORDYCE, J. S., and BAUM, R. L. (1966). J. Chem. Phys. 44, 1159.

FØRLAND, T. (1964). In "Fused Salts" (B. R. Sundheim, ed.), pp. 63–164. McGraw-Hill, New York.

FØRLAND, T., and DIJKHUIS, C. G. M. (1961). Discussions Faraday Soc. 32, 161.

FØRLAND, T., and OSTVOLD, T. (1966). Acta Chem. Scand. 20, 2086.

FOSTER, M. S. (1967). Regenerative E.M.F. cells. Advan. Chem. Ser. 64, 136–149.

FRENKEL, J. (1946). "Kinetic Theory of Liquids," p. 441. Oxford Univ. Press, London and New York.

GRAVES, A. D., and INMAN, D. (1968). In "EMF Measurements in High Temperature Systems" (C. B. Alcock, ed.), pp. 183–197. Am. Elsevier, New York.

GRAVES, A. D., HILLS, G. J., and INMAN, D. (1966). Advan. Electrochem. Electrochem. Eng. 4, 117–185.

GREENBERG, J., and HALLGREN, L. J. (1960). Rev. Sci. Instr. 31, 444.

GRUEN, D. M. (1964). In "Fused Salts" (B. R. Sundheim, ed.), pp. 301–339. McGraw-Hill, New York.

GRUEN, D. M., and McBETH, R. L. (1959). J. Phys. Chem. 63, 393.

GUGGENHEIM, E. A. (1952). "Mixtures." Oxford Univ. Press, London and New York.

HAMBY, D. C., and SCOTT, A. B. (1968). J. Electrochem. Soc. 115, 704.

HERSH, L. S., and KLEPPA, O. J. (1965). J. Chem. Phys. 42, 1309.

HERSH, L. S., NAVROTSKY, A., and KLEPPA, O. J. (1965). J. Chem. Phys. 42, 3752.

HILDEBRAND, J. H., and SALSTROM, E. J. (1933). J. Am. Chem. Soc. 54, 4257.

HILLS, G. J., and POWER, P. D. (1968). Trans. Faraday Soc. 64, 1629.

INMAN, D. (1968). *In* "EMF Measurements In High Temperature Systems" (C. B. Alcock, ed.), pp. 163–181. Am. Elsevier, New York.

JAMES, D. W. (1964). *In* "Molten Salt Chemistry" (M. Blander, ed.), pp. 507–534. Wiley (Interscience), New York.

JANZ, G. J. (1967). "Molten Salts Handbook." Academic Press, New York.

JANZ, G. J., and SAEGUSA, F. (1961). *J. Electrochem. Soc.* **108**, 663.

KLEMM, A. (1964). *In* "Molten Salt Chemistry" (M. Blander, ed.), pp. 535–607. Wiley (Interscience), New York.

KLEPPA, O. J. (1965). *Ann. Rev. Phys. Chem.* **16**, 187–211.

KRONENBERG, M. L. (1965). *Denki Kagaku* **33**, 203.

LAITINEN, H. A., and OSTERYOUNG, R. A. (1964). *In* "Fused Salts" (B. R. Sundheim, ed.), pp. 255–301. McGraw-Hill, New York.

LAITINEN, H. A., TISCHER, R., and ROE, D. K. (1960). *J. Electrochem. Soc.* **107**, 546.

LAITY, R. W. (1957). *J. Am. Chem. Soc.* **79**, 1849.

LAITY, R. W. (1959). *J. Chem. Phys.* **30**, 682.

LAITY, R. W. (1960). *Ann. N. Y. Acad. Soc.* **79**, 997.

LAITY, R. W. (1961). *In* "Reference Electrodes" (D. J. G. Ives and G. J. Janz, eds.), pp. 524–606. Academic Press, New York.

LEONARDI, J., and BRENET, J. (1965). *Compt. Rend.* **261**, 116.

LEVY, H. A., and DANFORD, M. D. (1964). *In* "Molten Salt Chemistry" (M. Blander, ed.), pp. 109–127. Wiley (Interscience), New York.

LEWIS, G. N., and RANDALL, M. (1923). "Thermodynamics." McGraw-Hill, New York.

LIU, C. H., JOHNSON, K. E., and LAITINEN, H. A. (1964). *In* "Molten Salt Chemistry" (M. Blander, ed.), pp. 681–735. Wiley (Interscience), New York.

LUKS, K. D., and DAVIS, H. T. (1967). *Ind. Eng. Chem. Fundamentals* **6**, 194.

MAYER, J. E., and MAYER, M. G. (1940). "Statistical Mechanics." Wiley, New York.

MAYER, S. W. (1964). *J. Chem. Phys.* **40**, 2429.

MONSE, E. U. (1957). *Z. Naturforsch.* **12a**, 526.

MURGULESCU. I. G., and STERNBERG, S. (1957). *Rev. Chim. Acad. Rep. Populaire Roumaine* **2**, 251.

NEIL, D. E., CLARK, H. M., and WISWALL, R. H. (1965). *J. Chem. Eng. Data* **10**, 21.

NOTZ, K., and KEENAN, A. G. (1966). *J. Phys. Chem.* **70**, 662.

OSTVOLD, T. (1966). *Acta Chem. Scand.* **20**, 2320.

PANISH, M. B., BLANKENSHIP, F. F., GRIMES, W. R., and NEWTON, R. F. (1958). *J. Phys. Chem.* **62**, 1325.

REISS, H., MAYER, S. W., and KATZ, J. L. (1961). *J. Chem. Phys.* **35**, 820.

REISS, H., KATZ, J. L., and KLEPPA, O. J. (1962). *J. Chem. Phys.* **36**, 144.

SALSTROM, E. J. (1934). *J. Am. Chem. Soc.* **56**, 1272.

SCHWARZ, K. (1941). *Z. Elektrochem.* **47**, 144.

SENDEROFF, S., and MELLORS, G. W. (1958). *Rev. Sci. Instr.* **29**, 151.

SENDEROFF, S., and MELLORS, G. W. (1966). *Science* **153**, 1475.

SENDEROFF, S., MELLORS, G. W., and BRETZ, R. I. (1960). *Ann. N. Y. Acad. Sci.* **79**, 878.

SHAMS EL DIN (1961). *Electrochim. Acta* **4**, 242.

SMITH, G. P. (1964). *In* "Molten Salt Chemistry" (M. Blander, ed.), pp. 427–507 Wiley (Interscience), New York.

STERN, K. (1956). *J. Phys. Chem.* **60**, 679.

STERNBERG, S., and HERDLICKA, C. (1966). *Rev. Roumaine Chim.* **11**, 29.

STILLINGER, F. H. (1961). *J. Chem. Phys.* **35**, 1581.

STILLINGER, F. H. (1964). *In* "Molten Salt Chemistry" (M. Blander, ed.), pp. 1–109. Wiley (Interscience), New York.

STILLINGER, F. H., KIRKWOOD, J. G., and WOJTOWICZ, P. J. (1960). *J. Chem. Phys.* **32**, 1837.

SUNDHEIM, B. R. (1957). *J. Phys. Chem.* **61**, 485.

SUNDHEIM, B. R. (1964). *In* "Fused Salts" (B. R. Sundheim, ed.), pp. 165–254. McGraw-Hill, New York.

TEMKIN, N. (1945). *Acta Physicochim. U.R.S.S.* **20**, 411.

UKSHE, E. A., and BUKUN, N. G. (1961). *Zh. Fiz. Khim.* **35**, 2689.

VANARTSDALEN, E. R., and YAFFE, I. S. (1955). *J. Phys. Chem.* **55**, 118.

WAGNER, C. (1966). *Advan. Electrochem. Electrochem. Eng.* **4**, 1–47.

WILMSHURST, J. K., and SENDEROFF, S. (1958). *J. Chem. Phys.* **35**, 1078.

YAFFE, I. S., and VANARTSDALEN, E. R. (1956). *J. Phys. Chem.* **60**, 1125.

YAO, Y. Y., and KUMMER, J. T. (1967). *J. Inorg. Nucl. Chem.* **29**, 2453.

Chapter 11

Bioelectrochemistry*

J. WALTER WOODBURY, STEPHEN H. WHITE,
MICHAEL C. MACKEY, WILLIAM L. HARDY,
AND DAVID B. CHANG

I.	Introduction	904
	A. What Is Bioelectrochemistry?	904
	B. Scope of the Chapter	904
	C. Aspects of Bioelectrochemistry Not Included	905
	D. Units	907
II.	Electrolytes in Living Organisms	908
	A. Tissues and Cells	908
	B. Cells	909
III.	Membranes: Steady-State Ion Transport	914
	A. The Membrane's Role in Cell Function	914
	B. Ion Transport through Neutral, Homogenous Membranes	917
	C. Transmembrane Potential	923
IV.	Membrane Structure	927
	A. Chemical Composition of Membranes	928
	B. Physical and Morphological Characteristics of Membranes	929
	C. Architectural Models of the Membrane	936
V.	Excitability: Voltage-Dependent Permeability	939
	A. The Nerve Impulse: Initiation and Propagation	940
	B. Cable Properties of Cylindrical Cells	944
	C. Voltage Clamping	948
	D. Analysis of Membrane Ionic Current	950
	E. Prediction of Excitable Behavior: The Hodgkin–Huxley Equations	953
VI.	Ion Channel Characteristics	955
	A. Special Properties of Na^+ and K^+ Channels	955
	B. Selectivity Sequences of Na^+ and K^+ Channels	958

* Sections IV, VI,B, and VII describe original contributions by the authors. These were supported in part by Research Grant NB–01752 and Training Grant 2 Tl GM 739 from the National Institutes of Health, US Public Health Service.

903

VII. Membrane Properties: Some Theoretical Approaches 961
     A. Homogeneous Membranes 961
     B. Eyring Rate Kinetics and Ion Penetration 967
     References 977

I. Introduction

A. What Is Bioelectrochemistry?

Bioelectrochemistry is the application of the theories, principles, and techniques of electrochemistry to the solution of biological problems. Many of the chapters in this book deal with aspects of electrochemistry having direct applicability to biological processes or techniques of biological experimentation. It is not possible to give in one chapter complete coverage of the possible uses of electrochemistry in biology; it is not even possible to cover all aspects of bioelectrochemistry in which the authors are competent. By choice, this chapter presents general coverage of the electrical and ionic aspects of cell membrane properties with occasional excursions in depth into areas of particular interest to us to give the "flavor" of that area and to indicate the frontiers.

Since considerable time is required for new knowledge in one field to "diffuse" into another, it is not surprising that the more "classical" parts of electrochemistry have found the widest application in biology. Some modern concepts have been partially applied, but there is little doubt that much of more modern physical chemistry and electrochemistry are applicable to biology; the problem is to be sufficiently knowledgeable in both fields. Perhaps this chapter will stimulate some graduate students in physical chemistry or electrochemistry to also become biologists and thus to facilitate the diffusion of knowledge.

B. Scope of the Chapter

This chapter describes the factors which affect and/or control the movements of ions through the membranes which bound all animal cells. No previous knowledge of biology is assumed; brief descriptions of necessary background are provided together with appropriate references.

At the boundary of an animal cell is a thin membrane (50–100 Å) of highly organized lipids and proteins called the *surface* or *plasma* membrane. The cell's contents and the surrounding medium are aqueous electrolyte solutions and fair conductors, with resistivities of 20–200 ohm cm. Surface membranes have resistivities in the semiconductor range, 10^7–10^{10} ohm cm. Since the membrane is only 50–100 Å thick, the surface resistance is low (10–10^3 ohm cm²) and capacitance is high (10^{-6} F/cm²). In the normal cell there is a potential difference of 30–100 mV (inside negative) across the membrane. The amplitude depends on the cell type: e.g., muscle, -90 mV; nerve, -70 mV; liver, -40 mV. The potential difference across the membrane arises from the differential permeability of the membrane to inorganic ions, mainly Na^+ and K^+ and the oppositely directed concentration gradients of Na^+ and K^+. The $[Na^+]$ is low inside cells and high outside; the reverse is true of $[K^+]$. Excitable cell (muscle and nerve) membranes have special ion permeability properties which can give rise to regenerative changes in the voltage across the membrane: A decrease in the size of the transmembrane voltage causes a rapid (0.1 msec) increase in Na^+ permeability and a consequent increase of Na^+ influx and reversal of the voltage across the membrane. This is followed by a slower increase in K^+ permeability (1 msec) which restores the original potential difference. This potential change is propagated as a wave along the surface membranes of the elongated cylindrical nerve and muscle cells.

This chapter is devoted primarily to describing what is known of the factors responsible for the unique electrochemical behavior of excitable cells: (1) The properties of the electrolyte solutions which make up the intracellular (inside cells) and extracellular (outside cells) fluids of animals; (2) the factors responsible for the development and maintenance of a potential difference across cell membranes; (3) the membrane changes responsible for regenerative behavior in excitable cells; and (4) what is known of the molecular bases of these intriguing phenomena. Insights provided by physical and electrochemical theories, knowledge, and techniques will be included at the appropriate places. Cole's (1968) book and Hodgkin's (1964) monograph are excellent sources of information on these phenomena by pioneers in the field. Katz (1966) has written a lucid introduction which is helpful to the beginning student. Shorter coverage has been given by Woodbury (1962, 1965).

C. ASPECTS OF BIOELECTROCHEMISTRY NOT INCLUDED

Many aspects of cellular and animal function illuminated by electrochemical principles cannot be covered here due to lack of space and expert knowledge. A few of the more important ones are:

1. *Biological Oxidation**

The energy derived from the oxidation of glucose and other substances is converted into usable form (usually the "high-energy" phosphate compound adenosine triphosphate, ATP) by the sequential oxidation and reduction of a series of intermediate enzyme complexes. Each cycle produces one ATP molecule at the expense of an electron being passed onto the next complex at lower energy. This energy-yielding electron-transport process is being studied intensely, since it lies at the heart of the whole cellular metabolic process. This process is carried out in mitochondria, special membrane bound structures which are scattered throughout the cell fluid (cytoplasm).

2. *Gross Biopotentials, ECG, EMG, EEG*†

The variations in transmembrane voltage generated in excitable tissues give rise to current flow in the extracellular fluid. This electrolyte solution is a volume conductor and the transient potential changes across cell membranes (frequently referred to as *action potentials* or simply as *electrical activity*) generate currents which spread throughout the volume. The spatial and temporal patterns of the resulting potential differences at the body surface tell something of the electrical activity going on within the body, but there is no way of deducing uniquely the nature of a generator within a volume conductor solely from the surface-potential distribution. Hence interpretation of surface potentials must be based on direct experimental measurements of the charactèristics of the generator. The time sequence of surface potentials generated by electrical activity in the heart is called an *electrocardiogram* (ECG); this temporal pattern of changes precedes and controls the contraction of the heart. Similarly, the *electromyogram* (EMG) refers to the time-varying surface potentials generated by electrical activity in skeletal muscle which precede and control the muscle's contraction. The *electroencephalogram* (EEG) is the time-varying potential sequence which can be recorded between any two points on the scalp. These changes are due to the partially synchronous activities of the billions of nerve cells contained in the brain.

3. *Effects of Current on Organisms*‡

Since activity in excitable cells causes current flow in the extracellular fluid, it follows that externally applied currents affect the behavior of

* See White *et al.* (1964).
† See Ruch and Patton (1965), Chapters 2, 5, 22, 30.
‡ See Toman (1965).

organisms. The most dramatic example is the convulsion or fit induced by a short (0.5 sec), strong (30 mA) pulse of alternating current through the brain. Less dramatic but more lethal is the disruption of the pumping action of the heart by ventricular fibrillation, a disorganization of the normally synchronous contraction of all the cells surrounding the ventricles. Currents of 10–100 mA through the trunk are likely to cause fibrillation. These effects are understood in general, but not in detail.

4. *Irreversible Thermodynamics**

Irreversible-thermodynamic techniques for dealing with flows and driving forces have been used to describe water and neutral solute and ion fluxes through cell membranes. Although these techniques have clarified these problems to some extent, we feel that irreversible thermodynamics is not the method of choice for attacking the problems of ion transport through membranes. Since much evidence indicates that specific ions penetrate membranes through specific regions or channels of molecular dimensions, theories of ion transport must be molecular. Thermodynamics is macroscopic, and thus not as well suited to dealing with specific molecular models of ion penetration as is Eyring-type rate kinetics. Further, since absolute rate theory reduces to irreversible thermodynamics near equilibrium (Eyring *et al.*, 1964), absolute rate theory is more general.

5. *Bioelectrodes†*

As a rough approximation, an animal can be regarded as a bag of electrolyte solution. Hence reversible electrodes should be used in recording potentials from or applying currents to living organisms. Techniques for making and using such electrodes are standard electrochemistry. However, the requirements on reversibility and matching of electrode pairs are usually not as demanding for biological as for electrochemical recording.

D. Units

Every field has its own tradition of unit usage. Convenient size of numbers is the usual determinant, but there are idiosyncrasies. The meter-kilogram-ohm-second system is used here with modifications, the most consistent modification being the use of centimeters instead of

* See Katchalsky and Curran (1965).
† See Feder (1968).

meters (e.g., resistivities are given in ohm-centimeters rather than ohm-meters). Cell dimensions are given in micrometers (microns, μm) and membrane dimensions in ångstroms (Å), 10^{-10} m. The idiosyncrasies arise in the units of concentration. Concentrations of ions in extracellular fluids are usually given as millimoles (mM) (rather than milliequivalents) per liter of solution (millimolar) rather than as millimolal. Since the total solute concentration in extracellular fluids is about 300 mM and nearly all of the dissolved particles are monovalent, inorganic electrolytes, the difference between molar and molal is not large, and use of equivalents is not strongly indicated. On the other hand, since intracellular fluid is only about 80% water, it is customary to give concentrations of electrolytes in intracellular fluid as millimoles per kilogram of cell water (millimolal). These units are annoying in flux calculations, where concentrations must be expressed in moles per centimeter cubed. Hence concentrations will be given in micromoles per centimeter cubed, μM/cm^3 (= millimoles per liter, mM/l).

II. Electrolytes in Living Organisms

A. TISSUES AND CELLS

1. *Tissue Structure*

A tissue consists of an aggregation of one or several types of cells organized in a fashion meaningful for the function of the tissue and held together by intercellular binding substances. Skeletal muscle is one of the simplest tissues, consisting almost entirely of muscle cells interspersed with the blood vessels which carry nutrients (O_2, glucose) to the muscle cells and take away wastes (CO_2, water). A skeletal muscle fiber (cell) is cylindrical, having a roughly circular cross section with an average diameter of about 70 μm. The length of the fiber is usually that of the muscle (up to tens of centimeters). Tension is exerted along the axis when a muscle fiber contracts, i.e., it tends to shorten and get fatter. The cylindrical cells are closely packed, with their axes parallel; 85–90% of total volume of the whole muscle is occupied by muscle cells and about half of the remaining 10–15% by blood vessels. The volume occupied by the cells is termed *intracellular space*, the total space outside cells is *extracellular space*. The fraction of extracellular space outside of blood vessels and directly in contact with cells is *interstitial space*.

2. Extracellular Space

a. Blood Plasma and Interstitial Fluid. Extracellular space is comprised of interstitial space and blood plasma, the nonred-cell fraction of blood. The distinction is meaningful because blood plasma contains about 7.5% of protein (valence ≈ -18). There is a Gibbs–Donnan equilibrium between blood plasma and interstitial fluid: Monovalent cation concentrations are about 5% higher in plasma than in interstitial fluid. The concentrations of materials in the solution directly bathing cells are the important quantities here, so electrolyte concentrations of interstitial fluid rather than of blood plasma are given. Since it is difficult to obtain samples of pure interstitial fluid and simple to obtain blood samples, concentrations of substances in interstitial fluid are calculated from their values in blood plasma using the Gibbs–Donnan ratio.

b. Composition of Interstitial Fluid. Interstitial fluid is a water solution of electrolytes. The major component is NaCl, but K^+, Ca^{++}, Mg^{++}, HCO_3^-, $H_2PO_4^-$, and HPO_4^{--} are also present in appreciable concentrations. The concentrations of organic substances, ionized and neutral, are low, but detectable. The concentrations of ions in mammalian interstitial fluid, $[S]_o$, and in intracellular fluid, $[S]_i$, of mammalian skeletal muscle are given in Table I; here S stands for an ion species.

There is nothing particularly remarkable about this electrolyte solution. The composition of the interstitial fluid varies from time to time in the same animal (about 5%) and from animal to animal in the same species. The above figures, while typical for most mammals, are different from those for lower vertebrates, e.g., frog interstitial ion concentrations are roughly 80% of the values in the table. Invertebrates living in sea water have interstitial fluids little different from sea water: 0.5 M NaCl with much lesser concentrations of K^+, Ca^{++}, Mg^{++}. In general, $[Na^+]_o$ is about 40 times $[K^+]_o$ in interstitial fluid, regardless of absolute values.

B. CELLS

1. Structure*

A reasonably accurate picture of cell structure is provided by electron micrographs taken of thin (500 Å) stained sections of a cell (Fig. 1). In such a picture membranes appear as two dense lines separated by a less-dense line (like a divided highway). This triple-layer structure is 50–100 Å thick. This structure is almost certainly the image of the barrier that

* See De Robertis *et al.* (1965).

910   J. Walter Woodbury *et al.*

gives rise to the selective permeability of the cell surface as revealed by isotope and electrical measurements. In addition to the surface (or plasma) membrane, the cell contains other membrane structures: The *endoplasmic reticulum* and *mitochondria*. The *endoplasmic* or *cytoplasmic reticulum* (ER), a tortuously folded membrane structure, is the probable site of synthesis of proteins and sterols. The ER is widely distributed

TABLE I

TYPICAL CONCENTRATIONS OF IONS IN MAMMALIAN INTERSTITIAL FLUID AND MUSCLE INTRACELLULAR FLUID[a]

Ion	Concentration of interstitial fluid (o) (μM/cm^3)	Concentration of intracellular fluid (i) (μM/cm^3 cell water)	$\frac{[S]_o}{[S]_i}$	$V_S = \frac{61}{z} \ln \frac{[S]_o}{[S]_i}$ (mV)
Na$^+$	145	12	12.1	66
K$^+$	4.1	150	1/36.6	-96
Ca^{++}	1.7	2		
Mg^{++}	0.7	17		
H$^+$ (pH)[b]	4.0×10^{-5} (7.40)	9.6×10^{-5} (7.02)	1/2.4	-23
Cl$^-$	116	3.9[c]	30	-90
HCO$_3^-$ [b]	29	12	2.4	-23
Others	9/z[d]	94[e]		
Resting potential V_r	0	-90 mV		

[a] After Ruch and Patton (1965), Chapters 1 and 45. $T = 37°C$.
[b] pH and [HCO$_3^-$] are maintained at these values by dissolved CO$_2$ at a partial pressure $P_{CO_2} \approx 40$ mm Hg. The P_{CO_2} values in tissues are somewhat higher.
[c] Calculated as [Cl$^-$]$_i$ = 116 × 10$^{-90/61}$. Direct measurements are inaccurate.
[d] z is average valence of remaining ions.
[e] A mixture of mostly organic anions (see text), average valence $Z \approx 2.1$ to make internal solution neutral.

intracellularly in muscle, and plays an important function in controlling muscle contraction. In nerve cells the ER is found only in the cell body (neuron) and not in the long, thin, electrically active cylindrical portions (axons and dendrites). Mitochondria are enclosed oval structures about 1 μm long and are the probable sites of most of a cell's chemical machinery for oxidative energy metabolism. Since neither the reticulum, mitochondria, nor nucleus (found in the cell body) plays a direct role in the electrical activity of nerve, these will not be considered further; they are not shown in Fig. 1, a schematic diagram of the features of cell structure involved in the unique electrochemical behavior of nerve and muscle fibers.

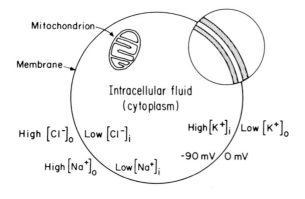

Membrane

Intracellular fluid
(cytoplasm)

High $[Cl^-]_o$ Low $[Cl^-]_i$

High $[K^+]_i$ Low $[K^+]_o$

High $[Na^+]_o$ Low $[Na^+]_i$

-90 mV 0 mV

INTERSTITIAL FLUID

Fig. 1. Schematic representation of the cell structures of immediate importance in determining the electrochemical properties of excitable cells. Relative inorganic ion concentrations and the transmembrane potential are shown to indicate directions of electrochemical gradients. Active outward transport of Na^+ and inward transport of K^+ balance the passive flows down the respective electrochemical gradients. The remainder of cellular anions are mostly organic.

2. *Intracellular Fluid (Cytoplasm)*

The fraction of intracellular volume occupied by mitochondria, reticulum, and nucleus is small. The remaining volume is filled with *cytoplasm*, a complex aqueous solution of electrolytes, enzymes, and other proteins in a matrix of fibrous protein. In the normal state the cytoplasm is a gel. Mammalian cells are only 5–100 μm in diameter, but the squid has two nerve axons 300–1000 μm (0.3–1 mm) in diameter; hence these are termed giant axons. The large size of these axons makes them well suited for experimental investigations of the properties of excitable membranes. Most knowledge of excitability has come from such investigations (Cole, 1968; Hodgkin, 1964).

Water is the principle intracellular constituent, comprising approximately 80% of the contents in mammalian muscle cells and 86% in the squid giant axon. Intracellular water content can be changed rapidly (seconds to minutes) by varying the outside concentrations of solutes. Dick (1966) has carefully reviewed the properties of cell water.

As shown in Table I, there are striking differences between interstitial and intracellular fluids of mammalian skeletal muscle: $[Na^+]_o$ is about 10 times $[Na^+]_i$, while $[K^+]_i$ is about 30 times $[K^+]_o$; $[Cl^-]_o$ is about 30 times $[Cl^-]_i$. Although the absolute values of these concentrations vary greatly throughout the animal kingdom, the ratios of external and internal concentrations are rather constant. Brinley (1965) gives data on intracellular inorganic-ion concentrations in squid and other invertebrate axons. Another feature Table I shows is that most of the internal anion is organic (see below).

3. *Physical State of Cytoplasm*

There are two principle views of the physical state of the cytoplasm (Hechter, 1964). The dominant view (Hodgkin, 1958, 1964; Katz, 1966) is that the cytoplasm is in much the same physical state as the interstitial fluid, i.e., a water solution of electrolytes and organic substances (enzymes, etc.). The bulk of the intracellular water and univalent cations are believed to be nearly as mobile as in the extracellular environment. In this view the asymmetry of electrolyte concentrations is due to the properties of the surface membrane. The membrane theory has been greatly strengthened recently by studies on squid giant axons with the cytoplasm replaced by a simple electrolyte solution. Even in these radically-changed conditions the membrane can develop normal, regenerative electrical potential changes (action potentials).

The second major view of cytoplasm (Ling, 1962) is that the interior of a cell exists as a highly ordered latticelike structure with a large fraction of the ions bound to specific sites. Cytoplasmic ion concentrations depend on the concentrations of available sites. Excitability is viewed as a change in the availability of sites. This theory is, in our opinion, unable to account for the action potential.

Squid giant axon cytoplasm has a weak birefringence, indicating that the fibrous proteins are axially oriented (Bear *et al.*, 1937). A small fraction of intracellular water is intimately associated with these fibrous proteins and membrane systems of the cell (Hechter, 1964). This introduces some uncertainty into the estimation of intracellular electrolyte concentrations and activities. Water is not completely free to diffuse inside cells. According to Dick (1959), the diffusion coefficient of water in cytoplasm is of the order of 10^{-8}–10^{-10} cm^2/sec, as compared with 10^{-6} cm^2/sec for similar, artificial protein–water systems.

a. State of Intracellular Ions. The principle cation in cells, K^+, is freely diffusable. Keynes and Lewis (1951) and Keynes (1951) used radioactive K^+ to show that at least 90% of the total intracellular K^+ in the cuttlefish (*Sepia*) giant axon exchanges as one compartment with extracellular K^+. Hodgkin and Keynes (1953) loaded a small segment of the cytoplasm of giant axons with radioactive K^+, applied an electric field along the axis, and then measured radioactivity as a function of distance and time. From these data they estimated mobility (μ_K) and diffusion constants (D_K) of intracellular K^+ (at 18°C) as $\mu_K = 4.9 \times 10^{-4}$ $cm^2/V\text{-sec}$ and $D_K = 1.5 \times 10^{-5}$ cm^2/sec. Both values are close to those of a 500 $\mu M/cm^3$ KCl solution. They estimated that K^+ accounts for 60–70% of axoplasm (cytoplasm of an axon) conductivity, leaving no doubt that most K^+ ions are free to diffuse. Hinke (1961) measured the Na^+ and K^+ concentrations of extruded squid axoplasm (by flame photometry) and sodium and potassium activities with cation-selective glass microelectrodes. He found (at 18°C):

$$[Na^+]_i = 91.5 \pm 4 \quad \mu M/cm^3 \quad \text{cell water,}$$

$$a_{Na} = 37 \pm 2 \quad \mu M/cm^3 \quad \text{cell water,}$$

$$[K^+]_i = 370 \pm 19 \quad \mu M/cm^3 \quad \text{cell water,}$$

$$a_K = 203 \pm 5 \quad \mu M/cm^3 \quad \text{cell water.}$$

Assuming most intracellular potassium is unbound, the activity coefficient for potassium is $203/370 = 0.549$. Measurements on the axoplasm after extrusions gave a potassium activity coefficient of 0.605. The activity coefficient for 400 $\mu M/cm^3$ KCl is 0.666 (Robinson and Stokes, 1959).

The $[Cl^-]_i$ of most cells vary inversely with the voltage across the membrane, as expected from electrochemical equilibrium. Keynes (1963) has shown that the activity coefficient of chloride ions in extruded squid axoplasm (0.7) is sufficiently close to that of sea water to preclude substantial binding of this ion.

Koechlin (1959) isolated and identified the major anion in squid giant axoplasm as isethionic acid and estimated its concentration as 200 $\mu M/cm^3$ cell water. The squid axon membrane is relatively impermeable to this anion (Brinley and Mullins, 1965).

In other tissues a large fraction of the anion constituent has been found to be organic phosphates and charged proteins (see Conway (1957)).

In contrast to Na^+, K^+, and Cl^-, Ca^{++} appears to be tightly bound in squid axoplasm. Hodgkin and Keynes (1957) estimate that Ca^{++} diffusivity is less than 1/10 and mobility less than 1/45 of the same values in free solution, and conclude that the intracellular concentration of ionized calcium is less than 0.01 $\mu M/cm^3$ as compared to a total Ca^{++} concentration of 0.4 $\mu M/cm^3$.

b. Conductivity of Axoplasm. Cole and Hodgkin (1939) estimated the resistivity of squid axoplasm as about 1.4 times that of sea water. Although others (cf. Hodgkin and Rushton (1946)) have obtained resistivity values ranging from one to six times sea water, 1.4 seems to be the best estimate.

Axoplasmic conductivity can be estimated from Koechlin's (1955) values for the ion concentrations in squid axoplasm and their equivalent conductances at the total ion strength of axoplasm. This estimate gives 38.2 millimhos/cm at 25°C. A similar estimate for sea water gives 50.4 millimhos/cm (resistivities of 26.2 and 19.85 ohm cm, respectively). The ratio $26.2/19.85 = 1.32$ is sufficiently close to measured values as to preclude any large structural or protein binding effects on the mobilities of the major axoplasmic ions. The only correction for protein was to allow for the volume it occupies. Table II gives values for internal and external resistivities of various tissues and their ratios. It can be seen that the ratio in most tissues is higher than it is in squid axons.,

TABLE II

RESISTIVITIES OF EXTRACELLULAR AND INTERSTITIAL FLUIDS
OF SOME ANIMAL TISSUES[a]

Tissue	ρ_i (ohm cm)	ρ_o (ohm cm)	ρ_i/ρ_o
Frog sciatic nerve fibers	110	87	1.27
Frog sartorius muscle fibers	250	87	2.9
Lobster leg nerve fibers	61	22	2.8
Sepia nerve fibers	63	22	2.9
Sheep heart (Purkinje fiber)	154	60	2.6

[a] After Shanes (1958).

III. Membranes: Steady-State Ion Transport

A. THE MEMBRANE'S ROLE IN CELL FUNCTION

1. *The Cell: An Open Steady-State System*

Thermodynamically, a metabolizing cell is a complex open system in an approximate steady state: Oxygen and fuel enter through the membrane at a steady rate, and CO_2, water, and other waste products leave at a steady rate; heat production and loss are equal. In the steady state the fluxes of materials used or produced are nonzero constants; all other fluxes are zero, but the system is not in equilibrium. Evidently, the surface membrane is strategically placed to control the flow of materials into and out of the cell to meet its needs. This control is exercised in at least two ways: (1) The membrane is an effective barrier to the diffusion of nearly all materials; the diffusion constants of ions in membranes range from 10^{-3} to 10^{-8} of their values in water. Un-ionized materials, especially lipid-soluble substances, diffuse through membranes more rapidly than ions, but with rates still several orders of magnitude slower than in water. (2) Energy-consuming mechanisms transport materials against their electrochemical gradients into or out of cells. The nature of the processes which thus *actively transport* ions and organic compounds such as amino acids and simple sugars are rapidly being worked out (Stein, 1967; Bolis *et al.*, 1967). It is well established that proteins and other macromolecules also cross the membrane, but little is known of the mechanisms (Ryser, 1968).

2. *Determinants of Ion Fluxes*

The remainder of this chapter is devoted to a consideration of the forces which act to drive ions through the membrane and the mechanisms of their transit. The forces acting on ions are concentration gradients, potential gradients, and active transport processes. Other gradients are negligibly small. The diffusional barrier offered by the membrane can also be thought of as an equivalent frictional force.

The steps in the development are: (1) The passive forces (electrochemical gradient, frictional) acting on the various ionic species are used to calculate fluxes through a homogeneous membrane; (2) the properties of the active transport processes necessary to maintain a steady state of ion distribution between cells and their surroundings are deduced and the experimentally determined properties of active Na^+-K^+ transport are briefly described; (3) the inadequacy of current theories to describe membrane ion permeation leads to a consideration of the molecular structure of the membrane in a search of a better theoretical basis; (4) some of the experimentally determined special properties of ion permeation in excitable membranes are described; these properties must be explained by any satisfactory theory; and (5) some current theories are described to show the present state of the field and to indicate where electrochemists may be able to contribute.

3. *Intracellular Recording of Transmembrane Potentials*

One of the most generally useful experimental methods for studying the properties of cell membranes is the intracellular microelectrode. This technique is illustrated in Fig. 2. Here a frog sartorius muscle (a triangular muscle of the upper leg about 4 cm long and 1 cm wide at the base) is dissected free and pinned to a wax-bottomed chamber (Fig. 2, right) filled with an artificial fluid (Ringer's solution) having the same ionic composition as frog interstitial fluid. A capillary microelectrode is held in position over muscle with a micromanipulator (not shown). Electrical connection is made to the chamber and microelectrode by Ag–AgCl electrodes. The potential difference between the tip of the microelectrode and the bathing medium is amplified and displayed as a function of time by the recorder (upper left). The microelectrode is

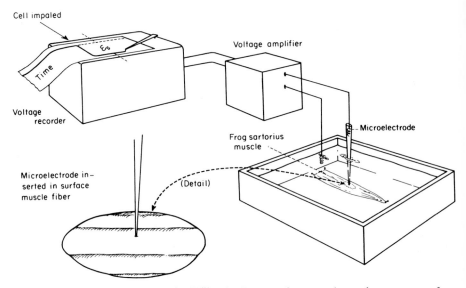

FIG. 2. Intracellular recording. Schematic diagram of an experimental arrangement for measuring transmembrane potentials of living cells. The detail at the lower left is an enlarged view of the electrode inserted through the membrane to show that the electrode tip (0.5 μm) is much smaller than diameter of muscle fiber (100 μm) [from Ruch and Patton (1965)].

advanced toward the surface with the micromanipulator. When the electrode penetrates the cell membrane (arrow on record at left), the pen is suddenly deflected to a new position representing the transmembrane potential $\mathscr{E}_s$. The key element in this technique is the microelectrode, a piece of glass capillary tubing drawn down to a fine tip. If the tip diameter is less than about 0.5 μm, it can be inserted transversely through a cell membrane without causing appreciable damage. Since a steady potential can be obtained, the membrane appears to "seal" to the glass. A gap between the electrode and the membrane allows large currents to flow from adjacent, intact regions, and the potential measured by the electrode is reduced to near zero. This sealing property of electrodes is absent or imperfect if the electrode tip is much greater than 1 μm in diameter.

The electrode is filled with 3 M KCl solution to reduce liquid junction potentials and electrode resistance. Saturated KCl ($\sim 5\ M$) cannot be used because of crystal formation near the tip. It is desirable to make electrode resistance as low as possible because the series resistance shunted by the capacity across the microelectrode wall (1 pF/mm) limits the high-frequency response of the electrode and thus distorts the transient potential changes of excitable cells (Woodbury, 1952). Resistances range from 5–100 megohms, depending both on tip diameter and taper.

The calculated difference in the liquid junction potential between microelectrode and bathing fluid and microelectrode and intracellular fluid is small (about 2 mV) compared with the transmembrane potential of a skeletal muscle cell of -90 mV. However, microelectrodes frequently have junction potentials as high as 30 mV; the electrode becomes negative with respect to the solution, as though the mobility of the anions in the tip were greatly restricted (Adrian, 1956). The known properties are: (1) The "tip" potential falls to normal if the tip is broken off; (2) lowering the pH of the KCl in the microelectrode to from two to three reduces the tip potential to near zero; (3) Th^{4+} in the external solution reduces tip potential (Holtzman, 1967).

B. Ion Transport through Neutral, Homogenous Membranes

1. Ions: Equilibrium Potentials and Membrane Permeabilities

The low ion permeability of membranes greatly simplifies experimental and theoretical considerations of ion movements across them. The rate-limiting step is at the membrane, so that concentration gradients of ions in the interstitial and intracellular fluids are nearly always negligibly small. Hence ion concentrations near the membrane closely approximate the bulk concentrations. Also, the membrane is thin compared with cell dimensions, so diffusion through the membrane is one dimensional.

Since concentrations near the membrane are close to bulk values, the electrochemical potential difference across the membrane for each ion species can be calculated from the transmembrane electric potential and the internal and external concentrations of the ion. This is done by comparing electrochemical equilibrium potential V_S for each ion species S with the measured transmembrane potential V_m as given in Table I. The potential V_S is the transmembrane potential at which the electrochemical potential difference ΔU_S across the membrane is zero:

$$\Delta U_S = RT \ln([S]_i/[S]_o) + FzV_S = 0;$$

F is the Faraday (9.65×10^4 C/equivalent) and z is valence. Hence

$$V_S = (RT/Fz) \ln([S]_0/[S]_i). \tag{3.1}$$

The equilibrium potentials for Na^+, K^+, H^+, Cl^-, and HCO_3^- ions are given in Table I. The Cl^- ion is the only ion listed in Table I whose equilibrium potential is equal to the steady or resting transmembrane potential V_r. If the cell were in thermodynamic equilibrium, the conclusion would be that the membrane is permeable to Cl^- and impermeable to all other ions. However, the cell is an open system and the results of measurements with tracers show that all inorganic ions listed in Table I can penetrate the membrane. The fluxes of Na^+, K^+, and Cl^- are many times larger than those of the other ions, and hence determine the transmembrane potential. The other ions will not be considered further. In muscle, membrane permeability to K^+ (P_K) is equal to P_{Cl} and both are about 100 times P_{Na}. Most ions are not equilibrated across the cell membrane, since they can penetrate it, and their equilibrium potentials are different than V_r. The inescapable conclusion (backed by a large amount of experimental data) is that part of the cell's metabolic energy is used to move nonequilibrated ions uphill (in the direction of their electrochemical gradients). Such a process is called *active transport*. Table I and Figure 1 show that electrochemical gradients ($V_r - V_S$) tend to force Na^+ and H^+ into and K^+ out of cells. In order to maintain a steady state, active transport processes must generate outward fluxes of Na^+ and of H^+ and an inward flux of K^+ equal to their leakage rates. Experimentally, it has been found that the active transport of Na^+ out of the cell and of K^+ into the cell are loosely coupled; in normal circumstances extrusion of Na^+ is accompanied one-for-one (or three-for-two) by active uptake of K^+. This active process is necessary and sufficient to maintain the cellular concentrations of Na^+, K^+, and Cl^- at their observed steady-state values. The biochemical nature of the active Na^+–K^+ pumping process is rapidly being discovered.

Even though the cell is not at equilibrium, thermodynamics can be used to set limits on the maximal rates of active ion transport. The power required for active transport must be less than the metabolic energy-production rate of the cell. Experiments have established that active Na^+ transport in noncontracting skeletal-muscle cells requires only about 10% of the total energy-production rate of the cell (Keynes and Maisel, 1954).

Unfortunately, lack of space precludes further consideration of active transport processes. The interested reader is referred to recent reviews, symposia, and books on this subject (Stein, 1967; Bolis *et al.*, 1967; Anon, 1968).

2. *Ion Fluxes and Permeability: The Goldman Equation*

Although the steady state is maintained by active ion transport, this process is usually electrically neutral. Thus the size of the transmembrane potential is determined solely by the passive ion fluxes; net ion flux depends on concentration, transmembrane potential, and membrane permeability. In the absence of external current the potential goes to the value necessary to make net membrane ionic current equal to zero. Calculation of transmembrane potential requires the relationship among passive (nonactive) net flux of an ion, its internal and external concentrations, and transmembrane potential. The usual approach to calculating passive fluxes is to solve the transport differential equation, a combination of Fick's and Ohm's laws for the appropriate boundary conditions. The total force acting on the ions of a given species in a unit volume is the sum of the diffusional (concentration gradient) and electrical (voltage gradient) forces. The flux is assumed proportional to the applied force, i.e., the mean velocity of an ion is directly proportional to the applied force (i.e., the frictional retarding force is proportional to velocity).

a. Diffusional Flux. The diffusional force on a mole of ions, F_M, is the negative of the chemical potential gradient:

$$F_M = -\mathbf{grad}\ U_S = -\mathbf{grad}\ RT \ln[S] = -(RT/[S])\ \mathbf{grad}[S].$$

Activities are approximated by concentrations, since concentrations appear only as ratios between inside and outside values and ionic strengths are about the same in the two media. Diffusion through a cell membrane is a one-dimensional problem; hence $\mathbf{grad}[S] = d[S]/dx$, where x is perpendicular to the membrane surface, outward positive. Ions migrate at a velocity proportional to force/ion $= F_M/N_0$, where N_0 is Avogadro's number. The proportionality constant is μ_S, the mobility. Hence average velocity $\bar{V} = \mu_S F_M/N_0$. Thus the flux M_S (moles/cm^2-sec) $= [S]\bar{V} = [S]\mu_S F_M/N_0 = -\mu_S kT\ d[S]/dx$. Fick's law of diffusion states that flux is proportional to $\mathbf{grad}[S]$, the proportionality constant being the diffusion constant D_S. Hence $M_S = -D_S\ d[S]/dx = -\mu_S kT\ d[S]/dx$ and $D_S = \mu_S kT$.

b. Electrical (Drift) Flux. A force $F_e = -ze\ dV/dx$ is exerted on each ion in a region where the electric field is $E = -dV/dx$, z is the valence, and e is the electronic charge. Hence these ions migrate with an average velocity $\bar{V}_S = \mu_S F_e = -\mu_S ze\ dV/dx$ and the flux is

$$M_S = \bar{V}_S[S] = \mu_S ze[S]\ dV/dx = -D_S(ez/kT)[S]\ dV/dx.$$

c. Transport Equation. The total steady-state passive flux of an ion species is the sum of diffusional and drift fluxes:

$$- M_S = D_S\{d[S]/dx + (Fz/RT)[S]\,dV/dx\}. \tag{3.2}$$

This equation is based on the assumption that the independence principle holds, i.e., the movements of any solute particle are independent of those of any other solute particle. This is a reasonable assumption in dilute solutions, but there are many known cases of nonindependent behavior in membrane transport processes.

Use of this equation to calculate ion movements through thin membranes requires some assumption about the variation of D_S with distance; the simplest is that the membrane is homogeneous and D_S is constant. The transport equation can be solved by itself only for the case where $M = 0$ (all permeant ions at equilibrium). The variables can be separated, x disappears, and integration for the boundary conditions $[S] = [S]_o$ and $V = 0$ on the outside and $[S] = [S]_i$ and $V = V_S$ on the inside gives Eq. (3.1), $V_S = (RT/Fz)\ln([S]_o/[S]_i)$.

When two or more ions are permeable and have different V_S, individual fluxes are nonzero and Eq. (3.1) is not directly solvable; there is one equations for each ion, but V is an extra dependent variable. The additional necessary relationship is Poisson's equation,

$$\nabla^2 V = -\frac{F}{\epsilon_0 \epsilon} \sum_i z_i[S]_i,$$

where ϵ_0 is the permittivity of free space and ϵ is the dielectric constant. Unfortunately, the complete equation set is difficult to solve. Hence it is usual to make an assumption about the variation of voltage or of concentrations with distance.

d. Constant-Field Assumption. Goldman (1943) made the assumption that the electric field in the membrane is constant [see also Hodgkin and Katz (1949)]. This implies that net charge within the membrane is zero and that the field is determined solely by charges accumulated at the interfaces between membrane and bathing media. Considering the high capacity of thin membranes ($1\ \mu F/cm^2$), this assumption seems reasonable, and has turned out to be the most successful general method for interpreting experimental data. The effects of the diffuse-double-layer charge distribution in the electrolyte just outside the membrane (Chapter 2; Davies and Rideal, 1963) are incorporated into the permeability coefficient.

The constant-field assumption permits the calculation of ion fluxes as a function of the total potential difference across the membrane, V_m. In a membrane x_m centimeters thick the electric field is V_m/x_m and the potential at any point in the membrane with respect to the outside solution $(V = 0)$ is $V(x) = V_m[1 - (x/x_m)]$; positive x is outward. The integrating factor of the right side of Eq. (3.2) is $\exp(FzV/RT)$:

$$-(M_S/D_S) \exp[FzV(x)/RT] = d\{[S] \exp[FzV(x)/RT]\}/dx. \quad (3.3)$$

The boundary conditions are: at $x = 0$ (inside), $V = V_m$ and $[S] = [S]_i$; at $x = x_m$ (outside), $V = 0$ and $[S] = [S]_o$. To integrate, set $V(x) = V_m[1 - (x/x_m)]$ and $P_S \equiv D_S/x_m$ (note that M_S does not depend on x);

$$M_S = \frac{P_S FzV_m}{RT} \frac{[S]_o - [S]_i \exp(FzV_m/RT)}{1 - \exp(FzV_m/RT)}. \quad (3.4)$$

This is the Goldman flux equation; it accurately describes the fluxes of many ion species through surface membranes in a variety of cells, e.g., chloride fluxes in striated muscle (Hodgkin and Horowicz, 1959b), sodium and potassium fluxes during activity in myelinated nerve fibers (Frankenhaeuser, 1960, 1962).

Permeabilities are low, ranging from 10^{-3} to 10^{-8} cm/sec, three to eight orders of magnitude smaller than corresponding values in water. It should be emphasized, however, that ion penetration through biological membranes frequently cannot be described by Eq. (3.4), but each exception seems to be a special case.

Equation (3.4) can be written in a more compact and easily visualizable form by setting $FzV_m/RT \equiv v$, factoring $([S]_o \cdot [S]_i)^{1/2}$ out of the numerator, defining $v_S \equiv \ln([S]_o/[S]_i)$, and converting to hyperbolic functions:

$$M_S = P_S \frac{v([S]_o[S]_i)^{1/2} \sinh[(v - v_S)/2]}{\sinh(v/2)}. \quad (3.5)$$

From this it can be seen that the current–voltage relationship is nonlinear and shows rectification; the greater the ion equilibrium potential, the greater the rectification. Physically, this means that a given driving force carries more charge into the membrane from the side of higher concentration than *vice versa*.

e. One-Way Fluxes and Their Ratio. Equation (3.4) can be used to define unidirectional fluxes. Influx, M_S^{in}, is given by Eq. (3.4) with $[S]_i = 0$, and efflux, M_S^{out}, with $[S]_o = 0$, V_m being held constant in both cases. The equivalent experiment is to add a radioactive tracer of S to the solution on one side of the membrane.

A useful general principle is obtained by taking the ratio of the unidirectional fluxes:

$$m \equiv M_S^{out}/M_S^{in} = ([S]_i/[S]_o) \exp(FzV_m/RT)$$
$$= \exp[Fz(V_m - V_S)/RT]. \tag{3.6}$$

This equation follows from the independence principle. Equation (3.6) can be obtained directly from Eq. (3.3) because the right-hand side is a perfect differential whose integrated value depends only on the boundary concentrations and voltages, and the coefficient of M_S on the left cancels when the flux ratio is taken. Therefore the constant-field assumption is not a necessary condition.

The flux ratio is a useful means of testing the independence of ion movements through a membrane. There are two ways in which the flux ratio for an ion species can deviate from the value predicted by Eq. (3.6): (1) Active transport of the ion causes a nonzero net flux, so that $m \neq 1$ at $V_m = V_S$; passive (downhill) ion movements could still be independent. (2) Ions may interact in the membrane; at $V_m = V_S$, $m = 1$, but Eq. (3.6) is not obeyed at other voltages. The flux ratio may be greater or less than expected when $V_m \neq V_S$. Nonindependence can result from the direct interaction of ions in a narrow pore, e.g., the "in-file" behavior of K^+ in sepia axons described by Hodgkin and Keynes (1955); the flux of K^+ in the downhill direction is bigger, that in the uphill direction smaller, and hence m is larger than predicted by Eq. (3.6). However, m is accurately described by Eq. (3.6) if $z \approx 3$, i.e., about three ions are "in file" in the membrane. The other category of nonindependent behavior is the "exchange diffusion" mechanism of Ussing: Ions can cross the membrane attached to a carrier (confined to the membrane) and the carrier can move only if combined with an ion. This results in a one-for-one exchange, and m is less than expected. Na^+ fluxes in frog skeletal muscle show this type of behavior; M_{Na}^{out} increases with $[Na^+]_o$ (Keynes and Swan, 1959).

3. Probable Nature of Ion Penetration

K^+ ions are generally the most permeable cations, a typical value being $P_K = 2 \times 10^{-6}$ cm/sec (Hodgkin and Horowicz, 1959b). This corresponds to $D_K = 2 \times 10^{-12}$ cm²/sec in a membrane 100 Å thick. The value of D_K in water is $\sim 10^{-5}$ cm²/sec. The low P_K must be attributed to the product of: (1) a low ion mobility and a relatively high concentration of ions in the membrane, or (2) a relatively high mobility and a low concentration, or (3) a low mobility and low concentration. The latter two conditions are compatible with a constant field in the membrane. A high mobility of ions in the membrane is indicated by the low variation of permeability with temperature. The Q_{10}'s are 1.0–2.5 (Stein, 1967; Hodgkin et al., 1952) (activation energies of 0–17 kcal/mole) for different cell types. Most values are about 1.3, near that of K^+ diffusion in free solution ($Q_{10} = 1.26$). Thus the mobility of K^+ in the membrane is probably not less than about 0.1 of the mobility in water. Mean membrane ion concentration $[S]_m$ can be estimated roughly from the relation $M_m/M_w = D_m/D_w = \mu_m[S]_m/\mu_w[S]_w$ for unity driving force; the subscripts m and w refer to membrane and water, respectively. The middle expression comes from Eq. (3.2) with $d[S]/dx = 1$ and $dV/dx = 0$, and the right hand one with $d[S]/dx = 0$, $dV/dx = 1$. The above data give $[K^+]_m/[K^+]_w = 2 \times 10^{-6}$. On the other hand, near-equal mobilities in water and membrane suggest that $[K^+]_m \approx 0.1[K^+]_w$. This discrepancy suggests that K^+ (and other) ions penetrate the membrane through rare, presumably specialized regions of high mobility ("pores" or "channels"). This view is supported by the recent experimental findings that artificial bilipid layer membranes have resistivities 10^5 times those of biological membranes (Tosteson et al., 1968); the lower resistivities of biological membranes are attributed to special, rare sites for ion penetration.

The density of ions in the membrane, estimated from the product of concentration and thickness, is 2×10^{-17} moles/cm² or one molecule per 8 µm². This may also be taken as a rough estimate of the density of pores. The amount of charge separated by the membrane (stored on membrane capacitance) due to V_m is about 10^4 times this value, so that surface charge largely determines the electric field in the membrane; the constant-field approximation is a good one on the average. However, the field in special ion-penetration regions may be variable due to high local-charge densities. Thus it is understandable that the fluxes of many ion species are adequately described by the constant-field flux equation, while those of many other species are not.

C. Transmembrane Potential

1. Steady-State Potential

Since the cell is an open system and the membrane is permeable to the major ionic constituents of the interstitial and intracellular fluids, it follows that the existence of a steady-state transmembrane potential must ultimately be due to active ion transport processes; without these, internal and external concentrations would equalize. However, the size of the potential is determined by the ratio of Na^+ to K^+ permeability, P_{Na}/P_K.

In nerve and muscle membranes P_{Na} is about 0.01 of P_K (Hodgkin and Katz, 1949; Hodgkin and Horowicz, 1959b). In some other tissues P_{Na}/P_K is higher, but always rather less than one. In the steady state the flux of each ion through the membrane is zero. For M_{Na} and M_K active fluxes must be included. Denoting these by M_{Na}^* and M_K^* and using Eq. (2.4),

$$M_{Na} = P_{Na}f(V_m)\{[Na^+]_o - [Na^+]_i e^v\} + M_{Na}^* = 0 \qquad (3.7)$$

and

$$M_K = P_K f(V_m)\{[K^+]_o - [K^+]_i e^v\} - M_K^* = 0, \qquad (3.8)$$

where $f(V_m) = v/(1 - e^v)$ and $v = FzV_m/RT$. Although there is some knowledge concerning the voltage and concentration dependencies of the active transport terms (Keynes and Swan, 1959; Keynes, 1965; Horowicz and Gerber, 1965a,b), this knowledge is not sufficient to completely specify M_{Na}^* and M_K^*. Nevertheless, the experimentally well-founded assumption that $M_{Na}^* = - M_K^*$, i.e., active efflux of Na^+ is accompanied one-for-one by active uptake of K^+, simplifies matters: If Eqs. (3.7) and (3.8) are added, the active terms cancel, as does $f(V_m)$. The steady-state voltage V_{ss} is

$$V_{ss} = \frac{RT}{F} \ln \frac{p[Na^+]_o + [K^+]_o}{p[Na^+]_i + [K^+]_i}, \qquad (3.9)$$

where $p = P_{Na}/P_K$. Since Cl^- is not actively transported, $[Cl^-]_i/[Cl^-]_o = \exp(FV_{ss}/RT)$. Thus V_{ss} is determined by P_{Na}/P_K and by $[K^+]_i$ and $[Na^+]_i$. The latter, in turn, are determined by the kinetics of the active transport process. Experimentally, it is found that $[Na^+]_i$ and $[K^+]_i$ are much the same in all animal tissues regardless of V_{ss} and p. One can obtain V_{ss} as a function of $[Na^+]_o$, $[K^+]_o$, and p if it is assumed that $M_{Na}^* = M_K^* = (const)[Na^+]_i^n$ (Woodbury, 1963).

The reason V_{ss} depends on p can be seen qualitatively from both steady- and transient-state arguments. In the steady state a coupled Na^+–K^+ pump requires that net active efflux of Na^+ = net passive influx of Na^+ = net active efflux of K^+ = net passive influx of K^+. In order for passive influx of Na^+ to equal passive efflux of K^+, there must be a much larger driving force on Na^+ to compensate for its much lower permeability: A high driving force on Na^+ times a low P_{Na} must equal a low driving force on K^+ times a high P_K. Since concentration gradients of Na^+ and K^+ are nearly equal and opposite, a large, negative V_m must build up to drive Na^+ into the cell and hinder K^+ exit. The charge separation process that builds up V_m can be grasped by supposing that

V_m is momentarily reduced to zero. In the first instant thereafter about 100 times more K^+ will exit from the cell than Na^+ enter because the driving forces are roughly the same and $P_{Na} \approx 0.01\, P_K$. The resulting exit of positive charges develops a V_m closer to V_K and further from V_{Na}. This process continues until current is zero at V_{ss}.

2. Quasisteady-State Potential

A reduction in V_m (inside solution becomes less negative) is called a depolarization, an increase is called an hyperpolarization. If any of the parameters determining the steady state are suddenly changed, there are two transients leading to a new steady state: (1) a fast transient of a few milliseconds for redistribution of charges separated by the membrane; and (2) a slow transient of minutes to hours for changes in internal ion concentrations. For example, if P_{Na} is suddenly doubled, Na^+ influx also doubles, and this depolarizes the membrane until a V_m is reached such that net transmembrane current is zero; depolarization increases K^+ efflux and Cl^- influx by moving V_m away from V_K and V_{Cl} until current is zero. A slow transient follows; there are net inward movements of Na^+ and Cl^- and a net outward movement of K^+. Hence $[Na^+]_i$ and $[Cl^-]_i$ increase and $[K^+]_i$ decreases until a new steady state is reached with each ion's net flux equal to zero. The time constant of this transient depends on cell size: The rate of change of $[S]_i$ in a cell of volume U and surface area A is given by $U\, d[S]_i/dt = AM_S$. For a cylindrical cell of 20 μm radius and a typical net efflux of 2×10^{-12} moles/cm²-sec (Hodgkin and Horowicz, 1959a) the time constant (ratio of $[S]_i$ to $d[S]_i/dt$) is 1100 min for K^+ and 25 min for Cl^-. During the short periods of time required to make measurements of V_m (less than a minute) internal ion concentrations usually can be regarded as constant at their original steady-state values.

In this slow transient or quasisteady-state situation V_m is determined by the condition that net current flow through the membrane is zero, but that individual ion currents are not. Assuming that Na^+, K^+, and Cl^- carry all of the current, then $M_{Na} + M_K - M_{Cl} = 0$. Each flux is given by Eq. (3.4). Substituting and solving for V_{qs}, the quasisteady-state voltage,

$$V_{qs} = \frac{RT}{F} \ln \frac{P_{Na}[Na^+]_o + P_K[K^+]_o + P_{Cl}[Cl^-]_i}{P_{Na}[Na^+]_i + P_K[K^+]_i + P_{Cl}[Cl^-]_o}. \quad (3.10)$$

This follows because $f(V_m)$, as defined under Eq. (3.7), is independent of the sign of z (but not the size), and thus cancels.

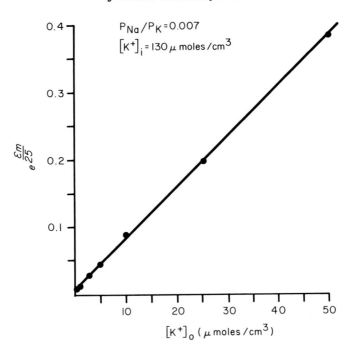

FIG. 3a. Effects of changes in $[K^+]_o$ on quasisteady-state transmembrane potentials of frog skeletal muscle fibers. The applicability of the Goldman relation is tested by plotting $\exp(\mathscr{E}_m/25)$ against $[K^+]_o$ (solid circles) with $[K^+]_o + [Na^+]_o = \text{const}$ ($\mathscr{E}_m \equiv V_m$ in this and subsequent figures). Since the data are well represented by a straight line, the slope and intercept can be interpreted using Eq. (3.10). The interpretation is simplified if the highly permeant Cl^- ions in the bathing solution are equilibrated at all voltages. This is approximately the case here. The slope and intercept give the values of P_{Na}/P_K and $[K^+]_i$ shown. Measured $[K^+]_i$ was 139 $\mu M/cm^3$ [after Adrian (1956)].

Experimental Determination of P_{Na}/P_K. Equation (3.10) can be used to interpret experimental data on the relationship between V_{qs} and $[K^+]_o$. In order to change $[K^+]_o$, a tissue must be placed in a bathing solution of the desired composition. Since cell membranes are highly permeable to water, the osmotic pressures of internal and external solutions must be kept constant to prevent cell shrinkage or swelling and consequent changes in internal concentrations. Too much swelling damages cells irreversibly. Thus if $[K^+]_o$ is increased, there should be a corresponding reduction of $[Na^+]_o$; $[K^+]_o + [Na^+]_o = \text{const}$. The interpretation of V_{qs}, $[K^+]_o$ data is simplified if Cl^- is replaced by an impermeant anion such as sulfate ($P_{SO_4} = 0$). In this case Eq. (3.10) shows that the plot of $\exp(FV/RT)$ against $[K^+]_o$ should be a straight line. Figure 3a

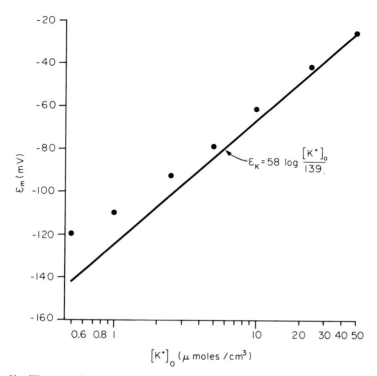

FIG. 3b. The same data as Fig. 3a with $\mathscr{E}_m$ plotted as a function of $[K^+]_o$ on a logarithmic scale. The straight line is $\mathscr{E}_K$ for $[K^+]_i = 139 \ \mu M/cm^3$. At high $[K^+]_o$ values experimental points approach the $\mathscr{E}_K$ line because the contributions of Na^+ and Cl^- become negligible in Eq. (3.10) and it reduces to the Nernst equation, Eq. (3.1) [from Ruch and Patton (1965)].

shows such a plot for skeletal muscle. The customary plot, V_{qs} vs. log $[K^+]_o$, is shown in Fig. 3b. At high $[K^+]_o$ values all but the $[K^+]$ terms are negligible, and V_{qs} approaches V_K as shown. In other tissues P_{Na}/P_K values are larger, e.g., $p \approx 0.1$ for thyroid cells (Williams, 1966).

IV. Membrane Structure

The foregoing section has dealt with the macroscopic or bulk theories of ion transport through membranes. These theories are adequate to explain the movements of some ions through some membranes, but there are so many exceptions [cf. Cole (1968), Hodgkin and Huxley (1952d)] that general theories must be deemed inadequate. Cells appear to have evolved special mechanisms of ion penetration to meet particular needs; hence detailed molecular theories are necessary, particularly to explain the bizarre, nonindependent, highly voltage-dependent Na^+ and K^+ ion currents of excitable cells. Therefore it is necessary to carefully consider the membrane's structure as revealed by electrical, chemical, and electron-microscopic analyses in search of clues to the molecular bases of the special mechanisms involved in ion transport. Unfortunately, the exact structure of membranes is not known despite much research, but considerable is known about the membrane's general architecture.

Insight into membrane architecture may be gained from study of the chemical, constituents and morphological and physical characteristics of membranes. The chemical composition is obtained from quantitative chemical, electrophoretic, and optical spectroscopic analyses of membrane fragments isolated from cells by various fractionation procedures; e.g., cells are lysed by immersing them in low-osmotic-pressure solutions; centrifugation separates the cell fragments into mitochondria, nuclear, microsomal (largely surface or plasma membranes), and supernatant (cell plasm) fractions. Information about the morphological and physical characteristics is obtained by electron-microscope, x-ray diffraction, surface-tension, and electrical measurements.

A. Chemical Composition of Membranes

The primary constituents of biological membranes are lipids and proteins. By weight, 25–80% of a membrane is lipid. For example, mitochondrial membranes are 25% lipid, erythrocytes 40–50%, and myelin 80% (O'Brien, 1967). However, a molecular model of membranes requires knowledge of the relative proportions of various molecular species: The molar ratios of cholesterol to phospholipid for several different membranes (Fig. 4) show that all membranes do not have the same lipid composition. These and other data also indicate that the primary lipids are phospholipids and cholesterol. The phospholipids are amphipathic molecules (dual affinity for water and hydrocarbon) and play a central role in theories of membrane architecture. These surface-active lipids possess polar or charged groups localized at one end of a long (~ 20Å) hydrocarbon chain (Hanahan, 1960; Bangham, 1963).

The protein portion of a membrane has at least two functions. Studies of mitochondrial (Green and Tzagoloff, 1966) and plasma membrane proteins [cf. Emmelot and Bos (1966)] demonstrate enzyme activities. Other proteins seem to play only a structural role, i.e., help maintain membrane integrity (Criddle *et al.*, 1961a,b,c; Richardson *et al.*, 1963;

Wolfgram, 1967). Rather surprisingly, membrane proteins are globular rather than sheetlike.

The relationship between protein and lipid is critical in membrane architecture. Both electrostatic and hydrophobic (van der Waal type forces between hydrocarbon moities of molecules) bonds between lipids

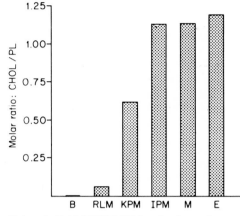

FIG. 4. Cholesterol/phospholipid (CHOL/PL) ratios for various membranes. Phospholipid and cholesterol are the major constituents of membranes, but the relative amounts vary. B, bacteria; RLM, rat liver mitochrondria; KPM, kidney plasma (surface) membrane; IPM, intestinal epithelium plasma membrane; M, myelin; E, erythrocyte ghosts (red-blood-cell plasma membrane). Note that mitochondria tend to have low ratios; myelin and erythrocyte ghosts have ratios greater than one; and other surface membranes have intermediate values [after O'Brien (1967), Korn (1966), Finean et al. (1966)].

and proteins are found. Much evidence suggests, however, that hydrophobic interactions are the more common (Criddle et al., 1961b; Lenard and Singer, 1968). Lipid–protein complexes are frequently found in membranes (Gent et al., 1964; Brown, 1963; Green and Tzagoloff, 1966) and it may be more appropriate to consider such complexes in architectural schemes rather than as separate proteins and lipids.

B. Physical and Morphological Characteristics of Membranes

1. Electrical Measurements

a. Static Membrane Capacitance.   Electrical measurements on cells gave the first evidence for the existence of a high-resistance surface membrane and some information about its structure. Fricke (1925) measured red-cell membrane capacitance with an ac bridge. He found that there was a surface structure (membrane) which behaved as a constant dielectric capacitor of 0.8 μF/cm^2, indicating that the membrane was of molecular dimensions. Assuming a lipid membrane with dielectric constant of 3, the thickness calculated from the parallel-plate-capacitor formula is 33 Å. This membrane capacity is something of a biological constant; values are always near 1 μF/cm^2 at 1 kHz. Table III gives the membrane capacities

TABLE III

MEMBRANE CAPACITIES AND LOSS ANGLES OF
VARIOUS CELLS[a]

Cell	Capacity (μF/cm^2)	Loss angle (deg)
RBC	0.8	90
Frog egg	2.0	86
Squid giant axon	1.07	76
Arbacia eggs	1.0	90

[a] Based on Cole and Curtis (1950).

and loss angles of a number of cell types; an extensive list can be found in the review by Cole and Curtis (1950).

b. Complex-Plane Impedance Locus. The complex impedance plot exploited by K. S. Cole is a relatively simple way to determine the component values of the equivalent electrical circuit of tissue impedance. Tissue reactance X_T is plotted against the tissue resistance R_T with angular frequency ω as the parameter. Tissue impedance $Z_T(\omega)$ is the sum of real, resistive—$R_T(\omega)$—and imaginary, reactive—$jX_T(\omega)$—components: $Z_T(\omega) = R_T(\omega) + jX_T(\omega)$, where $X_T = \omega L - (1/\omega C)$; L is inductance and C capacitance. This is the impedance of R, L, and C in series. To relate tissue impedance Z_T to various tissue elements (e.g., membrane capacity and resistance, electrolyte resistance, and intracellular resistance), an equivalent circuit for the tissue is assumed and the total impedance Z_E of the equivalent circuit calculated. By equating Z_E to Z_T, $R_T(\omega)$ and $X_T(\omega)$ may be expressed in terms of the elements of the equivalent circuit and the values of the elements can be calculated from experimental data, as illustrated in Fig. 5.

The tissue impedance Z_T is measured using a Wheatstone bridge (Fig. 5a). At a given frequency ω when the bridge is balanced (null at detector) and $R_1 = R_2$ values of R_s and C_s give Z_T directly, i.e.,

$$Z_T = R_s(\omega) - [j/\omega C_s(\omega)].$$

The equivalent circuit for a cell or tissue in bathing fluid is obtained by putting in a circuit element for each possible current pathway through the cell–electrolyte system (Fig. 5b). For a single cell the geometry and circuit are simple; for a cell suspension or tissue, geometry is complicated; in Fig. 5b r_m is membrane resistance, c_m membrane capacitance, r_{es} series extracellular electrolyte resistance, r_{ep} parallel extracellular electrolyte

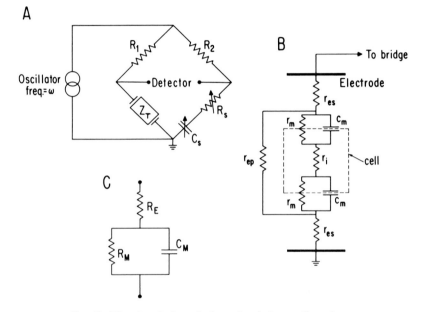

FIG. 5. The electrical equivalent circuit for a cell or tissue.

resistance, and r_i intracellular resistance. Interpretation of r_{es} and r_{ep} in terms of tissue geometry is possible only in special cases. The final equivalent circuit, Fig. 5c, "lumps" the elements of Fig. 5b into as few elements as possible. The impedance Z_E of the equivalent circuit is written in terms of the lumped elements. To determine the "tissue" circuit elements in terms of the known values R_s and C_s, Z_E is set equal to Z_T. Frequently some of the tissue elements cannot be uniquely determined, e.g., R_M and R_E represent a combination of r_i, r_m, r_{es}, and r_{ep}. If only resistive and capacitative elements are present in the tissue (usually the case), the plot of $X_T(\omega)$ (imaginary axis) against $R_T(\omega)$ (real axis) is a circular arc. If the circuit elements are ideal, a semicircle with center on the real axis is obtained, as may be shown by eliminating ω between the equations for $R_T(\omega)$ and $X_T(\omega)$.

Figure 6 shows a simple geometrical proof that the locus is a circle. The admittance of parallel RC is $Y_p = G_p + j\omega C_p$, where $G_p = 1/R_p$ (Fig. 6a). The locus of Y_p is shown in Fig. 6c by the vertical line in the upper quadrant starting at $Y_p = G_p$; the arrow indicates the direction of increasing angular frequency ω. At any ω the impedance of the circuit in Fig. 6a is the sum of resistive, $R_s(\omega)$, and reactive, $jX_s(\omega) = 1/j\omega C_s$,

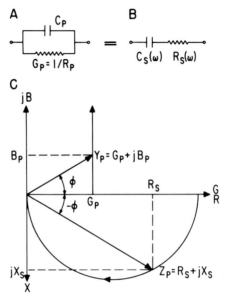

FIG. 6. Semicircular impedance-plane locus of an RC circuit. Geometrical relationships (c) are shown between (a) the admittance of a parallel RC circuit and (b) the impedance of an equivalent series RC circuit.

components (Fig. 6b). The values of R_s and X_s must vary with ω to match the impedance of the circuit in Fig. 6a. The impedance Z_p of the parallel circuit is $Z_p = 1/Y_p = R_s + jX_s$. Since $Y_p = |Y_p|e^{j\varphi}$,

$$Z_p = 1/(|Y_p|e^{j\varphi}) = |Y_p|^{-1}e^{-j\varphi},$$

where $\tan \varphi = B_p/G_p$. Hence the direction of Z_p is $-\varphi$. The real and imaginary components of Z_p are obtained from similar triangles: $R_s/G_p = (R_s^2 + X_s^2)^{1/2}/(G_p^2 + B_p^2)^{1/2}$ and from

$$1/|Y_p| = |Z_p| = 1/(G_p^2 + B_p^2)^{1/2} = (R_s^2 + X_s^2)^{1/2}.$$

The ratio then becomes $R_s/G_p = R_s^2 + X_s^2$. Since $G_p = 1/R_p$, $R_s^2 - R_s R_p + X_s^2 = 0$, or $(R_s - \frac{1}{2}R_p)^2 + X_s^2 = (R_p/2)^2$. This is the equation of a circle with center at $R_s = R_p/2$, $X_s = 0$ and radius of

$R_p/2$, as shown in Fig. 6c. As ω increases, the semicircle is traversed in the direction shown by the arrow. The addition of a resistance R_1 in series with the shunt R_p–C_p combination has the effect of shifting the whole circle to the right in the series impedance plane by a distance R_1 since Z_s is now $R_1 + R_s(\omega) + j\omega C_s(\omega)$. This permits interpretation of the locus of Fig. 7b, which shows how R_e, R_m, and C_m as defined in Fig. 5 can be quickly determined from the center and radius of the circle and the ω for maximum X_T.

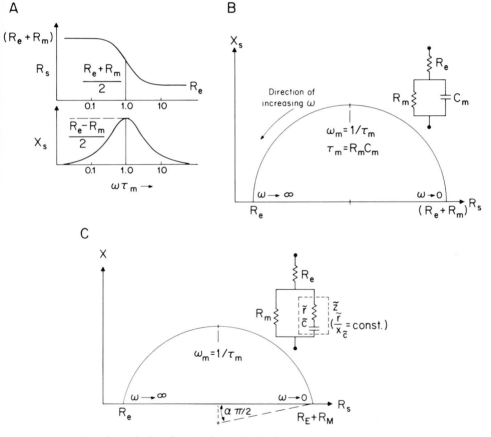

FIG. 7. Impedance loci and equivalent circuits for tissues. (a) Plots of R_s and X_s as functions of angular frequency ω for the circuit in (b). The abscissa is in multiples of $\omega\tau_m$, where $\tau_m = R_m C_m$. (b) Circular impedance plot of X_s vs. R_s obtained if circuit elements are ideal. Here R_e and R_m are determined from high- and low-frequency intercepts of the locus with the real axis, and C_m is determined from the time constant τ_m. (c) Experimentally, the center of the semicircular locus is depressed below the real axis as shown (described by the angle $\alpha\pi/2$).

In nearly all cases the center of the circular arc lies *below* the real axis because the circuit elements are not ideal. At least one of the elements, usually the capacity, must be represented by an impedance whose phase angle ($\phi \neq \pi/2$) is independent of frequency. Thus instead of a simple capacitor C_m with impedance $Z = 1/X_c$, a nonideal, "lossy" capacitor whose impedance is $\mathscr{E}\tilde{r} + jX_{\tilde{c}}$ such that $\tan \phi = \tilde{r}/X_{\tilde{c}} = \text{const} = (1 - \alpha)\pi/2 \leq \pi/2$ must be envisioned (Fig. 7c); $\tilde{r}$ and $\tilde{c}$ must change with frequency in a way which keeps ϕ constant: The amount of depression indicates the degree of nonideal behavior and is measured by the "loss" angle $(1 - \alpha)\pi/2$ (Fig. 7c, Table III).

This peculiar "loss" is an inherent property of the cell membrane. A similar phenomenon is seen in dielectrics (Cole and Cole, 1941; R. H. Cole, 1965) and electrode polarization (Schwan, 1966). The relation between the constant-phase-angle behavior in the three systems is not clear; Cole (1949) implies that the constant phase angle of membranes is due to the membrane's dielectric.

The circular arc with depressed center locus obtained when imaginary dielectric constant is plotted against real dielectric constant can be explained empirically by assuming the dipoles of the dielectric have a special, very broad distribution of relaxation times rather than a single one. There is no known molecular basis for such an assumption.

2. *Lipoid Solubility and Membrane Permeability*

The earliest evidence indicating the lipoidal nature of the membranes came from measurements of membrane permeability to neutral substances. In general, membranes are ten to 1000 times more permeable to water and to relatively small lipid-soluble organic compounds (below a certain size) than they are to even smaller ions. For many cells there is a direct relation between the permeability of the cell membrane to a series of organic substances and the oil/water partition coefficients of these substances. The evidence has been carefully reviewed by Davson and Danielli (1952). Thus many uncharged organic molecules, although much larger than hydrated ions, permeate the membrane quite easily; these behave as though they "dissolved" in the membrane. Water [see review by Dick (1966)] behaves anomalously in that it is distinctly lipid-insoluble but in many cases permeates just as well as organic compounds.

3. *Electron Microscopy of Cell Membranes*

Electron microscopy has contributed significantly to knowledge of membrane structure, but the effects of various fixation, embedding, and staining procedures on the tissue to be examined are not understood well enough to permit unambiguous conclusions about membrane structure. Electron-microscopy histological techniques have been summarized by Pease (1964).

Much data have now been accumulated on the appearance of membranes when prepared by fixation with $KMnO_4$ and OsO_4. With $KMnO_4$ fixation (preservation) a trilaminar structure (membrane) consisting of a pair of dark lines separated by a light inner space is always seen (Fig. 1). Typically, each band is about 25 Å thick, giving a total thickness of approximately 75 Å. This observation has profoundly influenced models of membrane structure.

Using both freeze-drying and $KMnO_4$ fixation, Sjöstrand (1963) showed that cytoplasmic membranes and mitochondria of some tissues have a granular appearance both in cross-section and frontal views. Sjöstrand interprets this as evidence for an orderly array of lipid globules or micelles embedded in a stabilizing protein matrix (dark staining). A number of other investigators have since reported similar findings using a variety of techniques. Nilsson (1965) found a similar globular appearance in some membranes, but found the common trilaminar configuration in many others.

Branton (1966) used the freeze-etching technique to show that cell membranes appear laminated as in OsO_4 or $KMnO_4$ fixations; in many electron micrographs the bilayer (85 Å thick) appears to have been split into halves during the fracturing procedure, so that an internal view of the membrane is obtained. Seen on the exposed internal surfaces of the membranes are numerous particles and depressions which have a diameter of about 85 Å; it is assumed that in these regions the membrane has a globular form.

4. Cell Membrane Thickness

Electron microscopy (EM), x-ray diffraction, and electrical measurements yield reasonably consistent values for the thickness of membranes. The direct visualization of the membrane with the electron microscope is the most convincing evidence that membrane thickness is 100 Å or less. Typically, EM measurements give 75 Å for thickness with a range of 50–100 Å. Mitochondrial membranes are thinner (50–70 Å) than plasma membranes (70–100 Å). However, it is valid to compare membrane thicknesses only when the membranes are in the same section. Yamamoto (1963) noted that in sections of bullfrog sympathetic ganglia the membranes fell into two groups: Mitochondrial, nuclear, and Golgi lamella membranes are 85–88 Å thick, while synaptic and Golgi vesicular, and plasma membranes are about 100 Å thick. These slight differences probably indicate differences in the molecular species making up the membrane; the same architectural principles can be followed without using identical construction materials.

There is an uncertainty in the values of membrane thickness calculated from electrical measurements using the parallel-plate-capacitor formula ($\delta = 8.85\ \epsilon A/C$; δ, thickness in Å; C/A in $\mu F/cm^2$), since the dielectric constant of the membrane is not accurately known. Assuming $C_m = 1$ $\mu F/cm^2$ and $\epsilon = 3$ (typical for nonpolar lipids), then $\delta = 27$ Å, a value much less than that obtained from EM measurements. On the other hand, a thickness of 75–100 Å and $C_m = 1\ \mu F/cm^2$ gives ϵ's between 8.5 and 11.5, unreasonably high values for pure lipids. Schwan (1957) indicated that ϵ's for proteins might be as high as 60; if so, an ϵ of 10 could be accounted for by assuming that the membrane is a 90%, 10% mixture of lipid and protein. Cole (1949) estimated, on the basis of crude dielectric arguments (independent of the parallel-plate-capacitor formula), that the dielectric constant could be as low as 10. Electrical estimates of membrane thickness thus range from 25 to 100 Å assuming dielectric constants of 3–12.

Small-angle x-ray diffraction of myelin indicates a repeat period of between 65 and 185 Å, depending upon the state of the myelin. (Myelin consists of up to several hundred double layers of membrane formed by wrapping an axon in a Schwann cell with all its cytoplasm squeezed out, i.e., like a jelly-roll.) These thicknesses are consistent with electron-microscope and electrical-thickness measurements. Finean *et al.* (1966) report fundamental repeat distances of 110 Å for packed red-blood-cell ghosts (membranes), 70 Å for intestinal epithelium, and 60 Å for liver cell membranes.

C. ARCHITECTURAL MODELS OF THE MEMBRANE

The data on the chemical and physical characteristics of membranes furnish clues to the general plan of membrane architecture. All membranes have similar kinds of lipids and similar thicknesses (50–100 Å), and hence presumably the same basic structure. Apparently, all membranes contain globular catalytic proteins, although the type of protein depends upon the metabolic activities of the membrane. It is possible, however, that membranes appear to be alike simply because we lack the techniques to discern the differences. For instance, 25–50 Å seems to be a typical size for biological molecules; almost any arrangement of these molecules which gives a structure of only a few molecules thickness might not be distinguishable by present techniques from any other arrangement of similar thickness. Nevertheless, it is worthwhile considering the most probable arrangements of molecules which might make up the membrane's "skeleton." There are three candidates for the backbone molecule: lipid, protein, and lipid–protein.

Figure 8 shows five different architectural schemes for biological membranes which use one of the three types of membrane "skeleton." Figure 8a shows the scheme of Davson and Danielli [cf. Davson and Danielli (1952)], which was the first detailed proposal for membrane structure and has had a profound influence on all subsequent models. The oriented lipid molecules were suggested on the basis of permeability and monolayer studies. The thickness of the membrane was not specified, although Danielli suggested that it must be at least two lipid molecules thick, but probably not greater than five. Monolayer studies indicated that 10–20 dynes/cm would be reasonable for the surface tension at the lipid–water interface. Since marine eggs were known to have a surface tension of only 0.1–1.0 dynes/cm and because the surface tension of protein-coated oil inclusions of some fish eggs also have a low surface tension, they suggested that the lipid membrane must be coated with denatured protein. However, recent studies show that thin lipid membranes which are believed to have the bilayer configuration have inherent surface tensions of 0.1–1.0 dyne/cm, and the necessity of postulating protein for lowering surface tension is negated.

X-ray diffraction polarized light and electron-microscope studies on myelin led Robertson (1960) to suggest the unit membrane hypothesis illustrated in Fig. 8b. The general principles of the Davson–Danielli model are retained. The experimental data indicate that the membrane is only two lipid molecules thick. The protein coats on the membrane are asymmetrical, i.e., the structure and/or composition of the protein on the inside and outside are different. Although the model is based primarily on studies of myelin, Robertson feels that all membranes have this construction because all membranes have the "railroad-track" appearance when fixed with $KMnO_4$. There is some question (Korn, 1966) about the general validity of this model, but recent studies suggest that it is probably correct for myelin (Napolitano et al., 1967; Stoeckenius, 1962).

There is other evidence indicating that the natural unit of membrane construction is a spherical lipid micelle approximately 40 Å in diameter. This led Lucy (1964) to propose the model shown in Fig. 8c. Typically, the micelles are held together by hydrogen bonds between cholesterol molecules in the micelles, and the micelles are embedded in a protein matrix. These are presumed to be in equilibrium with bilayer regions of the membrane. The natural interstices found between the micelles form

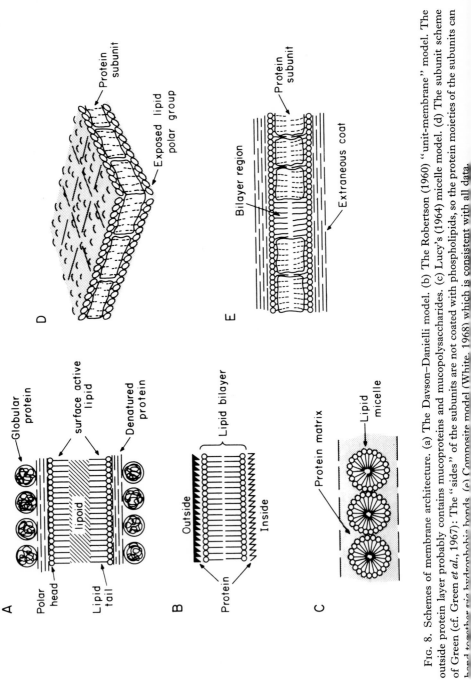

Fig. 8. Schemes of membrane architecture. (a) The Davson–Danielli model. (b) The Robertson (1960) "unit-membrane" model. The outside protein layer probably contains mucoproteins and mucopolysaccharides. (c) Lucy's (1964) micelle model. (d) The subunit scheme of Green (cf. Green *et al.*, 1967): The "sides" of the subunits are not coated with phospholipids, so the protein moieties of the subunits can bond together via hydrophobic bonds. (e) Composite model (White, 1968) which is consistent with all data.

"pores" about 4 Å in radius, which is consistent with much data for radii of red-blood-cell membrane pores. However, to obtain a correct value for membrane electrical resistance, it must be assumed that the "pore-forming" micelles occupy only a few (0.25–25) per cent of the surface area.

An entirely different model is that of Green and his associates (Kopaczyk *et al.*, 1966; Green and Perdue, 1966; Green and Tzagoloff, 1966) shown in Fig. 8d. Their model is based on numerous studies of mitochondria. The "skeleton" is composed of protein subunits which, when coated with phospholipid, may bond together to form a two-dimensional sheet. The lipid tails of the phospholipid are bonded only to the top and bottom surfaces of the protein subunits by nonelectrostatic (hydrophobic) forces. The subunits also bond together *via* hydrophobic forces. Since the tops and bottoms of the subunits are coated with polar groups, the subunits are restricted, when in aqueous solutions, to joining only at the sides. In the absence of phospholipid, three-dimensional (nonmembrane) structures are formed. This scheme has been demonstrated in a number of membrane preparations (Green *et al.*, 1967). However, Fleischer *et al.* (1967) have shown that although the inner mitochondrial membrane structure is unaffected by almost-complete lipid extraction (indicating a protein core), many membranes showing subunit construction are completely destroyed by lipid extraction. Thus not all membranes have exactly the same architecture.

Obviously, an adequate architectural model must be consistent with all of the chemical, physical, and morphological data; none of the above models are. A composite model which is consistent with all data is shown in Fig. 8e. In this case the "skeleton" consists of both protein subunits and lipid bilayer, the relative amounts depending upon the particular membrane and its function. Thus Robertson's unit-membrane hypothesis and Green's subunit hypothesis represent the extreme limits of this composite model. Metabolically inert myelin is all bilayer, while metabolically active mitochondrial membranes are composed only of subunits. Plasma membranes occupy the middle ground. This model is supported by the EM studies of Branton (1966), the permeability studies of Gainer (1967), and the high dielectric constant of membranes.

V. Excitability: Voltage-Dependent Permeability

In the nervous system information is transmitted in a pulse frequency modulation code; e.g., the strength of a sensory stimulus (light, sound, touch) is signaled to the brain by the combined number of impulses per second on all of the nerve fibers made active by the stimulus. As the strength of the stimulus is increased, the frequencies of firing in the active nerve fibers increases, and nearby, previously-inactive fibers are brought into activity. A nerve impulse is a stereotyped, phasic sequence of changes in transmembrane potential and associated ion fluxes, heat production, and other quantities, and propagates at constant speed along the fiber. Since transmembrane potential changes are the most easily measured, the nature of the impulse has been elucidated primarily from electrical measurements; ion tracer and other measurements have also contributed.

A. THE NERVE IMPULSE: INITIATION AND PROPAGATION

The sequence of potential changes accompanying a nerve impulse is called an *action potential* (AP). Figure 9a is a diagrammatic representation of how an AP can be initiated and recorded. A nerve trunk (a bundle of hundreds of individual nerve fibers) is dissected from a frog or cat. Such a trunk is 10–20 cm long. Two pairs of chlorided silver stimulating electrodes, S_1 and S_2, are applied to the nerve trunk at different distances from the recording microelectrode (far right of upper part of Fig. 9a). When a microelectrode is inserted in a fiber in the trunk (penetrate axon, lower part of Fig. 9a) the recorded potential changes abruptly from 0 to about -70 to -90 mV. Following stimulation at S_1 (indicated in the recording by stimulus "escape" or "artifact") an action potential (AP) is recorded after a short but definite delay. The AP rises rapidly to a peak (depolarization) and then recovers somewhat more slowly to the steady value (repolarization). At the peak of AP the transmembrane potential has reversed in sign.

Figure 9b shows the action potential of a mammalian myelinated nerve fiber. (A myelin sheath is formed about a nerve fiber by Schwann cells. A Schwann cell surrounds a 2-mm length of nerve fiber and wraps itself around the fiber about 100 times, forming an insulating sheath. In the wrapping process all the Schwann cell plasma is squeezed out, leaving the Schwann cell membranes closely opposed, which form a layer 170 Å thick. There is a gap of about 1 μm between adjacent Schwann cell wrappings where the axon membrane is in free communication with the interstitial fluid. This arrangement greatly increases conduction speed; cf. Hodgkin (1964).)

Several unique properties of the action potential are shown in Fig. 9: (1) The duration of the AP is about 0.7 msec (values range from 0.5 to a few milliseconds and are highly temperature-dependent); (2) the total

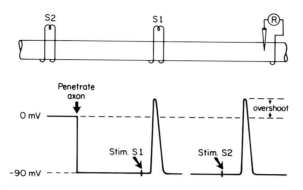

FIG. 9a. (Upper) A method for studying electrical activity of nerve fibers. (Lower) Sketch of sequence of potential changes recorded by microelectrode when it is inserted into a single fiber and when nerve trunk is stimulated by sufficiently strong, short (100 μsec) shocks applied first at S_1 and then at S_2. Abscissa: time (a few milliseconds); ordinate: transmembrane potential (about 100 mV) [from Ruch and Patton (1965)].

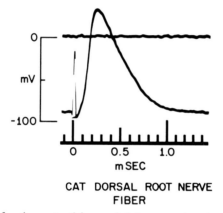

CAT DORSAL ROOT NERVE FIBER

FIG. 9b. Tracing of action potential recorded from cat dorsal root fiber. Conduction distance about 1 cm. Photograph is a double exposure, consisting of one sweep when microelectrode was in fiber and one sweep immediately after electrode was withdrawn from fiber and stimulus turned off [from Ruch and Patton (1965)].

potential change is about 120 mV, 30 mV greater than the resting potential; (3) when the stimulus is applied at S_2 rather than S_1, the sequence of events is exactly the same, except that the *latency*, the time between the delivery of the stimulus and the beginning of the AP, is longer. Further investigation shows that the latency is directly proportional to the conduction distance. Hence the action potential is a brief (0.5–1.0 msec) phasic change in membrane potential of fixed shape and amplitude traveling along the nerve fiber at constant speed, i.e., propagated wave of fixed shape, duration, and speed. The conduction speed of large (20 μm diameter) mammalian myelinated nerve fibers is 120 msec and that of the squid giant axon (500 μm diameter) is 20 msec. The faster speed of the mammalian nerve fiber is due to the myelin sheath and to higher temperature.

Another property of the AP not shown in Fig. 9 is *threshold*. If a stimulus is *suprathreshold*, a full-size action potential is initiated at the stimulating site and propagates along the fiber in both directions. If the stimulus is *subthreshold*, an AP is not initiated and the voltage changes occurring at the stimulus site die out in a few millimeters. This is called *threshold-all-or-nothing* behavior; a sufficient stimulus elicits a full response; an insufficient one, no regenerative response. These are two aspects of the same underlying process. A simple example of such a process is a brick standing on end; following a small displacement of the top of the brick in a horizontal direction, the brick will return to its original position. A suprathreshold stimulus tips the brick over.

1. *Nature of the Action Potential*

The action potential is an explosive or regenerative process. When the membrane is depolarized to threshold, an inward (depolarizing) current increases regeneratively and the depolarization proceeds to a limiting value. An analogous situation is the burning of a line of gunpowder. If a match brought near the line at one point raises the heat liberated by decomposing gunpowder above the heat losses to the surroundings, the process becomes self-sustaining and the flame travels rapidly in both directions.

a. Upstroke of the AP (Depolarization). In nerve a depolarization increases the permeability of the membrane to Na^+ ions which results in an increased inflow of Na^+ ions down their electrochemical gradient. In turn the increased entry of Na^+ neutralizes some of the negative charges on the inside of the membrane and thus further depolarizes it. If the Na^+ entry is large enough, the process is regenerative and continues until maximum Na^+ permeability is reached and the transmembrane potential V_m approaches the Na^+ equilibrium potential. This process, termed the Hodgkin cycle, can be diagrammed as shown in Fig. 10 (Hodgkin, 1951). This process accounts for the reversal of membrane polarity during the action potential; V_m approaches V_{Na} as P_{Na} becomes much larger than P_K and P_{Cl}.

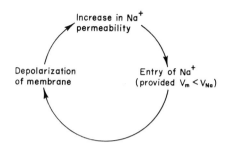

FIG. 10. The Hodgkin cycle (Hodgkin, 1951).

b. Repolarization. If the Hodgkin-cycle mechanism was a maintained process, V_m would remain near V_{Na} until the entering Na^+ came into a new steady state with the active Na^+ extrusion process. However, the increased P_{Na} induced by depolarization is transient; even if depolarization is maintained by external current flow, the inward Na^+ current decays to near zero in a few milliseconds. If this were the only mechanism operating to produce repolarization, it would occur at a rate given by the time constant (resistance times capacity) of the resting membrane; this would take several milliseconds in nerve and as much as 200 msec in skeletal muscle. Repolarization is speeded by a depolarization-induced but delayed increase in P_K. Thus the depolarization caused by the Hodgkin-cycle mechanism automatically initiates the development of an increasing outward K^+ current which restores V_m to its resting value. Following rapid repolarization and because of the persistence of the increased P_K, the membrane hyperpolarizes for a few milliseconds and then, as P_K decays, V_m approaches V_r. The ability of the membrane to increase its P_{Na} in response to depolarization (*via* the Hodgkin cycle) recovers only after a few milliseconds in the repolarized state, and another action potential can then be initiated.

2. Propagation

The AP propagates along a fiber at constant speed because a current flows through the cell plasm from an active (high P_{Na}), depolarized region (positive inside) to an adjacent, inactive (low P_{Na}), polarized region, out through the membrane, and back through the interstitial fluid. The outward current through resting membrane depolarizes it, and when threshold is reached the previously inactive membrane becomes active. This process is repeated successively and continuously at each point, and the action potential is propagated by means of this *local circuit* current. The local circuits in nerve serve the same function as heat conduction and radiation do in a gunpowder train; both set the adjacent regions on "fire." However, nerve recovers back to the original state so that another impulse can be generated shortly after the end of the previous one.

Since the action potential is a wave of fixed size and duration and propagates at a fixed speed u, the voltage across the membrane obeys the one-dimensional wave equation:

$$\partial^2 V_m/\partial t^2 = u^2\, \partial^2 V_m/\partial x^2, \tag{5.1}$$

where x is distance along the axis of the nerve fiber.

The functional relation between V_m and time or space during the action potential depends on the specific properties of the membrane: the Hodgkin cycle, the delayed increase in P_K, and the electrical equivalent circuit of the axon (cable properties).

B. Cable Properties of Cylindrical Cells

A nerve or muscle cell is an extremely extended cylinder. For example, a 20-μm nerve fiber may be as much as 2 m long in the human body, a length-to-diameter ratio of 10^6. Clearly, plasm resistance and membrane resistance will limit the spread of a depolarizing current (signal) applied to one point of the fiber. The situation is entirely analogous to a coaxial telegraph cable without inductance: the intracellular and interstitial fluids are conductors separated by the membrane, which is a leaky insulator. A regenerative process is required in the membrane because the intracellular fluid is such a poor conductor that a sudden depolarization of the membrane at one point will spread only a few millimeters and is greatly slowed, as illustrated in Fig. 11. Two microelectrodes are inserted a distance x apart into a skeletal muscle fiber (Fig. 11b). A step current (Fig. 11a) is applied through one electrode (indicated by the circled I in Fig. 10b) by connecting a battery whose voltage ($\mathscr{E}_B$) is hundreds of times larger than the transmembrane potential to the fiber through a resistor R_s whose resistance is much higher than the microelectrode resistance, so $I \approx \mathscr{E}_B/R_s$. The resulting change in transmembrane voltage $\Delta\mathscr{E}_m$ ($\equiv \Delta V_m$) is recorded at the other electrode (the circled $\mathscr{E}_m$ in Fig. 11b) as a function of time. The process is repeated with the recording electrode inserted at different distances. Figure 11c shows the results of recording $\Delta\mathscr{E}_m$ vs. time at distances of 0, 2.5, and 5.0 mm from the current electrode in a skeletal muscle fiber. The response gets slower and the final value smaller as the distance between electrodes is increased.

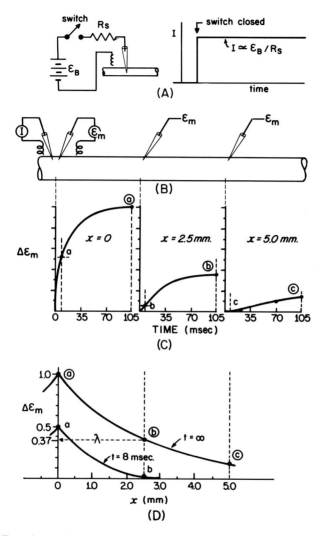

FIG. 11. Experimental measurement of cable properties in a skeletal muscle fiber. (a Left: Generation of a constant current and its application to a fiber *via* an intracellular electrode. Right: Applied current as a function of time; zero with switch open, and constant with switch closed. (b) Constant current is suddenly applied to a fiber at $x = 0$, and changes in transmembrane potential at several points along the fiber are measured with another intracellular electrode system. (c) Transmembrane potential changes as a function of time after switch closure are recorded at the distances indicated by the dashed upward extensions of ordinate lines. (d) Replot of data shown in (c) to show voltage changes as functions of distance along the fiber. Lettered points in (d) correspond to the same lettered points in (c). Spatial spread at early times is much less than at later times. Time constant of membrane τ_m is 35 msec. Space constant λ is 2.5 mm [from Ruch and Patton (1965)].

Figure 11d shows voltage as a function of distance (at the same scale as in Fig. 11b) at $t = 8$ msec (lower curve) and $t > 150$ msec (upper curve, labeled $t = \infty$). This behavior is accurately described by the cable equation.

1. *The Cable Equation*

The cable differential equation describes the variation of transmembrane voltage with distance and time, $V_m(x, t)$ [cf. Hodgkin and Rushton (1946), Taylor (1963)]. The equation is derived by equating membrane current density i_m calculated from the divergence of internal current to the current through the membrane's resistance and capacitance (Fig. 12).

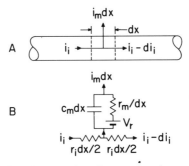

FIG. 12. Derivation of the cable equation. Current flow in a cylindrical cell with a high-resistance surface membrane. (a) Conservation of current: i_i decreases by an amount di_i over a distance dx, and this must exit through the membrane; hence $i_m\, dx = -di_i$ or $i_m = -\partial i_i/\partial x$. (b) Equivalent circuit of a cell dx centimeters long.

The derivation is based on the following assumptions: (1) Current flow in the cell plasm i_i (A) is axial and uniformly distributed; (2) current through the membrane (density i_m, A/cm) is perpendicular to the surface; and (3) the resistance of the external medium is negligible. Referring to the equivalent circuit of Fig. 12b, there is a resistance of r_i ohms/cm in the cell plasm; hence resistance of the segment is $r_i\, dx$; the voltage gradient is $\partial V_m/\partial x = -r_i i_i$; therefore $r_i i_m = \partial^2 V_m/\partial x^2$. Membrane current exits halfway along $r_i\, dx$ through an element of membrane consisting of a battery V_r and a membrane resistance r_m/dx in series shunted by a membrane capacity $c_m\, dx$; c_m is the capacity per unit length (F/cm) and r_m is the membrane resistance times unit length (ohm cm). (Membrane resistance is ohmic only for voltage changes less than 0.8 threshold.) Thus

$i_m \, dx = [c_m \, \partial V_m/\partial t + (V_m - V_r)/r_m] \, dx$. Equating membrane currents gives the cable equation,

$$i_m = \frac{1}{r_i} \frac{\partial^2 V_m}{\partial x^2} = c_m \frac{\partial V_m}{\partial t} + \frac{(V_m - V_r)}{r_m}, \tag{5.2}$$

where $c_m \, \partial V_m/\partial x$ is membrane capacitative current and $(V_m - V_r)/r_m$ is membrane ionic current. Multiplying through by r_m, making the definitions $\lambda^2 \equiv r_m/r_i$ and $\tau_m \equiv r_m c_m$ and defining $V = V_m - V_r$ converts the equation into the more usual form:

$$\lambda^2 \frac{\partial^2 V}{\partial x^2} = \tau_m \frac{\partial V}{\partial t} + V; \tag{5.3}$$

λ has the dimensions of length and is called the space constant; τ_m is the membrane time constant. A third intrinsic property of the cable is the input or characteristic resistance $R_c = V(0, \infty)/I_a$, where I_a is the applied current. These three quantities λ, τ_m, and R_c characterize the cable and can be measured experimentally (cf. Figs. 11c,d).

2. Space Constant, Time Constant, Characteristic Resistance

Equation (5.3) can be solved in closed form for an applied step of current applied at a point (Hodgkin and Rushton, 1946), but the functional relationship is complicated. The solution for $t = \infty$ is simple:

$$V(x, \infty) = R_c I_a \exp(-x/\lambda); \qquad R_c \equiv (r_i r_m)^{1/2}/2. \tag{5.4}$$

The experimental decline of V with x is shown in Fig. 11d.

The time course of V is S-shaped except at $x = 0$. Although the time constant $\tau_m = r_m c_m$ is the appropriate parameter, the time course is not exponential. In general, the greater the distance, the greater the time delay and the more slowly V rises (Fig. 11c). Physically, this behavior results because portions of the membrane capacitance at large distances from the current electrode must be charged by small currents flowing through the high resistance of the cell plasm, resulting in a large charging time constant. The S-shaped voltage–time characteristic arises because initially the applied current goes to charging nearby membrane capacitance; only when this is partially charged is current available to charge more distant capacitances [cf. Woodbury (1965)].

C. VOLTAGE CLAMPING

It is difficult to learn the detailed kinetics of an explosive process by studying the explosions. A better way is to control the critical variable so that threshold, all-or-nothing characteristics are eliminated. In the example of the brick given above the threshold behavior can be eliminated by applying sufficient external force to the brick (by holding it) so that its position is determined by the experimenter rather than the force of gravity and the brick's inertia. The kinetics of the threshold process can then be obtained in detail simply by measuring the force required to hold the brick in a particular position as a function of that position; the curve is smooth and shows no threshold behavior.

In nerve, Na^+ permeability depends on V_m; hence V_m is analogous to brick position, and the current necessary to keep voltage constant is analogous to applied force. Threshold behavior can be eliminated by "clamping" the voltage at a value set by the experimenter, and measuring the current as a function of time. The relation between membrane voltage and membrane ionic current at a particular time is obtained by clamping at different voltages. For sudden, small depolarizations an outward current must be supplied to carry K^+. At larger depolarizations the current must be inward to carry sodium ions, while at later times current must be outward, since P_{Na} has decreased and P_K increased (see Fig. 13). This procedure is equivalent to connecting an ideal battery (zero internal impedance) between the inside and the outside of the cell, so that V_m must equal the battery voltage, the battery supplying whatever current is demanded by the membrane at that voltage.

1. *Theory of Voltage Clamping*

Although the concept of voltage clamping is *a posteriori* simple, there are formidable theoretical and technical difficulties. The theoretical difficulties are embodied in a more general form of the cable equation [Eq. (5.2)], in which membrane ionic current is not approximated by an ohmic resistor. In a nerve membrane ionic current varies rapidly with time and voltage; $i_I = i_I(V_m, t)$ should replace the ohmic element $i_I = (V_m - V_r)/r_m$ in Eq. (5.2). Additionally, Eq. (5.2) is not a completely general description of membrane current, since the left-hand side includes only membrane current supplied from adjacent regions; current supplied by an internal electrode, i_a, must also be included. Thus Eq. (5.2) can be rewritten as

$$i_a + \frac{1}{r_i}\frac{\partial^2 V_m}{\partial x^2} = c_m \frac{\partial V_m}{\partial t} + i_I(V_m, t). \qquad (5.5)$$

The objective of a clamp experiment is to measure i_I, with V_m held constant to abolish regenerative behavior. This also eliminates the $\partial V_m/\partial t$ term in Eq. (5.5). In order to make $i_I = i_a$, and thus easily measured, $(1/r_i)\,\partial^2 V_m/\partial x^2$ must be made negligible. There are two ways to do this: (1) decrease r_i so that λ is much longer than the length of the dissected fiber; (2) isolate a short segment of fiber $(\ll\lambda)$ from adjacent regions by increasing the resistance of the external medium. The first method was originated by Cole [cf. Cole (1968), Marmont (1949)] and initially exploited by Hodgkin et al. (1952) and Hodgkin and Huxley (1952a,b,c,d) in their elegant (and Nobel Prize-winning) analysis of the ionic currents in the squid giant axon membrane. The method is to insert several centimeters of wire axially through the plasm of a giant axon. This wire, if properly prepared, reduces r_i by several orders of magnitude and increases the space constant. In other words, the axial wire connects all regions of the membrane by a low resistance, and hence the membrane is everywhere isopotential. With both spatial and temporal derivatives made negligible Eq. (5.5) reduces to $i_a = i_I$. Figure 13a illustrates the principle of the voltage clamp; the long internal electrode eliminates spatial variations in voltage, and connecting the internal and external electrodes to a battery holds the voltage constant in time. The experimental maneuver is to hold V_m constant at some value (usually the resting potential) and then suddenly switching to another value. Throwing the switch from 1 to 2 in Fig. 13a causes a surge of capacitive current followed by the membrane ionic current (Fig. 13b).

2. *Practical Aspects of Voltage Clamping*

The development of practical means of voltage clamping has been primarily an exercise in applied electrochemistry. The requirements on the axial electrode are severe. In order to suppress the strong negative-resistance characteristic of the squid axon membrane (increase in P_{Na} with depolarization), the surface resistance of the axial electrode must be low to prevent oscillations (Cole and Moore, 1960a; Taylor et al., 1960); the series resistance of the electrode and axoplasm must be less than the slope negative resistance of the membrane. Another viewpoint is to compare the space constant when Na^+ permeability is high, with fiber diameter. At the peak of activity the membrane resistance is 50 times lower than at rest (Cole and Curtis, 1939), and the space constant has thus fallen from about 2 mm to $2/(50)^{1/2} = 140$ μm, only about one-third of the fiber diameter. The axial electrode thus must reduce r_i by about three orders of magnitude to make active λ comparable with fiber length. These stringent requirements of the resistance of the axial electrode led

Cole and Kishimoto (1962) to develop a combination Ag–AgCl platinum black electrode to obtain low resistance at all frequencies. Other technical problems are discussed by Moore and Cole (1963).

3. *Internal Perfusion of Giant Axons*

It has been discovered that the plasm of squid giant axons can be either squeezed or washed out and replaced by artificial solutions (Baker *et al.*, 1961, 1962; Oikawa *et al.*, 1961). Resting and action potentials are close to normal when the inside of the axon is perfused with high $[K^+]$ solutions and the outside bathed in sea water (Tasaki *et al.*, 1965). These perfused axons can now be voltage clamped and the compositions of the internal and external solutions controlled [cf. Chandler and Meves (1965), Moore (1965), Lecar *et al.* (1967)]. Hence there should be a tremendous expansion of knowledge of membrane function in the next few years, setting more precise requirements on molecular theories of ion permeation.

D. Analysis of Membrane Ionic Current

1. *Membrane Ionic Current at Constant Voltage*

Figure 13b shows some early voltage-clamp results obtained by Hodgkin and Huxley (1952d). The ordinate is ionic current, the abscissa is time, and the parameter is change in membrane voltage from the resting value. Record 1 shows the sequence of current changes when the membrane is suddenly depolarized by 27 mV ($\Delta\mathscr{E}_m \equiv \mathscr{E}_m - \mathscr{E}_s \equiv V_m - V_r = 27$ mV). There is an initial outward current, too small to see on this scale, a rapid change to an inward current, and, finally, a slower change to a maintained outward current. The initial outward current is that expected from the resting membrane resistance and the change in voltage. The inward component is carried by Na^+ ions flowing down their electrochemical gradient and the late maintained outward current is carried by K^+.

Tracings 2–5 in Fig. 13b suggest strongly that the early phasic current is carried primarily by Na^+ ions. As larger and larger step depolarizations are applied the early inward current first gets larger (trace 2) then smaller (trace 3), disappears (trace 4), and finally reverses sign and becomes an outward current (trace 5) at a depolarization from the resting potential of 115 mV. In squid giant axons $V_r = 50$ mV and $V_{Na} = 60$ mV, so $V_{Na} - V_r = 110$ mV. Thus the early current reverses sign at about the voltage where the driving force on Na^+ reverses. Hodgkin and Huxley (1952a) found accurate quantitative agreement between the early current-reversal voltage and changes in V_{Na} by replacing part of the Na^+ ions with choline$^+$ ions in the artificial sea water bathing the axon. As described in Section VI,B, recent experiments have shown that the early current or Na^+ channel is 12 times more permeable to Na^+ than to K^+ (Chandler and Meves, 1965). The early current is large enough to account for the upstroke of the action potential in the unclamped axon:

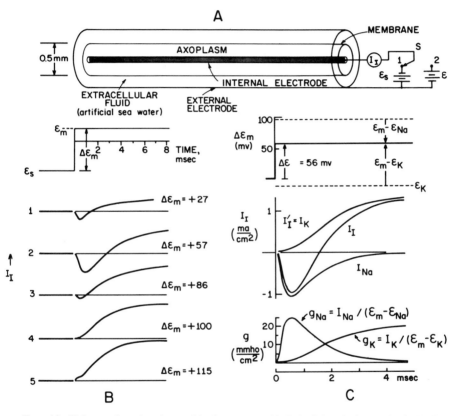

FIG. 13. Voltage clamping in squid giant axon. (a) Principle of the method. Ionic current I_I is measured as function of time with a cathode ray oscilloscope. (b) Transmembrane current flow as function of time after a sudden change in $\mathscr{E}_m$. Uppermost curve is $\mathscr{E}_m$ as a function of time. Traces 1–5 show membrane current following abrupt depolarizations by the amounts ($\Delta\mathscr{E}_m$, millivolts) shown at right. Time scale at the top applies to all records in (b). (c) Components of total membrane current and conductance. Top section: $\mathscr{E}_m$ as function of time; $\mathscr{E}_{\text{Na}}$ and $\mathscr{E}_{\text{K}}$ are indicated by dashed horizontal lines. Middle section: membrane ionic current is separated into its two components, I_{Na} and I_{K}, for $\Delta\mathscr{E}_m = 56$ mV. Bottom section: g_{Na} and g_{K} as functions of time for $\Delta\mathscr{E}_m = 56$ mV. Time scale at bottom applies to all records in (c). Part (b) after Hodgkin and Huxley (1952d), part (c) after Hodgkin (1958) [from Ruch and Patton (1965)].

The rate of rise of the action potential is about 750 V/sec at the inflection point where $\partial^2 V_m/\partial t^2 = 0$, and thus $c_m \, \partial V_m/\partial t + i_I = 0$ [Eq. (5.5), $i_a = 0$]. The ionic current required to change the membrane potential at the observed rate is calculated to be 0.75 mA/cm², a value close to 1 mA/cm² measured under voltage clamp conditions at about the same voltage (trace 2, Fig. 13b).

2. *Separation of Current into Specific Ion Components*

It is generally accepted that the early current is normally carried by Na⁺ and the late current by K⁺. In order to quantitatively analyze these currents, Hodgkin and Huxley (1952a) developed an interpolation method of separating ionic current into Na⁺, K⁺, and leakage components. First, total measured membrane ionic current is converted into current density, I_I (A/cm²) by dividing by membrane surface area. (Capital I's refer to current densities.) The separation is based on two assumptions: (1) I_K is unaffected by changes in $[Na^+]_o$, and (2) I_K does not start to change for a period sufficient to establish the rate of rise of I_{Na}, i.e., early current changes are due entirely to I_{Na}. The principle of the analysis is illustrated in Fig. 13c. Separation was made (in principle) by measuring I_I, then reducing $[Na^+]_o$ to a value at which a depolarization of 56 mV made $\mathscr{E}_m = \mathscr{E}_{Na}$. Since $I_{Na} = 0$ under these conditions, total ion current I_I' is equal to I_K (middle portion of Fig. 13c). Thus $I_{Na} = I_I - I_I' = I_I - I_K$. Actually, I_K includes a leakage component, but this is small and is separated from I_K by other means (Hodgkin and Huxley, 1952b).

3. *Specific Ion Conductances*

The separation of ionic current into Na⁺ and K⁺ moieties permits the calculation of membrane permeabilities to Na⁺ and K⁺ in terms of the chord conductance, defined as $g_S = I_S/(V_m - V_S)$ (mhos/cm²). Since V_m is constant throughout any record of Fig. 13c, individual ionic conductances are exactly the same shape as the separated currents, I_K and I_{Na}, and of magnitude obtained by dividing the current by the driving force (shown in the upper portion of Fig. 13c). The lowest portion of Fig. 13c shows that g_{Na} increases rapidly to a peak in less than 1 msec and declines to near zero in about 4 msec; g_K stays low for about 0.5 msec and then increases fairly rapidly to a plateau in about 4 msec. The currents I_{Na} and I_K can be separated and g_{Na} and g_K calculated for other depolarizations. The pattern of conductance changes shown in Fig. 13c is found for other depolarizations: As the amount of depolarization increases, the rate of rise of the conductances and the peak g_{Na} and final g_K values also increase (Hodgkin and Huxley, 1952a).

E. Prediction of Excitable Behavior: The Hodgkin–Huxley Equations

Hodgkin and Huxley (1952d) successfully predicted all the essential features of normal action-potential generation and propagation from an empirical mathematical description of the results of the voltage-clamp analysis of membrane ionic currents. This analysis and synthesis was an order-of-magnitude increase in our knowledge of nerve behavior and resulted in their receipt of the Nobel Prize in 1963. The question is whether or not the information obtained from analyzing voltage-clamp experiments is sufficient to describe the normal behavior of a nerve fiber. Their methods of fitting the voltage-clamp data and means for calculating normal nerve behavior have been described adequately in the literature (Hodgkin and Huxley, 1952a,b,c,d; Hodgkin, 1958, 1964; Hodgkin et al., 1952; Woodbury, 1962, 1965; Katz, 1966), and hence will be only briefly outlined here to serve as an introduction to the terminology Hodgkin and Huxley (1952d) used.

The total ionic current through the membrane is written in the form

$$I_I = I_{Na} + I_K + I_l, \qquad (5.6)$$

where I_l is an unspecified leakage current. Similarly, the current due to each ion species is $I_S = g_S(V_m - V_S)$, so that

$$I_I = g_{Na}(V_m - V_{Na}) + g_K(V_m - V_K) + g_l(V_m - V_l). \qquad (5.7)$$

The dependencies of g_{Na} and g_K on voltage and time were ingeniously represented in the Hodgkin–Huxley treatment by defining

$$g_{Na} = \bar{g}_{Na}m^3h; \qquad g_K = \bar{g}_K n^4; \qquad g_l = \text{const}, \qquad (5.8)$$

where $\bar{g}_{Na}$ and $\bar{g}_K$ are constants. The variables m, h, and n, functions of V_m and t and having values between zero and one, are defined by

$$dp/dt = \alpha_p(V_m)(1 - p) - \beta_p(V_m)p, \qquad (5.9)$$

where p represents h, m, or n and $\alpha_p(V_m)$ and $\beta_p(V_m)$ depend *only* on transmembrane voltage and not on time. The functional relationships between the α's and β's and voltage are complicated and suggest no obvious physical interpretation; β_h, α_m, and α_n increase with increasing depolarization, and α_h, β_m, and β_n decrease. The first group is described by equations of the form $(V_m/V_0)/\{\exp[(V_m + V_0)/10] - 1\}$, and the second by simple exponential functions of voltage. Under voltage clamp, V_m is constant and the solution of Eq. (5.9) is exponential in time. The rate constants α and β instantly assume their new values when V_m is changed. The solutions to Eq. (5.9) can be substituted in Eqs. (5.7) and

(5.8) to give the calculated total ionic current under voltage clamp. The agreement with experiment is good except that time courses of the calculated I_K's are not sufficiently delayed before rising rapidly. This could have been circumvented by defining $g_K = \bar{g}_K n^6$ or some higher power, but high-speed digital computers were not available in 1952 and the computation problems were already formidable.

The fraction of Na$^+$ (early current) channels open is $m^3 h$ and the fraction of K$^+$ (late current) channels open is n^4. Following a sudden depolarization m increases rapidly because α_m is increased and β_m decreased by the voltage change; m is exponential; m^3 is inflected. At the same time h starts to decrease exponentially, but with time constant ten times slower than that for m. The product $m^3 h$ thus increases rapidly and decreases more slowly. Similarly, n^4 increases slowly along a highly inflected time course.

When voltage is not controlled by the experimenter, the solution of the Hodgkin–Huxley equations, Eqs. (5.5) and (5.7)–(5.9), is difficult, since this set of equations contains a nonlinear, partial differential equation. The simplest case is the "space clamp"; the internal axial wire is in place but is not connected to a low-internal-impedance voltage source. This makes $\partial^2 V/\partial x^2 = 0$ and Eq. (5.5) becomes

$$c_m \, dV_m/dt + I_I = I_a. \tag{5.10}$$

Equations (5.7)–(5.10) now form a complete set of simultaneous, ordinary differential equations which can be solved by standard numerical means for any applied current I_a. Hodgkin and Huxley also obtained solutions for their equations for normal, nonspace-clamped conditions by assuming that there is a propagating wave solution. In this case the $\partial^2 V_m/\partial x^2$ term in Eq. (5.5) can be replaced by $u^{-2} \, \partial^2 V_m/\partial t^2$ [Eq. (5.1)] and the resulting ordinary second-order equation can be solved. However, the conduction speed u is now an unknown parameter and its values must be varied arbitrarily to find a value making the solution $V_m(t)$ finite for all t.

Hodgkin and Huxley (1952d) calculated action potential by both methods and obtained curves closely resembling recorded AP's. In addition, they predicted numerous other phenomena associated with nerve impulse conduction, e.g., conduction speed (calculated, 18.8 m/sec; measured, 21.2 m/sec), refractory period, conductance change during activity, anodal break excitation, ion exchange during activity, and threshold behavior. Thus voltage clamping is a powerful tool for analyzing nerve membrane behavior and has permitted specification of membrane properties with sufficient accuracy to challenge the abilities of theoreticians to propose workable molecular models.

Frankenhaeuser and Huxley (1964) did a voltage-clamp analysis of ionic currents in myelinated nerve fibers and found that equations quite similar to Eqs. (5.6)–(5.9) adequately describe the currents and accurately predict excitable behavior.

VI. Ion Channel Characteristics

A. Special Properties of Na$^+$ and K$^+$ Channels

Although many of the pertinent experimental findings about nerve membrane ionic currents have been described above, these and other facts are summarized here to put in compact form the data which must be encompassed by an adequate theory of membrane ion penetration.

1. *Variation of Ion Permeability with Voltage*

The peak value of g_{Na} and the final value of g_K (Fig. 13c) depend dramatically on the size of the step depolarization under voltage-clamp conditions. The plot of conductance as a function of voltage is S-shaped, with the inflection point at a depolarization of about 30 mV from the resting potential for both g_{Na} and g_K. For lesser depolarizations the relationship is approximately exponential; g_{Na} varies as $\exp(V_m/4)$ and g_K as $\exp(V_m/5)$.

The significance of this finding can be seen from the following argument of Hodgkin and Huxley (1952d): Suppose that ions penetrate the membrane through special channels specific for each ion species. A depolarization could cause these channels to open because some negatively charged molecules previously at the outside of the membrane now can move to the inside. (There must be relatively few such channels and many ions must move through each channel because the inward migration of the negatively charged molecules immediately following a step depolarization constitutes an outward current flow at early times. This is not observed experimentally even at depolarizations to V_{Na}, where it would be particularly obvious.) Suppose that there are N molecules of valence z per unit area in the membrane and that a specific ion permeability is proportional to the number N_i on the inside surface of the membrane, the remainder, $N_o = N - N_i$, being on the outside surface. If N_i is 0.5 when $V_m = 0$, then $N_i/N_o = \exp(-Fz V_m/RT)$, and hence

$$N_i = 1/[\exp(-Fz V_m/RT) + 1]. \tag{6.1}$$

For large negative values of $Fz V_m/RT$ this expression reduces to $\exp(Fz V_m/RT)$. Experimentally, g_{Na} varies with $\exp(V_m/4)$; hence

$z = 25/4 = 6$. Thus six charges (or a triply charged dipole) must move through the full transmembrane potential to account for the observed variation of sodium conductance with voltage. For potassium the magnitude of the valence is 4–5. If the charges or dipoles do not move completely through the membrane, the number of required charges is correspondingly increased. However, the observed dependences of g_{Na} and g_K on V_m are not accurately described by Eq. (6.1); the more complicated relationship given in Section V,E is required.

2. Linear Instantaneous Current–Voltage Relationship

The rise of I_{Na} following a step depolarization is S-shaped (Fig. 13c), the peak current being reached in something under 1 msec. The time course of I_K is even more S-shaped; the time to the inflection point is nearly 2 msec in Fig. 13c. This delay is most likely due to the time required for the molecular rearrangements which result in increases in g_{Na} and g_K. More generally, any change in permeability (as distinguished from the driving force) must require a finite amount of time, since such a change implies a physical or chemical rearrangement of the molecules in the membrane.

An experimental consequence is that the currents carried through the membrane by an ionic species just before and just after a sudden change in V_m have the same ratio as the driving voltages, because g has not had time to change. In the squid giant axon the instantaneous Na^+ current–voltage relationship is linear over a 200-mV range: $I_{Na} = g_{Na}(V_m - V_{Na})$, where g_{Na} has the same value at $t = -0$ and $t = +0$ (Hodgkin and Huxley, 1952b). This result is rather surprising, since the large transmembrane concentration and potential differences indicate that the instantaneous current–voltage relationships could be more accurately described by the Goldman constant-field equation [Eqs. (3.4) and (3.5)] than by Ohm's law. The Goldman equation does accurately describe the instantaneous current–voltage relation in voltage-clamped myelinated nerve fibers (Frankenhaeuser, 1960, 1962). Further, the linear current–voltage relation found in squid axons holds only at normal $[Na^+]_o$ values. At lower $[Na^+]_o$ the relationship is curved in a manner predicted by assuming the independence principle holds (Hodgkin and Huxley, 1952b).

3. Time Characteristics of I_K

The time delay before I_K starts increasing rapidly following a step depolarization is long and the maximum rate of rise is large compared with other mechanisms (such as diffusion) showing delay behavior (Fig.

13c). The delay times of S-shaped curves can be compared by calculating the dimensionless number $f = \dot{y}_{max} t_i / y_{max}$, where $\dot{y}_{max}$ is the slope at the inflection point, t_i is the time to the inflection point, and y_{max} is the final value of the variable y. For the K^+ channel $f = 1$ in some circumstances, while f is less than 0.1 in planar diffusion. Cole and Moore (1960b) made a detailed study of the kinetics of I_K turn-on. They found that the time to the inflection point (and the value of f) increased with the initial value of the membrane potential; if the membrane is hyperpolarized for 10 or more milliseconds preceding a depolarization to V_{Na}, the resulting curve is delayed with respect to one obtained with no preceding hyperpolarization. More importantly, they found that the shape of the I_K vs. t curve is independent of the initial voltage and dependent only on the final voltage. The I_K–t curves for various initial voltages can be superimposed by shifts along the time axis. The only effect of increasing the initial hyperpolarization is to increase the time to the inflection point. The term $I_K(t)$ is accurately described by an equation of the form

$$I_K(t) = I_{K,0}[1 - \exp(t - t_0)/\tau]^n,$$

where n is about 30. The curves for different initial hyperpolarizations can be fitted by altering t_0. As Cole and Moore (1960b) point out, this finding means that V_m is a state variable; the time course of I_K does not depend on past history (except to determine the starting value), but does depend only on the present value of V_m. Both $I_{K,0}$ and t_0 are, of course, functions of initial V_m. As mentioned in Section III,B,2, K^+ influxes and effluxes interact as though K^+ ions have to traverse a long, narrow channel (Hodgkin and Keynes, 1955). This "in-file" behavior must be explained by any theory of I_K [cf. Macey and Oliver (1967)].

4. Temperature Dependence of g_{Na} and g_K

The delay in the rise of g_{Na} following a step depolarization is a few tenths of a millisecond (Fig. 13c), depending on final V_m and on temperature. The rate of rise of g_{Na} (and g_K) depends sensitively on temperature; Q_{10}'s are about 3 in squid giant axons (Hodgkin et al., 1952). The Q_{10}'s of the final values of g_{Na} and g_K are much lower, about 1.3. The Q_{10}'s of the α's and β's in toad myelinated nerve fibers range from 1.7 to 3.2 (Frankenhaeuser and Moore, 1963).

5. Effects of Calcium Concentration on g_{Na} and g_K

Calcium is well known to have large effects on the electrical properties of excitable cells (Frankenhaeuser and Hodgkin, 1957; Huxley, 1959;

Blaustein and Goldman, 1966; Hille, 1968). The effects are complex and manifold. The effects of $[Ca^{++}]_o$ on the parameters describing g_{Na} and g_K are well approximated by shifting the voltage-dependent parameters (α, β) along the voltage axis by amounts proportional to $\log[Ca^{++}]_o$ but the coefficient is only about half of RT/zF. This implies that Ca^{++} binds to the membrane and alters the electric field near Na^+ and K^+ penetration sites. The Na^+ channels are more affected by $[Ca^{++}]$ than are K^+ channels.

6. *Separability of* Na^+ *and* K^+ *Channels*

The findings that the early current following a step depolarization of a squid axon membrane is carried largely by Na^+ and the late current largely by K^+ indicates that these currents flow through separate channels. This is also indicated by the completely different ion selectivity sequences of the early and late currents (Section VI,B). However, the action of the puffer fish poison, tetrodotoxin, is the most convincing evidence that there are separate channels. This substance selectively blocks the early (Na^+) current in concentrations as low as 10^{-8} M in the external bathing solution, but has no effect on the late current (K^+) channel [cf. Narahashi and Moore (1968)]. Tetrodotoxin has no effect when it is added to the solution bathing the inside of the membrane. Moore *et al.* (1967) found that only about 15 molecules of tetrodotoxin per μm^2 of membrane are required to block the early current. These effects are strong evidence that (1) Na^+ and K^+ channels are separate entities, (2) the inside and outside entrances to the Na^+ channel are different, and (3) the fraction of surface area occupied by Na^+ channels is very small.

The rough calculation made in Section III,B,3 indicated that there is only one ion channel per 8 μm^2 ($0.12/\mu m^2$) in the resting state. This can be compared with the number of Na^+ channels in the active state by taking the ratio of resting and active membrane conductances. Resting conductance is about 1 mmho/cm^2, peak g_{Na} is about 100 mmho/cm^2, and their ratio is $1/100$. Thus the expected density of channels in the active state is $0.12/\mu m^2 \times 100 = 12/\mu m^2$, a number in good agreement with $15/\mu m^2$ inferred from tetrodotoxin experiments.

B. SELECTIVITY SEQUENCES OF Na^+ AND K^+ CHANNELS

Most natural membranes are much more permeable to cations than anions (red blood cells are an exception). Furthermore, membranes differentiate between cations. There are two types of theories to explain this behavior: (1) Cations traverse membranes *via* discrete channels or pores with an ease determined solely by the ion's size (Mullins, 1956). (2) There are negative fixed charges on the membrane which confer cation-selective properties by cation exchanger action (Tobias, 1964; Lettvin et al., 1964; Eisenman, 1968; Ling, 1962). The bulk of the chemical, biochemical, and physiological evidence supports the second theory, and gives clues to the possible chemical identity of the fixed negative charges.

1. Cation Selectivity and Binding Sequences

a. Selectivity Sequences.   Cation selectivity sequences are best known for the squid giant axon, since internal perfusion makes possible more reliable and extensive measurements. The relative effects of various cations on the resting potential led Baker et al. (1962) to the sequence $P_K > P_{Rb} > P_{Cs} > P_{Na} > P_{Li}$ for the resting (nonexcited) membrane. The same sequence of relative cation permeabilities is found in frog sartorius muscle, lobster muscle, red blood cells, yeast cells, and E. coli. There are exceptions; Hodgkin (1947) found the sequence $P_K > P_{Cs} > P_{Rb} > P_{Na} > P_{Li}$ for crab nerve.

Clearly the cation selectivity sequences of Na^+ and K^+ channels differ greatly. Unfortunately the "single-file" behavior of K^+ channels (Hodgkin and Keynes, 1955) makes it difficult to determine uniquely the selectivity sequence of the K^+ channel. Sequences are obtained by replacing all or part of the internal or external Na^+ or K^+ with the test cation and measuring the resulting changes in transmembrane potential and/or current–voltage relations. The interactions of the two cation species in K^+ channels makes interpretation of the results equivocal [cf. Armstrong and Binstock (1965), Binstock and Lecar (1967)]. This difficulty does not apply to the Na^+ channel since it obeys the independence principle. The cation selectivity sequence of Na^+ channels has been measured quantitatively in the perfused squid axon:

$$P_{Li}:P_{Na}:P_K:P_{Rb}:P_{Cs} = 1.1:1:0.083:0.025:0.016$$

[cf. Meves (1966)]. Tasaki et al. (1966) added $P_{Na} > P_{NH_4} > P_K > P_{Guan} > P_{Rb}$, and the combined result is $P_{Li} > P_{Na} > P_{NH_4} > P_K > P_{Guan} > P_{Rb} > P_{Cs}$ for the sodium channel.

b. Binding Sequences.   If cation selectivity is related to the binding affinity of some charge groups on the membrane for cations, then there should be a one-to-one relationship between binding and selectivity sequences. Bundenberg de Jong (1949) examined the cation binding ability of a number of molecules that contain negatively charged groups.

Egg lecithin, soya bean phosphatide, and Na nucleate all contain phosphate groups ionized to some extent at pH $= 7$. He measured affinity A of the phosphate group for some small monovalent cations and found the sequence $A_{Li} > A_{Cs} > A_{Rb} > A_{Na} > A_{NH_4} > A_K$. He also found that binding affinities of monovalent cations for the ionized carboxyl group are in the sequence $A_{Cs} > A_{Rb} > A_{Guan} > A_K > A_{NH_4} > A_{Na} > A_{Li}$.

2. *Relationship between Selectivity and Binding*

Examination of the selectivity and binding data reveals a consistent pattern: The relative cation permeability sequence of the Na^+ channel is exactly the reverse of the binding affinity sequence of ionized carboxyl groups. This supports the hypothesis that selectivity sequences of cations for the Na^+ channels are a direct consequence of the existence of ionized carboxyl groups in or on the membrane. More specifically, the ease with which a cation passes through a Na^+ channel is inversely related to the binding affinity of ionized carboxyl groups for that cation. Considerable evidence gives qualitative support for this hypothesis; there are insufficient quantitative data to make a convincing test.

a. Charged Groups of Membranes. If cation selectivity is due to specific anion groups, then one or both surfaces of the membrane must have a negative charge. More specifically, changes which affect surface charge should affect excitability mechanisms. Experiments on internally-perfused squid axons provide strong evidence for negative charge on the internal surface. Tasaki and Shimamura (1962), Baker *et al.* (1962), Narahashi (1963), and Baker *et al.* (1964) found that perfusing squid axons with solutions diluted with sucrose (low ionic strength) reduced the resting potential to near zero, but did not greatly affect action-potential generation. This behavior is not predicted by the Hodgkin–Huxley equations; the kinetics of $g_{Na} = \bar{g}_{Na} m^3 h$ are such that $h \to 0$ for V_m's maintained near zero, and thus excitability should disappear. Maintained excitability is due to both a horizontal shift of the peak g_{Na} vs. V_m curve and Na^+ inactivation (h) vs. V_m curve. Baker *et al.* (1964) and Chandler *et al.* (1965) concluded that these shifts are probably due to the decrease in ionic strength of the internal perfusing solution rather than the concomitant decrease in the internal $[K^+]$. They obtained a good description of these effects by assuming that there is net excess negative charge of -2.23 $\mu C/cm^2$ on the inner surface of the membrane. Rojas and Atwater (1968) have used a more direct approach.

If the surface charge is uniform, the g_K vs. V_m curve should be shifted by approximately the same amount as the g_{Na} curves in low-ionic-strength solutions. Unfortunately, the available data are equivocal. Moore *et al.* (1964) found approximately the expected change, while Chandler *et al.* (1965) obtained shifts varying from zero to about half the shifts of the $g_{Na}-V_m$ curve. Thus the effects of low-ionic-strength solutions on the g_K-V_m curve are difficult to reconcile with the $g_{Na}-V_m$ curve, unless it is supposed that the surface charges are not uniformly distributed; there may be more charges near Na^+ channels than K^+ channels.

VII. Membrane Properties: Some Theoretical Approaches

The foregoing sections on electrolyte distribution, transmembrane potentials, membrane structure, and the permeability changes during activity have been aimed at giving the present state of knowledge of membrane properties. This section outlines a few of the current attempts to develop a detailed theory of the molecular mechanisms of ion penetration through membranes and of related phenomena. Such attempts can be lumped into two general categories: (1) Statistical-mechanical or non-equilibrium-thermodynamical calculations of the properties of homogeneous membranes with specified characteristics are made, e.g., selectivity properties of a fixed charge lattice [cf. Eisenman *et al.* (1968), Lindley (1967), Essig *et al.* (1966), Sandblom and Eisenman (1967)]. (2) Special and specific molecular mechanisms aimed at predicting observed behavior are postulated [cf. Tobias (1964), Goldman (1964), Hoyt (1963)]. Since this is the "frontier," one approach is *a priori* as likely to be useful as any other. Hence some of our individual efforts in this area are given to illustrate the state of the art.

A. Homogeneous Membranes

Although nearly all the experimental data indicate that ions penetrate the membrane through specialized regions, it is nevertheless probably useful to explore the statistical-mechanical properties of ions driven by voltage and concentration gradients through a homogeneous region containing fixed scatterers. The scattering groups can be assigned various properties (e.g., charged, dipolar, induced dipolar, or neutral) to determine the effects on ionic conductance. This type of calculation comes from plasma physics and appears valid, since the molecules making up the special regions where ions do penetrate behave as fixed scatterers and may have properties close to those of a homogeneous membrane of the same composition.

Another statistical-mechanical approach is to calculate the behavior of a sheet of dipoles in response to variations in the applied field. Some of the surface groups of membrane molecules are dipolar, and consideration of the cooperative properties of such groups may shed some light on membrane behavior.

1. *Ion Movements through Charged Scatterers*

K. S. Cole (1965) has summarized past work on electrodiffusion models of transport through biological membranes including transient behavior. In this approach hydrodynamic equations are used. An ionic current is set proportional to the sum of an electrical potential and a concentration gradient [Eq. (3.2)] and combined with Poisson's equation. In Cole's model singly-charged positive ions of density $n(x)$ ($\equiv$ [S]) move through immobile negative ions of uniform concentration $\bar{m}$. The particle conservation equation

$$\partial n/\partial t + \partial M/\partial x = 0 \tag{7.1}$$

relates the rate of change of the density to the divergence of the particle current $M(x)$ (flux). The current is given by the transport equation [Eq. (3.2)]. Rewriting Eq. (3.2) in terms of mobility μ gives

$$M = -\mu(kT \, \partial n/\partial x + ne \, \partial v/\partial x).$$

Poisson's equation gives the remaining necessary relationship,

$$\partial^2 V/\partial x^2 = -(e/\epsilon_0\epsilon_m)(\dot{n} - \bar{m}), \tag{7.2}$$

where ϵ_m is the dielectric constant of the medium and ϵ_0 is the permittivity of free space, 8.85×10^{-12} F/m. Combining Eqs. (7.1), (3.2), and (7.2) gives

$$\frac{\partial n}{\partial t} = \mu kT \frac{\partial^2 n}{\partial x^2} + \mu e \frac{\partial V}{\partial x}\frac{\partial n}{\partial x} - \frac{ne^2}{\epsilon_0\epsilon_m}\mu(n - \bar{m}). \tag{7.3}$$

If V is everywhere constant, Eq. (7.3) reduces to a diffusion equation with a diffusion time of the order of

$$\tau_D \approx \delta^2/\mu kT, \tag{7.4}$$

where δ is a characteristic distance over which n changes appreciably, in this case about membrane thickness. Similarly, if n is spatially nearly constant, the right side of Eq. (7.3) reduces to the rightmost term; n is nearly equal to $\bar{m}$ and the time variation of $n - \bar{m}$ is an exponential with a relaxation time constant

$$\tau_e = \epsilon_0\epsilon_m/ne^2\mu. \tag{7.5}$$

K. S. Cole (1965) used Eqs. (7.3)–(7.5) to analyze the sequence of events following the sudden application of a current to a membrane: Surface charges build up until a sufficient field exists in the membrane for the ions to carry the current. The charging process occurs on the time scale of Eq. (7.5) (membrane time constant). After the initial charging process the ions gradually redistribute themselves on the time scale of Eq. (7.4). Although Cole concludes that the hydrodynamic (electrodiffusion) equation [Eq. (7.3)] cannot account for the K^+ current in the squid axon membrane, it is worthwhile determining if more general statistical models are useful in predicting some membrane properties, e.g., dependence of conductance on voltage.

One of the assumptions in deriving Eq. (7.3) is that μ is independent of particle velocity. This gives rise to a single relaxation time inversely proportional to the particle mobility, Eq. (7.5). In general, a particle's mobility depends on its speed, because mobility varies inversely with collision frequency. Collision frequency depends on the type of ion–scatterer interaction. Hence in general there is a spectrum of relaxation times rather than one relaxation time. A more general analysis is based on the particle distribution function $f(\mathbf{r}, \mathbf{v}, t)$ rather than on particle density (concentration) $n(\mathbf{r}, t)$. The function $f(\mathbf{r}, \mathbf{v}, t) \, dr^3 \, dv^3$ is defined as the number of particles in the velocity interval dv^3 at v and position interval dr^3 at r at time t; $n(\mathbf{r}, t) = \int f(\mathbf{r}, \mathbf{v}, t) \, dv^3$.

A conservation equation for $f(\mathbf{r}, \mathbf{v}, t)$ similar to that for $n(r)$, Eq. (7.1), may be written by relating the time rate of change of $f(\mathbf{r}, \mathbf{v}, t)$ to the divergence of its current in r, $\partial \cdot (f \, d\mathbf{r}/dt)/\partial \mathbf{r}$, the divergence of its current in $\mathbf{v}$, $\partial \cdot (f \, d\mathbf{v}/dt)/\partial \mathbf{v}$, and to the rate of change due to collisions, $(\partial f/\partial t)_c$:

$$\frac{\partial f}{\partial t} + \frac{\partial}{\partial \mathbf{r}} \cdot \left(f \frac{d\mathbf{r}}{dt} \right) + \frac{\partial}{\partial \mathbf{v}} \cdot \left(f \frac{d\mathbf{v}}{dt} \right) = \left(\frac{\partial f}{\partial t} \right)_c. \tag{7.6}$$

Using Hamiltonian derivatives,

$$\frac{\partial}{\partial \mathbf{r}} \cdot \left(\frac{d\mathbf{r}}{dt} \right) + \frac{\partial}{\partial \mathbf{v}} \cdot \left(\frac{d\mathbf{v}}{dt} \right) = 0,$$

and so Eq. (7.6) reduces to

$$\frac{\partial f}{\partial t} + \mathbf{v} \cdot \frac{\partial f}{\partial \mathbf{r}} + \frac{d\mathbf{v}}{dt} \cdot \frac{\partial f}{\partial \mathbf{v}} = \left(\frac{\partial f}{\partial t} \right)_c. \tag{7.7}$$

For scattering with cross section $\sigma(v)$ from stationary scatterers of density N the Boltzmann (1964) collision integral is

$$\left(\frac{\partial f}{\partial t} \right)_c = -N\sigma v \left[f - \int \frac{d\Omega'}{4\pi} f(v, \theta', \varphi') \right], \tag{7.8}$$

where $d\Omega'$ is solid angle in velocity space and v, θ', and φ' are spherical coordinates in velocity space.

Ion movements through fixed scatterers having various properties, e.g., charge and dipole, can be approximated by retaining only the first two terms of an expansion of f in spherical harmonics in velocity space. If (1) the electric field is constant, (2) there are no spatial gradients in particle density n_0, and (3) the collision frequency of the ions with the fixed scatterers is given by $\nu_i = \nu v^n$, where v is an ion's speed, ν is a constant, and n depends on the nature of the scatterer as shown in Table IV, then a solution of Eqs. (7.7) and (7.8) for ionic current I_I as a function of electric field E is

$$I_I = G(E)E; \qquad G(E) = \frac{n_0(ez)^2}{3m\bar{\nu}} \frac{\int_0^\infty U^{n+4} e^{-W}\, dU/(U^{2n} + \bar{E}^2)}{\int_0^\infty U^2 e^{-W}\, dU}; \qquad (7.9)$$

where $\bar{\nu} = \nu v_t^n$; $v_t = (3kT/m)^{3/2}$, the thermal velocity of the ions, is a constant; $\bar{E}$ is a dimensionless electric field given by $\bar{E} = ezE/mv_t\nu_i\eta^{1/2}$, where m is ion mass and η is the fractional energy loss per collision; and W is defined by

$$W = 3 \int_0^U \frac{x^{2n+1}\, dx}{x^{2n} + \bar{E}^2}.$$

TABLE IV

<small>COLLISION FREQUENCY ν_i OF IONS WITH VARIOUS TYPES OF FIXED SCATTERING PARTICLES[a]</small>

Ions interacting with	n	$\nu_i = \nu v^n$
Neutral particles	1	νv
Induced dipoles	0	ν
Fixed dipoles	−1	ν/v
Fixed charges	−3	ν/v^3

[a] v is velocity of an ion; ν is a constant.

Numerical evaluation of Eq. (7.9) shows that when the scatterer is a neutral particle $G(E)$ decreases as $\bar{E}$ increases, as shown in Fig. 14. This relation resembles that found in squid axon membranes bathed in iso-tonic KCl [cf. Cole (1968)]. However, the calculated variation in $G(E)$ with E is only about half as fast as observed. Nevertheless, this suggests that neutral scatterers may occur in K^+ channels. Conductance decreases as the field increases because mobility is inversely proportional to collision frequency, and collision frequency increases with ion speed and hence with E.

Neutral scatterers in the K^+ channels cannot account for all of its properties; the actual changes in g_K following a step change in $\bar{E}$ occur on a time scale several orders of magnitude slower than the calculated changes.

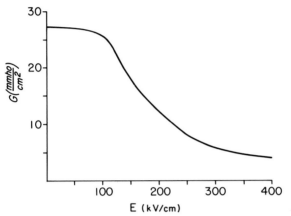

FIG. 14. Steady-state conductance vs. electric field, calculated from Eq. (7.9) for ions moving through fixed, neutral scattering particles ($n = 1$). The conductance scale was obtained by assuming (1) K^+ moves through channels 60 Å in length and 100 Å² in cross-sectional area; (2) the average number of ions in a channel is three; (3) the distance between neutral scatterers is 15 Å; and (4) there are 150 channels/μm². $T = 300°K$. The voltage (mV) across a 60-Å membrane is 0.6 times the value of the field (kV/cm).

2. Statistical Behavior of Dipole Sheets

Goldman (1964) proposed a structural model for axon membranes in which the dipole heads of some membrane phospholipid molecules change their orientation and combining properties under the influence of an electric field. In this model the flow of ions through the bulk of the membrane is given by the Nernst–Planck (hydrodynamic) equation, with the dipolar phosphate groups acting as ion-exchange gates at the surfaces. The affinity of ion-exchange sites for cations is assumed to depend markedly on the configuration of the dipolar complexes; the configuration in turn depends on electric field.

Debye (1945) described the behavior of a collection of electric dipoles in a solid with a Fokker–Planck equation, assuming that the rotation of the dipoles under the influence of an electric field is limited by collisions:

$$I\nu \frac{\partial f}{\partial t} = \frac{kT}{\sin \theta} \frac{\partial}{\partial \theta} \left(\sin \theta \frac{\partial f}{\partial \theta} \right) + \frac{\mu E}{\sin \theta} \frac{\partial f}{\partial \theta} (f \sin^2 \theta). \qquad (7.10)$$

In this equation $f(\theta, t) \sin \theta \, d\theta$ denotes the number of dipoles in the polar angle interval $d\theta$ at θ at time t; E is the electric field acting on a dipole; I is the dipole's amount of inertia; μ is the electric dipole moment; and ν is the collision frequency. Debye solved this equation for the case of no equilibrium field, and found that $f(\theta, t)$ responded to a small applied field with a single relaxation time inversely proportional to the temperature.

In the membrane there is an equilibrium field E_0; it is convenient to expand $f(\theta, t)$ in terms of Legendre polynomials, $P_n(\cos \theta)$:

$$f(\theta, t) \approx \exp\left[\frac{\mu E_0 \cos \theta}{kT}\right]\left[1 + \sum_{n=0}^{\infty} C_n(t)P_n(\cos \theta)\right]. \qquad (7.11)$$

Differential-difference equations for the C_n are obtained by substituting this expansion in Eq. (7.10). If only first-order terms E_1 in E are retained, $E = E_0 + E_1$. The field E_1 is the sum of the applied field E_a and a response field E_r:

$$E_1 = E_a + E_{r,0}\left(\frac{\mu kT}{E_0}\right)^{1/2} \sum_{n=0}^{\infty} \frac{C_n}{2n+1}\left[nI_{n-(1/2)} + (n-1)I_{n+(3/2)}\right]; \qquad (7.12)$$

$I_n \equiv I_n(\mu E_0/kT)$ is a modified Bessel function. The term E_r depends on the geometrical arrangement and density of the dipoles through the constant $E_{r,0}$. The series for E_r shows that the presence of an equilibrium electric field results in a spectrum of relaxation times in contrast to the single relaxation time when $E_0 = 0$. The peak of the relaxation time spectrum, τ_{rel}, shifts from being inversely proportional to temperature for small equilibrium fields, $\tau_{rel} = I\nu/2kT$, to being inversely proportional to the field for a large field, $\tau_{rel} = I\nu/2\mu E_0$.

The average dipole moment $\bar{\mu}$ at equilibrium is

$$\bar{\mu} = \mu[ctnh(\mu E_0/kT) - kT/\mu E_0]. \qquad (7.13)$$

In general, E_0 consists of a portion due to the dipoles, $\bar{\mu}$, and a portion E_e resulting from other charges:

$$E_0 = E_e + \gamma\bar{\mu}. \qquad (7.14)$$

The field E_e in a membrane is due to an applied voltage and to charges separated at the electrolyte–membrane boundary. Since $\bar{\mu}$ must depend on E_0 through the S-shaped Langevin function of Eq. (7.13) and depend linearly on E_0 through Eq. (7.14), it can be seen that $\bar{\mu}$ can suddenly change sign as E_e is varied. Provided μ is large enough, there are three points of intersection of Eqs. (7.13) and (7.14); the middle one at $E_0 = 0$ is unstable and the other two are stable, oriented one way or the other way. This flip-flop characteristic may enable membrane dipoles to act as "open-or-shut" gates for ion penetration channels. This mechanism cannot explain the behavior of the K^+ channel, since V_m is a state variable (the final g_K depends only on V_m, not on past history) and dipole orientation depends on the present field and previous values, but there are permeability changes in some tissues, e.g., heart, which may be the result of this type of behavior. Insufficient data are available to make quantitative tests.

B. Eyring Rate Kinetics and Ion Penetration

Heretofore membrane penetration by ions has been treated theoretically as a continuous diffusion process through a homogeneous medium. However, it is at least as realistic to treat ion movement through a 75-Å membrane as successive jumps over a few potential-energy barriers. There are unlikely to be more than five to ten potential-energy valleys encountered by an ion as it traverses a membrane, and the low temperature coefficients of Na^+ and K^+ ion penetration indicate activation energies of about 4800 cal/mole when Na^+ and K^+ channels are open (membrane depolarized). Hence the Eyring absolute rate theory of transport processes is well suited to dealing with ion penetration. Since there are several different kinds of ion channel, the potential energy contours through the membrane for each channel and for each ion species will be different. One complication is that penetration rates of Na^+ and K^+ depend on transmembrane voltage, and thus the absolute heights of the potential-energy barrier, as well as the driving force, depend on V_m. Alternatively, the fraction of channels open could depend on voltage; this is the interpretation of Hodgkin and Huxley (1952d). The problem is to see if the potential-energy contours through the membrane can be deduced from the unique properties of excitable membranes which were listed above.

1. Ion Flux through a Four-Barrier Membrane

The experimental finding in squid axons that the instantaneous current–transmembrane-voltage relationship is linear (constant chord conductance g) yields considerable information about the potential-energy barriers in the membrane. The relationship expected from the Goldman flux equation [Eq. (3.4)] for a homogeneous membrane is nonlinear unless internal and external concentrations are equal. There are no simple modifications of the transport equation [Eqs. (3.2) and (3.3)] which can give rise to linear I_{Na} vs. V_m curves over the observed range of $(V_{Na} - 150)$ to V_{Na} (in mV) (Hodgkin and Huxley, 1952b). However, an appropriate choice of potential-energy barrier heights does give linearity.

Figure 15 shows a possible potential-energy diagram for a membrane.

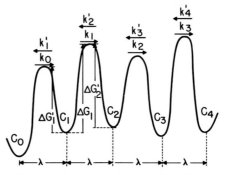

FIG. 15. Four-potential-energy-barrier model of an ion channel through a membrane. Concentrations of the ion in internal and external solutions are C_0 and C_4, respectively. Concentrations at minima are C_1, C_2, and C_3. Forward rate constants are unprimed, backward are primed. The ΔG's are similarly labeled and are measured from minima.

For simplicity the membrane is represented by four barriers, but the results are not greatly different for a five- or six-barrier model. Eyring *et al.* (1949, 1964) give the steady-state flux over successive potential energy barriers as

$$
\begin{aligned}
M &= C_0 \lambda k_0 - C_1 \lambda k_1{}' \\
M &= C_1 \lambda k_1 - C_2 \lambda k_2{}' \\
M &= C_2 \lambda k_2 - C_3 \lambda k_3{}' \\
M &= C_3 \lambda k_3 - C_4 \lambda k_4{}',
\end{aligned}
\tag{7.15}
$$

where M is flux (moles/cm²-sec), λ is the width of each barrier, and the remaining quantities are defined in Fig. 15. The absolute rate constants k_i are given by Eyring [cf. Eyring *et al.* (1964)]:

$$
k_i = \kappa_i(kT/h) \exp(-\Delta G_i/RT).
\tag{7.16}
$$

Eyring *et al.* (1949, 1964) give the general solution to Eq. (7.15). The result is

$$
M = \frac{\lambda(k_0 k_1 k_2 k_3 C_0 - k_1{}' k_2{}' k_3{}' k_4{}' C_4)}{k_1 k_2 k_3 + k_1{}' k_2 k_3 + k_1{}' k_2{}' k_3 + k_1{}' k_2{}' k_3{}'},
\tag{7.17}
$$

where C_0 and C_4 are internal and external ion concentrations (activities), respectively.

Since the internal and external solutions are of nearly equal ionic strengths and total particle concentrations, the free-energy difference between the two solutions is zero when $C_0 = C_4$, $V_m = 0$, and $M = 0$. This places one constraint on the k_i. When $V_m \neq 0$ half the trans-membrane voltage adds to the activation energy in the forward direction and half subtracts in the reverse direction: $\Delta G_i = \Delta G_{i0} + \frac{1}{2}FV_\lambda z$ and $\Delta G_i' = \Delta G_{i0} - \frac{1}{2}FV_\lambda z$, where V_λ is the voltage across the ith barrier. The rate constants thus can be written as $k_i = k_{i0} \exp(FV_\lambda z/2RT)$, where $k_{i0} = \kappa_i(kT/h) \exp(-\Delta G_i/RT)$. Similarly, $k_i' = k_{i0}' \exp(-FV_\lambda z/2RT)$. Since M must equal zero when $C_0 = C_4$ and $V_m = 0$, then

$$k_{00}k_{10}k_{20}k_{30} = k_{10}'k_{20}'k_{30}'k_{40}'$$

and (7.18)

$$\Delta G_{00} + \Delta G_{10} + \Delta G_{20} + \Delta G_{30} = \Delta G_{10}' + \Delta G_{20}' + \Delta G_{30}' + \Delta G_{40}'.$$

In Section III,B it was shown that the electric field in the membrane is constant, to a good approximation. If this is true in the Na^+ and K^+ channels, then $V_\lambda = V_m/4$. Using this relation, Eq. (7.18), and the definition $FzV_m/RT \equiv v$, Eq. (7.17) becomes

$$M = \frac{\lambda(C_0 e^{v/2} - C_4 e^{-v/2})}{(k_{00})^{-1}e^{3v/8} + (k_{40}')^{-1}e^{-3v/8} + k_{10}'(k_{00}k_{10})^{-1}e^{v/8} + k_{30}(k_{30}'k_{40}')^{-1}e^{-v/8}}.$$

(7.19)

A more compact form is obtained by defining

$$\exp(v_S) \equiv C_4/C_0, \qquad v_S \equiv V_S z F/RT,$$

$$\exp(3v_1/8) \equiv k_{00}/k_{40}, \qquad \exp(v_2/8) \equiv k_{20}'/k_{20} = k_{10}'k_{30}'k_{40}'/k_{00}k_{10}k_{30},$$

(7.20)

where V_S is the ion equilibrium potential. Substitution into Eq. (7.19) and rearranging gives the compact form

$$M = \frac{\lambda(C_0 C_4 k_{00} k_{40}')^{1/2} \sinh[(v - v_S)/2]}{\cosh[3(v - v_1)/8] + [k_{10}'k_{30}/(k_{10}k_{30}')]^{1/2} \cosh[(v - v_2)/8]}.$$ (7.21)

2. Linear Current–Voltage Relationship

Analysis of Eq. (7.21) shows that it can be made nearly linear over the range $-150 < (V - V_S) < 150$ mV if v_1 and v_2 are set equal to v_S and $k_{10}'k_{30}/k_{10}k_{30}' \equiv a^2$ is chosen appropriately. The nature of these constraints on the k_{i0} are best illustrated by a specific example, the linear relationship between instantaneous I_{Na} and V_m in the squid giant axon at normal $[Na^+]_o$ (Hodgkin and Huxley, 1952b). In this case $C_4 = [Na^+]_o = 480$ and

$C_0 = [\text{Na}^+]_i = 60\ \mu\text{M}/\text{cm}^3$; $C_4/C_0 = [\text{Na}^+]_o/[\text{Na}^+]_i = 8$; $V_{\text{Na}} = 52$ mV. To make $v_1 = v_2 = 52$ mV ($RT/F = 25$ mV at $T = 15°\text{C}$) requires that $k_{00}/k_{40}' = 8^{3/4} = 4.75$ and $k_{20}'/k_{20} = 8^{1/4} = 1.68$. For these conditions if $a = 0.6$ the deviation from linearity is less than 2%.

The linearity requirement puts three constraints on the rate constants, and Eq. (7.18) is a fourth. There are eight rate constants, but fortunately the four constraints specify the ratios of rate constants and thus the differences between the heights of the four maxima (Eyring *et al.*, 1949). Temperature variation of I_{Na} gives the absolute heights of each barrier. The barrier height differences are obtained by substituting the values of the k_{i0} and k_{10}' into Eqs. (7.18) and (7.21), assuming that the transmission coefficients κ_i and κ_i' out of the ith well are equal, and solving for the corresponding differences in the ΔG's.

The depths of the potential-energy wells cannot be determined from these conditions. The reason is that these are steady-state kinetics, and the concentrations in membrane potential-energy wells build up to whatever values are necessary to make the flux over internal barriers equal to the incoming fluxes, and thus play no role in determining steady-state fluxes.

Expressed in terms of ΔG's, the three linearity conditions and Eq. (7.18) become:

$$\ln(k_{00}/k_{40}') = (\Delta G_{40}' - \Delta G_{00})/RT = \tfrac{3}{4}\ln(C_4/C_0),$$
$$\ln(k_{20}'/k_{20}) = (\Delta G_{20} - \Delta G_{20}')/RT = \tfrac{1}{4}\ln(C_4/C_0), \qquad (7.22)$$
$$\ln(k_{10}'k_{30}/k_{10}k_{30}') = (\Delta G_{10} + \Delta G_{30}' - \Delta G_{10}' - \Delta G_{30})/RT = \ln(a^2),$$
$$\Delta G_{00} + \Delta G_{10} + \Delta G_{20} + \Delta G_{30} = \Delta G_{10}' + \Delta G_{20}' + \Delta G_{30}' + \Delta G_{40}'.$$

Reference to Fig. 15 shows that a barrier $i + 1$ is $\Delta G_{i0} - \Delta G_{i0}' \equiv RT\,\delta_{i+1,i}$ higher than the barrier i to the immediate left. Similarly, defining $\delta_{14} \equiv \Delta G_{00} - \Delta G_{40}'$, Eq. (7.22) becomes

$$
\begin{aligned}
-\,\delta_{14} &= \tfrac{3}{4}\ln(C_4/C_0) = 1.56,\\
\delta_{32} &= \tfrac{1}{4}\ln(C_4/C_0) = 0.52,\\
\delta_{21} \qquad -\,\delta_{43} &= 2\ln a = -1.02,\\
\delta_{21} + \delta_{32} + \delta_{43} + \delta_{14} &= 0.
\end{aligned}
\qquad (7.23)
$$

Equation (7.23) is easily solved for the δ's. The results are shown in Fig. 16. It can be seen that the highest barrier is on the side of higher $[\text{Na}^+]$. The difference between the right- and left-hand barriers is $\delta_{41} = -\delta_{14} = 1.56RT$; the corresponding ΔG is 930 cal/mole, a modest difference. In the general case of N barriers this difference is $\delta_{N,1} =$

$[(N - 1)/N]\ln(C_N/C_0)$. Thus to produce a linear current–voltage relation, it is necessary to have barriers of progressively increasing heights going from the low- to the high-concentration side of the membrane.

Clearly, the current–voltage is linear only for one value of C_4/C_0. Hodgkin and Huxley (1952b) found this to be true in the squid axon;

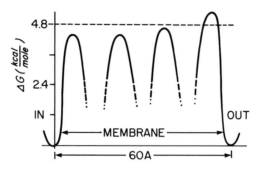

FIG. 16. Four-barrier potential-energy profile of Na^+ channel in squid giant axon membrane. The heights of the barriers are chosen to make the I_{Na} vs. V_m relation linear over the range $-150 \leq (V-V_{Na}) \leq 150$ mV for normal internal and external Na^+ concentrations $[Na^+]_o = 480$, $[Na^+]_i = 60$ μM/cm³. The heights of the minima are not constrained by the linearity requirement and are not shown. The absolute height of the barrier peaks was obtained from a $Q_{10} = 1.3$ for I_{Na} ($\Delta G = 4800$ cal/mole). Out: outside solution; In: inside solution.

the current–voltage relationship is linear only at normal $[Na^+]_o$. If the I_{Na}–V_m relationship is known at normal $[Na^+]_o$, the current I'_{Na} at any other value $[Na^+]_o'$ can be predicted from the independence principle. Since Eq. (7.21) is based on this principle, I'_{Na}/I_{Na} is obtained by dividing Eq. (7.21) for an altered $[Na^+]_o$ by Eq. (7.21) for normal $[Na^+]_o$. The denominators cancel, and the result is

$$\frac{I'_{Na}}{I_{Na}} = \frac{M'_{Na}}{M_{Na}} = \left(\frac{[Na^+]_o'}{[Na^+]_o}\right)^{1/2} \frac{\sinh[F(V_m - V'_{Na})/2RT]}{\sinh[F(V_m - V_{Na})/2RT]}. \quad (7.24)$$

Equation (7.24) is a slightly different form of Hodgkin and Huxley's Eq. (12) (1952a).

3. Transmission Coefficient and Its Interpretation

a. Absolute Barrier Height. The height of the potential-energy barrier between the bathing solutions and the membrane can be estimated from the variation of maximum I_{Na} with temperature. Hodgkin *et al.* (1952) estimate that the Q_{10}'s of the I_{Na} and I_K are about 1.3, but experimental difficulties make this only an approximation. Since $\Delta G =$

41,000 $\log_{10}(Q_{10})$ at room temperatures, $\Delta G = 4800$ cal/mole for $Q_{10} = 1.3$. Equation (7.21) shows that the principal temperature dependence of the flux is contained in the term $(k_{00}k'_{40})^{1/2}$. Ignoring the kT/h term ($Q_{10} = 1.03$) and $k'_{10}k_{30}/k_{10}k'_{30}$, where variations with temperature nearly cancel, then $(\Delta G_{10} + \Delta G'_{40})/2 = 4800$ cal/mole. This value was used to set the heights of the barriers in Fig. 16.

b. Value of $\lambda(\kappa_0\kappa_4')^{1/2}$. The linearity condition and the temperature coefficient of I_{Na} thus lead to an expression for the absolute rate of Na$^+$ movement through the membrane with all constants determined except those in the term $\lambda(\kappa_0\kappa_4')^{1/2}$. One constraint can be put on these by comparing Eq. (7.21) with k_{i0}'s chosen to make it linear with experimental results in the form $I_{Na} = g_{Na}(V - V_{Na})$. The linear expansion of Eq. (7.21) for the Na$^+$ channel is

$$I_{Na} = \frac{170 \times 10^{-6} F \lambda (\kappa_0\kappa_4')^{1/2} (kT/h) \exp(-4800/RT)}{1.6 \times 50 \times 10^{-3}} (V_m - 0.052),$$

$$(7.25)$$

where V is now in volts, I_{Na} is in A/cm^2, and F is a Faraday. Hence the coefficient of $(V_m - 0.052)$ is equal to g_{Na}. This equality can be solved for $\lambda(\kappa_0\kappa_4')^{1/2}$.

The general expression for a four-barrier membrane is

$$\lambda(\kappa_0\kappa_4')^{1/2} = \frac{2g_S[1 + (k'_{10}k_{30}/k_{10}k'_{30})^{1/2}]RT}{F^2(C_0C_4)^{1/2}(kT/h) \exp[-(\Delta G_{00} + \Delta G'_{40})/2RT]} \quad (7.26)$$

if the constraints on the k_{10}'s are met. Here g_S is the conductance of ion S.

Hodgkin and Huxley (1952d) give $g_{Na} = 0.12$ mhos/cm^2 when all sodium channels are open (large depolarization preceded by a hyperpolarization). However, the uncertainty in the value of the Q_{10} makes the accuracy of Eq. (7.25) not much better than an order of magnitude. Solving Eq. (7.26) for the Na$^+$ channel gives

$$\lambda(\kappa_0\kappa_4')^{1/2} = 4.2 \times 10^{-13}. \quad (7.27)$$

The form of Eq. (7.26) is relatively independent of the number of barriers, since the only term dependent on barrier number is $a^2 = k'_{10}k_{30}/k_{10}k'_{30}$ and values of a for two, three, four, and five barriers are one or less. Thus Eq. (7.26) is likely independent of the number of barriers within a factor of two.

c. Interpretation of $\lambda(\kappa_0\kappa_4')^{1/2}$. The small value of $\lambda(\kappa_0\kappa_4')^{1/2}$ is due to the small value of λ (~ 10 Å) and the small fraction of the membrane surface area occupied by Na^+ channels. Steric factors may also be important, i.e., there may be restrictions on a Na^+ ion's direction of approach to a channel if the Na^+ is to enter. The transmission coefficient $(\kappa_0\kappa_4')^{1/2}$ can be thus taken as the product of three factors: the size of the channel entrance, A; the number of channels per unit area, N; and a steric factor, $\Omega/2\pi$, the fractional solid angle describing approach-angle restrictions. Thus

$$(\kappa_0\kappa_4')^{1/2} = NA\Omega/2\pi. \tag{7.28}$$

The number of Na^+ channels has been estimated from the effects of tetrodotoxin to be about 15 channels/μ^2 (Moore et al., 1967). Hence $N = 15 \times 10^8$ channels/cm². The size of the opening to the Na^+ channel must be about the size of the largest ion that can penetrate the Na^+ channel; this is taken as 100 Å². A four-barrier model for a 60-Å membrane gives $\lambda = 15$ Å. Substituting these values in Eqs. (7.27) and (7.28) gives $(\kappa_0\kappa_4')^{1/2} = 2.8 \times 10^{-6}$ and $\Omega/2\pi = 0.2$. Considering the uncertainties in the various quantities, particularly the Q_{10} of I_{Na}, this value could be as small as 0.002 and as large as one. Thus entry into a Na^+ channel may not be unduly restricted.

This theory thus gives an accurate description of the Na^+ current through an open Na^+ channel. The same type of calculation can be made for K^+ channels, and the results are similar. However, this approach must be used with caution, since there is evidence that K^+ ion movements through a K^+ channel are not independent of each other. Hodgkin and Keynes (1955) analyzed the K^+ interactions and conclude that there are about three ions in each channel simultaneously. The independent behavior of Na^+ ions in Na^+ channels suggests that any particular Na^+ traverses the channel before another one enters from either direction, and thus that the potential-energy wells in the membrane are shallow. Similarly, the non-independent behavior of K^+ in K^+ channels suggests that the wells are deep and mean transit time is about three times slower than mean entry time.

d. Myelinated Nerve Fibers. The instantaneous $I_{Na} - V_m$ and $I_K - V_m$ relationships of nodes of frog myelinated nerve fibers obey the Goldman flux equation [Eqs. (3.4) and (3.5)] (Frankenhaeuser and Huxley, 1964). This implies that the membrane barriers are all about the same height, since Eq. (7.21) reduces to Eq. (3.5) when the number of barriers becomes large and all k_{i0}'s are equal except $k_{00} = k'_{N0}$. Since the barrier-height differences required to give linearity are small, the type of analysis used above for squid axons can be applied to frog nodes with only slight modifications.

4. *Approximate Description of a* Na$^+$ *Channel*

The information presented above on cell membrane architecture, the nature of ion selectivity mechanisms, and the potential-energy profile of a Na$^+$ channel given in Fig. 16 permit a fairly detailed but hypothetical description of a Na$^+$ channel and the surrounding region of membrane: A Na$^+$ channel consists of a hole or "pore" passing through the lipid–protein matrix of the membrane. The "pore" is formed in a region where several lipoprotein subunits of the membrane meet. In traversing the membrane an ion successively encounters an outer layer of protein (glycoprotein), the lipid coating of the lipoprotein subunits, the protein core of the membrane, and the lipid coating and glycoprotein again. This pore or channel is about 6 Å in radius and is likely guarded at one or both ends by carboxyl groups which may be part of a sialic acid group in the glycoprotein coat of the membrane. These groups constitute only a small fraction of the total negative surface charge. The specific blockage of Na$^+$ channels by tetrodotoxin indicates that the outer and inner approaches are different; tetrodotoxin is effective only on the outside; the guanidinium group of the toxin appears to combine with carboxyl groups of the outer pore entry region (Narahashi and Moore, 1968).

When a Na$^+$ channel is entered by an ion it presumably interacts with a carboxyl group in a way that acts to produce the observed selectivity. However, this interaction is not simple binding of the ion to a carboxyl group at the channel entrance. A tightly bound ion has the same probability of traversing the membrane as a loosely bound one because the concentration of the ion in that particular potential-energy minimum builds up to a level to keep fluxes equal over all barriers, i.e., the maxima, not the minima, determine fluxes (Section VI,B; Eyring *et al.*, 1964). On a statistical basis this means that a more tightly bound ion spends a larger fraction of the time bound to a carboxyl group than does a more loosely bound ion. Since Na$^+$ ions cross the membrane independently of each other, the mean time spent in the bound state is probably less than the mean time between effective collisions of Na$^+$ with the entrance. However, it is not known whether or not the movements of the more tightly bound ions, e.g., K$^+$ and Cs$^+$, through the Na$^+$ channel obey the independence principle.

In order to be selective, the pore entrance must in some way modify an ion's binding energy to carboxyl groups to produce a potential-energy barrier whose height is approximately the negative of the binding energy. One possible mechanism is to suppose that an ion must lose some of its hydration shell in order to enter the pore (Mullins, 1956, 1959; Lindley, 1967). Although the molecular mechanism is not apparent, the exact one-to-one relationship between the Na^+ channel selectivity sequence and the reversed carboxyl binding affinity sequence is convincing evidence of a causal relationship between the two.

The steric factor, $\Omega/2\pi = 0.2$ (range from 0.002 to 1) in the transmission coefficient is the probability that a Na^+ ion will enter a channel if it hits the entrance with sufficient kinetic energy. Once in the channel the ion progresses over a series of barriers until it exits. The wells may be negative charges (carboxyl groups) occurring at 10–15-Å intervals. Alternatively, the barriers may be neutral scatterers (hydrophobic groups of the lipid or protein). In either case these groups are probably positioned by loops of protein encasing the channel. The lumen of the channel is presumably filled with water. The higher potential-energy barriers toward the outer face of the membrane could be due to slight changes in the size of the lumen or closer positioning of charged groups to the lumen.

5. Voltage Dependence of Na^+ Conductance

The essence of the excitable process in nerve is that depolarization causes a large, rapid, and transient increase in Na^+ conductance. The preceding section dealt with the current–voltage relations of an open Na^+ channel. The next problem is to consider the possible mechanisms whereby changes in voltage open and close Na^+ channels. The kinetics of the opening of the channel following a sudden depolarization are third order (m^3), while the channel closes with first-order (h) kinetics. For both processes the rate constants depend on voltage in a rather complicated manner (Hodgkin and Huxley, 1952d). Several mechanisms for this process have been suggested (Mullins, 1959; Goldman, 1964; Hoyt, 1963). The concept of ion penetration through protein regions suggests that permeability or conductance changes are due to voltage-induced changes in the shape or conformation of the protein. It is well known that changes in the concentrations of certain constituents can cause conformation changes in proteins. The most accurately known example is that of the oxygen-carrying protein of the blood, hemoglobin. Hemoglobin consists of two pairs (α and β) of chains cross linked in a roughly tetragonal array, with members of a pair oppositely placed. Each chain has a molecular weight of about 16,700. When hemoglobin is oxygenated the heme group in each chain takes up an O_2 and there is a concomitant decrease in the distance between marked regions of the β chains from 40 to 33 Å. The oxygenation of one chain increases the association constants of the remaining chains for oxygen, so that the uptake process is facilitated. This leads to an S-shaped curve of oxygen content of hemoglobin vs. oxygen partial pressure.

It is attractive to suppose that the opening of a Na^+ channel (or a K^+ channel) is the result of a conformation change in the channel protein due to depolarization of the membrane. An increase in chain spacings of 7 Å could easily change a channel from closed to open. It seems reasonable to suppose that a change in electric field acting on a "channel" protein molecule could change its conformation in much the same way as the uptake of a O_2 by a hemoglobin molecule [cf. Hill (1967)]. A change in the voltage across the protein molecule, ΔV, is, at least, roughly equivalent to a change in concentration of an ion from C_0 to $C_0 \exp(\alpha \Delta V F z / RT)$, where α is the fraction of the voltage acting on charges in the protein of total charge z.

The nature of protein conformation changes is not well understood, so it is difficult to put this model of permeability changes on a quantitative basis. A simple model based on the types of conformation change found in hemoglobin gives a satisfying (but not necessarily quantitatively satisfactory) picture of the permeability changes of the Na^+ channel: Suppose that the pore proteins like hemoglobin consist of four chains extending through the membrane and that there are two possible equilibrium positions for each chain N (near) and F (far). A Na^+ channel is open when all four chains are in the F position, but closed when one or more chains are in the N position. When the membrane is polarized three of the chains are in the N position and the fourth is in the F position. When the membrane is depolarized the three chains in the N position quickly change to the F position in a cooperative manner. The fourth chain's equilibrium is in the N position, but its kinetics are about ten times slower. Thus a depolarization produces a rapid opening of the channel as the three N chains separate, followed by a slower closing of the channel as the F chain moves to the N position. Movements of only a few Ångstroms would suffice to close or open the channel.

This molecular model is formally identical with the original model of Hodgkin and Huxley (1952d) and suffers from the same defect: The voltage variations of the rate constants describing changes in g_{Na} are too complex to be explained on the simple charge or dipole hypothesis outlined in Section VI,A,1. The possibility of cooperative interaction between the various parts of the molecule might, however, lead to the complicated functions needed to describe the rise of I_{Na}. Further, the voltage dependence of the inactivation of I_{Na} (decline of g_{Na} following a sustained depolarization) is of the form for a simple charge or dipole movement.

REFERENCES

ADRIAN, R. H. (1956). The effect of internal and external potassium concentration on the membrane potential of frog muscle. *J. Physiol. (London)* **133**, 631.

ANON. (1968). Cell membrane biophysics. *J. Gen. Physiol.* **51**, 1s.

ARMSTRONG, C. M., and BINSTOCK, L. (1965). Anomalous rectification in the squid giant axon injected with tetraethylammonium chloride. *J. Gen. Physiol.* **48**, 859.

BAKER, P. F., HODGKIN, A. L., and SHAW, T. I. (1961). Replacement of the protoplasm of a giant nerve fibre with artificial solutions. *Nature* **190**, 885.

BAKER, P. F., HODGKIN, A. L., and SHAW, T. I. (1962). The effects of changes in internal ionic concentrations on the electrical properties of perfused giant axons. *J. Physiol. (London)* **164**, 355.

BAKER, P. F., HODGKIN, A. L., and MEVES, H. (1964). The effect of diluting the internal solute on the electrical properties of a perfused giant axon. *J. Physiol. (London)* **170**, 541.

BANGHAM, A. D. (1963). The physical structure and behavior of lipids and lipid enzymes. *Advan. Lipid Res.* **1**, 65.

BEAR, R. S., SCHMITT, F. O., and YOUNG, J. Z. (1937). Ultrastructure of nerve axoplasm. *Proc. Roy. Soc.* **B123**, 505.

BINSTOCK, L., and LECAR, H. (1967). Ammonium ion substitutions in the voltage clamped squid axon. *Ann. Meeting Biophys. Soc., Houston, 11th, 1967*, Abstr. WC 4, Biophysical Soc.

BLAUSTEIN, M. P., and GOLDMAN, D. E. (1966). Competitive action of calcium and procaine on lobster axon. *J. Gen. Physiol.* **49**, 1043.

BOLIS, L., CAPRARO, V., PORTER, K. R., and ROBERTSON, J. D. (1967). *Symp. Biophys. Physiol. Biol. Transport, Frascati, 1965.* Springer, New York.

BOLTZMANN, L. (1964). "Lectures on Gas Theory." Univ. of California Press, Berkeley, California.

BRANTON, D. (1966). Fracture faces of frozen membranes. *Proc. Natl. Acad. Sci. U.S.* **55**, 1048.

BRINLEY, JR., F. J. (1965). Sodium, potassium, and chloride concentrations and fluxes in the isolated giant axon of *Homarus. J. Neurophysiol.* **28**, 742.

BRINLEY, JR., F. J., and MULLINS, L. J. (1965). Ion fluxes and transference numbers in squid axons. *J. Neurophysiol.* **28**, 526.

BROWN, A. D. (1963). The peripheral structures of gram-negative bacteria, IV. The cation-sensitive dissolution of the cell, membrane of the Halophilic Bacterium, *H. Halobium. Biochim. Biophys. Acta* **75**, 425.

BUNDENBERG DE JONG, H. G. (1949). *In* "Colloid Science" (H. R. Kruyt, ed.), Vol. II, pp. 284–289 Am. Elsevier, New York.

CHANDLER, W. K., and MEVES, H. (1965). Voltage clamp experiments on internally perfused giant axons. *J. Physiol.* **180**, 788.

CHANDLER, W. K., HODGKIN, A. L., and MEVES, H. (1965). The effect of changing the internal solution on sodium inactivation and related phenomena in giant axons. *J. Physiol.* **180**, 821.

COLE, K. S. (1949). Some physical aspects of bioelectric phenomena. *Proc. Natl. Acad. Sci. U.S.* **35**, 558.

COLE, K. S. (1965). Electrodiffusion models for the membrane of squid giant axon. *Physiol. Rev.* **45**, 340.

COLE, K. S. (1968). "Membranes, Ions and Impulses." Univ. of California Press, Berkeley, California.

COLE, K. S., and COLE, R. H. (1941). Dispersion and sbsorption in dielectrics, I. Alternating current characteristics. *J. Chem. Phys.* **9**, 341.

COLE, K. S., and CURTIS, H. J. (1939). Electric impedance of the squid giant axon during activity. *J. Gen. Physiol.* **22**, 649.

COLE, K. S., and CURTIS, H. J. (1950). *In* "Medical Physics" (O. Glasser, ed.), Vol. II, pp. 82–90. Year Book Publ., Chicago, Illinois.

COLE, K. S., and HODGKIN, A. L. (1939). Membrane and protoplasm resistance in the squid giant axon. *J. Gen. Physiol.* **22**, 671.

COLE, K. S., and KISHIMOTO, U. (1962). Platinized silver chloride electrode. *Science* **136**, 381.

COLE, K. S., and MOORE, J. W. (1960a). Ionic current measurements in the squid giant axon membrane. *J. Gen. Physiol.* **44**, 123.

COLE, K. S., and MOORE, J. W. (1960b). Potassium ion current in the squid giant axon: Dynamic characteristic. *Biophys. J.* **1**, 1.

COLE, R. H. (1965). Relaxation processes in dielectrics. *J. Cellular Comp. Physiol. Suppl.* **2**, 13.

CONWAY, E. J. (1957). Nature and significance of concentration relations of potassium and sodium ions in skeletal muscle. *Physiol. Rev.* **37**, 84.

CRIDDLE, R. S., BOCK, R. M., GREEN, D. E., and TISDALE, H. D. (1961a). Specific interaction of mitochondrial structural protein (SP) with cytochromes and lipid. *Biochem. Biophys. Res. Commun.* **5**, 75.

CRIDDLE, R. S., BOCK, R. M., GREEN, D. E., and TISDALE, H. D. (1961b). The structural protein and mitochondrial organization. *Biochem. Biophys. Res. Commun.* **5**, 81.

CRIDDLE, R. S., BOCK, R. M., GREEN, D. E., and TISDALE, H. D. (1961c). Isolation and properties of the structural protein of mitochondria. *Biochem. Biophys. Res. Commun.* **5**, 109.

DAVIES, J. T., and RIDEAL, E. K. (1963). "Interfacial Phenomena." Academic Press, New York.

DAVSON, H., and DANIELLI, J. F. (1952). "The Permeability of Natural Membranes." Harvard Univ. Press, Cambridge, Massachusetts.

DEBYE, P. (1945). "Polar Molecules." Dover, New York.

DE ROBERTIS, E. D. P., NOWINSKI, W., and SAEZ, F. (1965). "Cell Biology." Saunders, Philadelphia, Pennsylvania.

DICK, D. A. T. (1959). Osmotic properties of living cells. *Intern. Rev. Cytol.* **8**, 387.

DICK, D. A. T. (1966). "Cell Water." Butterworths, Washington, D.C.

EISENMAN, G. (1968). Ion permeation of cell membranes and its models. *Federation Proc.* **27**, 1249–1251.

EISENMAN, G., BATES, R., MATTOCK, G., and FRIEDMAN, S. M. (1968). "The Glass Electrode." Wiley, New York.

EMMELOT, P., and BOS, C. J. (1966). On the participation of neuraminidase-sensitive sialic acid in K^+-dependent phosphohydrolysis of p-nitrophenyl phosphate by isolated rat-liver plasma-membranes. *Biochim. Biophys. Acta* **115**, 244.

ESSIG, A., KEDEM, O., and HILL, T. L. (1966). Net flow and tracer flow in lattice and carrier models. *J. Theoret. Biol.* **13**, 72.

EYRING, H., LUMRY, R., and WOODBURY, J. W. (1949). Some applications of modern rate theory to physiological systems. *Record. Chem. Progr.* (*Kresge-Hooker Sci. Lib.*) **10**, 100.

EYRING, H., HENDERSON, D., STOVER, B. J., and EYRING, E. M. (1964). "Statistical Mechanics and Dynamics." Wiley, New York.

FEDER, W., ed. (1968). Bioelectrodes. *Ann. N. Y. Acad. Sci.* **148**, 1.

FINEAN, J. B., COLEMAN, R., and GREEN, W. A. (1966). Studies of isolated plasma membrane preparations. *Ann. N. Y. Acad. Sci.* **137**, 414.

FLEISCHER, S., FLEISCHER, B., and STOECKENIUS, W. (1967). Fine structure of lipid-depleted mitochondria. *J. Cell Biol.* **32**, 193.

FRANKENHAEUSER, B. (1960). Quantitative description of sodium currents in myelinated nerve fibres of *Xenopus laevis. J. Physiol. (London)* **151**, 491.

FRANKENHAEUSER, B. (1962). Potassium permeability in myelinated nerve fibres of *Xenopus laevis. J. Physiol. (London)* **160**, 54.

FRANKENHAEUSER, B., and HODGKIN, A. L. (1957). The action of calcium on the electrical properties of squid axons. *J. Physiol. (London)* **137**, 218.

FRANKENHAEUSER, B., and HUXLEY, A. F. (1964). The action potential in the myelinated nerve fibre of *Xenopus laevis* as computed on the basis of voltage clamp data. *J. Physiol. (London)* **171**, 302.

FRANKENHAEUSER, B., and MOORE, L. E. (1963). The effect of temperature on the sodium and potassium permeability changes in myelinated nerve fibers of *Xenopus laevis. J. Physiol. (London)* **169**, 431.

FRICKE, H. (1925). The electric capacity of suspensions with special reference to blood. *J. Gen. Physiol.* **9**, 137.

GAINER, H. (1967). Plasma membrane structure: Effects of hydrolases on muscle resting potentials. *Biochim. Biophys. Acta* **135**, 560.

GENT, W. L. G., GREGSON, N. A., GAMMOCK, D. B., and ROPER, J. H. (1964). The lipid-protein unit in myelin. *Nature* **204**, 553.

GOLDMAN, D. E. (1943). Potential, impedance and rectification in membranes. *J. Gen. Physiol.* **27**, 37.

GOLDMAN, D. E. (1964). A molecular structural basis for the excitation properties of axons. *Biophys. J.* **4**, 167.

GREEN, D. E., and PERDUE, J. F. (1966). Membranes as expressions of repeating units. *Proc. Natl. Acad. Sci. U.S.* **55**, 1295.

GREEN, D. E., and TZAGOLOFF, A. (1966). Role of lipids in the structure and function of biological membranes. *J. Lipid Res.* **7**, 587.

GREEN, E. C., ALLMANN, E. W., BACHMANN, E., BAUM, H., KOPACZYK, K., KORMAN, E. F., LIPTON, S., MACLENNAN, D. H., McCONNEL, D. G., PERDUE, J. F., RIESKE, J. S., and TZAGOLOFF, A. (1967). Formation of membranes by repeating units. *Arch. Biochem. Biophys.* **119**, 312.

HANAHAN, D. J. (1960). "Lipid Chemistry." Wiley, New York.

HECHTER, O. (1964). Intracellular water structure and mechanisms of cellular transport. *Ann. N. Y. Acad. Sci.* **128**, 625.

HILL, T. L. (1967). Electric fields and the cooperativity of biological membranes. *Proc. Natl. Acad. Sci. U.S.* **58**, 111.

HILLE, B. (1967). The selective inhibition of delayed potassium currents in nerve by tetraethylammonium ion. *J. Gen. Physiol.* **50**, 1287.

HILLE, B. (1968). Charges and potentials at the nerve surface. *J. Gen. Physiol.* **51**, 221.

HINKE, J. A. M. (1961). The measurement of sodium and potassium activities in the squid axon by means of cation-selective glass microelectrodes. *J. Physiol. (London)* **156**, 314.

HODGKIN, A. L. (1947). The effect of potassium on the surface membrane of an isolated axon. *J. Physiol. (London)* **106**, 319.

HODGKIN, A. L. (1951). The ionic basis of electrical activity in nerve and muscle. *Biol. Rev. Cambridge Phil. Soc.* **26**, 339.

HODGKIN, A. L. (1958). The Croonian Lecture: Ionic movements and electrical activity in giant nerve fibers. *Proc. Roy. Soc.* **B148**, 1.

J. Walter Woodbury *et al.*

HODGKIN, A. L. (1964). "The Conduction of the Nervous Impulse." Thomas, Springfield, Illinois.

HODGKIN, A. L., and HOROWICZ, P. (1959a). Movements of Na and K in single muscle fibers. *J. Physiol. (London)* **145**, 405.

HODGKIN, A. L., and HOROWICZ, P. (1959b). The influence of potassium and chloride ions on the membrane potential of single muscle fibres. *J. Physiol. (London)* **148**, 127.

HODGKIN, A. L., and HUXLEY, A. F. (1952a). Currents carried by sodium and potassium ions through the membrane of the giant axon of *Loligo*. *J. Physiol. (London)* **116**, 449.

HODGKIN, A. L., and HUXLEY, A. F. (1952b). The components of membrane conductance in the giant axon of *Loligo*. *J. Physiol. (London)* **116**, 473.

HODGKIN, A. L., and HUXLEY, A. F. (1952c). The dual effect of membrane potential on sodium conductance in the giant axon of *Loligo*. *J. Physiol. (London)* **116**, 497.

HODGKIN, A. L., and HUXLEY, A. F. (1952d). A quantitative description of membrane current and its application to conduction and excitation in nerve. *J. Physiol. (London)* **117**, 500.

HODGKIN, A. L., and KATZ, B. (1949). The effect of sodium ions on the electrical activity of the giant axon of the squid. *J. Physiol. (London)* **108**, 37.

HODGKIN, A. L., and KEYNES, R. D. (1953). Mobility and diffusion coefficient of potassium in giant axon from Sepia. *J. Physiol. (London)* **119**, 513.

HODGKIN, A. L., and KEYNES, R. D. (1955). The potassium permeability of a giant nerve fibre. *J. Physiol. (London)* **128**, 61.

HODGKIN, A. L., and KEYNES, R. D. (1957). Movements of labelled calcium ions in squid giant axons. *J. Physiol. (London)* **138**, 253.

HODGKIN, A. L., and RUSHTON, W. A. H. (1946). The electrical constants of a crustacean nerve fiber. *Proc. Roy. Soc.* **B133**, 444.

HODGKIN, A. L., HUXLEY, A. F., and KATZ, B. (1952). Measurements of current-voltage relations in the membrane of the giant axon of *Loligo*. *J. Physiol. (London)* **166**, 424.

HOLTZMAN, D. (1967). The effect of low ionic strength extracellular solutions on the resting potential in skeletal muscle fibers. *J. Gen. Physiol.* **50**, 1485.

HOROWICZ, P., and GERBER, C. J. (1965a). Effects of external potassium and strophanthidin on sodium fluxes in frog striated muscle. *J. Gen. Physiol.* **48**, 489.

HOROWICZ, P., and GERBER, C. J. (1965b). Effects of sodium azide on sodium fluxes in frog striated muscle. *J. Gen. Physiol.* **48**, 515.

HOYT, R. C. (1963). The squid giant axon. Mathematical models. *Biophys. J.* **3**, 399.

HUXLEY, A. F. (1959). Ion movements during nerve activity. *Ann. N. Y. Acad. Sci.* **81**, 221.

KATCHALSKY, A., and CURRAN, P. F. (1965). "Nonequilibrium Thermodynamics in Biophysics." Harvard Univ. Press, Cambridge, Massachusetts.

KATZ, B. (1966). "Nerve, Muscle and Synapse." McGraw-Hill, New York.

KEYNES, R. D. (1951). The ionic movements during nervous activity. *J. Physiol. (London)* **144**, 119.

KEYNES, R. D. (1963). Chloride in the squid giant axon. *J. Physiol. (London)* **169**, 690.

KEYNES, R. D. (1965). Some further observations on the sodium efflux in frog muscle. *J. Physiol. (London)* **178**, 305.

KEYNES, R. D., and LEWIS, P. F. (1951). The sodium and potassium content of cephalopod nerve fibers. *J. Physiol. (London)* **144**, 151.

KEYNES, R. D., and MAISEL, G. W. (1954). The energy requirements for sodium extrusion from a frog muscle. *Proc. Roy. Soc.* **B142**, 383.

KEYNES, R. D., and SWAN, R. C. (1959). The effect of external sodium concentration on the sodium fluxes in frog skeletal muscle. *J. Physiol. (London)* **147**, 591.

KOECHLIN, B. A. (1955). On the chemical composition of the axoplasm of squid giant nerve fibers with particular reference to its ion pattern. *J. Biophys. Biochem. Cytol.* **1**, 511.

KOECHLIN, B. A. (1959). The isolation and identification of the major anion fraction of the squid giant nerve fibers. *Proc. Natl. Acad. Sci. U.S.* **40**, 60.

KOPACZYK, K., PERDUE, J. F., and GREEN, D. E. (1966). The relation of structural and catalytic protein in the mitochondrial electron transfer chain. *Arch. Biochem. Biophys.* **115**, 215.

KORN, E. D. (1966). Structure of biological membranes. *Science* **153**, 1491.

LECAR, H., EHRENSTEIN, G., BINSTOCK, L., and TAYLOR, R. E. (1967). Removal of potassium negative resistance in perfused squid giant axons. *J. Gen. Physiol.* **50**, 1499.

LENARD, J., and SINGER, S. J. (1968). Structure of membranes: Reaction of red blood cell membranes with phospholipase C. *Science* **159**, 738.

LETTVIN, J. Y., PICKARD, W. F., McCULLOCH, W. S., and PITTS, W. (1964). A theory of passive ion flux through axon membranes. *Nature* **202**, 1338.

LINDLEY, B. D. (1967). Membrane solvation as a basis for ionic selectivity. *J. Theoret. Biol.* **17**, 213.

LING, G. N. (1962). "A Physical Theory of the Living State: The Association-Induction Hypothesis." Ginn (Blaisdell), Boston, Massachusetts.

LUCY, J. A. (1964). Globular lipid micelles and cell membranes. *J. Theoret. Biol.* **7**, 360.

MACEY, R. I., and OLIVER, R. M. (1967). The time dependence of single file diffusion. *Biophys. J.* **7**, 545.

MARMONT, G. (1949). Studies on the axon membrane. I. A new method. *J. Cellular Comp. Physiol.* **34**, 351.

MEVES, H. (1966). Experiments on internally perfused squid giant axons. *Ann. N. Y. Acad. Sci.* **137**, 807.

MEVES, H., and CHANDLER, W. K. (1965). Ionic selectivity in perfused giant axons. *J. Gen. Physiol.* **48**, 31.

MOORE, J. W. (1965). Voltage clamp studies on internally perfused axons. *J. Gen. Physiol.* **48**, 11.

MOORE, J. W., and COLE, K. S. (1963). *In* "Physical Techniques in Biological Research" (W. L. Nastuk, ed.), Vol. VI, Pt. B, pp. 263–321. Academic Press, New York.

MOORE, J. W., NARAHASHI, T., and ULBRICH, W. (1964). Sodium conductance shift in an axon internally perfused with a sucrose and low-potassium solution. *J. Physiol. (London)* **172**, 163.

MOORE, J. W., NARAHASHI, T., and SHAW, T. I. (1967). An upper limit to the number of sodium channels in nerve and membrane? *J. Physiol. (London)* **188**, 99.

MULLINS, L. J. (1956). *In* "Molecular Structure and Functional Activity in Nerve Cells" (R. C. Grenell and L. J. Mullins, eds.), pp. 123–156. Am. Inst. Biol. Sci., Washington, D.C.

MULLINS, L. J. (1959). An analysis of conductance changes in squid axons. *J. Gen. Physiol.* **42**, 1013.

NAPOLITANO, L., LABANON, F., and SCALETTI, J. (1967). Preservation of myelin lamellar structure in the absence of lipid. *J. Cell Biol.* **34**, 817.

NARAHASHI, T. (1963). Dependence of resting and action potentials on internal potassium in perfused giant axons. *J. Physiol. (London)* **169**, 91.

NARAHASHI, T., and MOORE, J. W. (1968). Neuroactive agents and nerve membrane conductances. *J. Gen. Physiol.* **51**, 93S.

NILSSON, S. E. G. (1965). The ultrastructure of the receptor outer segments in the retina of the leopard frog (*Rana pipiens*). *J. Ultrastruct. Res.* **12**, 207.

O'BRIEN, J. S. (1967). Cell membranes—Composition; Structure; Function. *J. Theoret. Biol.* **15**, 307.

OIKAWA, T., SPYROPOULOS, C. S., TASAKI, I., and TEORELL, T. (1961). Methods for perfusing the giant axon of *Loligo pealii*. *Acta Physiol. Scand.* **52**, 195.

PEASE, D. C. (1964). "Histological Techniques for Electron Microscopy." Academic Press, New York.

RICHARDSON, S. H., HULTIN, H. D., and GREEN, D. E. (1963). Structural proteins of membrane systems. *Proc. Natl. Acad. Sci. U.S.* **50**, 821.

ROBERTSON, J. D. (1960). The molecular structure and contact relationships of cell membranes. *Progr. Biophys. Biophys. Chem.* **10**, 343.

ROBINSON, R. A., and STOKES, R. H. (1959). "Electrolyte Solutions," 2nd ed. Academic Press, New York.

ROJAS, E., and ATWATER, I. (1968). An experimental approach to determine membrane charges in squid giant axons. *J. Gen. Physiol.* **51**, 131S.

RUCH, T. C., and PATTON, H. D., eds. (1965). "Physiology and Biophysics." Saunders, Philadelphia, Pennsylvania.

RYSER, H. J.-P. (1968). Uptake of protein by mammalian cells: An underdeveloped area. *Science* **159**, 390.

SANDBLOM, J. P., and EISENMAN, G. (1967). Membrane potentials at zero current. *Biophys. J.* **7**, 217.

SCHWAN, H. P. (1957). *Advan. Biol. Med. Phys.* **5**, 148–209.

SCHWAN, H. P. (1966). Alternating current electrode polarization. *Biophysik* **3**, 181.

SHANES, A. M. (1958). Electrochemical aspects of physiological and pharmacological action in excitable cells. *Pharmacol. Rev.* **10**, 59.

SJÖSTRAND, F. S. (1963). A new ultrastructural element of the membranes in mitochrondria and of some cytoplasmic membranes. *J. Ultrastruct. Res.* **9**, 340.

STEIN, W. D. (1967). "The Movements of Molecules across Cell Membranes." Academic Press, New York.

STOECKENIUS, W. (1962). Some electron microscopical observations on liquid-crystalline phases in lipid-water systems. *J. Cell Biol.* **12**, 221..

TASAKI, I., and SHIMAMURA, M. (1962). Further observations on resting and action potential of intracellularly perfused squid axon. *Proc. Natl. Acad. Sci. U.S.* **48**, 1571.

TASAKI, I., SINGER, I., and TAKENAKA, T. (1965). Effects of internal and external ionic environment on excitability of squid giant axon. *J. Gen. Physiol.* **48**, 1095.

TASAKI, I., SINGER, I., and WATANABE, A. (1966). Excitation of squid giant axons in sodium free external media. *Am. J. Physiol.* **211**, 746.

TAYLOR, R. E. (1963). *In* "Physical Techniques in Biological Research" (W. L. Nastuk, ed.), Vol. IV, Pt. B, pp. 219–262. Academic Press, New York.

TAYLOR, R. E., MOORE, J. W., and COLE, K. S. (1960). Analysis of certain errors in squid axon voltage clamp measurements. *Biophys. J.* **1**, 161.

TOBIAS, J. M. (1964). A chemically specific molecular mechanism underlying excitation in nerve: a hypothesis. *Nature* **203**, 13.

TOMAN, J. E. P. (1965). *In* "The Pharmacological Basis of Therapeutics" (L. S. Goodman and A. Gilman, eds.), pp. 215–236. Macmillan, New York.

TOSTESON, D. C., ANDREOLI, T. C., TIEFFENBERG, M., and COOK, P. (1968). The effects of macrocyclic compounds on cation transport in sheep red cell and thin and thick lipid membranes. *J. Gen. Physiol.* **51**, 373S.

WHITE, S. H. (1968). Unpublished work.

WHITE, A., HANDLER, P., SMITH, E. L., and STETTEN, JR., D. (1964). "Principles of Biochemistry," 3rd ed. McGraw-Hill, New York.

WILLIAMS, J. A. (1966). Effect of external K^+ concentration on transmembrane potentials of rabbit thyroid cells. *Am. J. Physiol.* **211**, 1171.

WOLFGRAM, F. (1967). The amino acid composition of some non-neural proteolipid proteins. *Biochim. Biophys. Acta* **147**, 383.

WOODBURY, J. W. (1952). Direct membrane resting and action potentials from single myelinated nerve fibers. *J. Cellular Comp. Physiol.* **39**, 323.

WOODBURY, J. W. (1962). *In* "Handbook of Physiology, Section 2, Circulation" (W. F. Hamilton, ed.), Vol. I, pp. 237–286. Am. Physiol. Soc., Washington, D.C.

WOODBURY, J. W. (1963). Interrelationships between ion transport mechanisms and excitatory events. *Federation Proc.* **22**, 31.

WOODBURY, J. W. (1965). *In* "Physiology and Biophysics" (T. C. Ruch and H. D. Patton, eds.), Chapters 1 and 2, pp. 1–58. Saunders, Philadelphia, Pennsylvania.

YAMAMOTO, T. (1963). On the thickness of the unit membrane. *J. Cell Biol.* **17**, 413.

Chapter 12

Kinetics of Reactions with Charge Transport

V. G. Levich

I. Reactions with Charge Transport in Solutions 986

II. Physical Pattern of Reaction with Electron Transfer in Polar Solvents without the Breaking of Chemical Bonds 989

III. Electrode Reactions and Electrode Current 998

IV. Model of a Polar Solution 1004

V. Adiabatic Approximation and Quantum Transition Probability in a System Consisting of Heavy and Light Particles 1010

VI. Probability of Electron Transfer 1015

VII. Transition Probability in Reactions without Chemical Bond Breaking . . 1019
  A. High Temperature Case 1020
  B. Low Temperature Case 1020

VIII. Flow of Current through the Metal–Solution Interface 1024

IX. Flow of Current through the Semiconductor–Solution Interface 1031

X. Physical Pattern of Charge Transfer Reaction in a Polar Medium, Which Is Accompanied by the Breaking or Forming of Chemical Bonds 1039

XI. Quantitative Description of a Charge Transfer Reaction with Chemical Bond Breaking 1041

XII. Transition Probability 1046

XIII. Current Density 1049
  A. Low Overvoltage 1051
  B. "Normal" Overvoltage 1053
  C. High Overvoltage 1054

XIV. Other Cases of Electrode Reactions 1060

XV. Critical Consideration of Theoretical Concepts about the Mechanism of the Elementary Act of the Electrode Process 1063

XVI. Conclusions 1070
  References 1072

I. Reactions with Charge Transport in Solutions

An idea which was developed in our researches and which we shall stress here, implies that the mechanism of the elementary act in electrode processes is closely connected with bulk reactions involving charge transfer (the charge is carried either by the electron or the proton).

At present it proves to be possible to formulate a quantitative theory for this wide class of reactions, this theory being in satisfactory agreement with the experiments.

In the course of presentation of problems of chemical reactions in solution, this theory is usually used to distinguish strictly between the case of homogeneous and of heterogeneous (electrode) reactions with charge transfer. The theory shows (see below) that the elementary act of all reactions with charge transfer, both bulk reactions and those at the electrode, are common in nature. Of course, this does not mean full identity of bulk and electrode reactions. Electronic reactions always require a reservoir of electrons—the electrode, which has its own important specificity, which should be taken into account.

When speaking about reactions with charge transfer we mean the reactions accompanied by the transport of an electron or a proton. By redox reactions (homogeneous or electrode ones) the transfer of one or several electrons takes place. In some cases, as for example, in the following reactions,

$$[Fe(CN)_6]^{3-} \rightleftarrows [Fe(CN)_6]^{4-}$$
$$MnO_4^- \rightleftarrows MnO_4^{2-} \qquad (1.1)$$
$$[Mo(CN)_8]^{3-} \rightleftarrows [Mo(CN)_8]^{4-}$$

(reactions of this kind we shall call the simplest ones), no chemical bond breaking occurs. In these reactions the relation between the redox reaction and the electron transfer appears in a most direct way. In other reactions, e.g.,

$$Fe^{2+} \rightleftarrows Fe^{3+}, \qquad Ce^{3+} \rightleftarrows Ce^{4+}, \qquad (1.2)$$

though no chemical bond breaking is occurring, the quasichemical bond between the ions and the surrounding water molecules is broken. It appears, e.g., that Fe ions exist as complexes $[Fe(H_2O)_6]^{3+}$ and $[Fe(H_2O)_6]^{2+}$ with various parameters (e.g., distances between the iron and the oxygen). By redox reaction a corresponding transformation of a complex occurs.

In more complicated cases the redox transformations are connected not with electron transfer but, e.g., with oxygen atom transfer as it takes place in the reaction (Frölich, 1954)

$$ClO^- + NO_2^- \rightarrow Cl^- + NO_3^-.$$

Reactions of this kind will not be considered here.

Reactions with proton transfer proceed according to the general scheme

$$AH^+ + B^- \rightleftarrows A + BH. \tag{1.3}$$

BH, as well as A, may be charged or neutral particles.

Bulk reactions with electron transfer may be compared to the electrode reactions, e.g.,

$$[Fe(CN)_6]^{3-} + e(Me) \rightleftarrows [Fe(CN_6)]^{4-}$$

where e(Me) is the electron delivered by a metal. Electrode reactions with proton transfer,

$$H_3O^+ + e(Me) \rightarrow H_{ads} + H_2O, \tag{1.4}$$

represent the slow stage of hydrogen ion discharge and belong to the most important reactions of electrochemistry.

The theory of reaction kinetics in solutions is faced with the following problems:

(1) The clarification of the mechanism of reactions with charge transfer, in particular the explanation of the role of the polar solvent in the kinetics of these reactions.

(2) The derivation of the theory to explain empirical laws and the determinations of their range of applicability.

(3) The calculation of reaction velocity constants depending on the properties of the medium and the reacting particles; this calculation must include calculation of the activation energy as well as the pre-exponential factor (the latter at least to the order of magnitude).

There are some general empirical laws and facts that have been established for a very wide class of bulk reactions in solution. These are the following:

(1) The Arrhenius law, according to which the velocity constant k depends on the temperature T by an exponential law:

$$k = A \exp[-(E_a/kT)]. \tag{1.5}$$

The preexponential factor A depends comparatively weakly on the temperature, and differs essentially from the pair collision number between the reacting particles. It means that only a few particles possessing energy, which exceeds E_a, really react in the course of every collision. At low temperatures, in some cases, deviations from the Arrhenius law are observed.

The existence of an activation energy means that at small distances there are effective repulsive forces which act between the reacting particles. These forces may be of various kinds. In the case of identically charged ions, one of these forces is the electrostatic repulsion. However, even in the absence of electrostatic forces the bringing together of particles to the reaction distance is hindered in polar solvent by the solvation shells which surround the particles.

(2) For a series of reactions of similar type the activation energy proves to be connected with the reaction heat Q by the following relation:

$$\alpha = -(\partial E_a / \partial Q) \tag{1.6}$$

where α is a slowly varying function of Q. Thus in a sufficiently narrow interval of variation of the reaction heat, a linear relation between E_a and Q may be used:

$$E_a = -\alpha Q + E_0. \tag{1.7}$$

The value of α varies from unity to zero when Q varies over a sufficiently wide interval. Relation (1.6), known as the Brönsted rule, was originally established for bulk reactions with proton transfer, which represents the main stage of base–acid catalysis. However, a similar relation is valid also for reactions with electron transfer. The special importance of the Brönsted relation is determined by the fact that it connects two characteristics of a reacting system, which appear at first glance to be independent, namely, the activation energy having the kinetic sense, and the reaction heat representing a thermodynamic quantity.

(3) The last fact which is basic to any theory of reactions in solutions is the following one: the reacting particles interact strongly with a polar solvent. This follows immediately from comparing the reaction velocities between particles which are similar or different in nature, e.g.,

$$Cr^{3+} + e \rightleftarrows Cr^{2+}$$

or

$$Fe^{3+} + e \rightleftarrows [Fe(OH)]^{2+}.$$

If one compares these reactions with analogous reactions in gases, one immediately finds out an essential difference between these reactions. The velocity of exchange reactions in gases, which occurs between particles of similar nature ($A^- \rightleftarrows A$), exceeds by many orders of magnitude the velocity of the same reaction between various particles ($A \rightleftarrows B$). The velocity is connected with the possibility of resonant electron transition between similar particles. If particles A and B are different, so are the energy levels for electrons in both particles. The energy difference must pass to other degrees of freedom or turn into radiation. This sufficiently lowers the probability of the relevant process. In the case of similar particles A^- and A, this difficulty is removed and a resonance transition occurs with large probability.

No such resonance phenomena are observed in solutions. Velocities of electron exchange reactions do not differ. The absence of resonance phenomena means that all the systems of ion energy levels are displaced by interaction with the solvent, which is different for ions which are similar in nature but differ in charge.

In Section II we shall put in correspondence with these empirical laws of homogeneous reactions analogous laws for electrode reactions.

The establishment of a mechanism of reaction with charge transfer and the obtaining of general laws (1.5) and (1.6) for them proved to be possible (at least, for simple reactions of this class) on the basis of the quantum-mechanical theory to be presented.

II. Physical Pattern of Reaction with Electron Transfer in Polar Solvents without the Breaking of Chemical Bonds

Before passing to a systematic presentation of theory, it is useful to imagine qualitatively the pattern of phenomena occurring in the process of redox reactions in polar solvents. It is necessary to emphasize that we mean the general scheme of phenomenon, which is characteristic of all reactions with charge transfer. However, one must have in mind that various reactions, as will be seen from the following, may possess rather essential peculiarities, which have to be taken into account in quantitative theory.

To simplify the arguments, we start with the consideration of a reaction with electron transfer, which goes on without the breaking or forming of chemical bonds. Let us consider two similarly charged ions, which may enter into a redox reaction. If this reaction is not accompanied by the breaking or forming of chemical bonds, all the changes are reduced to an electron transfer from the oxidizing ion to the reducing one. Electrostatic interaction does not permit the ions to approach to a distance of less than r_0, where r_0 is determined by the condition

$$\frac{e^2}{\epsilon_0 r_0} \sim kT$$

with the static dielectric constant of the medium ϵ_0.†

† In a strong molecular field, $\epsilon_0 \sim 45$ or less.

We may write for the velocity of reaction k:

$$k = c_1 c_2 \int_{r_0}^{\infty} W(R) p(R) \, dR \qquad (2.1)$$

where c_1, c_2 are the average ion concentrations in solution, $p(R)$ is the probability that ions approach to a distance R, and $W(R)$ is the probability (per unit of time) that an electron performs the transition between the ions separated by the distance R. The transition probability $W(R)$ depends not only on distance but also on a number of other parameters. Since the transition probability decreases sharply with increasing R (see below), whereas $p(R)$ generally speaking increases with R, some most probable distance R_{cr} can be found which corresponds to the maximum of

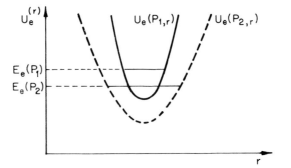

FIG. 1. Electron energy as a function of the distance to the ion nucleus at different values of solvent polarization.

the integrand. Then for the velocity constant one may write the following expression:

$$k = c_1 c_2 p(R_{cr}) W(R_{cr}). \qquad (2.2)$$

The main problem is to find the transition probability $W(R)$. Let $\mathbf{R}_1$ and $\mathbf{R}_2$ be the coordinates of ion nuclei in the oxidized and the reduced forms, so that $R = |\mathbf{R}_1 - \mathbf{R}_2|$. Let us represent schematically the potential energy of electrons in the vicinity of one of the ions (Fig. 1). The figure shows the energy levels occupied by electrons as horizontal lines. In solution the interaction changes the potential energy and displaces the energy levels, so that the curve shown in Fig. 1 differs from the potential theory of an electron on the same ion in vacuum. Thus, as the molecules of solvent possess dipole moments (we should recall that only the case of polar liquids is being considered), there is an electrostatic interaction between the charged ion and the molecules surrounding it. We may say that the potential energy of electron U_e in an ion introduced into a solvent depends on the polarization $\mathbf{P}$ (dipole moment in unit volume) of the latter near the ion, i.e., $U_e = U_e(\mathbf{P})$; by this $U_e(\mathbf{P})$ differs from the potential energy in vacuum $U_e(0)$. Accordingly, the energy levels of electrons in an ion will depend on the polarization $\mathbf{P}$, i.e., $U_e = U_e(\mathbf{P})$ and $\epsilon = \epsilon(\mathbf{P}) \neq \epsilon(0)$, where $\epsilon(0)$ is the energy level of a given ion in vacuum.

Molecules of solvent are in continuous thermal movement, and the polarization $\mathbf{P}$ and the potential energy $U_e(\mathbf{P})$ change in time. The solid and dotted lines of Fig. 1 may be said to represent an instant photograph corresponding to various moments of time when the polarization had the values $\mathbf{P}_1$ and $\mathbf{P}_2$.

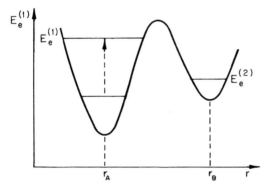

FIG. 2. Electron energy as a function of the distance between the nuclei of two ions.

If we compare the possibilities of transition of an electron from the reduced to the oxidized form of the same ion in vacuum and in solution, they prove to be quite different. In vacuum the transition of electron may go from the level $\epsilon(0)$ on the vacant level with the same energy. The energy levels of the reduced and oxidized form of the same ion do not coincide. Due to strong interaction with the solvent (which varies with the ion charge), the energy level $\epsilon(\mathbf{P})$ in the reduced form appears to be displaced into some position ϵ_{red} and in the oxidized form into another position ϵ_{ox}. Thus, even in the case of transition between ions of similar nature, the resonant transition in solution is impossible. In solution the transition of electron between ions is always accompanied by a change of the electronic energy of the system, giving $\Delta\epsilon = \epsilon_{red} - \epsilon_{ox}$ (Fig. 2). The law of conservation of energy requires that this energy must pass to some external system. Thus, three kinds of transition are possible in principle.

(1) *Transition with radiation.* The energy difference $\Delta\epsilon$ is radiated. Quantum-mechanical calculation shows immediately that such a transition has a low probability W.

(2) *Transition through excited states.* Electrons in reduced states as a result of thermal agitation pass into the region of high energies, where the energy spectrum is almost continuous. In this region the energy spectrum of the oxidized form always exists in a level with the same energy, as in the reduced form. After agitation the transition between states of equal energy occurs. The probability of thermal agitation is given by the Gibbs factor

$$\rho(\epsilon) = \text{const} \exp \Delta\epsilon/kT.$$

Electronic excitations are of the order of a few electron volts, whereas $kT \sim 25 \cdot 10^{-3}$ eV. It is clear that the transition according to such a scheme is of low probability. However, we shall see that there are some reactions in which excited states play an important role.

(3) *The only remaining possibility is closely connected with the existence of interaction of ions and solvent.* In Figs. 3A and 3B the potential energy and electron energy levels of two ions are plotted as functions of distance. Below, the dipoles surrounding ions are represented schematically by arrows. As a result of fluctuation (see the dipole's position), the

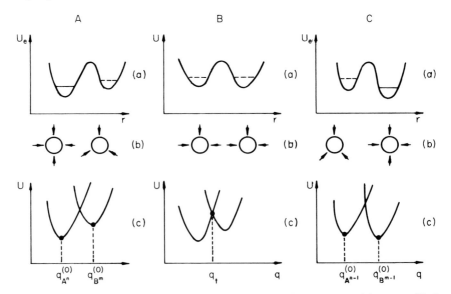

FIG. 3. The comparison (schematic) between (a) the electron potential energy, (b) the polarization of solvent, and (c) the electron term in the (A) initial, (B) transition, and (C) final states of the system.

position of level ϵ_{red} is changed in such a way that it was displaced into position ϵ_{ox}. In this case the transition of an electron without radiation becomes possible. The probability of such a transition, which we shall call a radiationless transition, is comparatively great if interaction of the ion with the medium is strong enough. After the transition of an electron there arises a completely nonequilibrium situation—polarization of the solvent by ions with different charges turns out to be the same. This is the reason why fluctuation of polarization decays and the system goes over to a final state represented in Fig. 3C. It is clear that the probability W of the electronic transition (or the velocity of the redox reaction) depends not only on the height of the energetic barrier which is overcome by the electron traveling from one ion to another but also on the width of this barrier determined by the distance between ions. This probability depends as well upon the velocity of the level with energy ϵ_{red} passing by the level with energy ϵ_{ox} (it would be equally right to speak about ϵ_{ox} level passing by the level ϵ_{red}). If this velocity of traveling is arbitrarily high, the electron has no time to perform the transition from one ion to the other, i.e., the probability of transition tends to zero. If, on the contrary, the velocity of passing is arbitrarily small, so that the system stays for an indefinitely long time in the state represented in Fig. 3B, the electron may occupy with equal probabilities either the left or the right potential hole. In this case, the transition probability is equal to unity. In actual cases the movement of levels goes on with finite velocity, and so the transition probability lies between zero and unity.

It is useful for what follows to compare this transition with the usual tunneling transition. Let the constant in time potential field $U(x)$ be fixed, this field having the form of a barrier presented in Fig. 4. If the free

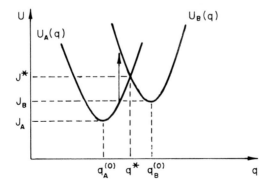

FIG. 4. Two electron terms of the system composed of two ions, corresponding to different solvent states. Abscissa, generalized coordinates characterizing the solvent state. Ordinate, values of the electron terms.

particles having energy E smaller than the barrier height falls on the barrier from the left, then the particle will penetrate this barrier with a probability given by Gamov's formula

$$W \sim \exp\{2/\hbar) \int_{x_1}^{x_2} [2m(E - U(x))]^{1/2} \, dx\}.$$

It is assumed that the form of the barrier is given and is independent of the time and remains constant during the penetration of the particle. It is clear that such a penetration differs qualitatively from the one considered above. In the course of a radiationless transition the form of the barrier is continuously changing during the very process of transition (compare the initial state, Fig. 3A, and the final state, Fig. 3C), whereas the probability of transition depends on the velocity of traveling of level ϵ_{red} by level ϵ_{ox}, i.e., on the time of the existence of the situation corresponding to Fig. 3B. The whole process of transition has an essentially nonstationary nature. The attempt to describe the radiationless transition with the aid of Gamov's formula makes no sense and may cause considerable theoretical as well as practical errors.

While describing reactions with electron transfer which is treated as radiationless transition we have made use of the potential energy curves for the electrons as a function of the distance from the nuclei. It is more convenient to introduce the concept of an electronic term on the surface of the potential energy of the system considered as a whole.

Let us consider for this purpose a plot of an electron term. (A term is known to be the total potential energy of a system for given values of the parameters on which this energy depends.) In a diatomic molecule, for instance, the electron term is the total potential energy of the molecule for a fixed distance between nuclei. A change in this distance involves a change in the potential energy of the molecule, which is thus dependent on the internuclear distance as a parameter.

In our case, the electron term is the total energy of the whole electron system in the field of the ion and of the surrounding medium, the energy of the interaction between ions, and the energy of the polar solvent. Thus, in our case, the electron term refers to the total energy of the system (electron + ions + solvent) with the exception of the kinetic energy of the heavy particles (ions and water dipoles).

Since polarization of the solvent in the vicinity of the ion depends on the electric field of the latter, the electron terms of ions with different charges will be different. Figure 4 shows two electron terms $U(q)$ for two polarization states of the solvent. The left-hand curve is the term of the system when the electron is localized near ion A^n; the right-hand curve is the term of the system when the electron is localized near ion B^m.

The solvent state is graphically represented by one coordinate q. In reality, the solvent state should be characterized by a set of coordinates corresponding to the number of its degrees of freedom. Therefore, $U(q_1, q_2, \ldots, q_N)$ is a surface in a multidimensional space and Fig. 4 should be considered as a schematic representation of the true picture.

The value of the "solvent coordinate" $q_A^{(0)}$ corresponds to the equilibrium state of the solvent near ion A^n; $q_B^{(0)}$ has a similar meaning. Electron transfer corresponds to the transition of the system from the state with the energy $\mathscr{T}_A = U(q_A^{(0)})$ to that with the energy $\mathscr{T}_B = U(q_B^{(0)})$. The intersection point of the terms† corresponding to the "coordinate" q^* is designated as U^*.

We confine ourselves to nonradiative transitions (type 1). Later, it will be proved that nonradiative transitions can be of two kinds, as it is shown in Figs. 5 and 6.

(a) The system can pass from the point $q_A^{(0)}$ to the point q^* (Fig. 5) moving continuously along the potential curve $U_A(q)$. At the intersection point q^* the system passes to the term $U_B(q)$, and again moves continuously along the curve $U_B(q)$ to the point $q_B^{(0)}$. The path of the system is shown in Fig. 5 by thick lines and arrows. This is a usual, classical transition of the system, which is in keeping with the theory of the transition state. Point q^* corresponds to the transition state of the system.

It is evident that the difference

$$\Delta E^* = \mathscr{T}^* - \mathscr{T}_A$$

should be considered to be the activation energy of the process. The transition of the system from point $q_A^{(0)}$ to point q^* requires a supply of energy from without, in this case from the surrounding medium—the solvent. Therefore, some thermal fluctuation in the surrounding solvent should correspond to the transition

$$q_A^{(0)} \rightarrow q^*.$$

This shows that the classical method that is described corresponds to sufficiently high temperatures. Quantitative estimates for the transition probability will be given in Section VII.

(b) The system passes from point $q_A^{(0)}$ to point q^t, moving continuously along the curve $U_A(q)$. Then it follows a horizontal curve, passing from

---

† In the case under consideration the wavefunctions of the electron in both states are not orthogonal. Therefore, the curves of the terms in Fig. 4 are not identical to the usual curves of the terms for diatomic molecules.

term $U_A(q)$ to term $U_B(q)$, and arrives directly at point $q_B^{(0)}$ (Fig. 6). The horizontal section of the path does not correspond to the motion along the term and, therefore, cannot be described by the classical method. In this case, the system of nuclei and electrons is not in a state with a definite energy. If we wished to describe the system using the usual nomenclature, such a transition should be called a "tunnel

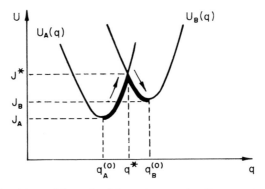

FIG. 5. Schematic picture of the path of the system passing from one term to the other. Abscissa, generalized coordinates. Ordinate, electron terms.

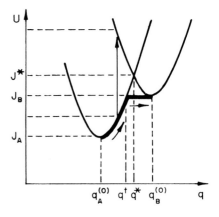

FIG. 6. Schematic picture of the path of the system passing from one term to the other (low-temperature transition). Abscissa, generalized coordinates. Ordinate, electron terms.

transition of the system." In contrast to the classical path of the transition through point q^*, in the case of the "tunnel transition of the system," the activation energy is equal to

$$\varDelta E^t = \mathscr{T}^t - \mathscr{T}_A$$

and less than $\varDelta E^*$. Therefore, such transitions need a smaller thermal fluctuation and are important at lower temperatures. Naturally, it does not mean that at lower temperatures the probability of the "tunnel transition of the system" is greater than that of a transition following the classical path. The probability of a "tunnel transition of the system" will be determined in Section VII.

"Tunnel transitions of the system" should not be confused with the usual tunnel transition of the electron. It should be borne in mind that the curves of the potential energy of the whole system (terms) in Figs. 4–6 show graphically the state of the whole system and not that of the electron. Therefore, the curves in Figs. 4–6 do not represent the behavior of the electron during the change in the state of the system.

It is important to understand correctly the correspondence between the curves of the terms $U(q)$ and the energies of electrons $U_e(r)$ in the process of transition. The polarization q^* corresponds to equal electron energy E_e in both holes. Simultaneously with the electron energy, the values of both terms $U_A(q^*)$ and $U_B(q^*)$ are equalized at the point $q = q^*$, which corresponds to the intersection point. It is clear that a tunnel-electron transition may take place only in a situation corresponding to the position of the system in the point q^* (Fig. 5). In the case of a transition shown in Fig. 6 and called a "tunnel transition of the system," the usual tunnel transition is impossible.

When the system moves along the horizontal section in Fig. 6, the potential energy of the whole system is constant, but the electron energy is not. The change in the electron energy from $E_e^{(1)}$ to $E_e^{(2)}$, for the scheme in Fig. 6, is shown by a vertical arrow. In fact, when the position of all heavy particles is fixed, i.e., at a constant value of q, the change of energy during the transition between the terms is equal to that of the electron energy. The relationship between the curves of the terms and the electron energy is graphically shown in the scheme of Fig. 3. Each top diagram represents $U_e(\mathbf{r})$ for two ions, and the arrows below it schematically show molecules of the polar solvent surrounding both ions. Different values of the coordinate q correspond to different polarization values. The bottom diagram shows the corresponding term, the points on it representing the state of the system. The polarization near ion A is large after transition; the energy level of the electron near ion A is lower than that near ion B. The transition $q_A^{(0)} \rightarrow q^t$ corresponds to the equalization of polarization of both ions and to the equality of energy levels. The electron is transferred to ion B at point q^t and the fluctuation of polarization disappears. In the final state, ion B has a larger polarization and the energy of the electron near ion B is lower than it would be if the electron were near ion A.

In conclusion, it should be noted that it would be wrong to consider the "tunnel transition of the system" as the tunnel transition of the nuclei. The tunnel transition, as we understand it, presumes the motion of the particle to occur in an external field, forming a barrier. Incidentally, heavy nuclei themselves form a field. Their motion, whether classical or nonclassical, changes the field distribution, and cannot be considered on the basis of the scheme of tunnel transition.

III. Electrode Reactions and Electrode Current

The special feature of electrode processes is that all these processes are accompanied by charge transfer through the electrode–electrolyte boundary surface. The electrode serves as an electron reservoir in these processes.

For many years the physical properties of this reservoir were not dealt with by electrochemists. However, the development of semiconductor electrochemistry several years ago demonstrated the importance of a correct knowledge of the electrode properties, in particular, of the electron energy distribution (Myamlin and Pleskov, 1967; Vdovin et al., 1949).

This chapter deals only with the proper stage of the electrode process, i.e., the charge transfer through the boundary interface electrode solution. We shall not be concerned with the complications connected with processes of diffusion and migration as well as with such processes as crystallization, oxidation, or passivating, surface film formation, or emission of gases. We shall consider first the redox reactions as well as the discharge reaction of hydrogen ions.

Let us consider a big electrode, a metallic or semiconducting one, having the interface boundary with solution which contains ions— cations or anions. Let us assume that the electrode reaction is connected with the transfer of one electron:

$$A^+ + e(\mathscr{E}l) \rightleftarrows A.$$

Here $e(\mathscr{E}l)$ designates the electron which is delivered into the reaction by the electrode. The problem of many-stage reactions will be discussed later. The general expression for electrode current density may be represented as a difference between cathode and anode currents

$$i = i_k - i_a. \tag{3.1}$$

Let us consider every item independently. We may write the following general expression for the cathode current:

$$i_k = e \int c(x; \varphi) W(x, \epsilon; \varphi) n(\epsilon) \rho(\epsilon) \, d\epsilon \, dx. \tag{3.2}$$

Here $c(x; \varphi)$ denotes the concentration of reacting particles A^+, which are removed a distance x from the electrode surface (the latter is taken as $x = 0$ surface). The ion at a distance x is subjected to the action of an electric field with the potential $\varphi(x)$. $n(\epsilon)$ is the number of electrons in the electrode volume and having the energy in the interval ϵ, $\epsilon + d\epsilon$. $\rho(\epsilon)$ is the number of energy levels of the electrode in the same interval. $W(x, \epsilon; \varphi)$ represents the probability (related to the unit of time) that the electron occupying the level ϵ in the bulk of the electrode will go over to the ion removed from the electrode to the distance x. The probability of this transition generally speaking depends on the potential of the electric field acting on the ion. The integration is carried over all possible distances, i.e., $0 \leq x < \infty$, and the whole energy spectrum of the electrons. Correspondingly, the anode current density from the electrolyte into the electrode may be represented as follows:

$$i_a = e \int c(x; \varphi)[1 - n(\epsilon)]W(x, \epsilon; \varphi)\rho(\epsilon) \, d\epsilon \, dx. \qquad (3.3)$$

The factor $[1 - n(\epsilon)]$ shows that the transition may occur only to an unoccupied level in full accordance with the Pauli exclusion principle. The electron distribution function $n(\epsilon)$ in the metal conductors is well known. This function is represented by the Fermi distribution law

$$n(\epsilon) = \{\exp[(\epsilon - \epsilon_F)/kT] + 1\}^{-1} \qquad (3.4)$$

where the Fermi energy ϵ_F is a function which varies slowly with temperature. In the quasi-noninteracting approximation for electrons

$$\epsilon_F = \epsilon_0 - \tfrac{1}{6}\pi^2(kT)^2\left[\frac{\partial}{\partial\epsilon} \ln \rho(\epsilon)\right]_{\epsilon = \epsilon_F} \qquad (3.5)$$

where the constant ϵ_0 is about 1 to 2 eV for metals. At room temperatures the following relation is usually valid for metals:

$$\epsilon_F \simeq \epsilon_0. \qquad (3.6)$$

The quasi-noninteracting approximation accounts well for properties of the function $n(\epsilon)$ when the energy values are near the Fermi energy. The electrons energy distribution appears to be similar to a step function: for $\epsilon < \epsilon_F$ every state is occupied by one electron, for $\epsilon > \epsilon_F$ the Fermi distribution turns into the classical Maxwell distribution.

The density of electron levels in metals represents a slowly varying function of energy. In the same quasi-noninteracting (quasi-free) electron approximation we have

$$\rho(\epsilon) \simeq (mV/4\pi^2\hbar^3)[2n(\epsilon - E_0)]^{1/2} \tag{3.7}$$

where E_0 is the depth of the potential box in the metal and V is the volume of the metal.

The characteristic feature of the energy spectrum in semiconductors with intrinsic conductivity is the presence of a wide gap in the density of states as compared to kT. All the states are divided into two zones—the valence zone filled with electrons and the conductivity zone. There are only a few electrons in the latter zone, and for these electrons the Fermi distribution again turns into the Maxwell distribution

$$n(\epsilon) \simeq \exp[-(\epsilon - \epsilon_F)/kT]. \tag{3.8}$$

In the valence zone the distribution of holes also has the form of a Maxwell distribution. Consequently the probability of finding an electron removed from the Fermi level by an amount $|\epsilon - \epsilon_F|$ above this level equals the probability of finding a hole removed the same amount $|\epsilon - \epsilon_F|$ below it. The Fermi level in an intrinsic semiconductor lies in a forbidden zone. Densities of electron and holes near zone edges are given by the expression

$$\rho_e(\epsilon) = (m_e V/4\pi^2\hbar^3)[2m_e(\epsilon - E_0)]^{1/2} \tag{3.9}$$

and

$$\rho_h(\epsilon) = (m_h V/4\pi^2\hbar^3)[2m_h(\epsilon - E_0)]^{1/2} \tag{3.10}$$

for the conductivity zone ρ_e and the valence zone ρ_h, where m_e and m_h are the effective masses of electron and hole, respectively.

In the case of impurity semiconductors one may also write the general formulas for $n(\epsilon)$ and $\rho(\epsilon)$.

The electron and energy level distributions are not only calculated but also determined experimentally. Thus, those factors in the general expressions (3.2) and (3.3) which depend on electrode properties may be considered in some cases as quantitatively known. In any case, their qualitative behavior is always known. The behavior of the function $c(x; \varphi)$ is known less exactly. Numerous electrochemical data indicate that the distribution of potential (and correspondingly, the concentrations of reacting particles in the electric double layer near the boundary surface) affects essentially the rate of the charge transfer process. These topics are discussed in detail in other chapters of this volume. Here we indicate only some characteristic features of the function $c(x; \varphi)$. First, there exists a plane $x = x^*$ of minimal possible distance to which hydrated ions can approach the electrode surface. In fact, there is a plane at which the electron transfer in the course of the electrode reaction occurs. In the case of high concentration of the solution the quantity $c^* \equiv c(x^*, \varphi^*)$ is near the bulk concentration c_0. However, for low concentrations the difference between c^* and c_0 may be a significant one. For this we have

$$c^* \simeq c_0 e^{-(e\psi_1/kT)} \tag{3.11}$$

where ψ_1 is the potential at the boundary between the diffuse and the Helmholtz parts of the external covering of the double layer. If the transfer process takes place between ions which are charged similarly to the charge near the electrode surface, then $c(x; \varphi)$ decreases rapidly with the increase in the distance from the electrode. We do not intend to discuss the methods of calculation and measuring of the quantity c^* in this chapter. The central problem of the theory consists of determining the transition probability $W(x, \epsilon; \varphi)$. Anticipating the results to be derived later, we make use of one of the general features of $W(x, \epsilon; \varphi)$, namely that this function falls off rapidly with increasing distance between the electrode and that particle on which (or from which) the electron is transferred, since this electron transfer at great distances into the bulk of the solution is not very probable. This property of the function $W(x, \epsilon; \varphi)$ enables us to simplify essentially the general expressions for i_k and i_a. Namely we may write approximately

$$i_k \simeq ec^* \int W(x^*, \epsilon; \varphi^*)n(\epsilon)\rho(\epsilon) \, d\epsilon, \tag{3.12}$$

and an analogous relation exists for i_a. Here φ^* denotes the potential in the plane of maximal approach.

If the surface potential is far from the equilibrium one, one of the following inequalities holds: $i_k \gg i_a$ or $i_k \ll i_a$. Here we may introduce the rate constants of cathode and anode electrode reactions:

$$k = \frac{i_k}{ec^*} = \int W(x^*, \epsilon; \varphi^*)n(\epsilon)\rho(\epsilon) \, d\epsilon \tag{3.13}$$

$$k' = \frac{i_a}{ec^*} = \int W(x^*, \epsilon; \varphi^*)[1 - n(\epsilon)]\rho(\epsilon) \, d\epsilon. \tag{3.14}$$

This is a convenient form for comparing the rates of the bulk and the electrode reactions involving charge transfer.

The rates of electrode reactions, i.e., the rates of the processes with charge transfer through the boundary surface between the electrode and the solution, are given by the empirical formula of Tarel. In the general case, when any essential concentration of the neutral electrolyte is absent, the electrode current density is expressed by Tafel's empirical formula

$$i = i_0 \left[\exp\left\{ \frac{\alpha(\eta) e \eta}{kT} \right\} - \exp\left\{ -\frac{(1 - \alpha(\eta)) e \eta}{kT} \right\} \right]$$

$$\cdot \exp\left\{ -\frac{\alpha(\eta)[e\psi_1(\eta) - \psi_1(eq)]}{kT} \right\}. \quad (3.15)$$

Here i_0 is the exchange current,

$$i_0 = (i_k)_{eq} = (i_a)_{eq}. \quad (3.16)$$

Overvoltage is defined as $\eta = \varphi - \varphi_{eq}$, where φ and φ_{eq} are the true potential and the equilibrium potential of the electrode and $\psi_1(\eta)$ is the potential at the boundary between the dense and the diffuse parts of the electric double layer and generally depends upon the overvoltage. The quantity $\alpha(\eta)$, called the transport coefficient, is the function of over-voltage (it is not a constant, contrary to what is usually accepted in the electrochemical handbooks). The variation of $\alpha(\eta)$ lies in the interval

$$0 < \alpha(\eta) < 1. \quad (3.17)$$

Although $\alpha(\eta)$ is a rather slowly varying function of overvoltage, the very fact of this variation is of fundamental importance for some processes. If a considerable concentration of indifferent electrolyte is present, then $\psi_1 \simeq \psi_1(eq)$ and the expression (3.15) is slightly simplified. With some modifications which are due to the necessity of accounting for the potential distribution in the bulk of the semiconductor, Tafel's formula may be written and is valid also for reactions on the semiconducting electrode. The fact that the voltage–current characteristic for the large majority of charge transfer processes is represented by the empirical relation (3.15) over a wide range of the overvoltage changes can by no means be accidental. On the contrary, it gives evidence that the mechanism of these processes have a common nature, although they include such different reactions as, e.g.,

$$Cl^- \rightarrow Cl + e(\mathscr{E}l)$$

or

$$H_3O^+ + e(\mathscr{E}l) \rightarrow H_2O + H_{ads}.$$

The purpose of the theory is to establish this common mechanism of electrode reactions. Another important problem of the theory is to establish the similarity and difference between the electrode and the bulk reactions with charge transfer. The fact that the mechanism of these processes is very similar in nature was established rather long ago (Frumkin, 1932). Particularly, it is not difficult to see that the Tafel law

(3.15) represents nothing else but a combination of the activation formula with the Brönsted rule. Really, the caloric effect of the electrode reaction Q may be represented in the following form:

$$Q = Q_0 + e\,\Delta\varphi = Q_0 + e\eta. \tag{3.18}$$

This relation shows that a part of the caloric effect is equal to the variation of the energy with charge transfer from the solution (at potential $\varphi = 0$) to the electrode (at potential φ). Then from (3.18) and (1.13) the Tafel's law follows:

$$i = A\exp[-(E_a/kT)] = \text{const}\,\exp[-(\alpha Q/kT)]$$
$$= \text{const}\,\exp[-(\alpha e\eta/kT)].$$

This reasoning is somewhat rough, and it will be made more precise subsequently.

Thus, we see that the electrode reactions as well as the bulk reactions obey in general the same experimental laws. On the other hand, while comparing the general theoretical formulas for the rates of the electrode and the bulk reactions, we see that the main factor determining the mechanism of the process, i.e., the probability of the charge transfer process, is common for any reaction of this kind. The difference arises from the necessity of accounting for the electrode properties, and first of all for the presence of a continuous energy spectrum of the electrode. This circumstance is frequently of comparatively minor importance. Thus from the most general considerations it follows that the nature of the bulk reactions with charge transfer do not differ from the electrode processes of the same kind. The latter are, however, more amenable to theoretical analysis as well as to experimental investigation for a series of features. From the theoretical point of view, the convenience of electrode reactions is due to the fact that microscopic properties of metallic and semiconducting electrodes are known comparatively well. As will be seen from the following, this circumstance may be used in the experimental verification of the theory. From the experimental point of view, the study of electrode reactions has some essential advantages as compared to that of bulk reactions. For example, in the case of electrode reactions the activation energy may vary over a wide range as the potential jump varies. For this reason the investigator may easily vary the reaction rate over a wide range of up to 5–6 orders. Moreover, the measurements of the current and the potential present many more possibilities for variations in studying the kinetics. However, in order to avoid confusion it should be recalled that for a series of electrode pro-cesses, which are, e.g., connected with deposit formation, or gas emission, it is impossible to make any analogy with bulk reactions. Furthermore, the voltage–current characteristics of electrode processes affect the reagent concentration near the electrode, and the experimental and theoretical determination of this concentration causes considerable difficulties.

IV. Model of a Polar Solution

As we stressed previously, the solvent plays the main role when reactions with charge transfer occur. Therefore an adequate description of a polar solvent constitutes a necessary preliminary for the development of the theory.

A consistent statistical theory of the liquid does not exist at the present time. This is particularly true in the case of such a complex liquid as a polar solvent. Nevertheless, for our limited purposes, a very simple solvent model will do; viz., we shall be interested in the changes in the liquid state occurring in the electron transfer times, i.e., time intervals of the order of 10^{-15} sec. Liquid molecules perform motions of two kinds. On one hand, they make diffusion jumps from one temporary equilibrium position to another. On the other hand, the molecules take part in the thermal motion of the liquid as a whole, vibrating about the temporary equilibrium positions.

The time intervals for diffusion jumps are approximately 10^{-9} sec. Apparently, this time interval is large compared to those of interest to us. Therefore, diffusion jumps may be neglected and the molecules should be considered as fixed in equilibrium positions. In other words, temporary equilibrium positions may be considered as true equilibrium positions for the time intervals of interest to us.

We shall consider the thermal motions of liquid molecules as not differing from those of atoms in a crystal lattice of a solid. We shall confine ourselves to the case of small displacement when the vibrations are of a harmonic nature. In fact, when dealing with a liquid, it is quite inadmissible in a number of cases to neglect the anharmonicity. A harmonic approximation, however, appears to be sufficient when considering thermal motion of the liquid with a view of using this approximation to estimate the probability of transitory and hardly probable processes of liquid polarization fluctuation. Therefore, in what follows we shall make use of the relevant results of the theory of polar crystals.

Two kinds or branches of vibrations, acoustic and optic, are known to exist in a polar crystal. In the case of acoustical vibrations, contiguous particles vibrate, on the average, in one direction. In the case of optical vibrations, contiguous particles move in opposite directions. This accounts for acoustical vibrations being responsible for variations in the crystal density. Optical vibrations, involving a considerable mutual displacement of contiguous particles, change the dipole moment at the corresponding place in the crystal. Optical vibrations determine the optical behavior of crystals, as well as other phenomena associated with their electrical polarization. We shall only be interested in the optical branch of vibrations in this treatise.

The main characteristic of the optical branch is that in the limiting case of long waves, i.e., at $\lambda \to \infty$, the frequency tends to approach a certain limiting value, ω_0.

We are really interested in the displacements of atoms, which cause noticeable polarization in the vicinity of the ion under consideration. This, in turn, changes the potential energy of the ion and may bring it into the reaction state, as described earlier. Let us consider the hydrated ion, which is surrounded by polar molecules of solvent. Somewhat schematically, one may separate all the molecules of the solvent into two groups: first, molecules, which are arranged most closely to the ion and are in quasichemical interaction with it—the so-called inner hydrate sphere; second, molecules forming the outer hydrate sphere and having an electrostatic interaction with the ion. The existence of the quasi-chemical interaction follows from very large values of the hydration energy. It is natural to assume that the thermal movement of molecules of the inner coordination sphere comprises small oscillations about some equilibrium position of the ion. Moreover, there are intramolecular oscillations in molecules (independent of the distance of the molecules from the ion), which occur with frequencies $\omega_{mol} \sim 10^{13}$ to 10^{14} sec^{-1}. Frequencies of these molecular oscillations ω_{mol} are of the same order of magnitude as are the intramolecular oscillations, so that the following inequality holds:

$$\hbar\omega_{mol} \gg kT.$$

The thermal movement of dipole molecules in the outer coordination sphere presents libration oscillations about equilibrium orientations. The frequencies of these oscillations seem to be essentially lower than the frequencies of oscillations in the inner sphere, because the inter-molecular attractions of the outer molecules are also much smaller. As the distance between the outer coordination sphere and the ion exceeds the interatomic distance, the whole solvent with the exception of the inner sphere may be to a sufficiently good approximation described as a dielectric continuum completely neglecting its molecular structure. By such a macroscopic description the dynamical state of the solvent is specified by a specific polarization vector, $\mathbf{P}(r, t)$, and its derivative with respect to time, $\dot{\mathbf{P}}(r, t)$. In this continuum approximation the description of the liquid does not differ from the description of a solid dielectric. Vibrations of dipole molecules and their relative displacements correspond to variations in the polarizability of the dielectric medium: these types of oscillations with some characteristic frequencies form the whole spectrum of the continuous medium. Before proceeding further, we stress that the separation of the solvent into inner and outer coordination spheres and the procedure of separating out the frequency ω_{mol} from the whole frequency spectrum is a rather rough approximation. However, as we shall verify, we have reason to consider this approximation as quite sufficient for our aims. But, as soon as this rough approximation is accepted, there is no meaning in attempting to specify the dielectric continuum further. We make use of this circumstance in what follows. We have to start by considering the behavior of pure liquid solvent, which contains no impurity ions. A more precise description has been given elsewhere (Dogonadze et al., 1965).

In order to realize which wavelengths are of interest, let us note that a rather large number of ions take part in creating a reaction situation. The reaction situation is connected with a change of energy of the system (ion + hydrate cover) by an amount of the order of the activation energy E_a. It is clear, however, that every dipole may contribute to E_a an amount of energy not exceeding its own energy $\hbar\omega_0$. Therefore N dipole particles take part in creating the reaction situation:

$$N \gtrsim E_a/\hbar\omega_0 \sim E_a/kT.$$

If $E_a \sim 0.5$ eV, then $N \gtrsim 10^3$. The dimension L of the region, where the polarization fluctuation causing the reaction situation takes place, is of the order

$$L \sim (M_0 N/\gamma)^{1/3}.$$

Here M_0 is the mass of a dipole molecule and γ its density. For water we have $E_a \sim 0.5$ eV and $L \sim 20$ Å. The scale of the solvent region, where the relevant change of polarization takes place, is large compared to the interatomic spacing d. This shows that if we forget the role of the inner coordination sphere (we consider it in what follows) we may with sufficient accuracy limit ourselves to considering only wavelengths for which $\lambda \gg d$, or (what is in fact the same, with frequencies ω equal to ω_0). All other frequencies do not contribute to the energy of the liquid in the continuum approximation. As we pointed out earlier, the state of the polarization of the liquid may be characterized by a polarization vector $\mathbf{P}(\mathbf{v}, t)$ and its time derivative $\dot{\mathbf{P}}(\mathbf{r}, t)$ describing the rate of the polarization change in the course of time.

Expanding $\mathbf{P}$ and $\dot{\mathbf{P}}$ in a Fourier series, one may write

$$\mathbf{P}(\mathbf{r}, t) = \sum_{\mathbf{k}} (\mathbf{P}_{\mathbf{k}} e^{ikz} + \mathbf{P}_{-\mathbf{k}} e^{-ikz})$$

$$\dot{\mathbf{P}}(\mathbf{r}, t) = \sum_{\mathbf{k}} (\dot{\mathbf{P}}_{\mathbf{k}} e^{ikz} + \dot{\mathbf{P}}_{-\mathbf{k}} e^{-ikz}).$$

The following equalities must hold simultaneously due to the physical nature of the vectors $\mathbf{P}$ and $\dot{\mathbf{P}}$:

$$\mathbf{P}_{\mathbf{k}} = \mathbf{P}_{-\mathbf{k}}; \qquad \dot{\mathbf{P}}_{\mathbf{k}} = \dot{\mathbf{P}}_{-\mathbf{k}}.$$

Thus, the polarization of the liquid may be represented as a set of standing polarization waves, this set being formed as a result of small displacements of the dipole molecules from their equilibrium positions. As is well known, the energy of such wave sets may be presented in the following form:

$$E = \int \sum \beta(\omega_{\mathbf{k}})[P_{\mathbf{k}}^2 + \omega_{\mathbf{k}}^{-2}\dot{P}_{\mathbf{k}}^2]\, dV. \tag{4.1}$$

Here $\beta(\omega_{\mathbf{k}})$ is a set of constants, characterizing the properties of the particles of the medium and their interaction. Formula (4.1) represents the expression for the energy of the system of particles performing small oscillation where this energy is expressed using normal coordinates. In order to realize the degree of accuracy with which the thermal movement of a liquid is described by expression (4.1), we note that we may obtain the dielectric permeability from this expression. The latter appears to have the following form:

$$\epsilon = \epsilon_{re}(\omega) = \sum \delta(\omega - \omega_i); \qquad \epsilon_{im} = 0$$

where ϵ_{re} and ϵ_{im} are the real and imaginary parts of the dielectric permeability. This means that the dielectric medium performing the small polarization oscillations does not absorb light and has the refractive coefficient which possess sharp maxima on proper frequencies of the medium. In polar liquids—e.g., in water—the real part of the dielectric permeability has maxima, which are, however, like rather wide peaks. The imaginary part of the dielectric permeability also has its maxima in this region. A nonzero value of ϵ_{im} signifies the existence of the damping of waves with appropriate frequencies.

Thus the consideration of thermal movement in the liquid as the system of undamped polarization waves which arise as a result of small displacements of molecules from the equilibrium positions is only a rough approximation of the real situation. The existence of sufficiently distinct maxima of dielectric permeability shows that this approximation nevertheless shows the general character of thermal movement. Here we restrict ourselves to this approximation. A more precise description has been presented by us elsewhere (Dogonadze *et al.*, 1965). The characteristic frequencies corresponding to peak values are $\omega_0 = 10^{11}$ sec^{-1} and $\omega_1 = 10^{14}$ sec^{-1}. The first frequency accounts for orientation oscillations (vibrations) of water molecules as a whole, while the second one corresponds to water molecule oscillations along the O—H or H—H bonds. At room temperature only oscillations with frequency ω_0 are excited. This frequency corresponds to a wave having a rather long wavelength as compared to interatomic spacings.

It is possible therefore to write down, approximately, the energy of a pure solvent for *m*:

$$E \simeq \beta(\omega_0) \int \sum_k [P_k{}^2 + \omega_0^{-2} \dot{P}_k{}^2] \, dV. \qquad (4.2)$$

If one calculates the macroscopic potential energy of a polarizable medium consisting of light (electrons) and heavy (nuclei) particles, one obtains (Levich, 1965; Pekar, 1951):

$$\mathscr{V}_{\text{pot}} = 2\pi\left(\frac{1}{n^2} - \frac{1}{\epsilon_0}\right) \int P^2 \, dV = \frac{2\pi}{C} \int P^2 \, dV \qquad (4.3)$$

where ϵ_0 is the static and n^2 the optical value of the dielectric permittivity. From (4.3) we find for $\beta(\omega_0)$

$$\beta(\omega_0) \sim 2\pi/C. \qquad (4.4)$$

We emphasize once more that relation (4.4) permits the estimation of the order of magnitude of the quantity $\beta(\omega_0)$. However, the essentially molecular characteristic $\beta(\omega_0)$ cannot depend on such macroscopic parameters as density or temperature.

Below we shall pass from the classical description of the medium to the quantum-mechanical one. Following the usual procedure, we introduce dimensionless and real quantities p_k and q_k defining them by the formulas:

$$P_k = (\hbar\omega_0/4L^3)^{1/2}(iq_k + q_{-k})$$

$$\dot{P}_k = (\hbar\omega_0{}^3/4L^3)^{1/2}(ip_k + p_{-k})$$

and substitute the expansion of **P** and **Ṗ** into Eq. (5.2). After some simple transformations, we obtain the following expression for the energy over the whole volume:

$$E = \sum (\hbar\omega_0/2)(p_\mathbf{k}^2 + q_\mathbf{k}^2). \tag{4.5}$$

The quantities $p_\mathbf{k}$ and $q_\mathbf{k}$ play the role of dimensionless, normal co-ordinates and momenta of the oscillator. A set of oscillators with various wave vectors $\mathbf{k}$ is equivalent to a complete set of standing waves in the volume $V = L^3$.

The transition from Eq. (4.5) to the quantum-mechanical Hamiltonian of the pure solvent $\mathscr{H}$ is accomplished by the usual substitution of coordinates and momenta by the corresponding quantum operators:

$$q_\mathbf{k} \rightarrow q_\mathbf{k}, \qquad p_\mathbf{k} \rightarrow \frac{1}{i}\frac{\partial}{\partial q_\mathbf{k}}.$$

This yields the expression

$$\mathscr{H} = \sum \frac{\hbar\omega_0}{2}\left(q_\mathbf{k}^2 - \frac{\partial^2}{\partial q_\mathbf{k}^2}\right). \tag{4.6}$$

Formula (4.6) is the sought for quantum-mechanical Hamiltonian of the polar solvent. It is identical to the Hamiltonian obtained by Pekar (1951) and Fröhlich (1954) in their studies of the theory of polar crystals.

The Hamiltonian in Eq. (4.6) shows that the optical vibrations of the solvent can be substituted by a set of phonons with the frequency ω_0 and the wave vectors $\mathbf{k}$. Equation (4.6) shows that in our approximation optical oscillations may be represented as a set of oscillators or phonons. The wavefunction of such a system is well known. Namely,

$$\Psi = \prod \psi_\mathbf{k}$$

where $\psi_\mathbf{k}$ is the wave function of the harmonic oscillator. Correspondingly, the energy of the system is equal to

$$E = \sum \hbar\omega_0(n + \tfrac{1}{2}).$$

If we know the Hamiltonian of the pure solvent, we may proceed to a discussion of the problem of how this Hamiltonian changes under the influence of dissolved ions. As these ions are few in number it is quite natural to suggest that their presence does not change the spectrum of oscillation of the liquid as a whole.

Let us consider the hydrated ion surrounded by dipole molecules of the hydrated solvent. We may divide rather schematically all the molecules of the solvent into two groups: (1) molecules which are arranged closely around the ion and form the so-called inner hydrate sphere; (2) all the remaining molecules, forming the outer hydrated sphere. As the energy of hydration is large enough, molecules of the inner sphere in most cases are in some quasichemical interaction with ions. Molecules of the outer sphere are subjected to electrostatic interaction with the ion, which is up to some degree shielded by the dipoles of the inner sphere.

It is natural to suggest that the thermal movement of the solvent molecules in inner sphere is reduced to small oscillations with frequencies ω_1 along the bonds A—H_2O. Because of strong coupling between the ion and the molecules of the inner sphere, these frequencies appear to possess the same order of greatness as intramolecular oscillations, so that $\hbar\omega_1 \gg kT$, and these oscillations are weakly excited at room temperature.

The consideration of transitions accompanied by considerable change in the inner coordination sphere does not differ from the consideration of reactions going with chemical bond breaking. We postpone the discussion of this problem until Section XI.

However, for some ions (relevant points were considered in Section I) the interaction with the solvent is weaker, and the change in the charge of the ion is not accompanied by an inner sphere rearrangement. The change introduced by ions on outer hydrate sphere is reduced to a displacement of the equilibrium positions of the oscillating ions and to the creating of some equilibrium polarization.

If we assume that the displacements of heavy molecules of solvent, which are caused by the electric field of the ion, are sufficiently small, then the oscillation frequencies do not change. Hereafter we shall use this qualitative picture for the quantitative consideration of the behavior of the system (solvent plus ions).

V. Adiabatic Approximation and Quantum Transition Probability in a System Consisting of Heavy and Light Particles

In what follows we shall consider the quantum-mechanical behavior of a system consisting of electrons, protons, and solvent molecules. As we have seen previously, from various solvent oscillations, one movement of dipole solvent molecules is of special interest for us, namely, the orientation oscillations with characteristic frequency $\omega_0 \sim 10^1$ sec^{-1} or a characteristic time $\tau_0 \sim 10^{-11}$ sec. Intramolecular oscillations, e.g., the proton oscillations in molecules of the type AH, have essentially higher frequencies, $\omega_1 = 10^{14}$ sec^{-1} or characteristic times, $\tau_1 \sim 10^{-14}$ sec.

Finally, the characteristic times, corresponding to electronic movement, are of the order $\tau_2 \sim 10^{-16}$ sec. We see that according to the character of the movement in a reacting system all the particles of this system may be divided into three groups or subsystems—rapid subsystem (electrons), slow subsystem (nuclei in molecules), and the slowest subsystem (solvent molecules moving as a whole). Characteristic times or velocities of movements of these three subsystems differ by 2 to 3 orders of magnitude. In quantum mechanics, when considering the behavior of the system, which may be divided into rapid and slow subsystems, one makes use of a special method called the adiabatic separation. The idea of this separation is as follows: during the time needed for the configuration of heavy particles to change noticeably, electrons pass these particles many times, creating in effect an electron cloud smeared out in space. It is clear, that by this movement one may consider rapid and slow particles as independent to some degree. In terms of wavefunctions this independence means that the full wavefunction of the particle system may be approximately represented by multiplication of wavefunctions of the slow part, χ depending on the corresponding coordinates R, and the wavefunction of the rapid subsystem (electrons) $\varphi(r, R)$,

$$\Psi(r, R) = \chi(R)\varphi(r, R) \tag{5.1}$$

where r is the electron coordinate. The wavefunction $\Psi(r, R)$ satisfies the Schrödinger equation

$$H\Psi = E\Psi \tag{5.2}$$

where E is the full energy, and H the full Hamiltonian of the system,

$$H = -\sum (\hbar^2/2M)\, \Delta_R - \sum (\hbar^2/2m)\, \Delta_r + U(r, R). \tag{5.3}$$

Here $U(r, R)$ is the full potential energy of the system consisting of nuclei and electrons.

If one substitutes the wavefunction in the form (5.1) into the Schrödinger equation (5.2) the latter will not be satisfied. It appears, however, that if we neglect some small terms, we obtain two independent equations for the wavefunctions of the rapid and the slow subsystems. The wavefunction of the slow subsystem obeys the following Schrödinger equation:

$$\left[-\frac{1}{2m_i}\, \Delta_r + u(r_i; R) \right]\varphi = \epsilon_{el}(R)\varphi \tag{5.4}$$

where $u(r_i; R)$ is the potential energy of the rapid subsystem (electrons) whose value is taken for fixed positions of the heavy particles. The energy $\epsilon_{el}(R)$ of the rapid subsystem depends upon the heavy particles positions as well as upon parameters. The wavefunction of the slow subsystem obeys another equation,

$$\left\{ -\sum_k \frac{1}{2M_k} \Delta_{R_k} + \epsilon_{el}(R_k) \right\} \chi(R_k) = E\chi(R_k) \qquad (5.5)$$

where E is the full energy and $\epsilon_{el}(R_k)$ the full energy of the light subsystem. The omitted terms possess a specific structure. They contain expressions of the following type:

$$(\hbar^2/M) \int \varphi^*(\partial/\partial R_i)\varphi \, dV \qquad (5.6)$$

which has the mass of the heavy particles in the denominator. Moreover, the derivatives of wavefunctions of the rapid subsystem with respect to the slow subsystem coordinates are usually small: the wavefunction of the light particles is rather strongly "smeared out" in the whole of the space occupied by the system. Therefore the change of position of the comparatively very exactly localized heavy particles does not cause any noticeable change in the wavefunction of the light particles.

Approximate separation of the system into rapid and slow parts constitutes the basis of the whole theory of molecules and also of the solid state theory. The widely known principle of Frank and Condon is the special expression of this adiabatic separation; according to this principle, radiative transition of the electrons (rapid subsystem) in molecules occurs at fixed positions of the nuclei (slow subsystem).

As will be clear from what follows, the character of the changes in the state of the system in the course of chemical transformations appears to be closely connected with the Frank–Condon principle.

We present now the general formula for the transition probability of the system from the nth into the mth state. In quantum mechanics it is shown that if the system which is in the nth state characterized by the wavefunction Ψ_n is subjected to the action of some sufficiently small external perturbation, then the system with a certain probability may go over to another state—e.g., the mth (characterized by the wavefunction Ψ_m). The general quantum-mechanical formula for the $n \to m$ transition probability (according to first order perturbation theory) is

$$W_{nm} = (2\pi/\hbar) \sum_m \left| \int \Psi_m^* v \Psi_n \, dV \right|^2 \delta(E_n - E_m). \qquad (5.7)$$

Here v is the energy of the perturbation, causing the transition, δ is the

Dirac delta function,

$$\delta(x) = \frac{1}{2\pi} \int\limits_{-\infty}^{\infty} e^{ikx}\, dx = \begin{cases} 0, & x \neq 0 \\ \infty, & x = 0 \end{cases}, \qquad \int\limits_{-\infty}^{\infty} \delta(x)\, dx = 1.$$

The δ function of the argument $E_m - E_n$ expresses the energy conservation law in the transition. The summation in (5.7) is carried over all final states with the given energy E_m. Thus, the expression (5.7) gives the probability that the system will leave the state n (per unit of time). In every particular problem the perturbation v has its own meaning. However, it may be stated in the most general case that v varies essentially with the dimensions of the whole system and therefore depends comparatively weakly upon the coordinates of the strongly localized heavy particles. If one presents the wavefunction in the form (5.1) one obtains the following expression:

$$W_{nm} = \frac{2\pi}{\hbar} \sum_m \left| \int \varphi_m^*(r, R)\chi_m^*(R)v(r, R) \right.$$

$$\left. \cdot \chi_n(R)\varphi_n(r, R)\, dr\, dR \right|^2 \delta(E_m - E_n) \qquad (5.8)$$

where the integration is carried over all the coordinates r and R of the light and heavy particles. In formula (5.8) it is possible to make an essential simplification in the same approximation when the principle of Frank–Condon is valid. Namely, as the wavefunction of the heavy particles is localized in space, whereas the wavefunction $\varphi(r, R)$ and the interaction energy $v(r, R)$ are distributed over the whole volume of the system, the integral in (5.8) has the following form:

$$\mathscr{T} = \int f(x)F(x)\, dx \qquad (5.9)$$

where $f(x)$ and $F(x)$ are functions which vary slowly and sharply with their arguments, respectively. In this case, as is well known, one may take out the slowly varying function from the integrand, assigning a value of this function at some average point

$$\mathscr{T} = f(\bar{x}) \int F(x)\, dx. \qquad (5.10)$$

According to this transformation, one may present the transition probability in the following form:

$$W_{nm} = (2\pi/\hbar) \sum \left| \int \varphi_m^* v\varphi\, dr \right|^2 \left| \int \chi_m^*\chi_n\, dR \right|^2 \delta(E_m - E_n). \qquad (5.11)$$

The simplification arising from this procedure is that now the wavefunctions for the rapid and the slow subsystems are well separated in formula (5.11). This is especially significant when calculating the thermal average of the transition probability, in which we are really interested. Namely, such an averaged transition probability is contained in the formulas for the chemical reactions rate constant and for the electrode current. By definition,

$$\overline{W} = \text{Av } W_{nm} = \sum W_{nm}\rho_n$$

$$= \frac{2\pi}{\hbar} \text{Av} \sum_m \left| \int \varphi_m^* v \varphi_n \, d\mathbf{r} \right|^2 \left| \int \chi_m^* \chi_n \, d\mathbf{R} \right| \delta(E_m - E_n). \quad (5.12)$$

Averaging over the thermal motion or, speaking more correctly, statistical averaging, is carried out with the Gibbs distribution function

$$\rho_n = e^{-(\epsilon_n/kT)} \Big/ \sum e^{-(\epsilon_n/kT)} \quad (5.13)$$

where ρ_n represents the normalized probability that the system in statistical equilibrium occupies the state with energy E_n. The reaction we shall assume is sufficiently slow so that no essential distortion of the statistical equilibrium state of the system could arise. If the rapid subsystem is represented by electrons (and accordingly energy levels of the rapid subsystem are the electronic ones), then spacings between its levels comprise several tenths of an electron volt (i.e., a few thousands degrees). Therefore the rapid subsystem is in the ground state at room temperature. This means that by the averaging process we have to retain only one term in the sum over the initial states of the rapid subsystem; this term refers to the ground state.

Thus, one may write down the following expression:

$$\overline{W} = (2\pi/\hbar) \left| \int \varphi_{m_0}^* v \varphi_{n_0} \, d\mathbf{r} \right|^2 \text{Av} \sum_m \left| \int \chi_m^* \chi_n \, d\mathbf{R} \right|^2 \delta(E_m - E_n) \quad (5.14)$$

where the index zero will label quantities referring to the ground state. Formula (5.14) containing the averaging only over all possible states of the slow subsystem for a fixed electronic state constitutes the basis of the following theory. The dependence of $\overline{W}$ on temperature is completely contained in the second factor of (5.14),

$$\text{Av} \left| \int \chi_m^* \chi_n \, d\mathbf{R} \right|^2 = \frac{\sum e^{-(\epsilon_n/kT)} \left| \int \chi_m^* \chi_n \, d\mathbf{R} \right|^2}{\sum e^{-(\epsilon_n/kT)}}. \quad (5.15)$$

On the contrary, the first factor in (5.14) contains only the wavefunctions of the rapid subsystem in the ground state and the interaction energy and does not depend on the temperature.

VI. Probability of Electron Transfer

We start our consideration with the case of the simplest redox reactions which occur without chemical bond breaking.

We shall assume that there are no ions in the solution which may serve as charge carriers, i.e., ligands or bridges. In the examples given in Section I the charge transfer occurs between similarly charged ions. Electrostatic repulsion prevents the direct contact of the ions in the course of their mutual approach. We have to bear in mind, however, that the ions and some neutral molecules are strongly solvated. Their coupling with the solvate layer is so strong that the energy of thermal motion is not sufficient to "throw off" the solvate layer. When such strongly solvated particles approach one another, the exchange of an electron takes place at some distance. In order to obtain a quantitative expression for the transition probability of the electron we make use of the method of calculation described in Section V.

One may separate the reacting system (ion A, ion B, and the solvent molecules) into slow and rapid subsystems. The first subsystem includes the electron performing the transition for fixed positions of the particles of the slow subsystem (which contains A and B nuclei and solvent molecules). We shall assume that ions A and B are atomic and do not possess any internal degrees of freedom. The states of their internal electronic levels are considered as unchangeable and the transition of the valence electron goes without excitation (from the ground state into the ground state again). According to formula (5.14) the average transition probability is determined by two factors. The first of these factors contains only electronic wavefunctions, and because the electronic states do not change with temperature this factor may be carried outside the overall averaging and takes the form

$$L = \left| \int \varphi_f{}^* v \varphi_i \, d\mathbf{r} \right|. \tag{6.1}$$

Then formula (5.14) acquires the following form:

$$\overline{W}_{\alpha\beta} = L^2 \, \mathrm{Av} \sum_{\beta} \left| \int \Phi_\beta{}^*(q) \Phi_\alpha(q) \, d\mathbf{R} \right|^2 \delta(E_f - E_i). \tag{6.2}$$

Here the indices α and β label the initial and final states, respectively. $\Phi(q)$ is the wavefunction of the independent oscillators. The perturbation energy v† represents the interaction energy of the electron in the

---

† The perturbation energy v includes the nonadiabaticity operator which is small compared with $Ze^2/|\mathbf{r} - \mathbf{R}|$.

reduced ion with oxidized ion:

$$v = Z_{ox}e^2/|\mathbf{r} - \mathbf{R}| \tag{6.3}$$

for the direct transition and

$$v = Z_{red}e^2/|\mathbf{r} - \mathbf{R}| \tag{6.4}$$

for the inverse one. Evidently v represents a rather slowly varying function of the spatial coordinates as compared to the wavefunctions of the electrons and especially those of the heavy particles. The wavefunction of the electrons for complex ions is known much less exactly. We shall make use of some of the quantum-chemical expressions for electronic wavefunctions in order to obtain very crude estimations of the appropriate values. The main interest lies in the second factor, determined by the temperature dependence of W. The wavefunctions of the heavy subsystem entering into the second factor have to be determined from the solution of the Schrödinger equation (5.5) which in our particular case acquires the following form:

$$H_\alpha \Phi_\alpha(q) = E_\alpha \Phi_\alpha(q)$$

where $\alpha = i, f$, E_α is the full energy of the slow subsystem, and H according to (4.6) and (5.5) has the following form (we omit the index α):

$$H = H_s + \epsilon_{el}(q) = \frac{\hbar\omega_0}{2} \sum \left(q_k{}^2 - \frac{\partial^2}{\partial q_k{}^2} \right) + \epsilon_{el}(q). \tag{6.5}$$

The quantity $\epsilon_{el}(q)$ is the energy of the rapid subsystem (electron) which depends on the coordinates of the slow subsystem q_k. The problem becomes defined if the dependence of the electron energy ϵ_α on the coordinates of the phonons q_k is found, i.e., if the change of energy due to the interaction of the electron with the lattice vibrations is obtained. We shall suppose the interaction with the electron has no effect on the character and frequencies of the vibrations of the solvent atoms. The only change will be the displacement of equilibrium positions of the atoms, which, in the presence of an electron, will oscillate around new equilibrium positions. We shall consider that the change in the energy $\epsilon_\alpha(q_k)$ can be written as a linear function $(q_k - q_{k\alpha}^{(0)})$, where $q_{k\alpha}^{(0)}$ corresponds to the equilibrium configuration of the solvent. Thus,

$$\Delta\epsilon_\alpha = \epsilon_\alpha(q_k) - \epsilon_\alpha(q_{k\alpha}^{(0)}) = \sum \left(\frac{\partial \epsilon_\alpha}{\partial q_k} \right)_{q_k = q_k^{(0)}} \cdot (q_k - q_{k\alpha}^{(0)}). \tag{6.6}$$

The linear approximation in the expansion of $\epsilon_\alpha(q_k)$ shows that the displacement of the heavy molecules of solvent caused by the electric field of the electron is small enough. A similar approximation is used in all calculations in the theory of polar and ionic crystals (e.g., Born and Huang, 1957; Pekar, 1951).

The substitution of Eq. (6.6) into the Hamiltonian $\mathcal{H}_\alpha$ gives

$$\mathcal{H}_\alpha = \mathcal{H}_s + \epsilon_\alpha(q_k^{(0)}) + \sum \left(\frac{\partial \epsilon_\alpha}{\partial q_k}\right)_{q_k = q_k^{(0)}} \cdot (q_k - q_{k\alpha}^{(0)})$$

$$= \frac{\hbar \omega_0}{2} \sum \left\{ \left[q_k^2 - \left(\frac{\partial^2}{\partial q_k^2}\right) \right] + \sum \left(\frac{\partial \epsilon_\alpha}{\partial q_k}\right)_{q_k = q_k^{(0)}} \left(\frac{2}{\hbar \omega_0}\right) q_k \right\}$$

$$+ \epsilon_\alpha(q_{k\alpha}^{(0)}) - \sum \left(\frac{\partial \epsilon_\alpha}{\partial q_k}\right) \cdot q_{k\alpha}^{(0)}$$

$$= \frac{\hbar \omega_0}{2} \sum \left\{ \left[q_k + \left(\frac{1}{\hbar \omega_0}\right)\left(\frac{\partial \epsilon_\alpha}{\partial q_k}\right)_0 \right]^2 - \frac{\partial^2}{\partial q_k^2} \right\} + \epsilon_\alpha(q_{k\alpha}^{(0)})$$

$$- \sum \left\{ \left(\frac{\partial \epsilon_\alpha}{\partial q_k}\right)_0 \cdot q_{k\alpha}^{(0)} - \left(\frac{\hbar \omega_0}{2}\right) \cdot \left(\frac{\partial \epsilon}{\partial q_k}\right)_0^2 \right\}. \tag{6.7}$$

Evidently, since at the equilibrium points $q_k = q_k^{(0)}$, the potential energy of the solvent must have a minimum. Hence,

$$q_{k\alpha}^{(0)} = -\left(\frac{\partial \epsilon_\alpha}{\partial q_k}\right)_0 \cdot \frac{1}{\hbar \omega_0}. \tag{6.8}$$

Therefore Eq. (6.7) may be rewritten in the form

$$\mathcal{H} = \frac{\hbar \omega_0}{2} \sum \left\{ (q_k - q_{k\alpha}^{(0)})^2 - \frac{\partial^2}{\partial (q_k - q_{k\alpha}^{(0)})^2} \right\} + \epsilon(q_k^{(0)})$$

$$+ \frac{\hbar \omega_0}{2} \sum (q_{k\alpha}^{(0)})^2. \tag{6.9}$$

By introducing the notation

$$\mathcal{T}_\alpha = \epsilon(q_{k\alpha}^{(0)}) + \frac{\hbar \omega_0}{2} \sum (q_{k\alpha}^{(0)})^2, \tag{6.10}$$

we obtain

$$\mathcal{H}_1 = \frac{\hbar \omega_0}{2} \sum \left\{ (q_k - q_{k1}^{(0)})^2 - \frac{\partial^2}{\partial (q_k - q_k^{(0)})^2} \right\} + \mathcal{T}_1, \tag{6.11}$$

$$\mathcal{H}_2 = \frac{\hbar \omega_0}{2} \sum \left\{ (q_k - q_{k2}^{(0)})^2 - \frac{\partial^2}{\partial (q_k - q_k^{(0)})^2} \right\} + \mathcal{T}_2. \tag{6.12}$$

The quantity $(\hbar \omega_0 / 2) \sum (q_{k\alpha}^{(0)})^2$ is the energy of the solvent equilibrium polarization, corresponding to the presence of the electron near the nucleus α. According to Eq. (4.3) it can be written as

$$\tfrac{1}{2} \sum \hbar\omega_0 (q^{(0)}_{k\alpha})^2 = (C/8\pi) \int [D_\alpha^{(0)}]^2 \, dV = (2\pi/C) \int P^2 \, dV \quad (6.13)$$

where $D_\alpha^{(0)}$ is the vector of induction caused by the electron localized near the α nucleus. It should be recalled that all comments and results are valid when the macroscopic formula (4.3) is applied. We see that the Hamiltonian of the set of heavy particles, which form the slow subsystem, is of simple form both before and after the transition of the electron. The Hamiltonian is of the form appropriate for the system of independent harmonic oscillators, which oscillate about their equilibrium positions $q^i_{k_0}$ and $q^f_{k_0}$. Energies and wavefunctions of such a system are well known. The Hamiltonian (and accordingly the energy) both before and after the transition contains the quantities $\mathscr{T}_\alpha$. We recall once more their meaning: it is the sum of the energy of the electrons in the normal states of the ions A^m (i.e. $\epsilon^i_{el}(q)$) and B^{n+1} (i.e., $\epsilon^f_{el}(q)$) and the equilibrium polarization of the solvent $U_\alpha = \tfrac{1}{2} \sum \hbar\omega_0 (q^{(0)}_{k\alpha})^2$ (the solvent surrounds two ions before or after reaction). In order to estimate this energy to an order of magnitude one may use Eq. (4.3).

The case of an arbitrary frequency dispersion law, i.e., an arbitrary law $\omega(\mathbf{k})$ was also considered, and relations were obtained which generalize those cited in this chapter. It proved to be possible to establish the connection between the dispersion law and the correlation function, which correlates the average values of the polarizations $\langle \mathbf{P}(\mathbf{r})\mathbf{P}(\mathbf{r}')\rangle$ and the polarization currents $\langle \dot{\mathbf{P}}(\mathbf{r})\mathbf{P}(\mathbf{r}')\rangle$ at the neighboring points $\mathbf{r}$ and $\mathbf{r}'$ of the polar medium (Dogonadze et al., 1969). In the theory of radiationless transitions in semiconductors it was shown that the correlation functions method makes it possible to connect the transition probability with the properties of the imaginary part of the dielectric losses, corresponding to the maximum of $\epsilon_{im}(\omega)$ (at some frequency $\omega_0 = 10^{-11}$ sec^{-1}) leads to the value of the activation energy which coincides with the value calculated earlier on the basis of an assumed model.

Formula (6.1) determines the terms of the potential energy surfaces of the system, which were shown in Section II based on qualitative arguments; namely, since by definition the electronic term presents the full energy of the system minus the kinetic energy of the heavy particles. Formula (6.1) give

$$U_\alpha(q) = \mathscr{T}_\alpha + \frac{\hbar\omega_0}{2} \sum (q_k - q^{(0)}_{k\alpha})^2$$

$$= \epsilon(q^{(0)}_{k\alpha}) + \frac{\hbar\omega_0}{2} \sum (q^{(0)}_{k\alpha})^2 + \frac{\hbar\omega_0}{2} \sum (q_k - q^{(0)}_{k\alpha})^2. \quad (6.14)$$

A term of the potential energy surface in the initial and the final states are two paraboloids in $(N + 1)$-dimensional space (where N is the number of phonons in the liquid). The bottom of the latter paraboloid is displaced a distance $Q = \mathcal{T}_2 - \mathcal{T}_1$.

VII. Transition Probability in Reactions without Chemical Bond Breaking

Knowing wavefunctions Φ_n of the slow subsystem, one may with the aid of (6.2) calculate the electrons transition probability A → B.

Using for Φ_n the wavefunctions of the harmonic oscillator,

$$\Phi_n = A_n H_n(q_k - q_k^{(0)}) \exp\left[-\tfrac{1}{2}(q_k - q_{k0})^2\right]$$

where H_n is the Hermite polynomial, substituting for the δ function its Fourier expansion, we obtain from (6.2)

$$W_{\alpha\beta} = \frac{1}{\hbar^2} |L_{\alpha\beta}^{(2)}|^2 \text{ Av} \int_{-\infty}^{\infty} \sum_{n'} \left\langle \Phi_{\alpha n} \left| \exp\left[\frac{i}{\hbar} E_{\beta n'} t\right] \right| \Phi_{\beta n'} \right\rangle$$

$$\times \left\langle \Phi_{\beta n'} \left| \exp\left[-\frac{i}{\hbar} E_{\alpha n} t\right] \right| \Phi_{\alpha n} \right\rangle dt = \frac{1}{\hbar^2} |L_{\beta\alpha}^{(2)}|^2 \text{ Av} \int_{-\infty}^{\infty} \sum_{n'}$$

$$\times \left\langle \Phi_{\alpha n} \left| \exp\left[\frac{i}{\hbar} H_\beta t\right] \right| \Phi_{\beta n'} \right\rangle \cdot \left\langle \Phi_{\beta n'} \left| \exp\left[-\frac{i}{\hbar} H_\alpha t\right] \right| \Phi_{\alpha n} \right\rangle dt$$

$$= \frac{1}{\hbar^2} |L_{\beta\alpha}^{(2)}|^2 \int_{-\infty}^{\infty} \left\langle \Phi_{\alpha n} \left| \exp\left[\frac{i}{\hbar} (H_\beta - H_\alpha) t\right] \right| \Phi_{\alpha n} \right\rangle dt. \quad (7.1)$$

The averaging of the matrix element was made using Feynman's (1951) method. The following result is obtained for $W_{\alpha\beta}$ (for $\alpha = 1$, $\beta = 2$) after rather cumbersome and time-consuming calculations:

$$W_{12} = (2\pi/\hbar^2 \omega_0)|L_{21}^{(2)}|^2 I_m(z) \exp\left\{-(\hbar\omega_0 m/2kT) - z \text{ ch}(\hbar\omega_0/2kT)\right\} \quad (7.2)$$

where I_m is the Bessel function of the mth order and m and z are defined by

$$m = (\mathcal{T}_2 - \mathcal{T}_1)/\hbar\omega_0$$

$$z = \tfrac{1}{2} \text{csch}[\hbar\omega_0/2kT] \sum_k (q_{k1}^{(0)} - q_{k2}^{(0)})^2.$$

The probability of the reverse transition, in accordance with the general principles of quantum mechanics, is given by the formula

$$W_{21} = \exp[\hbar\omega_0 m/kT]W_{12} = \exp[(\mathcal{T}_2 - \mathcal{T}_1)/kT]W_{12}. \quad (7.3)$$

We see that the transition probability is expressed by a rather complicated formula (7.2). The latter equation contains, besides a simple parameter, the argument z; the latter depends in turn on an important quantity:

$$E_s = (\hbar\omega_0/2) \sum (q_{k1}^{(0)} - q_{k2}^{(0)})^2, \qquad (7.4)$$

The quantity E_s represents the main parameter of the theory. We shall call it the energy of reorganization or the energy of repolarization. In the macroscopic continuum approximation E_s may be presented in the following form [making use of (6.13)]:

$$E_s = \frac{C}{8\pi} \int (\mathscr{D}_1 - \mathscr{D}_2)^2 \, dV. \qquad (7.5)$$

Formula (7.5) is convenient in two aspects. It permits the visualization of E_s and at the same time leads to a rough estimate. We see that the quantity E_s represents the variation of the solvent energy in the transition from the initial state (with induction $\mathscr{D}_1$) into the final (with induction $\mathscr{D}_2$) state. Let us consider two limiting cases, for which formula (7.2) simplifies essentially.

A. HIGH TEMPERATURE CASE

It is not difficult to see that the high temperature case is the main one, which is realized in the case of the redox reactions. When the following inequalities are fulfilled,

$$kT \gg \hbar\omega_0,$$

$$kT \gg \hbar\omega_0 \frac{Q}{E_s} = \hbar\omega_0 \frac{\mathscr{I}_2 - \mathscr{I}_1}{E_s} \qquad (7.6)$$

(these inequalities we shall call the conditions for the high temperature approximation), formula (7.2) is amenable to further important simplification; namely as the argument of the Bessel function $z \gg 1$, one may use the asymptotic expansion of the Bessel functions. This gives

$$W_{12} = |L_{21}|^2 \cdot \left(\frac{\pi}{\hbar E_s}\right)^{1/2} \exp\left\{-\frac{(\mathscr{I}_2 - \mathscr{I}_1 + E_s)^2}{4E_s kT}\right\}. \qquad (7.7)$$

B. LOW TEMPERATURE CASE

At low temperatures, when the inverse inequalities are fulfilled formula (7.2) is again amenable to simplification. Particularly because in this case

$z \ll 1$, we expand the Bessel functions in series and this gives

$$W_{12} = |L_{12}|^2 \, \frac{2\pi}{\hbar^2 \omega_0 [(\mathscr{T}_2 - \mathscr{T}_1)/\hbar\omega_0]!} \left(\frac{E_s}{\hbar\omega_0}\right)^{(\mathscr{T}_2 - \mathscr{T}_1)/\hbar\omega_0}$$
$$\cdot \exp\left[-\frac{E_s}{\hbar\omega_0}\right] \cdot \exp\left[-\frac{\mathscr{T}_2 - \mathscr{T}_1}{kT}\right]. \tag{7.8}$$

The expressions thus obtained are of principal significance in the subsequent presentation.

The expressions obtained above, just as the formula of absolute rate theory, contain two factors: the preexponential factor (in our first approximation this factor does not depend on temperature), and the exponential one. Thus, expressions of the Arrhenius type are obtained, which include an activation energy

$$E_a = (E_s + \mathscr{T}_2 - \mathscr{T}_1)^2/4E_s \quad \text{(high-temperature approximations),} \tag{7.9}$$

$$E_a = \mathscr{T}_2 - \mathscr{T}_1 \qquad \text{(low temperature approximations).} \tag{7.10}$$

Contrary to the absolute reaction rate theory where the preexponential factor is calculated but no attempt is usually made to calculate the activation energy, we have calculated (at least, in principle) both of these quantities.

Before we proceed to a discussion of these formulas, we shall discuss the question of their range of applicability; this range is connected with some assumptions adopted in the course of the derivation. The main assumption is that we may use the perturbation theory. The condition of validity of this approximation is the following inequality:

$$\overline{W}_{12} \cdot t \ll 1 \tag{7.11}$$

where t is the time of action of the perturbation. The inverse inequality (7.11) represents nothing else but the condition for the diffusion approximation. Therefore, in those cases where the reaction occurs in the kinetic region and the probability of reaction is sufficiently small to exclude the existence of any diffusion limitations, the application of perturbation theory is justified.

Formulas (7.9) and (7.10) allow a simple interpretation in terms of the electronic terms. In particular for the case of redox reactions we shall discuss in detail only the high temperature approximation. The frequency ω_0 representing the frequency of orientation of the oscillations of the solvent dipole molecules always satisfies the first inequality (7.6). The heats of reaction Q are always less than E_s. In the high temperature approximation the slow subsystem (solvent) moves classically, i.e.,

$\Delta\epsilon = \hbar\omega_0$ and is small as compared to the thermal energy kT. We emphasize once more that polarization fluctuations have nothing in common with temperature fluctuations. The change of energy of the system is connected only with the electrostatic interaction of ion with the solvent field. In the harmonic approximation, when the terms represent a paraboloid, they intercept when

$$\mathcal{T}_1 + \frac{\hbar\omega_0}{2} \sum (q_k^* - q_{k1}^{(0)}) = \mathcal{T}_2 + \frac{\hbar\omega_0}{2} \sum (q_k^* - q_{k2}^{(0)})^2.$$

The corresponding full potential energy at the point $q = q^*$ is equal to

$$E_a = \frac{(E_s + \mathcal{T}_1 - \mathcal{T}_2)^2}{4E_s} = \frac{(E_s + Q)^2}{4E_s}. \qquad (7.12)$$

We arrive at an equation connecting the activation energy E_a with the reaction heat Q, i.e., to the Brönsted rule.

The Brönsted rule in its particular form (7.12) is to some extent connected with the use of a one-frequency harmonic approximation. For real reactions with electron transfer, E_s appears to be of the order of 1 eV. This value is obtained by comparing the theory with experiment and also from rough estimates according to formula (7.5). The value of Q for all the redox processes is rather small as compared to this value of E_s. It is seen, e.g., from the equilibrium relation between the concentrations of the oxidized or reduced forms in the volume of the solution, that

$$\ln(C_{ox}/C_{red}) = e^{-Q/kT}. \qquad (7.13)$$

If $Q \gg kT$, the concentration of one of the forms would have an unreasonably small value. Therefore there is always

$$E_a \simeq \tfrac{1}{4}E_s + \tfrac{1}{2}Q \qquad (7.14)$$

and the transport coefficient α equals $\tfrac{1}{2}$. Reactions possessing the transport coefficient near to $\tfrac{1}{2}$ will be referred to below as reactions occurring in the normal region. We shall explain further the meaning of this term. The inverse limiting case, i.e., the low temperature approximation, corresponds to the transition pattern, shown in Fig. 6. Having achieved the point q^*, the system as a whole performs quantum transitions immediately in the state with minimal energy. We return to a discussion of these transitions in Section XII where there will be a discussion of the internal coordination sphere. Here we formulate the result of this consideration. It appears that account of this interaction leads only to a change of the repolarization.

Up to now we have considered only the behavior of the exponential factor in the general formula for the transition probability.

The preexponential factor is also of considerable interest. In formula (7.8) it was calculated to the first order of perturbation theory. This approximation is, as a rule, sufficient. If the interaction energy between particles in the point of terms intersection is small, as is shown in Fig. 4, then the system crosses over from one potential surface to the other. Following the terminology accepted in chemical kinetics (this terminology is not quite approximate), one refers to such transitions as nonadiabatic. If, however, the interaction in the intersection point is large, then the situation shown in Fig. 7 is possible, as is well known from quantum

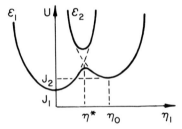

FIG. 7. The case of the nonintersecting terms.

mechanics. Strong interaction leads to repulsion terms and the system crosses over from the initial state to the final one, moving continuously on a smooth potential energy surface. This is the so-called adiabatic transition.

In the case of strong interaction, which may apparently occur in some cases in electron transfer phenomena, in order to calculate the transition probability one has to employ another method of calculation which is not connected with the use of the first order perturbation theory. The motion of the rapid subsystem (the electron) and the slow systems (the solvent) may be considered to be independent. In the so-called semiclassical approximation the solvent may be considered as a purely clasical system, whereas the description of the electron preserves its quantum character. In this approximation one obtains for the probability of adiabatic transition

$$W^a = (\hbar/2\pi kT)e^{-(E_a/kT)} \tag{7.15}$$

instead of (7.8).

VIII. Flow of Current through the Metal–Solution Interface

We can now calculate directly the current density for the oxidation–reduction reaction occurring on a metal electrode. The density of the solution–metal current can be written as

$$i_{ms} = e \int c_{n-1}(x)\, dx \int W_{sm}(x, \epsilon_f)[1 - n(\epsilon_f)]\rho(\epsilon_f)\, d\epsilon_f$$

$$= e \int c_{n-1}(x)\, dx \int W_{ms} \exp\left[\frac{\mathscr{T}_s - \mathscr{T}_m}{kT}\right][1 - n(\epsilon_f)]\rho(\epsilon_f)\, d\epsilon_f \quad (8.1)$$

In this case, we express W_{sm} in terms of W_{ms} according to Eq. (7.3). Consequently, the total current flowing through the metal–solution interface is given by the equation

$$i = e \int dx\, d\epsilon_f\, \rho(\epsilon_f) n(\epsilon_f) c_n(x)$$

$$\times \left\{\frac{c_{n-1}(x)}{c_n(x)} \cdot \frac{1 - n(\epsilon_f)}{n_f} \cdot \exp\left[\frac{\mathscr{T}_s - \mathscr{T}_m}{kT}\right] - 1\right\}. \quad (8.2)$$

To continue our calculation we have to introduce some assumptions concerning the concrete mechanism of electron transfer and the properties of the double layer on the metal–solution interface. We shall consider first that the flow of current does not interfere with the equilibrium distribution of ions in the double layer, and second, that the charge of the ion being discharged is of the same sign as the metal surface. The first assumption holds for a low current, when the number of ions participating in the electrode process is small compared to the total number of ions in the double layer. This assumption can be considered to be valid under the actual conditions of the passage of the direct current. This problem was discussed in more detail by Levich (1959). The second assumption permits us to consider the transfer to occur at a relatively large distance between the electrode and the metal. In this case we can use, for the transfer probability, the equation of the nonadiabatic approximation.

Because of the equilibrium for the ion distribution, we can write

$$\left[\frac{c_{n-1}(x)}{c_n(x)}\right] = \frac{[c_{n-1}]}{[c_n]} \exp\left[\frac{e[\varphi(x) - \varphi_0]}{kT}\right] \quad (8.3)$$

where $[c_n]$ and $[c_{n-1}]$ are the corresponding concentrations, and φ_0 the potential in the bulk of the solution at $x \to \infty$. In addition, for the

Fermi–Dirac distribution, we have

$$\frac{1 - n(\epsilon_f)}{n(\epsilon_f)} = \exp\left[\frac{\epsilon_f - \epsilon_F}{kT}\right] \tag{8.4}$$

where ϵ_F is the Fermi level of the metal. Evidently, for the equilibrium term of the electron, localized near ion A^{n-1} in the solution, we have

$$\mathcal{T}_s = -\epsilon_{n-1} - [(\epsilon_0 - 1)/8\pi\epsilon_0] \int (\mathcal{D}_{n-1})^2 \, dV + (n-1)e\varphi(x) \tag{8.5}$$

where $\varphi(x)$ is the potential of the self-consistent electrostatic field of the double layer in the solution at the distance x from the electrode, ϵ_{n-1} the energy of the electron in ion A^{n-1} (if the ion is in a vacuum), and $\mathcal{D}_{n-1}$ the induction of this ion.

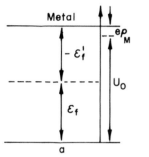

Fɪɢ. 8. Scheme of the localization of characteristic levels in the metal.

Similarly, we have for the system (ion A + electron in the metal)

$$\mathcal{T}_m = -[(\epsilon_0 - 1)/8\pi\epsilon_0] \int (\mathcal{D}_n)^2 \, dV + (\epsilon_f - e\varphi_M - U_0) \tag{8.5'}$$

where ϵ_f is the energy level of the electron in the metal, φ_M the inner electrostatic potential, which has a constant value in the metal, and U_0 the total energy of the electron bond in the metal (see Fig. 8). The difference of the terms contained in the transfer probability will be written as

$$\mathcal{T}_s - \mathcal{T}_m = (-\epsilon_f + e\varphi_M + U_0) - \epsilon_{n-1}$$
$$- [(\epsilon_0 - 1)/8\pi\epsilon_0] \int [(\mathcal{D}_n)^2 - (\mathcal{D}_{n-1})^2] \, dV - e\varphi(x).$$

According to Born, the quantity

$$\Delta\alpha = [(\epsilon_0 - 1)/8\pi\epsilon_0] \int (\mathcal{D}_n{}^2 - \mathcal{D}_{n-1}^2) \, dV$$

is the difference of hydration energies of ions with the charges n and

$n - 1$ (Conway and Bockris, 1959). Hence, we obtain, finally,

$$\Delta \mathcal{T} = \mathcal{T}_s - \mathcal{T}_m = (e\varphi_M + U_0 - \epsilon_f) - \epsilon_{n-1} + \Delta\alpha - e\varphi(x). \quad (8.6)$$

By substituting Eqs. (8.3), (8.4), and the difference $\mathcal{T}_s - \mathcal{T}_m$ from Eq. (8.6) to Eq. (8.2), we find

$$i = e \int dx \, d\epsilon_f \, \rho(\epsilon_f) n(\epsilon_f) c_n(x) W_{ms}(x, \epsilon_f)$$

$$\times \left\{ \exp\left[\frac{-\epsilon_F - kT \ln ([c_n]/[c_{n-1}]) + \Delta\alpha - u_0 + e(\varphi_M - \varphi_0) - \epsilon_{n-1}}{kT} \right] - 1 \right\}.$$

All the quantities between braces do not depend on the variable x and ϵ_f, so that

$$i = \left\{ \exp\left[\frac{\Delta\alpha - u_0 - \epsilon_{n-1} - \epsilon_F - kT \ln ([c_n]/[c_{n-1}]) + e(\varphi_M - \varphi_0)}{kT} \right] - 1 \right\} i_{sm}.$$

$$(8.7)$$

The difference $\varphi_M - \varphi_0$ between the inner potential of the metal and an infinitely removed point in the solution can be written in the form

$$(\varphi_M - \varphi_0) = (\varphi_M - \varphi_0)_{eq} + \eta \quad (8.8)$$

where $(\varphi_M - \varphi_0)_{eq}$ refers to the equilibrium state and η is the usual definition of the overvoltage. The equilibrium value of $(\varphi_M - \varphi_0)_{eq}$ is obtained from Eq. (8.2), if we assume $i = 0$. Thus,

$$(\varphi_M - \varphi_0)_{eq} - \epsilon_{n-1} + (\epsilon_F - u_0) + kT \ln ([c_{n-1}]/[c_n]) - \Delta\alpha. \quad (8.9)$$

Therefore, Eq. (8.7) can be represented as

$$i = (\exp[(e\eta/kT)] - 1)i_{sm}. \quad (8.10)$$

As the relation

$$i_{sm} = i_{ms} \exp[-(e\eta/kT)] \quad (8.11)$$

holds, Eq. (8.11) can be rewritten in the alternative form

$$i = i_{ms}(1 - \exp[-(e\eta/kT)]). \quad (8.12)$$

Our next task is to calculate i_{sm}. For this purpose, we have to integrate Eq. (8.2) over the coordinate x and over the energy spectrum. We have

$$i_{ms} \approx ec_n(\delta) \, \delta \int_0^\infty n(\epsilon_f) \rho(\epsilon_f) W_{ms}(\delta, \epsilon_f) \, d\epsilon_f \quad (8.13)$$

where δ is some distance from the electrode, at which the probability of the electron transfer is greatest, $W_{ms}(\epsilon_f, \delta)$, and $c_n(\delta)$ are the values of the transfer probability and the concentration of ions A^n in the plane $x = \delta$, respectively. Then by using Eq. (7.7), we obtain

$$i_{ms} = ec_n(\delta) \cdot \delta \cdot \bar{p} \int_0^{\epsilon_F} W_{ms}(\delta, \epsilon_f) n(\epsilon_f) \, d\epsilon_f$$

$$= ec_n(\delta)\rho \, \delta|L_{ms}^{(2)}(\delta)|^2 \left[\frac{\pi}{\hbar^2 k T E_s}\right]^{1/2} \int_0^{\epsilon_F} n(\epsilon_f)$$

$$\times \exp\left[-\frac{(\mathcal{T}_s - \mathcal{T}_m + E_s)^2}{4E_s k T}\right] d\epsilon_f. \tag{8.14}$$

The value of the difference $\mathcal{T}_s - \mathcal{T}_m$ should be taken in the plane $x = \delta$. On the basis of Eqs. (8.6), (8.8), and (8.9), we have

$$\mathcal{T}_s - \mathcal{T}_m = e\eta + (\epsilon_F - \epsilon_f) + e(\varphi_0 - \varphi(\delta)) + kT \ln ([c_n]/[c_{n-1}]). \tag{8.15}$$

We shall denote, as usual, the value of the potential in the plane $x = \delta$, which is identified with the Helmholtz plane, by ψ_1. Then, finally,

$$i = gc_n(\delta)I \tag{8.16}$$

where the constant factors are brought together in the form

$$g = e \, \delta\bar{p} \left[\frac{\pi}{\hbar^2 k T E_s}\right]^{1/2} |L_{ms}^{(2)}(\delta)|^2$$

and I denotes the integral over the energies,

$$I = \int \exp\left\{-\frac{\left[(\epsilon_F - \epsilon_f) + E_s + e\eta + e(\varphi_0 - \psi_1) + kT \ln ([c_n]/[c_{n-1}])\right]^2}{4E_s k T}\right\} n(\epsilon_f) \, d\epsilon_f. \tag{8.17}$$

Integration is carried out over all the energies, starting from zero; that is, the energy of the electron in the vacuum up to the energy of the electron in the lower level of the band. Let us rewrite I as

$$I = \int \exp\left\{-\frac{\left[(\epsilon_F - \epsilon_f) + E_s + e\eta + e(\varphi_0 - \psi_1) + kT \ln ([c_n]/[c_{n-1}])\right]^2}{4E_s k T} + \ln n(\epsilon_f)\right\} d\epsilon_f$$

$$= \int e^{f(\epsilon_f)} \, d\epsilon_f. \tag{8.18}$$

The integral should be calculated by the Laplace method, expanding the exponential into a series of the powers of ϵ_f near the maximum point, determined by the condition $f'(\epsilon_f{}^*) = 0$. Usually, the repolarization energy E_s is 1–5 eV, and is large compared to the overvoltage $e\eta \leq$ 1–2 eV, as well as to the values of $e(\varphi_0 - \psi_1)$ and $kT \ln[c_n]/[c_{n-1}]$. Therefore, we can write finally

$$i_{ms} = (8\pi)^{1/2}gkT[c_n] \cdot \exp\left\{-\frac{ne(\psi_1 - \varphi_0)}{kT}\right\}$$

$$\cdot \exp\left\{-\frac{[E_s + e\eta + e(\varphi_0 - \psi_1) + kT \ln([c_n]/[c_{n-1}])]^2}{4E_s kT}\right\}. \quad (8.19)$$

Here we have used Boltzmann distribution for the concentration $c_n(\delta)$.

In accordance with Eq. (8.12), the total current through the contact can be written in the form

$$i = g[c_n](1 - e^{-(e\eta/kT)}) \exp\left[-\frac{ne(\psi_1 - \varphi_0)}{kT}\right]$$

$$\cdot \exp\left\{-\frac{[E_s + e\eta + e(\varphi_0 - \psi_1) + kT \ln([c_n]/[c_{n-1}])]^2}{4E_s kT}\right\}. \quad (8.20)$$

The expression found is the final current–voltage characteristic for discharge. The potential ψ_1 in it is a function of the electrode potential. In the presence of a supporting electrolyte, all the equations obtained are valid. But in a binary solution, the dependence of ψ_1 upon φ_0 is different from that in the presence of a supporting electrolyte. We assumed in our calculations that the electrode surface has the same charge as the discharging ion. This permitted us to consider the electrode reaction to be of a nonadiabatic nature.

Let us consider in detail the last factor, writing it in the form

$$f_2(\eta, \psi_1) = \exp\left\{-\frac{[E_s + e\eta + e(\varphi_0 - \psi_1) + kT \ln([c_n]/[c_{n-1}])]^2}{4E_s kT} + e\eta\right\}$$

$$- \exp\left\{-\frac{[E_s + e\eta + e(\varphi_0 - \psi_1) + kT \ln([c_n]/[c_{n-1}])]^2}{4E_s kT}\right\}$$

$$= \exp\left[-\frac{E_a{}'}{kT}\right] - \exp\left[-\frac{E_a{}^2}{kT}\right]. \quad (8.21)$$

The quantities $E_a{}'$ play the part of effective energies of activation of the direct electron transfer. At the first glance, the result we obtained (i.e., the existence of the dependence of the current upon activation) is inconsistent with the properties of the electronic spectrum of the metal electrode. Since the energy spectrum of electrons in a metal is a continuous one, it would appear that the energy level of an ion could be compared to the same energy level in a metal. The transition between these levels should have to occur without activation. Actually, in a state of thermodynamic equilibrium and with small deviations from it, the energy levels of hydrated ions are not located arbitrarily with respect to the energy spectrum in the metal. They lie below the Fermi level and the electron transfer necessitates a fluctuation in the solvent polarization.

By using Eqs. (8.21), it is possible to write an expression for the change in the energy of activation during flow of current E_a, as compared to the equilibrium value of $(E_a)_{eq}$:

$$E_a - (E_a)_{eq} = \frac{[E_s + e\eta + e(\varphi_0 - \psi_1) + kT \ln([C_n]/[C_{n-1}])]^2}{4E_s}$$
$$- \frac{[E_s + e(\varphi_0 - \psi_1) + kT \ln([C_n]/[C_{n-1}])]^2}{4E_s}. \quad (8.22)$$

In the case of strongly hydrated ions, however, Eq. (8.22) can be simplified. In this case, the repolarization energy reaches several electron volts and is very large in comparison with other quantities in the square bracket, i.e., $E_s \gg e\eta + e(\varphi_1 - \psi_1) + kT \ln([C_n]/[C_{n-1}])$. Then, we shall have with good accuracy up to the second order of magnitude

$$\{E_1{}^* - (E_1{}^*)_{eq}\} \simeq (e\eta/2) + (e^2\eta^2/4E_s); \quad (8.23)$$

thus, the Brönsted's rule is obtained with the transfer coefficient equal to

$$\beta = \alpha = \tfrac{1}{2}. \quad (8.24)$$

In this case, formula (8.20) takes the form of Tafel's equation,

$$i = i_0 \exp\left[-\frac{ne(\varphi_0 - \psi_1)}{kT}\right]\left(\exp\left[\frac{e\eta}{2kT}\right] - \exp\left[-\frac{e\eta}{2kT}\right]\right), \quad (8.25)$$

where the exchange current is

$$i_0 = g([C_n][C_{n-1}])^{1/2} \exp\left[-\frac{E_s}{4kT}\right] \exp\left[\frac{e(\varphi_0 - \psi_1)}{2kT}\right]$$
$$= e\,\delta\bar{\rho}\left(\frac{\pi}{\hbar^2 kT E_s}\right)^{1/2} |L_{ms}^{(2)}(\delta)|^2([C_n][C_{n-1}])^{1/2}$$
$$\times \exp\left[-\frac{E_s}{4kT}\right] \exp\left[\frac{e(\varphi_0 - \psi_1)}{kT}\right]. \quad (8.26)$$

The dependence of the exchange current on the nature of the solvent is determined by the values of E_s and $\bar{\rho}$, $|L_{ms}^{(2)}(\delta)|^2$ and, in a conventional manner, by the concentrations of the oxidized and reduced ions. The transfer coefficient is of a universal character. It should be emphasized, however, that formula (8.25) is by no means universal, and appreciable deviations from it may be observed in practice. These deviations should be always present if the repolarization energy E_s is not large compared to the overvoltage $e\eta$.

The theoretical curve (which is not shown) was calculated on the density, does not depend on such quantities as the inner potential φ_M (or the work function) and the difference of hydration energies. According to Eq. (8.5), these quantities are contained in the transfer probability. By making use of the equilibrium condition, however, it is possible to exclude φ_M and $\Delta\alpha$ from the expression for the current density.

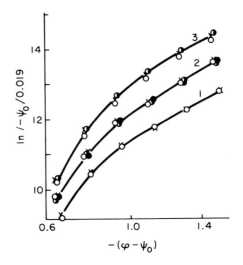

FIG. 9. Corrected Tafel plots of $Fe(CN)_6{}^{3-}$ reduction in solutions: (1) 10^{-3} N $Li_3Fe(CN)_6$ in the presence of LiCl in concentrations 10^{-3} N ($\times$), 3×10^{-3} N ($\bigcirc$); (2) 10^{-3} N $K_3Fe(CN)_6$ in the presence of KCl in concentrations 0 ($\bigcirc$), 5×10^{-4} N ($\times$), 10^{-3} N ($\mathbin{\text{◑}}$), 1.5×10^{-3} N ($\bullet$), (3) 10^{-3} N $Cs_3Fe(CN)_6$ in the presence of CsCl in concentrations 0 ($\bigcirc$), 3×10^{-4} N ($\times$), 5×10^{-4} N ($\mathbin{\text{◑}}$) (taken from Frumkin et al. (1963)).

There exists good quantitative agreement between the theory and experiment.

For the reaction of electroreduction of 10^{-3} N $Fe(CN)_6^{3-}$ on a dropping mercury electrode, an $(i - \varphi)$ curve was obtained (Frumkin et al., 1963) (Fig. 9).

The theoretical curve (which is not shown) was calculated on the basis of (8.20). In the case of ferricyanide, the energy of rearrangement E_s can be expected to be not too large. If E_s is evaluated by using Born's equation, it turns out that $E_s \simeq 0.65$ eV. Therefore, considerable deviations from Tafel's equation (8.25) should be expected at potentials sufficiently removed from the equilibrium potential, and it is necessary to use the more general Eq. (8.20).

We find that the theory is in quantitative agreement with experiment. The maximum deviation of the theoretical value of the current from the experimental one is near 10%.

In a precise measurement of the transfer coefficient in the reaction $Cr^{2+} \rightleftarrows Cr^{3+}$ for α, the result obtained (Parsonce and Passeron, 1966) was

$$\alpha = \tfrac{1}{2} + (e\eta/2E_s)$$

where $E_s = 0.26$ eV, in good agreement with the theory.

For comparison with experiment in detail, see Marcus (1964).

IX. Flow of Current through the Semiconductor–Solution Interface

We have seen that the current for electron exchange at a metal–solution interface depends only in a minor way on the properties of the metal. The flow of current at a semiconductor–solution interface should correspond to a somewhat different model. The concepts developed above, however, also apply to the latter type of interface.

Let us suppose that an oxidation–reduction reaction of type 2 occurs on a semiconductor electrode. The current through the interface is expressed by formulas (8.1)–(8.3). Actually, we did not imply in those equations any definite types of energy spectrum and distribution function, and these formulas are, therefore, of a general character. From the point of view of electron exchange of interest to us, the differences between the metal and the semiconductor are as follows:

(1) The electrons from the conduction and valence bands participate in the exchange at a semiconductor–solution interface.

(2) The electric field penetrates into the semiconductor, and this situation is responsible for a considerable part of the potential drop for the interface.

(3) Electrons in the semiconductor may be considered to be undegenerated in the conduction band.

Let us introduce the following notations (Fig. 10): The zero energy will be designated by U_0, as in the case of metals, the energy of particles (electrons and holes) by ϵ_f, and the Fermi level by ϵ_F. The half-width

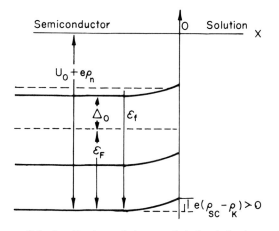

FIG. 10. Scheme of the localizations of characteristic levels in the semiconductor.

of the forbidden zone is equal to Δ_0; the inner potential in the semi-conductor in the equilibrium state is φ_{sc}^{eq}. The potential for the semicon-ductor–solution interface in the equilibrium state will be designated by φ_c^{eq}, so that in the range of depths of the order of $1/\kappa_{sc}$, a change in the potential from φ_{sc} to φ_c^{eq} occurs.

When current flows, the equilibrium distribution of potential in the semiconductor and in solution is disturbed. Thus, the potential for the nonequilibrium semiconductor is equal to

$$\varphi_{sc} = \varphi_{sc}^{eq} + \eta \tag{9.1}$$

where η is the overvoltage. We shall consider, however, the current flowing through the interface to be small, and the distribution of the carriers in it to be an equilibrium one. The change will amount only to the shift of potential in the equilibrium distribution of the carriers. We shall also take into consideration the change in the ψ_1 potential and in the

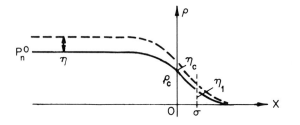

FIG. 11. Nonequilibrium potential distribution upon polarization. Abscissa, coordinate. Ordinate, potential near the semiconductor–solution interface.

potential at the interface φ_c during the flow of current. The following notation will be introduced:

$$\phi_c = \varphi_c^{eq} + \eta_c, \tag{9.2}$$

$$\psi_1 = \psi_1^{eq} + \eta_1. \tag{9.3}$$

On the basis of the foregoing, the potential distribution during the flow of current can be represented by the dashed line in Fig. 11. This shows the nonequilibrium potential distribution in the particular case of anodic polarization. Let us represent the current flowing through the interface by

$$i = [i^{(e)}_{sc-s} - i^{(e)}_{s-sc}] + [i^{(p)}_{sc-s} - i^{(p)}_{s-sc}] = i^{(e)} + i^{(p)} \tag{9.4}$$

where the subscript sc refers to the semiconductor and s to the solution; superscript (e) is used for the current component associated with the electron exchange between the ion and the conduction band, and (p) relates to the processes of electron exchange between the ion and the valence band. Since at the present time it is possible to measure the electron and the hole currents independently, we shall evaluate them separately.

In the same way as for a metal, we obtain the following general equations:

$$i^{(e)} = i^{(e)}_{s-sc}(\exp[e\eta/kT] - 1), \tag{9.5}$$

$$i^{(p)} = i^{(p)}_{s-sc}(\exp[e\eta/kT] - 1), \tag{9.6}$$

$$i^{(e)}_{s-sc} = i^{(e)}_{sc-s} \exp[-(e\eta/kT)], \quad i^{(p)}_{s-sc} = i^{(p)}_{sc-s} \exp[-(e\eta/kT)], \tag{9.7}$$

$$i = i^{(e)}_{s-sc}(\exp[e\eta/kT] - 1) + i^{(p)}_{s-sc}(\exp[e\eta/kT] - 1). \tag{9.8}$$

Thus, to obtain concrete relationships, we must calculate the electron and the hole components of the solution–semiconductor current $i^{(e)}_{s-sc}$ and $i^{(p)}_{s-sc}$.

First, let us write the expressions for the currents, similar to Eq. (8.2). Evidently, we have

$$i^{(e)}_{s-sc} = e \int_0^\infty c_n(x) \, dx \int_{(e)} W_{sc-s}(x, \epsilon_f) n^{(e)}(\epsilon_f) \rho^{(e)}(\epsilon_f) \, d\epsilon_f \tag{9.9}$$

$$i^{(p)}_{s-sc} = e \int_0^\infty c_n(x) \, dx \int_{(p)} W_{sc-s}(x, \epsilon_f) n^{(p)}(\epsilon_f) \rho^{(p)}(\epsilon_f) \, d\epsilon_f. \tag{9.10}$$

The probability for electron transfer from the semiconductor to the ion is given by the general equation (7.8).

In the conduction band we may put

$$n^{(e)}(\epsilon_f) \simeq \exp[-(\epsilon_f - \epsilon_F)/kT]$$

$$\rho^{(e)}(\epsilon_f) \simeq (1/4\pi^2)(2m^*/\hbar^2)^{3/2}(\epsilon_2 - \epsilon_f)^{1/2} \simeq \bar{\rho}^{(e)}.$$

Similarly, we have for the valence band

$$n^{(p)}(\epsilon_f) \simeq 1$$

$$\rho^{(p)}(\epsilon_f) \simeq (1/4\pi^2)(2m^*/\hbar^2)^{3/2}(\epsilon_f - \epsilon_1)^{1/2} \simeq \bar{\rho}^{(p)}$$

where ϵ_2 and ϵ_1 are the energies of the lower limit of the conduction band and of the upper limit of the valence band, respectively, $\bar{\rho}^{(e)}$ and $\bar{\rho}^{(p)}$ are the mean values of the level densities, and m^* is the effective mass of the electron in the semiconductor.

We can write, as for a metal, the approximate relationships

$$i^{(p)}_{s-sc} \simeq ec_n(\delta)\overline{\rho^{(p)}}\delta \int_{(e)} W_{sc-s}(\delta, \epsilon_f)\, d\epsilon_f, \tag{9.11}$$

$$i^{(e)}_{s-sc} \simeq ec_n(\delta)\overline{\rho^{(e)}}\delta \int_{(p)} W_{sc-s}(\delta, \epsilon_f)\, d\epsilon_f. \tag{9.12}$$

Let us consider now the range of integration over energies in Eqs. (9.11) and (9.12). The following circumstance should be taken into account in choosing the integration range in these bands: an electron transfer associated with the states located in the shaded area in Fig. 10 may occur, in principle, in the region where the bands bend. In the valence band, for example, the electron transfer in the range $0 \leq \epsilon_f \leq e(\varphi_{sc} - \varphi_c)$ may be of the type of a tunnel transfer through the barrier in the semiconductor. But the potential change inside the semiconductor from φ_{sc} to φ_c may occur at the depth equal to the Debye length in the semiconductor, i.e., of $1/\kappa_{sc}$. This depth corresponds to a very considerable barrier width. Since the barrier width is large, the subbarrier origin may be neglected. Under these conditions, the integration ranges have the width shown in Fig. 10. Thus, integration will be carried out over the ranges

$$u_0 + e(\varphi_{sc} - \varphi_c) \leq \epsilon_f \leq \epsilon_1 = \epsilon_F - \Delta_0 + e(\varphi_{sc} - \varphi_c)$$
$$\text{(valence band)}, \tag{9.13}$$

$$\epsilon_2 = \epsilon_F + \Delta_0 + e(\varphi_{sc} - \varphi_c) \leq \epsilon_f \leq u_0 + e(\varphi_{sc} - \varphi_c)$$
$$\text{(conduction band)}. \tag{9.14}$$

The integrands in Eqs. (9.11) and (9.12) converge rapidly. This indicates that it is only the states located at the limits of the bands (the states of electrons with a low energy corresponding to the bottom of the conduction band and those of the holes with the energies close to the upper edge of the valence band) which participate in setting up the current. Accordingly, the integration ranges may be extended, if we write

$$\int_{(e)} \cdots d\epsilon_f \simeq \int_{-\infty}^{\epsilon_F - \Delta_0 + e(\varphi_{sc} - \varphi_c)} \cdots d\epsilon_f, \tag{9.15}$$

$$\int \cdots d\epsilon_f \simeq \int_{\epsilon_F - \Delta_0 + e(\varphi_{sc} - \varphi_c)}^{\infty} \cdots d\epsilon_f. \tag{9.16}$$

The use of infinite integration ranges corresponds to a very small value of the integrand in the regions $\epsilon_f > \epsilon_F + \Delta_0 + e(\varphi_{sc} - \varphi_c)$ above the bottom of the valence band and $\epsilon_f < -[\epsilon_F - \Delta_0 + e(\varphi_{sc} - \varphi_c)]$ below the upper limit of the valence band.

Now pass on to the evaluation of the solution–semiconductor currents. We begin with the evaluation of the hole component of the current. We may use expression (8.17) for the evaluation of the hole component of the current by means of Eq. (9.10), by taking into consideration the integration ranges according to Eq. (9.15). We obtain from a calculation quite similar to that made in the preceding section for a metal

$$i_{s-sc}^{(p)} \simeq ec_n(\delta) \, \delta |L_{s-sc}|^2 \left(\frac{8\pi}{\hbar^2 E_s kT} \right)^{1/2} N_p^{(c)} \cdot \exp \left\{ \frac{\Delta_0 - e(q_{sc}^{eq} - \varphi_c^{eq})}{kT} \right\}$$

$$\times \exp \left\{ - \frac{\begin{matrix} [E_s + \Delta_0 + kT([c_n]/[c_{n-1}]) \\ + e(\varphi_0 - \psi_1) - e(\varphi_{sc}^{eq} - \varphi_c^{eq})]^2 \end{matrix}}{4E_s kT} \right\}. \tag{9.17}$$

Instead of $\overline{\rho^{(p)}}$ we introduce the equilibrium number of holes in a unit volume on the contact $N_p^{(c)}$; this quantity is

$$N_p^{(c)} = \int_0^{\epsilon_1} [1 - n(\epsilon_f)] \rho(\epsilon_f) \, d\epsilon_f \simeq \overline{\rho^{(p)}} \int_0^{\epsilon_F - \Delta_0 + e(\varphi_{sc}^{eq} - \varphi_c^{eq})}$$

$$\times \exp \left[\frac{\epsilon_f - \epsilon_F}{kT} \right] d\epsilon_f \simeq \overline{\rho^{(p)}} \exp \left\{ - \frac{\Delta_0 - e(\varphi_{sc}^{eq} - \varphi_c^{eq})}{kT} \right\}. \tag{9.18}$$

It should be emphasized first that no dependence on the overvoltage η is contained in Eq. (9.17). The significance is as follows. When the equilibrium is disturbed, the bottom edge of the valence band is somewhat lowered. But only the states located close to the upper edge of the band contribute to the hole component of the current. This becomes particularly apparent when we consider an electron transfer from the ion to the valence band of the semiconductor. Such a transfer is possible only into the vacant states at the upper edge of the band. For further simplification, let us consider the case when the rearrangement energy is large. In that case the squares of the values, which are small compared to E_s, may be neglected in the exponential. In addition, an equilibrium distribution for $c_n(\delta)$ can be used. Thus,

$$i_{s-sc}^{(p)} \simeq e\left(\frac{8\pi}{\hbar^2 E_s kT}\right)^{1/2} |L_{s-sc}|^2 \, \delta[C_n]^{\frac{1}{2} - [\Delta_0 - e(\varphi_{sc}^{eq} - \varphi_c^{eq})]/2E_s}$$

$$\times \, [C_{n-1}]^{\frac{1}{2} + [\Delta_0 - e(\varphi_{sc}^{eq} - \varphi_c^{eq})]/2E_s} \, \exp\left\{-\frac{[E_s + \Delta_0 - e(\varphi_{sc}^{eq} - \varphi_c^{eq})]^2}{4E_s kT}\right\}$$

$$\times \, N_p^{(c)} \exp\left\{\frac{\Delta_0 - e(\varphi_{sc}^{eq} - \varphi_c^{eq})}{kT}\right\}$$

$$\times \, \exp\left\{\left(-\frac{1}{2} + \frac{\Delta_0 - e(\varphi_{sc}^{eq} - \varphi_c^{eq})}{2E_s}\right)\left(\frac{e(\eta_c - \eta_1)}{kT}\right)\right\}$$

$$\times \, \exp\left[\frac{-ne\eta_1}{kT}\right] \tag{9.19}$$

where we introduced the quantities η_c and η_1, according to Eqs. (9.2) and (9.3).

In making a similar calculation for the electron component of the current $i_{s-sc}^{(e)}$, according to Eq. (9.9), taking into consideration (8.17) and (9.16), we find

$$i_{s-sc}^{(e)} \simeq e \, \delta |L_{s-sc}|^2 \left(\frac{8\pi}{\hbar^2 kTE_s}\right)^{1/2} \exp\left[\frac{\Delta_0 + e(\varphi_{sc}^{eq} - \varphi_c^{eq})}{kT}\right]$$

$$\times \, \exp\left[-\frac{[E_s + \Delta_0 + e(\varphi_{sc}^{eq} - \varphi_c^{eq})]^2}{4E_s kT}\right][c_n]^{\frac{1}{2} + [\Delta_0 + e(\varphi_{sc}^{eq} - \varphi_c^{eq})]/2E_s}$$

$$\times \, N_e[c_{n-1}]^{\frac{1}{2} - [\Delta_0 + e(\varphi_{sc}^{eq} - \varphi_c^{eq})]/2E_s}$$

$$\times \, \exp\left[-\left(\frac{1}{2} - \frac{\Delta_0 + e(\varphi_{sc}^{eq} - \varphi_c^{eq})}{2E_s}\right) \cdot \frac{e(\eta_c - \eta_1)}{kT}\right]$$

$$\times \, \exp\left[\frac{-ne\eta_1}{kT}\right] \exp\left[\frac{-e\eta}{kT}\right]. \tag{9.20}$$

In Eq. (9.20) the Boltzmann distribution for $c_n(\delta)$ has been taken into account. Unlike the hole component, the electron component of the current depends on the overvoltage η. When the equilibrium is disturbed, the bottom edge of the conduction band ϵ_2 is shifted by the value of η. Since the main contribution to the electron component is made by the states with an infinitesimal energy, the occupation of the states with large energies is exponentially small. This shift is directly reflected in Eq. (9.20).

Equations (9.19) and (9.20) can be somewhat simplified, however, if it is assumed that

$$E_s \gg \Delta_0 + e(\varphi_{sc}^{eq} - \varphi_c^{eq}).$$

This last inequality is not always valid.

Let us introduce the exchange current for the hole and the electron components of the current

$$i_0^{(p)} \simeq e\left(\frac{8\pi}{\hbar^2 kTE_s}\right)^{1/2} \delta |L_{s-sc}|^2([c_n][c_{n-1}])^{1/2} N_c^{(p)}$$

$$\times \exp\left[\frac{\Delta_0 - e(\varphi_{sc}^{eq} - \varphi_c^{eq})}{kT}\right] \exp\left\{-\frac{[E_s + \Delta_0 - e(\varphi_{sc}^{eq} - \varphi_c^{eq})]^2}{4E_s kT}\right\} \tag{9.21}$$

$$i_0^{(e)} \simeq e\,\delta\left(\frac{8\pi}{\hbar^2 kTE_s}\right)^{1/2} |L_{s-sc}|^2(c_n c_{n-1})^{1/2} N_c^{(p)} \exp\left[\frac{\Delta_0 + e(\varphi_{sc}^{eq} - \varphi_c^{eq})}{kT}\right]$$

$$\times \exp\left\{-\frac{[E_s + \Delta_0 - e(\varphi_{sc}^{eq} - \varphi_c^{eq})]^2}{4E_s kT}\right\}. \tag{9.22}$$

The ratio of the exchange currents of the electron and the hole components is equal to

$$\frac{i_0^{(e)}}{i_0^{(p)}} = \exp\left\{-\frac{e(\varphi_{sc}^{eq} - \varphi_c^{eq})}{kT}\left(1 + \frac{\Delta_0}{E_s}\right)\right\}. \tag{9.23}$$

By introducing the exchange currents, we can write the current components $i_{s-sc}^{(p)}$ and $i_{s-sc}^{(e)}$ in a more compact form:

$$i_{s-sc}^{(p)} = i_0^{(p)} \exp\left\{-\frac{e(\eta_c - \eta_1)}{2kT}\right\} \exp\left\{-\frac{ne\eta_1}{kT}\right\}. \tag{9.24}$$

$$i_{s-sc}^{(e)} = i_0^{(e)} \exp\left\{-\frac{e(\eta_c - \eta_1)}{2kT}\right\} \exp\left\{-\frac{ne\eta_1}{kT}\right\} \exp\left[-\frac{e\eta}{kT}\right]. \tag{9.25}$$

Equations (9.24) and (9.25) clearly show that by taking into consideration the potential drop in the electrolyte and in the semiconductor one can write the current components in terms of the factors of the form

$$\exp\left\{-\frac{e(\eta_c - \eta_1)}{2kT}\right\} \quad \text{and} \quad \exp\left\{-\frac{ne\eta_1}{kT}\right\}.$$

The total current flowing through the solution–semiconductor contact is given by the combination of Eqs. (9.9), (9.24), and (9.25). To avoid writing a cumbersome equation, let us write the expression for the density current in the following reduced form:

$$i = i_0^{(p)} \exp\left[-\frac{e\eta_c}{kT}\right] \exp\left[-(n - \tfrac{1}{2})\frac{e\eta_1}{kT}\right]\left(\exp\left[\frac{e\eta}{kT}\right] - 1\right)$$
$$+ i_0^{(e)} \exp\left[-\frac{e\eta_c}{kT}\right] \exp\left[-(n - \tfrac{1}{2})\frac{e\eta_1}{kT}\right]\left(1 - \exp\left[-\frac{e\eta}{kT}\right]\right). \quad (9.26)$$

The foregoing consideration dealt with semiconductors exhibiting intrinsic conductivity. The same relationships are obtained for impurity semiconductors. But in the latter case, the half-width of the band Δ_0 in the equations for the exchange currents $i_0^{(e)}$ and $i_0^{(p)}$ should be substituted by $\Delta_0^{(e)}$ and $\Delta_0^{(p)}$, respectively, where $\Delta_0^{(e)}$ and $\Delta_0^{(p)}$ are the distances from the Fermi level to the edge of the corresponding band.

The expressions for the electron and the hole components of the current, as well as the general expression for the current density derived in this section, clearly show how strong is the influence of the structure of the energy spectrum of the electrode upon the kinetics of the oxidation–reduction process. Because of the existence of a relatively wide forbidden zone, the electron-transfer processes occur mainly at the levels in the vicinity of the bottom of the conduction band or near the upper edge of the valence band. The current for each of these bands is

$$i^{(e)} = i_0^{(e)}\left(\exp\left[\frac{e\eta}{kT}\right] - 1\right) \exp\left[-\frac{e}{kT}(\eta_{c/2} + (n - \tfrac{1}{2})\eta_1)\right]. \quad (9.27)$$

$$i^{(p)} = i_0^{(p)}\left(1 - \exp\left[\frac{e\eta}{kT}\right]\right) \exp\left[-\frac{e}{kT}(\eta_{c/2} + (n - \tfrac{1}{2})\eta_1)\right] \quad (9.28)$$

for the conduction and the valence bands, respectively. Equations (9.27) and (9.28) differ essentially from a similar expression (8.20) for the current at the metal electrode. The essential difference between Eqs. (9.27) and (9.28) and the similar expression (8.19) for a metal electrode, is the absence of overvoltage, η, to the second power in the exponential of the former equations. If each of the current components is formally identified with Tafel's law, the transfer coefficients are

$$\alpha^{(p)} = 0, \qquad \alpha^{(e)} = 1,$$

respectively. We shall discuss later the problem of the determination of transfer coefficients at greater length.

The exchange currents for holes and for electrons do not coincide. But they are both proportional to the number of the electron or hole carriers in the semiconductor and to that of ions in the solution directly at the interface.

On the basis of a qualitative physical consideration, the author and collaborators (Myamlin and Pleskov, 1967; Vdovin *et al.*, 1949) and, independently, Green (1959), used expressions of the type (9.27) and (9.28), and took into consideration also the dependence of the exchange currents on the concentration of the carrier.

X. Physical Pattern of Charge Transfer Reaction in a Polar Medium, Which Is Accompanied by the Breaking or Forming of Chemical Bonds

In a previous section we have considered the reactions with electron transfer between two ions or an ion and a metal. These reactions, if not accompanied by breaking or forming of chemical bonds, form the class of simplest reactions. However, the reactions of this type are realized comparatively rarely. In an overwhelming number of cases the occurrence of reactions is accompanied by the breaking or forming of new chemical bonds. This refers both to the vast majority of redox reactions and to all the reactions with proton transfer. In particular, let us consider a mechanism for a process of the following type (Doganadze *et al.*, 1965, 1968b):

$$AH^+ + B^- \rightarrow A + HB$$

where A and B are arbitrary one-or-many atomic particles. We shall always assume the interaction of particles AH^+ and B^- with the solvent to be strong. That this is so is indicated by high values of the hydration energy, which may reach values up to 180 kcal/mol (i.e., about 8 eV).

The difference between the complex ion AH^+ and the simple ion of the type Fe^{2+} is first in the fact the complex ion also possesses the oscillatory degrees of freedom in addition to the electronic energy. Frequencies of intramolecular oscillations contribute values of about 0.2–0.3 eV (i.e., about 2–$3 \cdot 10^3$ degrees) and depend both on the stiffness of the corresponding bonds and on the particle masses. Thus, at room temperature the following inequality is always fulfilled:

$$\hbar\omega_i \gg kT.$$

Because of this the particles performing intramolecular oscillations behave as pure quantum particles independently of their masses.[†] The mass affects only the values of the intramolecular frequencies. Therefore, the

---

[†] In this connection we recall that the conditions of classical behavior for free particles are determined by their masses. This does not refer, however, to bound particles, present in molecules.

considerations made about the electron transfer are to some extent also valid for proton transfer. As there exists the reaction heat, which is emitted or absorbed, the energy levels of the initial and final system are displaced relative to one another by the amount of the reaction heat. In order to create the reaction situation it is necessary that the fluctuations of solvent should arise. However, as the gap between the oscillation levels is an order of magnitude less than the distance between the electronic levels, under some conditions there may be some significance to the transitions with thermal excitement of the lowest oscillatory states. In the language of terms, the pattern of the transition in a reaction appears to be the following: the surface of the potential energy of the molecules AH^+ and B^-H^+ comprise (in the approximation of harmonic oscillations) the paraboloids shown in Fig. 12, where R is the proton

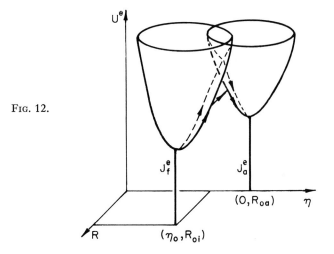

FIG. 12.

coordinate, and the solvent coordinate is denoted by q, i.e., the set of quantities characterizing the polarization of the solvent. When the solvent polarization fluctuates (i.e., its coordinate q changes values) the potential energy of the system in the initial state changes as well and the system moves continuously along the energy surface $U(q, R)$ preserving the constant value $R = R_0$. The oscillations along the bond $A-H^+$ remain unexcited. At the moment, when the solvent achieves the point q^*, the proton energy in the molecule AH^+, i.e., $U(R_{01}, q^*)$, proves to be equal to the proton energy in the molecule BH, i.e., $U(R_{02}, q^*)$, and the system performs a quantum transition, which is shown by an arrow in Fig. 12. For invariant values of the coordinates of the slow subsystem, $q = q^*$, the proton crosses over from the normal state in molecule AH

into the normal state of molecule B^-H^+. This transition takes place through the barrier, which changes continuously during the transition. Then the solvent moving continuously along the coordinate q leaves the fluctuation state, and the molecule B^-H^+ together with the solvent which surrounds it comes to the state with minimal energy (i.e., to the point R_{02}, q_{02}). This last transition also occurs for a fixed position of the proton. The calculation confirms the pattern of the process described above and allows us to calculate the rate of the corresponding reaction. Of course, we have chosen the example with an AH^+ molecule only for the sake of concreteness. The nature of the transported particle itself does not play any essential role. It is only the frequency that is important here and not the mass of the particle. With decreasing frequency the transitions through excited levels become increasingly important.

For all the coordinates satisfying the condition $\hbar\omega \ll kT$, the transition occurs classically. More precisely, the transition goes through the excited states and the saddle point of the potential energy surface. For all other coordinates which satisfy the inverse inequality, the transition has a quantum character. We note that the larger the particles mass, the smaller is the distance at which the quantum transition occurs. If we evaluate this distance for an electron to be about 5 to 7 Å, then for a proton (see below) it constitutes only about 0.5 Å.

XI. Quantitative Description of a Charge Transfer Reaction with Chemical Bond Breaking

After a qualitative discussion of the reaction which is accompanied by chemical bonds breaking, we may proceed to the quantitative description of such reactions. For the sake of generality we take as our examples the proton transfer reactions in a polar solvent,

$$AH^+ + B^- \rightarrow A + BH \tag{11.1}$$

where A and B are some molecules. In particular A or B may also be macroscopic bodies, i.e., a massive electrode surface. Then, if B is the metal electrode M, and A corresponds to the water molecule, reaction (11.1) reduces to

$$H_2O \cdot H^+ + M \rightarrow H_2O + H_{ads} + H \tag{11.1'}$$

where H_{ads} is the hydrogen atom adsorbed on the metal surface. One may neglect the change of one electron in the metal. We shall consider reaction (11.1) as sufficiently slow in the sense that the diffusion limitations may be eliminated. We shall assume that both in the initial and in the final state the proton forms sufficiently strong bonds. This means that oscillations of the proton along the bonds Me—H and H_2O—H^+ occur with frequencies which satisfy the inequality

$$\hbar\omega_{ads}, \qquad \hbar\omega_p \gg kT.$$

For bonds of the type C—N, N—H, or O—H, these frequencies comprise about 2600 to 3600 cm^{-1}. Frequencies corresponding to the bonds H^+—H_2O as well as to H_{ads} are not known exactly. However, the estimates from spectroscopic data (for H_2O^+) and also from the boundary energies of the corresponding hydrides of metals show $\omega_{ads} \sim 2 \cdot 10^{14}$ sec^{-1} so that inequality is fulfilled for both bonds.

We shall now assume the H_3O^+ ion discharge to occur directly from the outer Helmholtz plane. Therefore, instead of the space distribution of the discharging ions, it is necessary to know only their concentration at the surface C_S'. This assumption is physically justified since the transition probability decreases sharply with increasing distance from the electrode.

Actually, the ion distribution of an indifferent electrolyte affects only the potential distribution in the solution, i.e., it influences only the value of the initial energy of the system and the value of C_S. The statistical problem can be separated from the quantum-mechanical calculation and the current can be formally calculated and presented in closed form. However, for definiteness, we shall consider the concentration of the indifferent electrolytes to be large enough and the total potential drop to occur in the inner region of the double layer. In that case, the value of C_S will be close to that of the bulk concentration and the electrostatic energy of the discharging ion will be zero.

Since hydroxonium-ion discharge results in a proton transfer from the water molecule contained in the hydroxonium ion to the electrode, the complete quantum-mechanical Hamiltonian of the system can be conveniently written as

$$H(x_w, x_m, R, q) = H_i^0(x_w, x_m, R, q) + V_{pm}(x_m, R)$$
$$= H_a^0(x_m, x_w, R, q) + V_{pw}(x_w, R),$$

where x_w, x_m, and R are the coordinates of the electrons binding the proton in the hydroxonium ion, of the electron forming with the proton an adsorbed state, and of the proton, respectively; q is a set of normal coordinates of the solvent $\{q_k\}$; V_{pw} and V_{pm} are the energies of interaction of the proton with the water molecule in the hydroxonium ion and with the electrode, respectively. H_i^0 and H_a^0 are the unperturbed values of the

Hamiltonian of the system in the initial and final states of the system, respectively.

The electron terms of the initial and final states should be obtained using the adiabatic approximation. We shall consider the electrons to be the fast subsystem and the proton and the solvent the slow subsystem. The physical significance of this approximation is that in determining the wavefunction of the electron subsystem it is possible to ignore the kinetic energy of the slow subsystem, the proton and the solvent, owing to the slowness of their motion. In other words, the electron state can be determined accurately enough if the proton and the solvent are considered to be almost at rest. Naturally, the electron energy in the initial $\epsilon_f(R, q)$ and final $\epsilon_a(R, q)$ states depend on the coordinates of the slow subsystem as parameters.

The use of the adiabatic approximation for the slow subsystem means that the proton and the solvent interact with a cloud of fast-moving electron charge rather than with a point electron. In this case $\epsilon_f(R, q)$ and $\epsilon_a(R, q)$ act as the potential energy of interaction between the heavy particles and the electron cloud.

In an adiabatic approximation, the wavefunction of the total system in the initial and final states can be represented as

$$\Psi^0_{f,a}(x, R, q) = \Psi_{f,a}(x)\theta_{f,a}(R, q) \tag{11.2}$$

where $\Psi_{f,a}(x)$ are the wavefunctions of the electrons and $\theta_{f,a}(R, q)$ the wavefunction of the proton and the solvent, which are determined from

$$\left\{ -\frac{\hbar^2}{2M} \frac{\partial^2}{\partial R^2} - \tfrac{1}{2}\hbar\omega_0 \sum_k \frac{\partial^2}{\partial q_k{}^2} + U^e_{f,a}(R, q) \right\}\theta_{f,a}(R, q) = E^0_{f,a}\theta_{f,a}(R, q). \tag{11.3}$$

Here M is the proton mass and, $U^e_{f,a}(R, q)$ incorporates both $\epsilon_{f,a}(R, q)$ and the energy of direct interaction between the heavy particles. It is evident from (11.3) that the quantities $U^e_{f,a}(R, q)$ act as a complete potential energy for the slow subsystem; they are called electron terms.

In addition to the electron terms, the electron–proton terms can be introduced, owing to the existence of the relationship between the proton and polarization frequencies,

$$\omega_i, \qquad \omega_a \gg \omega_0. \tag{11.4}$$

When condition (11.4) is fulfilled, it is possible to use adiabatic perturbation theory, assuming the proton to be the fast subsystem and the solvent the slow one. Thus a complete physical picture is based on the use of a double adiabatic approximation: (1) the electron is considered as being fast compared to the proton and the solvent, (2) the proton is considered fast compared to the solvent.

In the double adiabatic approximation, the wavefunctions of the system in the initial and final states can be written

$$\Psi_{f,a}^0(x, R, q) = \Psi_{f,a}(x)\chi_{f,a}(R, q)\phi_{f,a}(q) \tag{11.5}$$

where Ψ and χ are the wavefunctions of the electrons and the proton.

The wavefunctions of the solvent ϕ are determined from

$$\left\{ -\tfrac{1}{2}\hbar\omega_0 \sum_k \frac{\partial^2}{\partial q_k^2} + U_{f,a}^{ep}(q) \right\}\phi_{f,a}(q) = E_{f,a}^0\phi_{f,a}(q) \tag{11.6}$$

where $U_{f,a}^{ep}$ is the potential energy of the solvent, i.e., the electron–proton term.

By expanding the electron term of the initial state in a power series in terms of small deviations from the equilibrium values of the coordinates of the proton and the solvent, we obtain

$$U_f^e = J_f^e + \tfrac{1}{2}\hbar\omega_0 \sum_k Q_k^2 + \tfrac{1}{2}\hbar\omega_i\xi^2 + \sum_k \gamma_k Q_k\xi_k \tag{11.7}$$

where $\xi = (M\omega_i/\hbar)^{1/2}(R - R_{0i})$ is the deviation of the dimensionless coordinate of the proton from the equilibrium value corresponding to the initial state; $Q_k = q_k - q_{k0}$ is the deviation of the normal coordinates of the solvent from the equilibrium values. The minimal value of the potential energy in the initial state can be written

$$J_f^e = \epsilon_f - e\varphi + E_{0i} \tag{11.8}$$

where φ is the potential of the metal and E_{0i} the minimal potential energy of the proton in the hydrated H_3O^+ ion, comprising both the electrostatic energy of hydration and the chemical binding energy.

The surface described by (11.7) is a paraboloid in $(N + 1)$-dimensional space (N is the number of oscillators). The presence of the last term in (11.7), which corresponds to the interaction between the proton vibrations and the water oscillators, leads to the turning of the main axes of the paraboloid about the axes of the coordinates in the $(N + 1)$-dimensional space. The angle of rotation is determined by

$$\gamma_k = \left(\frac{\partial^2 U_f^e}{\partial \xi \, \partial Q_k} \right)_{\xi = Q_k = 0}. \tag{11.9}$$

Although further calculations can be carried out taking into consideration the last term in (11.7) it is possible to simplify the formulas by taking advantage of the fact that the frequency of vibrations of the proton differs greatly from that of the solvent. An accurate calculation shows that the parameter determining the value of the interaction of these vibrations is

$$\frac{16 \sum_k \gamma_k^2}{\hbar^2(\omega_i^2 - \omega_0^2)} \approx 16 \frac{\omega_0}{\omega_i} \cdot \frac{\int (\mathbf{P}_0 - \mathbf{P}_0{}^0)^2 \, dv}{\int \mathbf{P}_0(\mathbf{P}_0 - \mathbf{P}_0{}^0) \, dv}. \tag{11.10}$$

The significance of the quantities $\mathbf{P}_0{}^0$ and $\mathbf{P}_0$ can be explained as follows. Let us denote by R_{0i}^0 the coordinate of the equilibrium position of the proton in the H_3O^+ ion in the gaseous phase. When the H_3O^+ ion is placed into the solvent, its equilibrium position alters due to polarization. Let us denote the new value of the coordinate of the equilibrium point by R_{0i}. The quantities $\mathbf{P}_0{}^0$ and $\mathbf{P}_0$ are equal to the equilibrium values of polarization due to the H_3O^+ ion, in which the equilibrium coordinate of the proton is R_{0i}^0 and R_{0i}, respectively. The estimation of the parameter (11.10) with a dielectric continuum approximation gives

$$\frac{16 \sum_k \gamma_k^2}{\hbar^2(\omega_i^2 - \omega_0^2)} \approx \frac{\omega_0}{\omega_i} \cdot \frac{|R_{0i} - R_{0i}^0|}{\tau_0} \ll 1 \tag{11.11}$$

where τ_0 is the radius of the H_3O^+ ion.

The inequality (11.11) shows that the last term in (11.7) can be dropped. Finally, $U_f{}^e$ can be written as

$$U_f{}^e = J_f{}^e + \tfrac{1}{2}\hbar\omega_0 \sum_k (q_k - q_{k0})^2 + \tfrac{1}{2}M\omega_i^2(R - R_{0i})^2. \tag{11.12}$$

Similarly, for the term of the final state $U_a{}^e$ we have

$$U_a{}^e = J_a{}^e + \tfrac{1}{2}\hbar\omega_0 \sum_k q_k^2 + \tfrac{1}{2}M\omega_a^2(R - R_{0a})^2 \tag{11.13}$$

where $J_a{}^e$ is the minimal potential energy corresponding to the adsorbed state of the hydrogen atom. R_{0a} is the equilibrium coordinate of the proton in the adsorbed state.

The theoretical analysis of our problem can be greatly simplified if the system of coordinates is turned in $(N + 1)$-dimensional space $\{q_k\}$ in such a way that in the new coordinates, η_k, the minimum of the initial state term, lies on one of the new axes η. Then the proton transfer will correspond to the change in the equilibrium coordinate of only one oscillator. Finally, the electron terms of the system can be written in the form

$$U_f{}^e = J_f{}^e + \tfrac{1}{2}\hbar\omega_0(\eta - \eta_0)^2 + \tfrac{1}{2}M\omega_i^2(R - R_{0i})^2 \tag{11.14}$$

$$U_a{}^e = J_a{}^e + \tfrac{1}{2}\hbar\omega_0\eta^2 + \tfrac{1}{2}M\omega_a^2(R - R_{0a})^2 \quad (\eta_0 = \sum_k q_{k0}^2). \tag{11.15}$$

Thus, in the variables η and R the electron terms prove to be two-dimensional (Fig. 12). This greatly simplifies the graphic analysis of the transition.

The electron–proton terms are obtained in a similar way,

$$U_{f,n}^{ep} = J_{fn} + \tfrac{1}{2}\hbar\omega_0(\eta - \eta_0)^2 \tag{11.16}$$

$$U_{a,n'}^{ep} = J_{an'} + \tfrac{1}{2}\hbar\omega_0\eta^2 \tag{11.17}$$

where

$$J_{fn} = J_f^e + \hbar\omega_i(n + \tfrac{1}{2}) \tag{11.18}$$

$$J_{an'} = J_a^e + \hbar\omega_a(n' + \tfrac{1}{2}). \tag{11.19}$$

XII. Transition Probability

The solutions of Schrödinger's equations of the unperturbed Hamiltonians H_i^0 and H_a^0 have been considered above. The dropped terms V_{pm} and V_{pw} cause transitions between the unperturbed states to occur (proton discharge and hydrogen ionization, respectively).

Now let us calculate the proton transfer probability from molecule AH to ion B$^-$. Let us use the general formula of the perturbation theory for the transfer probability (5.14). The wavefunctions of the final and the initial states of the system contained in the expression for the matrix element are solutions of Schrödinger's equation for the unperturbed Hamiltonian, $H_i^{(0)}$ and $H_a^{(0)}$, and the potential v contained in (5.14) describes the interaction of a proton, bound in the molecule H_3O^+, with the metal surface which leads to the proton jump. In calculating the matrix element, we shall use Condon's approximation, according to which the wavefunction of the fast subsystem depends slightly on the coordinates of the slow subsystem q. The wavefunction therefore can be taken at a point corresponding to the maximal contribution of the overlapping integral,

$$\langle \Psi_a^*(x, R, q)v\Psi_f(x, R, q)\rangle = \langle \Psi_a^*(x)v\Psi_f(x)\rangle\langle \chi_a^*\chi_f\rangle\langle \phi_a^*(q)\phi_f(q)\rangle$$
$$= L\langle \chi_a^*\chi_f\rangle\langle \phi_a^*\phi_f\rangle \tag{12.1}$$

where L is the overlapping integral of the electron wavefunctions. To a harmonic approximation, the wavefunction of the slow subsystem can be written as the product (11.5). To calculate the total probability of the proton transfer in unit time, it is necessary to carry out a statistical averaging over all possible initial states of the system and a summation over the final states. In this case we have

$$\bar{W} = \sum_{n,n'} \sum_{m,m'} \frac{\exp\{-(m + \tfrac{1}{2})\hbar\omega_\mathrm{p}/kT\}\cdot\exp\{-(n + \tfrac{1}{2})\hbar\omega_0/kT\}W_{n,n'}^{m,m'}}{Q_1\cdot Q_2} \delta(E)$$

(12.2)

where Q_1 and Q_2 are the corresponding statistical sums, and δE is the Kronecker symbol expressing the conservation of energy law. Calculating the matrix elements and carrying out the necessary summations, as well as making use of the fact that $\hbar\omega_\mathrm{p} \gg kT$ and $\hbar\omega_0 \ll kT$, we obtain the following expressions for the mean probability W:

$$\bar{W} = \frac{\omega_0}{2\pi} \frac{L^2}{M} \exp\left\{-\frac{E_\mathrm{p}}{\hbar\omega_\mathrm{p}}\right\} \sum_{l=-\infty}^{\infty} \left(\frac{E_\mathrm{p}}{\hbar\omega_\mathrm{p}}\right)^l \cdot \frac{1}{l!} \cdot \exp\left\{-\frac{\hbar\omega_\mathrm{p}}{2kT}(|l| + l)\right\}$$
$$\cdot \exp\left\{-\frac{(E_\mathrm{s} + \varDelta J - l\hbar\omega_\mathrm{p})^2}{4E_\mathrm{s}kT}\right\}$$

(12.3)

where

$$M = (\hbar\omega_0/2)(E_\mathrm{s}kT/\pi^3)^{1/2},$$

and

$$E_\mathrm{s} = \hbar\omega_0\eta_0^2/2$$

is the repolarization energy of the solvent upon proton transfer, and $E_\mathrm{p} = \hbar\omega_\mathrm{p}(\xi_0^2/2)$.

It should be noted in calculating the sum contained in (12.3) at different values of $\varDelta J$ that the main contribution to the sum is usually made by one term corresponding to a certain $l = l^*$, which can be determined from the exponential factor in (12.3). The factor $(1/l!)(E_\mathrm{p}/\hbar\omega_\mathrm{p})^l$ changes slower than the exponential factor and as it can be shown not to affect significantly the value of l^*,

$$l^* = \begin{cases} (\varDelta J - E_\mathrm{s})/E_\mathrm{s}, & E_\mathrm{s} \le \varDelta J \\ 0, & -E_\mathrm{s} < \varDelta J < E_\mathrm{s} \\ (\varDelta J + E_\mathrm{s})/E_\mathrm{s}, & \varDelta J \le -E_\mathrm{s}. \end{cases}$$

(12.4)

Thus, for simplicity, restricting ourselves to one term in the sum, we obtain

$$\bar{W} = \frac{\omega_0}{2\pi} |L|^2 \exp\left\{-\frac{E_\mathrm{p}}{\hbar\omega_\mathrm{p}}\right\} \frac{1}{M|l^*|!} \left(\frac{E_\mathrm{p}}{\hbar\omega_\mathrm{p}}\right)^l \exp\left\{-\frac{\hbar\omega_\mathrm{p}}{2kT}(|l^*| + l^*)\right\}$$
$$\cdot \exp\left\{-\frac{(E_\mathrm{s} + \varDelta J - l^*\hbar\omega_\mathrm{p})^2}{4E_\mathrm{s}kT}\right\}.$$

(12.5)

Expression (12.5) for the transfer probability has an Arrhenius form with the activation energy equal to

$$E_a = \frac{(E_s + \varDelta J - l^* \hbar \omega_p)^2}{4E_s} + \frac{\hbar \omega_p}{2} (|l^*| + l). \tag{12.6}$$

Using formula (12.4) we can rewrite (12.6) as

$$E_a = \begin{cases} \varDelta J, & E_s \leq \varDelta J \\ (E_s + \varDelta J)^2/4E_s, & -E_s < \varDelta J < E_s \\ 0, & \varDelta J \leq -E_s. \end{cases} \tag{12.7}$$

The parameter E_s contained in the expression for E_a is an important parameter of the theory, characterizing the effect of the solvent on the kinetics of a chemical reaction. A significant result of the present treatment is the fact that the activation energy in the proton transfer reaction is characterized not by the shape of the potential wells determining proton vibrations in H_3O^+ and H_{ads} as was assumed in the Horiuti–Polanyi (1935) theory, but by vibrational electron–proton terms of the solvent, this dependence being determined by a single parameter E_s, the repolarization energy of the solvent. The value of E_s can be estimated theoretically and is a quantity of the order of 1 to 2 eV. At the same time, as will be shown, the repolarization energy of the solvent can be determined from experimental data on the dependence of the reaction rate constant on the reaction heat.

Introducing the notion

$$\kappa^* = \frac{|L|^2}{M} \frac{1}{|l^*|!} \left(\frac{E_p}{\hbar \omega_p}\right)^l \exp\left\{-\frac{E_p}{\hbar \omega_p}\right\}. \tag{12.8}$$

we can finally rewrite formula (12.5) for the transfer probability $\bar{W}$ as follows:

$$\bar{W} = (\omega_0/2\pi)\kappa^* \exp(-E_a/kT). \tag{12.9}$$

The coefficient κ^* is usually called the effective transmission coefficient. It follows from the calculation carried out that κ characterized the probability of a transition of the system from one term to the other when passing through an activated state. Generally speaking, it should be noted that each value of κ^* corresponds to a particular term of the sum in (12.3). Therefore, the symbol κ^* corresponds to the term in (12.3) which makes the main contribution to the transition probability. In general, the transition probability with an allowance made for any initial and final states is of the form

$$W_{ia}(\epsilon_f) = Av_n \sum_{n'} W_{nn'}(\epsilon_f) \tag{12.10}$$

where

$$W_{nn'}(\epsilon_f) = \kappa_{nn'} \cdot \frac{\omega_0}{2\pi} \cdot \exp\left\{ -\frac{(J_{an'} - J_{fn} + E_s)^2}{4E_s kT} \right\}. \tag{12.11}$$

As has been shown in the theory of redox reactions, the transition of the solvent from the initial to the final state follows the classical path (through the intersection point of the electron potential energy). Naturally, the situation is the same in the case of a proton discharge, the only difference being that here the classical transition occurs on the electron–proton terms. The proton transition must be assumed to occur from a given level n to a definite level n'. The transition probability obtained above permits us to interpret the reaction path by means of the electron terms. For definiteness, the transition of the system at $n = n' = 0$ is shown in Fig. 13. It is clear from the figure that at first activation of

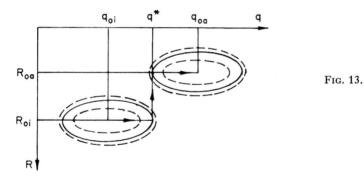

FIG. 13.

the solvent occurs at a fixed position of the proton. At the point q^*, at fixed coordinates of the solvent, a quantum proton transition occurs.

It should be stressed that in considering a quantum proton transition at the point q^*, it is impossible to calculate the transition probability using the Gamov formula.

XIII. Current Density

Knowing the transition probability it is possible to determine the cathodic cd by means of the general formula

$$i = 2eC_s \int w_{ia}(\epsilon_f)n(\epsilon_f)\rho(\epsilon_f)\, d\epsilon_f \tag{13.1}$$

where C_s is the surface concentration of hydroxonium ions (on the outer side of the double layer). $n(\epsilon_f)$ is the Fermi distribution,

$$n(\epsilon_f) = \left\{ \exp\left[\frac{\epsilon_f - \epsilon_F}{kT} \right] + 1 \right\}^{-1}, \tag{13.2}$$

$\rho(\epsilon_f)$ the electron-level density in the metal and $w_{ia}(\epsilon_f)$- the H_3O^+-ion discharge probability per unit time. The factor 2 is introduced since under steady state conditions there is one act of electrochemical desorption per one act of the H_3O^+ discharge.

In accordance with the general formula for the transition probability, all possible excited states of the proton both in the initial and the final states should be taken into consideration,

$$i = \mathrm{Av}_n \sum_{n'} i(n, n') \tag{13.2'}$$

where $i(n, n')$ is the current calculated assuming that in the initial state the proton is at the level E_n, and in the final state at $E_{n'}$. In the harmonic approximation,

$$E_n = \hbar\omega_i(n + \tfrac{1}{2}), \qquad E_{n'} = \hbar\omega_a(n' + \tfrac{1}{2}) \tag{13.3}$$

where the quantum numbers $n, n' = 0, 1, 2, \ldots$. First we shall calculate $i(n, n')$, and then find the expression for the mean current i.

Using the formula for the transition probability and the Fermi distribution, we write $i(n, n')$ as

$$i(n, n') = 2ec_s\rho^*(\omega_0/2\pi)\cdot\kappa(n, n') \int \frac{\exp\{-(\epsilon_f - \epsilon_{nn'}^0)^2/4E_s kT\}}{\exp\left[(\epsilon_f - \epsilon_F)/kT\right] + 1} \, d\epsilon_f \tag{13.4}$$

where

$$\epsilon_{nn'}^0 = \epsilon_F + E_s + (J_{an'} - J_{Fn}^0) - e\eta. \tag{13.5}$$

In this case we consider the density of the levels to be a smooth function of ϵ_f and therefore take it out of the integration sign. It is convenient to calculate this integral by the Laplace method (i.e., by the steepest descent method for the function of a real variable). For this purpose we equate to zero the derivative of the index of the exponent of the whole expression under the integral,

$$\frac{d}{d\epsilon_f} \left(\ln n(\epsilon_f) - \frac{E_a}{kT} \right) = 0. \tag{13.6}$$

The quantity

$$\alpha(\epsilon_f) = -\frac{dE_a}{d\epsilon_f} = \frac{dE_a}{d(J_{an'} - J_{fn})} \tag{13.7}$$

is readily seen to be the analog of the transfer coefficient in the Brönsted equation for electrode reactions. These quantities are not identical since the Brönsted equation relates the changes in the activation energy to changes in the reaction heat, which is a thermodynamic quantity. In our case, however, the quantity $(J_{an'} - J_{fn})$ has different values for electrons at different levels ϵ_f and has no thermodynamic significance. It will be shown below that for the electrochemical process the Brönsted equation can be generalized as

$$\alpha^* = \frac{dE_a}{d(J_{an'} - J_{fn})}\bigg|_{\epsilon_f = \epsilon*} \tag{13.8}$$

where ϵ^* corresponds to the electron level at which the expression under the integral for the current has a maximum. In other words, ϵ^* is the energy level from which the actual transfer of the electron participating in the electrode reaction is realized. ϵ^*, as pointed out above, is found from (13.6), which can be rewritten

$$n(\epsilon_f) = 1 - \alpha(\epsilon_f) \tag{13.9}$$

$$n(\epsilon_f) = \frac{1}{2} - \frac{J_{an'} - J_{fn}^0}{2E_s} + \frac{e\eta}{2E_s} + \frac{\epsilon_f - \epsilon_F}{2E_s}. \tag{13.10}$$

If the values of the parameters E_s and $(J_{an'} - J_{fn}^0)$ are known, the solution of this equation, ϵ^*, can be found numerically. However, we now give a method for the solution of this transcendental equation. We shall consider three regions of overvoltage.

A. Low Overvoltage

$$e\eta < (J_{an'} - J_{fn}^0) - E_s. \tag{13.11}$$

It is clear from Fig. 14 that the straight line, for which the corresponding equation coincides with the right-hand side of (13.10), intersects the

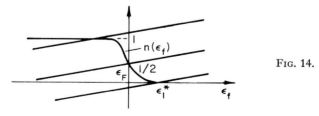

FIG. 14.

function $n(\epsilon_f)$ at such value of $\epsilon_1{}^*$ that $n^* \equiv n(\epsilon_1{}^*) \ll 1$. From (13.10) it is easy to find $\epsilon_1{}^*$, n^*, α^*, and E_a,

$$\epsilon_1{}^* \approx \epsilon_F - E_s + (J_{an'} - J_{fn}^0) - e\eta \qquad (13.12)$$

$$n^* \approx \exp\left\{\frac{[E_s - (J_{an'} - J_{fn}^0)]}{kT}\right\} \cdot \exp(e\eta/kT) \qquad (13.13)$$

$$E_a \approx E_s \qquad (13.14)$$

$$\alpha^* \approx 1 - n^*. \qquad (13.15)$$

The value of the activation energy thus found corresponds to the arrangement of the terms shown in Fig. 15, i.e., in this region of over-

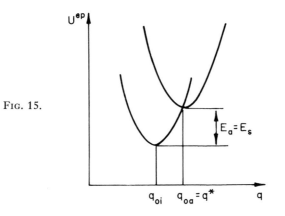

FIG. 15.

voltage the transition process is of a barrierless nature. For the current we have

$$i(n, n') = 2ec_s\rho^*\omega_0\kappa(n, n')\left(\frac{kTE_s}{\pi}\right)^{1/2}\exp\left\{\frac{J_{fn}^0 - J_{an'}}{kT}\right\}\exp\left(\frac{e\eta}{kT}\right). \qquad (13.16)$$

In the region of barrierless transitions the adsorption bond between hydrogen and the electrode is realized by an electron which before the transition had been a strongly excited state $(\epsilon_1{}^* - \epsilon_F > kT)$. Since the number of these electrons is very small $(n^* \ll 1)$, the current is small too. The slope of the polarization curve in this region in semilogarithmic scale is 1,

$$\frac{d \ln i(n, n')}{d(e\eta/kT)} \approx \alpha^* \approx 1. \qquad (13.17)$$

B. "Normal" Overvoltage

$$|e\eta - (J_{an'} - J_{fn}^0)| < E_s. \tag{13.18}$$

In this region the intersection takes place near the Fermi level and it can be readily shown that

$$\epsilon^* \approx \epsilon_F + \frac{2kT}{E_s}(J_{an'} - J_{fn}^0 - e\eta) \approx \epsilon_F \tag{13.19}$$

$$n^* \approx \frac{1}{2} - \frac{J_{an'} - J_{fn}^0 - e\eta}{2E_s} \approx \frac{1}{2} \tag{13.20}$$

$$E_a \approx \frac{E_s}{4} + \frac{J_{an'} - J_{fn}^0 - e\eta}{2} \approx \frac{E_s}{4} \tag{13.21}$$

$$\alpha^* \approx \frac{1}{2} + \frac{J_{an'} - J_{fn}^0}{2E_s} - \frac{e\eta}{2E_s} \approx \frac{1}{2}. \tag{13.22}$$

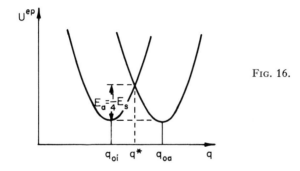

Fig. 16.

In this region the activation energy corresponds to the arrangement of the terms shown in Fig. 16. Accordingly, the current is of the form

$$i(n, n') = ec_s\rho^*\omega_0\kappa(n, n')kT \exp\left\{-\frac{E_s}{4kT} - \frac{J_{an'} - J_{fn}^0}{2kT} + \frac{e\eta}{2kT}\right\}. \tag{13.23}$$

We see that the transfer coefficient α^* is close to the value $\frac{1}{2}$. The exact value $\alpha^* = \frac{1}{2}$ is obtained at $e\eta_{1/2} = J_{an'} - J_{fn}^0$. However at large enough E_s, the transfer coefficient α^* is equal to $\frac{1}{2}$ in a considerable region of overvoltages. As we have emphasized before, the theory is based on the assumption that the reorganization energy is large, a quantity of the order of several electron volts.

In this region all the electrons participating in the reaction can be considered to have the energy which is very close to the Fermi level.

In the normal region the slope of the polarization curve on a semilogarithmic scale is close to $\frac{1}{2}$ in the range of overvoltage indicated.

In this region, as well as in the barrierless region, the value of α^* determined by means of (13.8) coincides with the determination of α as

$$\alpha = \frac{d \ln i(n, n')}{d(e\eta/kT)} \approx \alpha^*. \tag{13.24}$$

It should be noted that the theory provides the existence of (13.24) to a good approximation at arbitrary overvoltage values (in the absence of concentration effects).

Since in the metal the number of electrons near the Fermi level is large and the activation energy E_a four times as small as in the barrierless region, the current in the normal region is relatively large.

C. HIGH OVERVOLTAGE

$$e\eta > J_{an'} - J^0_{Fn} + E_s. \tag{13.25}$$

In this region

$$\epsilon^* \approx \epsilon_F + E_s + J_{an'} - J^0_{Fn} - e\eta \tag{13.26}$$

$$n^* \approx 1 - \exp\{(E_s + J_{an'} - J^0_{an} - e\eta)/kT\} \approx 1 \tag{13.27}$$

$$E_a \approx 0 \tag{13.28}$$

$$\alpha^* \approx 1 - n^* \approx 0. \tag{13.29}$$

The activation energy, which is zero, corresponds to the arrangement of the terms shown in Fig. 17. Since $E_a \approx 0$ at high enough overvoltage the discharge process must be of an activationless nature. Then the current is equal to

$$i(n, n') = 2ec_s\rho^*\omega_0\kappa(n, n')(kTE_s/\pi)^{1/2}. \tag{13.30}$$

Since in this region of overvoltage the discharge involves the participation of the electrons from practically all the occupied energy levels ($n^* \approx 1$), i.e., from the levels located below the Fermi level ($\epsilon^* < \epsilon_F$), and moreover the activation energy of the transition is zero, the current has a constant and very large value. This condition obviously offers great experimental difficulties in the way of the detection of activationless transitions.

The expressions for the current in three different regions of overvoltage given above can be combined into one formula in which the overvoltage in the explicit form is substituted by the transfer coefficient α^*, which depends on η,

$$
i(n, n') = \frac{2ec_s\rho^*\omega_0\kappa(n, n')}{[1 + (2E_s/kT)\alpha^*(1 - \alpha^*)]^{1/2}} \left(\frac{kTE_s}{\pi}\right)^{1/2}
$$

$$
\times (1 - \alpha^*) \exp\left(-(\alpha^*)^2 \frac{E_s}{kT}\right). \tag{13.31}
$$

The dependence of α^* on η can be found with good accuracy from

$$
\alpha^* = \begin{cases} 1 - \exp\left\{-\dfrac{J_{an'} - J_{fn}^0 - E_s - e\eta}{kT}\right\}; & e\eta < J_{an'} - J_{fn}^0 - E_s \\[2mm] \dfrac{1}{2} + \dfrac{J_{an'} - J_{fn}^0 - e\eta}{2E_s}; & \begin{array}{l} J_{an'} - J_{Fn}^0 - E_s < e\eta \\ \qquad < J_{an'} - J_{fn}^0 + E_s \end{array} \\[2mm] \exp\left\{\dfrac{J_{an'} - J_{fn}^0 + E_s - e\eta}{kT}\right\}; & e\eta > J_{an'} - J_{fn}^0 + E_s. \end{cases}
$$

$$\tag{13.32}$$

Figure 17 shows the polarization curve in a semilogarithmic plot for $E_s = 2$ eV. The parameters contained in the theory—the preexponential factor and $(J_{an'} - J_{fn}^0)$ are not defined concretely since, without changing shape, they can shift the curve only along the abscissa and ordinate axes. According to formula (13.2) the total cathodic current is

$$
i = \sum_{n,n'} \frac{\kappa(n, n')}{\kappa_0} i_0(e\eta - E_{n'} + E_n) \exp\left(-\frac{E_n - E_0}{kT}\right) \tag{13.33}
$$

where i_0 and H_0 are the current and the transmission coefficient calculated for the case when the proton passes from the nonexcited initial state to the nonexcited final state.

Strictly speaking, the measured current is given by (13.33), rather than by the expressions presented earlier. However, if the current is approximately represented as

$$
i \sim \sum_{n,n'} \kappa(n, n') \exp\{-\alpha(E_{n'} - E_0) - (1 - \alpha)(E_n - E_0)\}/kT \tag{13.34}
$$

by a simple analysis we can draw the following conclusions for the region of normal overvoltages. The main contribution to the total current is made by the transition from the ground initial state to the zero level of the final state.

In the regions of low and high overvoltage, the situation may be somewhat different. At low overvoltages $\alpha \approx 1$ and excited initial states

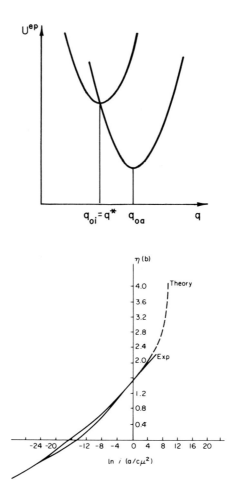

FIG. 17.

can contribute to the current. On the contrary, at high overvoltage, when $\alpha \approx 0$ the transitions to the final excited states can be realized.

It should be stressed, however, that the above effect may change the preexponential factor, but cannot alter the qualitative nature of the expression for the current/overvoltage characteristic.

In order to obtain the detailed comparison between various theories, a quantitative calculation according to formula (13.33) was carried out, account being taken of the possibility of transitions into excited states. In Fig. 17 the theoretical and experimental curves of the dependence of current on the overvoltage are presented. The experimental curve represents the summary of numerous measurements, and among these measurements we should especially note the new measurements of Nürnberg (1969).

The agreement between the theory and experiment proves to be quite satisfactory. The width of the normal region, where the value of the transport coefficient is near $\frac{1}{2}$, constitutes about 1.2 to 1.5 V. The most sensitive method to verify the theory is the comparison of this theory with measurements of isotopic effects.

The isotopic effect may be characterized by the ratio of reaction rates for two isotopes, e.g., protium and tritium:

$$S_{HT} = \frac{k_H}{k_T} = \frac{\int \kappa_H(R) \exp\{-U(R)/kT\}\, dR}{\int \kappa_T(R) \exp\{-U(R)/kT\}\, dR}. \tag{13.35}$$

The potential energy of the discharged ions $U(R)$ is reduced to the sum of the energy in the electric field of the double layer eER and the energy of repulsion, which prevents the ion from approaching close to the surface of the electrode $U(R)$:

$$U(R) = eER + U_r.$$

The dependence of κ on R is determined by the wavefunctions of the proton in the molecule, H_3O^+, and in the adsorbed state. In order to obtain the qualitative characteristics it is possible to limit oneself to the simplest model, a harmonic oscillator. Then a simple calculation yields

$$\kappa \sim \exp(-\alpha R^2)$$

where

$$\alpha \sim \omega m.$$

Because the integrals in (13.35) possess a sharp maximum, one may easily obtain the following expression for

$$S_{HT} = (S_{HT})_0 \exp(-\text{const } E) \tag{13.36}$$

where const $\sim \partial U_r/\partial R$ and $(S_{HT})_0$ is the coefficient of separation without the field. The latter is determined mainly by the difference of zero-point energies of the proton and triton,

$$\Delta J_H - \Delta J_T = (\epsilon_{iH}^{(0)} - \epsilon_{fH}^{(0)}) - (\epsilon_{iT}^{(0)} - \epsilon_{fT}^{(0)}), \tag{13.37}$$

in molecules H_3O^+, TH_2O^+ and in the adsorbed state.

It is necessary to stress that formula (13.36) has a qualitative but not a quantitative meaning. Contrary to the previously used concepts (see Section XV) the formula shows that the separation coefficient (1) depends on the field strength in the discharge region and (2) depends on the difference in the zero-point energies of the initial and final states.

For fixed values of the field E (or surface charge q) and $\Delta J_H - \Delta J_T$, the separation coefficient should depend neither on the overvoltage nor on the composition of solution, nor ψ_1-potential and should depend

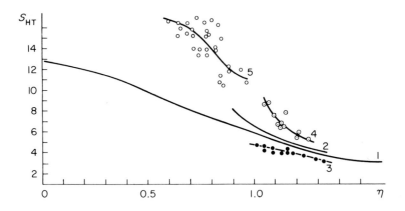

FIG. 18. KEY: Curve 1, Hg—HCl; curve 2, Hg—H_2SO_4; curve 3, am Tl—H_2SO_4; curve 4, Pb—H_2SO_4; curve 5, Ga—H_2SO_4.

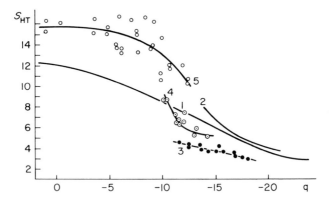

FIG. 19. See Fig. 18 for key.

only weakly on the cathode material. This conclusion of the theory was investigated in a specially designed experiment of Kryshtalik (1969).

In Figs. 18 and 19 are presented the values of the separation coefficient for various metals as a function of η and q. We find that the conclusions of the theory are completely confirmed by the experiment. Rough estimates lead to numerical values of $\Delta J_H - \Delta J_T$ that also are in agreement with the experiment. The importance of this result is connected with the fact that in all other theories of electrode processes the separation coefficient should depend upon the nature of the metal. We shall return to this problem in Section XV.

The qualitative difference between the two points of view, quantum-mechanical theory and theory of Horiuti–Polanyi, is even more pronounced in the case of electrolysis on mercury in alkaline solutions. Measurements have shown that in a tetramethyl–ammonium hydroxide solution the separation factor S electrode proves constant in a wide range of η (S varying significantly in this potential range in acid solution).

This result is totally at variance with main concepts of the Horiuti–Polanyi theory (see Section XV), which in principle does not differentiate between the H_3O^+ ion and the neutral H_2O molecule dischanges. In both cases activation is due to the stretching of the H—O bond and is reduced by the superimposed field. Therefore, both in acid and alkaline solutions S should drop with rising η.

In the quantum–mechanical theory presented above, where activation was associated with the solvent reorganization, there are no reasons to suppose that the distance from the electrode of the discharging neutral H_2O molecules should depend on η. Therefore, in the case of alkaline solutions the overlap of the heavy particle wave functions in H_2O and H_{ads} should not vary with η. In accordance with experiment, the separation factor should not depend on potential.

The following must be stressed: the phenomenon of electrons transfer in redox reactions is essentially quantum mechanical in nature. However, if this reaction proceeds without chemical bonds breaking, then the temperature dependence of the rate of the process is determined only by the behavior of the slow subsystem—the solvent. The latter is determined under usual conditions classically. Therefore, the quantum nature of the transport process affects only the value of the preexponential factor, and is therefore hidden to some extent. This enabled Marcus, without going deeply into the details of the true nature of the transition, to employ the theory of absolute reactions rates in order to obtain the correct expression for the probability of electronic transition. Of course, such a calculation formally contradicts the statements of the rate theory, where the transition along the reaction coordinate is considered. During the redox reactions the reaction coordinate is the coordinate of the electron. Substituting this coordinate for the classical coordinate of the solvent, Marcus succeeded in obtaining the correct result. The reason is quite clear: by the electron's transition the saddle point on the solvent potential energy surface corresponds to the minimal value of the activation energy.

The situation changes completely in the case of reactions which go on with forming (or breaking) of chemical bonds, at least in the case of those for which the coupling is sufficiently strong ($\hbar\omega \gg kT$).

The quantum nature of the process is clearly shown in this case. The minimal value of the activation energy does not correspond to the saddle point on the potential energy surface of any classical part of the system, as we saw in the case of a reaction with proton transfer.

Therefore the attempt of a formal application of the absolute rate theory to reactions of this kind cannot lead to a reasonable result. Quite similarly it is impossible [as was, e.g., done by Marcus and Cohen (1968) who do not yet seem to be acquainted with our work (Dogonadze *et al.*, 1968b)] to apply quite formally the expression for the reaction rate constant, which is correct for reactions with electron transfer without bonds breaking, to the general case of reactions in solution with chemical bonds breaking.

XIV. Other Cases of Electrode Reactions

The case of reactions with proton transfer considered in Section XII is a typical one, but it is not a single example of reaction accompanied by chemical bond breaking. This is a case of a reaction which goes according to scheme I:

$$
\begin{array}{llll}
\text{initial quantum} & & \text{final quantum} & \\
\text{degrees of freedom} & \hbar\omega_i \gg kT & \rightarrow \quad \text{degrees of freedom} & \hbar\omega_f \gg kT \\
& & & \hspace{2cm} \text{(I)} \\
\text{initial classical} & & \text{final quantum} & \\
\text{degrees of freedom} & \hbar\omega_i^{(0)} \ll kT \rightarrow & \text{degrees of freedom} & \hbar\omega_f^{(0)} \ll kT.
\end{array}
$$

In these reactions the quantum degrees of freedom remain quantum ones (e.g., oscillation frequencies H_2O—H^+ and H_{ads}—M) and the classical degrees of freedom remain classical ones (oscillations of the solvent in the same example). One may, however, cite examples of reactions which are performed according to other schemes where the "mixing" of classical and quantum degrees of freedom occurs. Some of these cases have already been considered by us quantitatively or qualitatively. Let us consider, e.g., the electrode reaction which proceeds according to scheme II:

$$
\begin{array}{llll}
\text{initial quantum} & & \text{final classical} & \\
\text{degree of freedom} & \hbar\omega_i \gg kT & \rightarrow \quad \text{degree of freedom} & \hbar\omega_f \ll kT \\
& & & \hspace{2cm} \text{(II)} \\
\text{initial classical} & & \text{final classical} & \\
\text{degree of freedom} & \hbar\omega_i^{(0)} \ll kT \rightarrow & \text{degree of freedom} & \hbar\omega_f^{(0)} \ll kT.
\end{array}
$$

As an example of reactions of this type the following electrode reaction may serve:

$$Cl^- + M \rightarrow Cl_{ads} + M. \tag{14.1}$$

M is a mercury electrode. The chlorine ion is strongly solvated so that the frequency of oscillations relative to the bond Cl^-—H_2O is large ($\hbar\omega_{Cl} \gg kT$). The bonding energy Cl_{ads}—M is small, and corresponds to the frequency $\hbar\omega_{Cl_{ads}} \ll kT$. The oscillations of the solvent before and after the transition are classical.

The calculation according to the general formula (5.14) (Kuznetsov, 1969b) leads to the activation energy

$$E_a = \frac{(E_s + E_r + \frac{1}{2}\hbar\omega_{Cl^-} + \Delta J)^2}{4E_s} \qquad \text{for} \quad \Delta J + \frac{\hbar\omega_{Cl^-}}{2} < E_s \tag{14.2}$$

$$E_a = E_r + \Delta J \qquad \text{for} \quad \Delta J + \frac{\hbar\omega_{Cl^-}}{2} > E_s. \tag{14.3}$$

Expressions (14.2) and (14.3) are amenable to the following interpretation of transition.

In the case of small reaction heats, to which formula (14.2) applies, the system moves along the classical coordinate (i.e., the fluctuation of the solvent occurs) with a constant value of the quantum coordinate $R_i^{(0)}$ (so that the particle remains in the ground quantum state with an unchanged length of the valence bond). At the point of intersection $(q^*, R_i^{(0)})$ of the final and initial potential curves when the corresponding energies become equal, the transition to the final potential curve occurs [this means reaction is completed (14.1)]. As in the initial state the particle possesses the zero point energy of $\hbar\omega_{Cl^-}/2$, so that it is necessary to raise the initial energy to the value of this energy in order to obtain the intersection of the potential curves. After such an increase is accomplished, the system moves classically along the initial curve and relaxes into the final state $(q_f, R_f^{(0)})$. In the final state when the particle is weakly connected with the surface, this particle performs classical oscillations along the bond Cl_{ads}—M. There is an essentially larger amplitude of the change of coordinate R, which corresponds to classical oscillations. On the other hand, as the radius of the hydrated ion Cl^- considerably exceeds the radius of the atom Cl, in the course of the electrode reaction the discharging particle very nearly approaches the electrode.

As was discussed in Section XII, the case of transition according to scheme I, the point $(q^*, R^{(0)})$ is not a saddle point on the potential energy surface, so that the reaction differs from that predicted by the absolute reaction rate theory. The case of larger reaction heats is of more interest, i.e., the case when E_a is given by formula (14.3). In this case the transport coefficient is equal to

$$\alpha = \partial E_a / \partial \Delta J = 1$$

and the process should be considered as a barrierless one. In fact, this is not quite correct, and the transition occurs as shown in Fig. 20.

The coordinate of intersection of the potential curves, q^*, coincides with the equilibrium final value $q_f^{(0)}$. However, the value of the quantum coordinate does not coincide with the final value of $R_f^{(0)}$. We see that the transition along the coordinate q from $q_i^{(0)}$ is really barrierless. However,

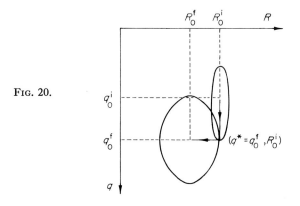

Fig. 20.

the transition into the final state which is the equilibrium state relative to coordinate R corresponds to activationless motion on the potential energy surface $R_i^{(0)} \rightarrow R_f^{(0)}$ for a fixed value of $q = q_f^{(0)}$. The transition as a whole (i.e., relatively to the set of coordinates q and R) is not a barrierless one. For the inverse transition of $f \rightarrow i$ an activation energy $E_a' = E_s$ is needed, though the transport coefficient of this transition is equal to zero.

Schemes I and II do not yet cover all the possible transitions involving chemical bonds breaking; e.g., the transition according to scheme III is possible:

initial quantum coordinate $\rightarrow$ final classical coordinate
initial classical coordinate $\rightarrow$ final quantum coordinate.   (III)

The theory of such a transition is now in a stage of development.

XV. Critical Consideration of Theoretical Concepts about the Mechanism of the Elementary Act of the Electrode Process

The first attempt to develop a theory of the electrode process was that of Gurney (1931). According to this theory, in the course of the electrode reaction, electrons in the metal intercept the inner phase boundary and add to ion acceptors after the usual tunnel transition in a given external potential force field. These concepts have been recently developed by Gerischer (1960), as well as by Bockris and Matthews (1965), and therefore deserve a more detailed discussion. According to Gurney's theory, the current density may be represented by formula (3.2), and the transition probability entering this formula according to Gurney is given by the known formula due to Gamov:

$$W \sim \exp\left\{ -\frac{2}{\hbar} \int_a^l p \, dx \right\} = \exp\left\{ -\frac{2}{\hbar} \int_a^l [2m(E - U(x))]^{1/2} \, dx \right\} \quad (15.1)$$

where p is the momentum of the electron, and U is the potential energy in the field at the interphase boundary between electrode and solution. Expression (15.1) gives the probability of tunneling by the prescribed form of the barrier. As the latter is unknown, it is usually assumed that the barrier has the simplest form, a rectangular or triangular one. It is usually assumed that the barrier has the simplest form, a rectangular or triangular one. It is clear from the preceding arguments that the use of the Gamov formula for the calculation of the transition probability for the system, which consists of a metal (semiconductor) and the solvated ion (molecule), is inadmissible. We also emphasized earlier that the transition of an electron from the electrode to the ion has the character of a nonstationary process, and in the course of this process the potential field itself changes. The calculation of the probabilities of such transitions according to Gamov's formula, which was derived using the assumption of the unchangeability of the spatial distribution of the potential field in the course of the transitions, is not allowed formally. Such a calculation may lead not only to quantitative errors but also to a qualitative distortion of the pattern of the whole phenomenon. The barrier penetration transition, as we have seen, occurs with some probability when the energy levels of the electron in the metal and in the ion coincide. However, the position of the energy level in the solvated particle changes in the time course together with the changes of the solvent polarization. One can say for the sake of clarity that the level in the ion passes the level in the metal at some velocity. If this velocity is large the transition probability proves to be small. In the limit the probability approaches $\frac{1}{2}$.

An electron with equal probability may occupy two potential boxes separated by the potential barrier (see, e.g., Landau and Lifshitz, 1964). On the other hand, the transition probability is rather sensitive relative to the unknown parameters of the barrier—its height and its width. Moreover, Gamov's formula only gives the probability of the passing of the plane wave through the barrier and not of the probability of the electron transition from the state described by the plane wave into the state described by the spherical wave directed to the center. If this were so, it would correspond to the transition of an electron from a quasi-free motion in the metal to the spherical potential box of the atom. All these arguments make the use of Gamov's formula unjustified in principle and useless in practice. It should be noted that for a long time the concepts of Gurney's theory were not employed, not because of the theoretical arguments presented above but due to quite different considerations; namely, as was pointed out by Frumkin (1933), the process

$$H_3O^+ + e \rightarrow \overbrace{H + H_2O}^{\text{volume}} \tag{15.2}$$

proves to be energetically unprofitable as compared to the process

$$H_3O^+ + e \rightarrow \overset{\text{surface}}{H_{ads}} + \overset{\text{volume}}{H_2O}. \tag{15.3}$$

In the last case the adsorption energy of the hydrogen atom H_{ads} on the electrode surface is gained. Therefore the further development of the theory of the electrode process was connected with a consideration of the process of transition of the proton from the hydroxinium molecule in the absorbed atom state. It was assumed that this process serves as the prototype of any electrode reaction. Horiuti and Polanyi suggested a theory of the elementary act of the electrode process of the type (15.3). This theory was called the theory of slow discharge.

The theory of slow discharge represents a particular case of the theory of absolute reaction rates. There are two physical ideas in the concepts of the Horiuti–Polanyi theory, which are practically unconnected with each other:

(1) The suggestion that the charge transfer process, in particular through the interphase boundary, presents the slow stage of the electrode process.

(2) The concepts concerning the mechanism of the process itself. These concepts form the ground for the quantitative formulation of this theory.

Although the Horiuti–Polanyi theory was originally formulated for the case of the hydrogen ion discharge, the main concepts of this theory were frequently carried over to other electrode reactions without any significant changes.

In a series of examples it has been established that after removing diffusion limitations, the stage of discharge is really the limiting one. Therefore, the critical analysis of the Horiuti–Polanyi theory will be concerned only with the second point.

The mechanism of reaction according to Horiuti and Polanyi is assumed to be the following one. The potential energy surface (electronic terms) corresponds to the proton in the hydroxonium molecule and also to the hydrogen atom adsorbed on the surface of the electrode. The thermal agitation results in the proton moving along the potential energy surface and stretching the valence bond H^+—H_2O. Through this, with a certain probability, the proton may get to the point of intersection of the potential energy surfaces of the initial and final states. Here the system comes into the activated state, and it may either return to the initial state or pass onto the new surface, corresponding to the neutral adsorbed atom. As the distance R^* corresponding to the intersection of the potential energy surfaces lies rather near the metal surface, it is assumed in the theory that the barrier penetration transition of an electron from the metal to the ion placed at the point R^* occurs with a probability which is close to unity. Therefore, the integral velocity of the process (15.3) is determined by the probability of the transition into the activated state. By changing the potential of the metal the potential energy surface of H_3O^+ is displaced, whereas the surface of the neutral atom H_{ads} does not change. The heat of the reaction and the activation energies change correspondingly, so that

$$Q = \Delta J = \Delta J^{(0)} - e\,\Delta\varphi \qquad (15.4)$$

$$E_a = E_a^{(0)} - \alpha e\,\Delta\varphi. \qquad (15.5)$$

The rate of the process is determined by the number of particles, which as a result of thermal agitation acquire an energy exceeding E_a, i.e.,

$$i \sim \exp[-(E_a^{(0)} - \alpha e\,\Delta\varphi)/kT]. \qquad (15.6)$$

Expression (15.6) represents the empirical Tafel law with the transport coefficient α. From simple geometrical considerations one finds that

$$\alpha = \tan\theta_1/(\tan\theta_1 + \tan\theta_2)$$

where θ_1 and θ_2 are the angles of inclination of the potential energy surfaces $U_{H_3O^+}$ and $U_{H_{ads}}$ at the point of their intersection P^*. Of course, neither the form of these surfaces nor the values of these angles follow immediately from the Horiuti–Polanyi theory. In order to obtain the values of the transport coefficients which were observed in the experiment, one should make further assumptions in the Horiuti–Polanyi theory, which concern the character of the intersection of the potential energy surfaces. For example, in order to obtain the value $\alpha = \frac{1}{2}$, it is necessary that in the region of intersection the inclination of both potential energy surfaces (for H_{ads} and H_3O^+) be equal to each other. There are the following principal objections, which may be put forward in connection with the Horiuti–Polanyi theory:

(1) The suggestion about the classical character of proton motion along the bond in the H_3O^+ molecule and along the bond with the metal surface does not correspond to the real situation. In fact at least one of these bonds or even both of them are sufficiently strong so that the distance between the levels of oscillation energy changes in the interval from some tenths of a volt up to one volt.

In the region of such slow excitations (two to four vibrational quanta) the motion along the reaction coordinate is of a purely quantum character. The quantum transition of the system from the normal to the excited state corresponds to this motion.

(2) The Horiuti–Polanyi theory avoids the discussion of the role of the solvent; even the existence of the solvent is not accounted for in this theory. Meanwhile, the most important role of the solvent follows directly from the arguments connected with the very nature of the theory. After neutralization at the point of intersections of the potential curves, the hydrogen atom still possesses a considerable amount of energy as compared to the adsorbed state. If the transition into the final, i.e., adsorbed state would occur into vacuum, the energy $E_a - E_{ads}$ should come out in the form of kinetic energy. If $E_a - E_{ads} \gg kT$, the particle with such an energy cannot remain on the surface at all. In fact, the difference between the energies of the activated and the adsorbed states is given over to the solvent. It is clear from these arguments that the interaction of the particles with the solvent has great significance.

(3) Furthermore, in the Horiuti–Polanyi theory the value of the transport coefficient is introduced as an arbitrary assumption, but it is not calculated.

(4) The constant value of the transport coefficient $\alpha = \frac{1}{2}$ in the language of the intersecting potential energy surfaces signifies that the angle with which the potential surfaces intersect has a constant value over a rather wide (of order 1 eV) interval of potential energy values. This seems quite unjustified from the point of view of the two intersecting potential energy surfaces.

(5) The Horiuti–Polanyi theory cannot explain the very fact of the

existence of barrierless and activationless discharge. If the transition goes on as described above and if the direct reaction is a barrierless one, then the inverse reaction will be activationless, and the rate of the latter would be rather large. In fact, the direct reaction under these conditions should not be observable. Previously we have seen in the example of the chloride ion how this problem is solved in quantum theory by taking account of the solvent role.

But the most convincing theoretical argument against the applicability of the Horiuti–Polanyi theory is the direct calculation.

(6) Using the general formula (3.1) it is possible to obtain the results for the Horiuti–Polanyi theory. According to the physical ideas of this theory, the solvent has no effect on the transition and its state remains unchanged, i.e., the overlapping of the wavefunctions of the solvent is equal to unity, $\langle \phi_e | \phi_{e'} \rangle = \delta_{e,e'}$. Therefore, in (3.1) it is necessary to calculate the exchange integral connected with the electron wavefunctions and the overlappings of the proton wavefunctions. The first quantity corresponds physically to the transmission coefficient and in the Horiuti–Polanyi theory should be substituted for unity. Thus, in the Horiuti–Polanyi approximation, the transition probability should be written as

$$w_{ia} = C_1 \cdot \text{Av}_n \sum_{n'} |\langle \chi_{n'} | \chi_n \rangle|^2 \, \delta(E_i^0 - E_a^0).$$

With a harmonic approximation, i.e., substituting the oscillator wavefunctions for χ, this expression can be accurately calculated and is equal to

$$w_{ia} = C_2 \exp\left\{ \frac{\Delta J}{2kT} - \frac{E_p}{\hbar\omega_p} \coth \frac{\hbar\omega_p}{2kT} \right\} \cdot I_{\Delta J/\hbar\omega_p}\left\{ \frac{E_p}{\hbar\omega_p} \operatorname{csch} \frac{\hbar\omega_p}{2kT} \right\} \quad (15.7)$$

where ΔJ is the reaction heat, E_p the proton state reorganization energy, ω_p the proton frequency, and $I_m(z)$ the Bessel function. In order to obtain the result of Horiuti and Polanyi, who used the activated complex method, it is necessary in (15.7) to assume $\hbar\omega_p \ll kT$. Then we obtain the activation formula

$$w_{ia} = C_3 \exp\{-(E_a/kT)\}, \qquad \hbar\omega_p \ll kT \quad (15.8)$$

where E_a is the activation energy (Fig. 21). In actual fact, a reverse relation to that used by Horiuti and Polanyi is valid, viz., $\hbar\omega_p \ll kT$. In this case (15.7) gives

$$w_{ia} = C_4 \exp\{-(\Delta J/kT)\}, \qquad \hbar\omega_p \ll kT. \quad (15.9)$$

Equation (15.9) has a quite obvious physical sense and shows that in the case of the higher vibrational levels being weakly excited ($\hbar\omega_p \gg kT$)

a quantum subbarrier transition takes place instead of the classical one, which is described by (15.9) (see Fig. 21). Thus, assuming the solvent to take no part in the discharge process, we inevitably arrive at formula (15.9), which can be easily seen to be in contradiction with the experimental data. In fact, according to (15.9) the activation energy should be equal to the reaction heat, i.e., the transfer coefficient α in the Brönsted equation should be unity. Thus, it follows necessarily from the analysis of the Horiuti–Polanyi theory that the reorganization of the solvent in the discharge process should be taken into consideration.

We have dwelt at some length on the analysis of the Horiuti–Polanyi theory for the reason that the main physical ideas of this theory have been used in one form or another in many later studies (Conway, 1958; Hush, 1958; Salomon and Conway, 1965).

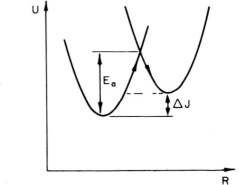

FIG. 21.

Together with these theoretical considerations the above-cited experimental data on the isotopic effect by the discharge of hydrogen ions indicate decisively the inapplicability of the Horiuti–Polanyi concepts. Really, according to the essence of their theory the value of the isotopic effect should be determined by the difference in zero-point energies of the proton in H_3O^+ and the adsorbed atom, H. The zero-point energy of the adsorbed atom depends essentially on the adsorbtion potential and therefore the isotopic effect has to depend on the nature of the metal. Meanwhile, as we have seen previously, the experiment shows that the isotopic effect does not depend on the nature of the metal and is not determined completely by the difference of zero-point energies. As the dependence of the isotopic effect on the nature of the metal is connected with the very essence of Horiuti and Polanyi, the measurements of isotopic effect proved to be some kind of *experimentum crucis* for the theory. The negative result of this experiment together with all theoretical considerations permits us to make the statement that the Horiuti–Polanyi theory is not able to describe the process of hydrogen ion discharge and cannot be generalized for other cases of electrode reactions.

Quite recently Bockris and Matthews (1965) have suggested a theory of the elementary act of the electrode process, this theory being to some extent a synthesis of the theories of Gurney and Horiuti–Polanyi.

In the theory of Bockris and Matthews the following scheme of the process is accepted: all the electrons in the metal are at the Fermi level. The energy level of an electron in the molecule is assumed to be lower than the Fermi level by the value ϵ_{cr}. Due to thermal excitement the bond H^+—H_2O is stretched, and the electronic energy level of the molecule is raised. By stretching the bond up to some length R^*, the level raises up to the value ϵ_{cr}. The probability W_1 of the bond stretching up to the length R is given by the formula

$$W_1 \sim \exp\{-(\epsilon_{cr}/kT)\}. \tag{15.10}$$

If the value ϵ_{cr} corresponds to the equality of the Fermi energy and the electronic energy level in the excited molecule H_3O^+, then at the moment of time when the excitation reaches the value ϵ_{cr}, the tunnel transition of the electron from the Fermi level to the excited molecule H_3O^+ will occur. Then the bond breaks, and the formed neutral hydrogen atom passes into the adsorbed state. By this passing the distance between the proton and the surface decreases from R^* to R_{ads}. The motion of a proton by excitation of the molecule H_3O^+ up to the activated state R^* and subsequent motions of the hydrogen atom formed in the activated form along the potential energy surfaces is assumed to be classical. Contrary to the Horiuti–Polanyi theory, Bockris and Matthews assume that the probability of an electron transfer from the metal to the proton placed at the distance R from the electrode surface differs from unity. They calculate this probability using the Gamov formula with the rectangular barrier:

$$W_2 = \exp\{-(2l/\hbar)[2m(E_F - E_R)]^{1/2} \tag{15.11}$$

where E_R is the electron energy in the plane $x = R$. The full probability W of the process of the hydrogen ion discharge is, according to Bockris and Matthews, determined by the relation

$$W = W_1 W_2.$$

Besides the general remarks about the inapplicability of Gamov's formula to the nonstationary transition under consideration, the meaning of the rectangular barrier itself is far from being clear: the electron goes over to the proton and the potential field of the proton by no means can be characterized by an infinitely wide rectangular potential surface. Even if such a transition were possible, one should assume that the electron is caught in a potential box of a spherical form. As is well known from quantum mechanics, in this case the transition probability becomes essentially smaller than the one calculated according to formula (15.11). In order to be caught by the spherical box the plane wave coming from the metal should be reduced to fit a region whose dimensions are of the order of the box width. Such a reduction is a highly improbable process. In other words, the wave function represented by the plane wave overlaps poorly with the wave function in the spherical box, especially in the case of underbarrier transitions. The transport coefficient in the Bockris–Matthews theory is determined exactly in the same way, as was done in the Horiuti–Polanyi theory—from the inclination of the potential energy curves at the intersection point. It is evident that all the above critical remarks concerning both the theory of Horiuti and Polanyi and the theory of Gurney refer also to the theory of Bockris–Matthews, the latter being a combination of the aforementioned theories. Besides this, there is another point in the Bockris and Matthews theory. Namely the concepts of the quantum state and the length of the H^+—H_2O bond (i.e., the distance of H^+ from the center of the molecule H_3O^+) have been confused. Formula (15.10) gives the probability of the system (molecule of H_3O^+) being in the state with the fixed energy but not of the proton having been removed from the center of the H_3O^+ molecule at the given distance. There is no unique correspondence between these two concepts in quantum mechanics, and the value of the activation energy of ϵ_{cr} of the molecule bears no relation to the position of the proton in this molecule. The proton, which performs oscillations in the molecule H_3O^+ may with a certain probability be at the distance R^* (more exactly, at the distance between R^* and $R^* + dR^*$) both in the normal state with energy ϵ_0 and in any excited state (with energy ϵ_n). This probability is determined by the corresponding value of $|\psi_0|^2 \, dV$ and $|\psi_n|^2 \, dV$, but not by the energy ϵ. In other words, the presence of a proton in the plane R is by no means connected with its energy. This conclusion makes the arguments of Bockris and Matthews especially difficult to understand.

XVI. Conclusions

In conclusion, we shall briefly discuss the historical background of this problem and the contribution made by various authors to its present state. The physical concept on which the calculations described above are based was first advanced by Libby (1952). He pointed out the great importance of thermal fluctuations in the polar solvent, which ensure the electron-exchange reactions. Libby, however, did not give a quantitative calculation of the transition probability.

The paper of Platzmann and Franck (1954) contains the very important idea that, in considering an electron transfer reaction, calculations similar to those used in the investigation of nonradiative transitions in a solid should be performed.

It was postulated in a number of papers that the probability W_{12} of the electron exchange reaction can be represented as a product

$$W_{12} = W_e \cdot W_{act}$$

where W_e is the electron-transfer probability and W_{act} the probability of formation of an activated complex.

In the paper of Zwolinski et al. (1955), W_e was calculated on the basis of the usual theory of the tunnel transfer of the electron.

In Weiss' paper (1959) the electron transfer was considered to be a collision of the second kind. According to Weiss, the electron transfer involves the formation of a transition state, which corresponds to some fixed distance between the ions. In reality, as we saw above, the transition state corresponding to the coincidence of the electronic energy levels of both ions may arise at any interionic distances, depending on the extent of the polarization fluctuation.

In a number of papers, Marcus (1956, 1957, 1959, 1960, 1961, 1963a–d) estimated the value of W_s. The physical concepts advanced in his papers are very close to those described above and had a great stimulating effect upon our work.

The first papers, where due account of the solvent role in adiabatic redox reactions not accompanied by chemical bonds breaking, belong to Marcus.

In the works of Levich (1965) and Levich and Dogonadze (1959) the quantum-mechanical theory of both adiabatic and nonadiabatic reactions with electron transfer not accompanied by the chemical bonds breaking was developed. This theory was generalized somewhat later for the cases of electrode redox reactions by Dogonadze and Chizmadzev (1963). Dogonadze and Kuznetsov developed a theory of redox electrode reactions for the case of semi-conducting electrodes. Marcus has independently generalized his works on bulk redox reactions for the case of electrode adiabatic reactions with metallic electrodes. In a paper of Dogonadze et al. (1968b) a quantum-mechanical theory of electrode reactions with chemical bonds breaking was developed. In a paper of Levich et al. (1968), the theory of bulk reactions with proton transfer was developed. General concepts of the theory of reactions, accompanied by chemical bond breaking, was set forth in a review by Levich. The detailed qualitative comparison of the theory of redox reactions with experimental data was carried out by Marcus (1964).

The experimental quantitative verification of redox reaction theory for the reaction $Fe(CN)_6^{3-} \rightleftarrows Fe(CN)_6^{4-}$ on a mercury electrode was carried out by Frumkin et al. (1963) and the reaction $Cr^{2+} \rightleftarrows Cr^{3+}$ on a mercury electrode by Parsonce and Passeron (1966). The barrierless electrode reactions were discovered by Kryshtalik (1969) for the hydrogen ion discharge on a mercury electrode.

The approximate model of a solvent with one frequency was suggested by Levich and Dogonadze. The dispersion of solvent frequencies was taken into account by Dogonadze and Kuznetsov (1967b).

The general quantum-mechanical theory of reactions in polar solvent which occurs with breaking and forming of chemical bonds with arbitrary oscillation frequencies was set forth in the paper by Dogonadze et al. (1968b), and its application to the ion discharge was made in the paper of Kuznetsov and (independently) by Kryshtalik. In connection with the quantum mechanical theory of hydrogen ion discharge special experiments were performed by Kryshtalik in order to determine the isotopic effect and in particular its dependence upon the nature of the metal (Kryshtalik, 1969).

At the present time there are review articles published on the theory of the elementary act of the redox processes: by Levich (1965), Marcus (1964), Dogonadze et al. (1965), and Dogonadze et al. (1968b).

REFERENCES

BOCKRIS, J. O'M, and MATTHEWS, D. B. (1965). J. Electroanal. Chem. **9**, 325; Proc. Roy. Soc (1966), **A292**, 479.

BORN, M., and HUANG, K. (1957). "Dynamical Theory of Crystal Lattices." Oxford Univ. Press, London and New York.

BRODSKY, A., LEVICH, V. G., and TOLMATCHEV, V. (1968). Dokl. Akad. Nauk SSSR **182**, 1036; (1968). **183**, 852.

CONWAY, B. E. (1958). Can. J. Chem. **37**, 178.

CONWAY, B. E., and BOCKRIS, J. O'M. (1959). "Modern Aspects of Electrochemistry," Vol. 1, pp. 47–102. Academic Press, New York.

DOGONADZE, R., and CHIZMADZEV, Yu. (1963). Dokl. Akad. Nauk SSSR **150**, 333.

DOGONADZE, R., and KUZNETSOV, A. (1967a). Electrochemia **3**, 380.

DOGONADZE, R., and KUZNETSOV, A. (1967b). Electrochemia **3**, 1324.

DOGONADZE, R., KUZNETSOV, A., and CHERNENKO, A. (1965). Usp. Chimi. **34**, 1779; Itogi Nauki (Advan. Sci.) Russ. Moskani Chem. (1967).

DOGONADZE, R., KUZNETSOV, A., and LEVICH, V. (1967). *Electrochemia* **3,** 739.
DOGONADZE, R., KUZNETSOV, A., and LEVICH, V. (1968a). *Electrochemia* **11,** 1025.
DOGONADZE, R., KUZNETSOV, A., and LEVICH, V. (1968b). *Electrochim. Acta* **13,** 1025.
DOGONADZE, R., KUZNETSOV, A., LEVICH, V., and CHAZKATZ, In. (1969). *Electrochem. Acta* **14,** 1001.
FEYNMAN, R. (1951). *Phys. Rev.* **84,** 108.
FRÖHLICH, H. (1954). *Advan. Phys.* **3,** 325.
FRUMKIN, A. N. (1932). *Z. Phys. Chem.* **A160,** 116.
FRUMKIN, A. N. (1933). *Z. Phys. Chem.* **A164,** 121.
FRUMKIN, A. N., PETZII, O., and NIKOLAEVA-FEDOROVICH, N. (1963). *Electrochim. Acta* **8,** 177.
GERISCHER, H. (1960). *Z. Phys. Chem. (NF)* **26,** 223.
GREEN, M. (1959). "Modern Aspects of Electrochemistry; p. 343. Academic Press, New York.
GURNEY, R. (1931). *Proc. Roy. Soc.* **134A,** 137; (1936). "Ions in Solution." Cambridge Univ. Press, London and New York.
HORIUTI, J., and POLANYI, M. (1935). *Acta Electrochem. USSR* **2,** 505.
HUSH, N. S. (1958). *J. Chem. Phys.* **28,** 962.
KRISHTALIK, L. (1960). *J. Fiz. Chim.* (Russian) **34,** 117.
KRISHTALIK, L. (1969). *Electrochimia* **5,** 120.
KUZNETSOV, A. (1968). *Electrochim. Acta* **13,** 1293.
KUZNETSOV, A. (1969a). *Electrochemia* **4.**
KUZNETSOV, A. (1969b). *Electrochem.* **5,** 479.
LANDAU, L., and LIFSHITZ, E. (1964). "Quantum Mechanics." Moscow.
LEVICH, V. G. (1949). *Dokl. Akad. Nauk SSSR* **67,** 309; (1959). **124,** 869.
LEVICH, V. G. (1965). *Advan. Electrochem. Electrochem. Eng.* **4.**
LEVICH, V. G. (1968). *Vestnik Akad. Nauk SSSR* **No. 7.**
LEVICH, V. G., and DOGONADZE, R. (1959). *Dokl. Akad. Nauk SSSR* **124,** 123.
LEVICH, V. G., DOGONADZE, R., and KUZNETSOV, A. (1968). *Dokl. Akad. Nauk SSSR* **179,** 137.
LIBBY, W. (1952). *J. Phys. Chem.* **56,** 863.
MARCUS, R. A. (1956). *J. Chem. Phys.* **24,** 966.
MARCUS, R. A. (1957). *J. Chem. Phys.* **26,** 867.
MARCUS, R. A. (1959). *Can. J. Chem.* **37,** 155.
MARCUS, R. A. (1960). *Discussions Faraday Soc.* **29,** 21.
MARCUS, R. A. (1961). *In* "Transactions of the Symposium on Electrode Processes" (E. Yeager, ed.), p. 239. Wiley, New York.
MARCUS, R. A. (1963a). *J. Phys. Chem.* **67,** 853.
MARCUS, R. A. (1963b). *J. Chem. Phys.* **38,** 1335.
MARCUS, R. A. (1963c). *J. Chem. Phys.* **38,** 1858.
MARCUS, R. A. (1963d). Paper presented at the Moscow CITCE Meeting.
MARCUS, R. A. (1964). *Ann. Rev. Phys. Chem.* **15,** 155.
MARCUS, R. A., and COHEN, A. (1968). *J. Phys. Chem.* **72,** 4249.
MYAMLIN, V., and PLESKOV, In. (1967). "Electrochemistry of Semiconductors." Plenum Press, New York.
NÜRNBERG, H. (1969). Studien Min. Modernen Technik Zur Kinetk Schneller Chemischer und Elektrochemisker Schzitte von Protonen Tzasfferprozessen. Bonn.
PARSONCE, R., and PASSERON, E. (1966). *J. Electroanal. Chem.* **12,** 524.
PEKAR, S. (1951). "Investigations of Electronic Theory of Crystals." Moscow.

PLATZMANN, R., and FRANCK, T. (1954). *Z. Phys.* **138,** 411.

SALOMON, M., and CONWAY, B. E. (1965). *Discussions Faraday Soc.* **39,** 223.

VDOVIN, Yn., LEVICH, V., and MYACULIN, V. (1949). *Dokl. Akad. Nauk SSSR* **124,** 350.

WEISS, J. (1959). *Proc. Roy. Soc.* **A222,** 128.

ZWOLINSKI, B., MARCUS, R. J., and EYRING, H. (1955). *Chem. Rev.* **55,** 157.

Author Index

Numbers in italics refer to the pages on which the complete references are listed.

A

Adrian, R. H., 917, 926, *977*
Agar, J. N., 790, *854*
Aggett, J., 759, *769*
Aikasjan, E. A., 826, *854*
Albert, A., 750, 767, *769*
Albert, L., 662, 666, 680, *729*
Albyshev, A. F., 876, *899*
Allmann, E. W., 938, 939, *979*
Almerini, A. L., 851, *857*
Amdur, I., *769*
Amelinckx, B., 680, *723*
Ammar, I. A., *569*
Anders, U., 897, *899*
Anderson, M. T., 750, *771*
Andreoli, T. C., 923, *982*
Antonov, A. Ya., 660, 661, *727*
Applegate, K., 760, 768, *771*
Aramata, A., 627, *725*
Arkharov, V., 670, 673, *723*
Armstrong, C. M., 959, *977*
Armstrong, R. D., 685, *723*
Arndt, F., 767, *769*
Asay, J., 763, *772*
Aten, A. H. W., 666, *723*
Atkinson, G., 763, *771*
Atwater, I., 960, *982*
Austin, L. G., 811, *854*
Avrami, M., 686, *723*
Awad, S. A., *569*
Ayabe, Y., 627, *728*
Azagami, S., 572, *609*

B

Bachmann, E., 938, 939, *979*
Bacon, A., 797, 806, *854*
Bagotzky (Bagotsky), V. S., 597, 599, 601, 602, 606, *607*, 812, 828, *854*, *856*
Baker, B. S., 833, *854*, *855*, 874, *899*
Baker, P. F., 950, 959, 960, *977*

Bakish, R., 717, *723*
Banerjee, B. C., 670, *723*
Banerjee, T., 706, *723*
Bangham, A. D., 928, *977*
Barker, G. C., 619, 627, *723*, 748, *769*
Barnes, S. C., 662, 666, 668, 669, *723*
Barrada, Y., 791, *856*
Barth, H., 579, *607*
Barton, J. H., 666, 691, *723*
Bartosik, D. C., 839, *855*
Basolo, F., 732, *769*
Bass, L., 739, *769*
Bates, R., 959, 961, *978*
Baticle, A. M., 627, *723*
Bauer, H. H., 627, *723*, *724*
Bauer, S. H., 860, *900*
Baum, H., 938, 939, *979*
Baum, R. L., 896, *900*
Bauman, H. F., 842, *855*
Baxendale, J. H., 741, *770*
Bazalski, B. T., 604, 606, *609*
Beacom, S. E., 655, *724*
Bear, R. S., 912, *977*
Bechtoldt, C., 696, *728*
Becker, R., 674, *724*
Behr, B., 627, *724*
Bell, R. P., 768, *769*
Bennion, D. N., 800, 802, 803, 804, 832, *855*
Benson, S. W., 732, *769*
Berl, W. G., 827, 830, *855*
Berndt, D., 597, 599, *609*
Berzins, T., 627, *724*
Bewick, A., 688, *724*
Bianti, G., 599, *607*
Bicelli, L. P., 669, 670, *729*
Biggs, A. I., 767, *769*
Biltz, W., 888, *900*
Binder, H., *607*
Bingulac, S., 629, 632, 633, 714, *725*
Binstock, L., 950, 959, *977*, *981*

1075

Blander, M., 868, 869, 870, 871, 875, 879, 884, 886, 887, *900*
Blankenship, F. F., 881, *901*
Bliznakov, G., 671, *727*
Blomgren, G. E., 862, 868, 869, 870, 885, 886, 887, *900*
Bloom, H., 862, *900*
Bock, R. M., 928, 929, *978*
Bockris, J. O'M., *569*, 584, 596, 600, 606, *607*, 612, 615, 616, 617, 619, 620, 621, 622, 624, 625, 626, 627, 628, 629, 630, 631, 632, 633, 634, 635, 637, 638, 639, 641, 642, 643, 645, 646, 647, 648, 650, 651, 652, 653, 654, 657, 658, 662, 666, 667, 668, 675, 677, 680, 681, 684, 685, 689, 690, 691, 692, 694, 695, 710, 711, 715, 716, 717, 718, *723, 724, 725, 727, 728, 729*, 793, *855*, 862, *900*, 1026, 1063, 1069, *1072, 1073*
Boerlage, L. M., 666, *723*
Bolis, L., 915, 918, *977*
Boltzmann, L., 963, *977*
Bondar, V. V., 699, 707, *729*
Bonhoeffer, K. F., 554, *607*
Born, M., 573, *607*, 1017, *1072*
Bos, C. J., 928, *978*
Bostanov, W., 678, *724*
Bowden, F. P., 790, *854*
Bozorth, R. M., 670, *724*
Bradley, C. L., 851, *857*
Branton, D., 935, 939, *977*
Braunstein, J., 868, 869, *900*
Breiter, M. W., 579, 580, 581, 598, 603, 604, 605, *607*, 826, *855*
Brenet, J., 881, *901*
Brenner, A., 698, *724*
Brettner, D. M., 842, *855*
Bretz, R. I., 877, 886, *901*
Briere, G., 748, *769*
Brinley, F. J., Jr., 912, 913, *977*
Brintzinger, H., 754, 755, 758, *769*
Brodsky, A., *1072*
Broers, H. J., 833, *855*
Brown, A. D., 929, *977*
Bruins, P. F., 851, *856*
Brummer, S. B., 603, 605, 606, *607*
Budevski, E., 666, 678, 679, 680, 690, *724*, 727
Bues, W., 873, *900*
Bukun, N. G., 892, *902*
Bundenburg de Jong, H. G., 959, *977*

Bunn, C. W., 674, 676, *724*
Burstein, R. (Burshtein, R. Kh.), 606, *609*, 639, *727*
Burton, W. K., 612, 674, 680, *724*, 793, *855*
Butler, J. A. V., 621, *724*
Butler, T. C., 767, *769*
Buttler, H. v., 579, *607*

C

Cabrera, K. N., 612, 674, 676, 680, *724*, 793, *855*
Cairns, E. J., 839, 845, *855, 856*
Caldin, E. F., *769*
Capraro, V., 915, 918, *977*
Carlson, C. M., 862, *900*
Castellan, G. W., 758, 763, *769, 772*
Cavasino, F. P., 761, *769*
Chance, B., 741, *771*
Chandler, W. K., 950, 951, 960, 961, *977, 981*
Chang, F. T., 591, 592, 594, *607*
Chazkatz, In., 1018, *1073*
Chein, C., 696, *730*
Chernenko, A., 1006, 1008, 1039, 1072, *1072*
Chernov, B., 676, *724*
Chizmadzev, Yu., 1071, *1072*
Christiansen, K. A., 639, *724*
Christiansen, G. A., 751, *770*
Christov, St. G., *569*
Ciarlarello, T. A., 844, *857*
Claesson, S., *769*
Clark, H. M., 877, *901*
Clark, M., 599, *607*, 828, *855*
Clippinger, E., 760, *772*
Cochrane, W., 670, *724*
Cohen, A., 1060, *1073*
Cohen, J. B., 696, *724*
Cole, D. L., 740, 748, 761, 763, 764, *769, 770, 771*
Cole, K. S., 905, 911, 913, 927, 930, 934, 936, 949, 950, 957, 962, 963, 964, *977, 978, 981, 982*
Cole, R. H., 934, *978*
Coleman, R., 929, 936, *979*
Collins, D. H., *855*
Connick, R. E., 753, 762, *770, 772*
Conway, B., 1026, 1068, *1072, 1074*
Conway, B. E., *569*, 584, 602, *607, 608*, 615, 616, 617, 631, 632, 689, *724*
Conway, E. J., 913, *978*

Cook, P., 923, *982*
Criddle, R. S., 928, 929, *978*
Crothers, D. M., 768, *769*
Curran, P. F., 907, *980*
Curtis, H. J., 930, 949, *978*
Czerlinski, G. H., 744, *769*

D

Damjanović, A., 641, 646, 647, 650, 651, 662, 666, 667, 668, 677, 680, 681, 684, 688, 690, 716, 717, *724, 725, 729*, 793, *855*
Danford, M. D., 861, *901*
Danielli, J. F., 934, 937, *978*
Dankov, P. D., 666, 696, *726*
Das, M. N., 748, *771*
Davies, G., 753, 759, *769*
Davies, J. T., 920, *978*
Davies, M. O., 599, *607*, 828, *855*
Davis, H. T., 863, 864, 871, 872, *900, 901*
Davison, P. F., 768, *769*
Davson, H., 934, 937, *978*
Davy, H., 774, *855*
Day, R. J., 599, *609*
De Béthune, A. J., *569*
Debye, P., 732, *769*, 965, *978*
de Groot, S. R., 691, *725*, 890, *900*
De Koranyi, A., 696, *728*
Delahay, P., 584, 599, *607, 609*, 627, 652, *724, 725, 730*
Delimarskii, Iu. K., 875, 880, 883, 887, 893, 894, 895, *900*
DeMaeyer, L., 740, 741, 745, 748, 751, 765, *769, 770*
De Robertis, E. D. P., 909, *978*
Despić, A. R., 619, 620, 624, 626, 627, 628, 629, 630, 632, 633, 634, 635, 637, 639, 641, 645, 646, 648, 650, 653, 654, 657, 658, 666, 667, 692, 694, 695, 714, 715, 716, 718, 720, 723, *724, 725, 729*
Devanathan, M. A. V., 597, *610*, 615, 710, 711, *724*
Dibble, E., 763, *772*
Dick, D. A. T., 911, 912, 934, *978*
Diebler, H., 732, *769*
Diggle, J., 653, 654, 657, 658, 666, 692, 694, 695, *725*
Dijkhuis, C. G. M., 875, 876, 877, 880, 881, 882, 883, 884, 885, *900*

Dijkhuis, R., 875, 876, 880, 881, 882, 883, 884, 885, *900*
Dirkse, T. P., 627, *725*
Dobrev, D., 670, *728*
Döring, W., 674, *724*
Dogonadze, R., 1006, 1008, 1018, 1039, 1060, 1071, 1072, *1072, 1073*
Dojlido, J., 627, *724*
Doto, H., 670, *726*
Dowden, D. A., 583, *607*
Dražić, D. M., 629, 630, 634, 635, 637, 638, 639, 678, 679, *724, 725*
Dribinsky, A. V., 660, *727*
Dugdale, I., 688, *725*
Duke, F. R., 879, *900*
Dulova, V. I., 748, *770*

E

Ebert, M., 741, *770*
Eckert, J., 605, *607*
Economou, N. A., 662, 666, 668, *725*
Edwards, J., 660, *730*
Efimov, E. A., 598, 599, *607*
Ehrenstein, G., 950, *981*
Eigen, M., 732, 739, 740, 741, 748, 751, 753, 760, 764, 765, 768, *769, 770*
Eischens, R. P., 603, *607*
Eisenberg, M., 777, 780, 791, 794, 795, 797, 835, 836, 839, 842, 843, 845, 851, *855, 857*
Eisenman, G., 959, 961, *978, 982*
Eley, D. D., 584, 586, *607*
Elving, P., 627, *723*
Emerson, M. T., 740, *770*
Emmelot, P., 928, *978*
Emmett, H., 674, 676, *724*
Engel, A., 706, *729*
Enke, C. G., 598, *609*
Enyo, M., 574, 578, 579, *607, 608*, 626, 627, 630, 633, 638, 641, 650, 667, *724, 727*
Epelboin, I., 638, *725*
Epstein, W. W., 750, 751, *770, 771*
Erdey-Grúz, T., 556, 557, 559, 561, 563, *608*, 612, 662, 688, *725*, 791, 826, *855*
Ergener, L., 767, *769*
Ershler, B. N., 627, *725*
Ertl, G., 748, *770*
Essig, A., 961, *978*
Euler, J., 800, 812, 813, *855*
Evans, D. F., 736, *771*

Evans, D. J., 670, *725*

Eyring, E. M., 740, 750, 751, 759, 761, 763, 764, *769, 770, 771, 772*, 907, 968, 974, *978*

Eyring, H., 544, 545, 546, 558, 574, *608, 609*, 732, *770*, 791, *855*, 862, *900*, 907, 968, 970, 974, *978*, 1071, *1074*

F

Fainberg, A. H., 760, *772*

Faircloth, R. L., 619, 627, *723*

Falk, S. U., 851, *855*

Farkas, A., 554, *608*

Farr, J. P., 841, *855*

Fatueva, T. A., 703, 705, *730*

Faust, F. W., 696, *725*

Feder, W., 907, *978*

Fedorowa, A. I., 826, *854*

Felder, E., 766, 767, *771*

Feynman, R., *1073*

Fiat, D., 762, *770*

Finch, G. I., 669, 670, *725*

Finean, J. B., 929, 936, *979*

Finholt, J. E., 759, *770*

Fischer, H., 662, 666, 667, 668, 669, 674, 680, 706, *725, 729*

Fleischer, A., 840, 841, *855*

Fleischer, B., 939, *979*

Fleischer, S., 939, *979*

Fleischmann, C. W., 604, *608*

Fleischmann, M., 649, 650, 667, 685, 686, 688, 710, *723, 725*, 793, *855*

Flood, H., 868, 869, 887, *900*

Florianovich, G. H., 639, *725*

Førland, T., 866, 868, 869, 877, 881, 887, *900*

Foerster, F., 698, *725*

Ford, J. I., 603, *607*

Fordyce, J. S., 896, *900*

Forty, A., 680, *725, 726*

Foster, M. C., 761, *771*

Foster, M. S., 874, *900*

Foulke, D. G., 659, *726*

Fox, J. J., 767, *770*

Franck, T., 1071, *1074*

Frank, F. C., 612, 668, 674, 676, 680, 681, 682, 683, *724, 725, 726*, 793, *855*

Frankenhaeuser, B., 921, 955, 956, 957, 973, *979*

Freedman, H. H., 751, *770*

French, T. C., 768, *770*

Frenkel, J., 688, *726*, 888, *900*

Fricke, H., 929, *979*

Friedman, H. L., 739, *770*

Friedman, S. M., 959, 961, *978*

Friess, S. L., *769*

Fröhlich, H., 1009, *1073*

Frost, A. A., 732, *770*

Frumkin, A. N., 578, 592, 593, 594, 602, 603, 605, *608, 609*, 639, 705, *726, 727*, 826, 828, *855*, 1002, 1030, 1064, *1073*

Fukuda, M., 577, *608*

Fukuda, S., 670, *726*

Fuoss, R. M., 753, *770*

G

Gainer, H., 939, *979*

Gammock, D. B., 929, *979*

Gardner, A. W., 619, 627, *723*

Gaspard, F., 748, *769*

Gavioli, G., 627, *726*

Gent, W. L. G., 929, *979*

Gerber, C. J., 924, *980*

Gerischer, H., 618, 627, 638, 642, 645, 716, 717, *726, 730*, 748, *770*, 1063, *1073*

Gibbs, W., 544, *608*

Gileadi, E., 606, *607*

Gilman, S., 598, 602, 603, 606, *608*, 834, *855*

Gilroy, D., 602, *608*

Giner, J., 605, *608*

Giron, I., 662, 696, *726*

Glasstone, S., 544, 545, 546, *608*, 732, *770*, 791, *855*

Glemser, O., 763, *770*

Glocker, R., 670, 673, *726*

Goldman, D. E., 920, 958, 961, 965, 975, *977, 979*

Good, A., 666, *727*

Gorbunova, K. M., 662, 666, 670, 671, 673, 696, 698, 699, 702, 705, 707, 708, 709, *726, 729, 730*

Gorgonova, E. P., 605, *609*

Gorodetzki, V. V., 629, *728*

Gorodskii, O. V., 895, *900*

Goswami, A., 670, *723*

Gottlieb, M. H., 604, *608*

Graf, L., 666, *726*

Graves, A. D., 893, 894, 895, 896, 898, 899, *900*

Grechukina, T. N., 670, *726*

Green, D. E., 928, 929, 939, *978*, *979*, *981*, *982*

Green, E. C., 938, 939, *979*

Green, M., 1039, *1073*

Green, W. A., 929, 936, *979*

Greenberg, J., 873, *900*

Greene, N. D., *729*

Greenhalgh, W. J., 739, *769*

Gregson, N. A., 929, *979*

Grens, E. A., 806, 807, 813, 815, 817, 819, 835, *855*, *856*

Grimes, W. R., 881, *901*

Grjotheim, K., 868, 869, 887, *900*

Grove, W. R., 827, *856*

Grubb, W. T., 833, *856*

Gruen, D. M., 873, *900*

Grunwald, E., 740, *770*

Guggenheim, E. A., 868, *900*

Gurevich, I. G., 812, *856*

Gurney, R. W., 559, *608*, 621, *726*, 1063, *1073*

H

Haber, F., 554, 557, *608*

Hallgren, L. J., 873, *900*

Halpern, J., 732, *770*

Hamby, D. C., 886, *900*

Hamilton, D. R., 692, 696, *726*

Hammes, G. G., 740, 753, 754, 755, 756, 759, 768, *769*, *770*

Hamori, E., 768, *770*

Hampson, N. A., 841, *855*

Hanahan, D. J., 928, *979*

Handler, P., 906, *982*

Hara, M., 670, *726*

Harris, G. M., 759, *770*

Haslam, J. L., 750, 751, *770*

Heath, C. E., 838, *856*

Hechter, O., 912, *979*

Henderson, D., 907, 968, 974, *978*

Herdlicka, C., 877, *901*

Hersh, L. S., 871, 886, *900*

Heusler, K. E., 639, *726*

Heynau, H. A., 761, *771*

Heyrovsky, J., 826, 827, *856*

Hildebrand, J. H., 867, *900*

Hill, T. L., 572, *608*, 961, 976, *978*, *979*

Hille, B., 958, *979*

Hills, G. J., 610, 712, *726*, 892, 893, 894, 895, 896, *900*

Hilson, P. J., 626, *726*

Hinke, J. A. M., 913, *979*

Hira, Lal, 605, *609*

Hirai, N., 862, *900*

Hirata, H., 670, *726*

Hirota, K., 544, 558, 559, 563, 566, 569, 574, 575, 586, *608*

Hiskey, C. F., *569*

Hoar, T. P., 634, 637, 639, 716, *726*

Hoare, J. P., 597, 599, 600, *608*

Hodgkin, A. L., 905, 911, 912, 913, 921, 922, 923, 924, 925, 927, 940, 942, 943, 946, 947, 949, 950, 951, 952, 953, 954, 955, 956, 957, 959, 960, 961, 967, 968, 969, 971, 972, 973, 975, 976, *977*, *978*, *979*, *980*

Hoeg, H., 639, *724*

Hoffmann, H., 741, 755, *771*

Hogen-Esch, T. E., 759, 760, *771*

Holmes, L. P., 740, 748, *771*

Holtje, W., 763, *770*

Holtzman, D., 917, *980*

Horiuti, J., 544, 545, 546, 548, 555, 557, 558, 559, 563, 564, 565, 566, 567, 569, 570, 571, 573, 574, 575, 576, 577, 578, 585, 586, *608*, *609*, 623, *726*, 1048, *1073*

Horowicz, P., 921, 922, 924, 925, *980*

Hoshi, M., 578, *607*

Hovorka, F., 599, *607*, 828, *855*

Howes, A., 662, *726*

Hoyt, R. C., 961, 975, *980*

Huang, K., 1017, *1072*

Hudson, R. G., 696, *730*

Hultin, H. D., 928, *982*

Hume-Rothery, W., 706, *726*

Hunt, J. P., 761, *771*

Huntington, J., 662, *726*

Huq, A. K. M. S., 596, 600, *607*

Hurlen, T., 627, 634, 637, 639, *726*, 727

Hurwitz, P. A., 732, 763, *771*

Hush, N. S., 623, 727, 1068, *1073*

Huxley, A. F., 923, 927, 949, 950, 951, 952, 953, 954, 955, 956, 957, 967, 968, 969, 971, 972, 973, 975, 976, *979*, *980*

I

Ibl, N., 655, *727*, 791, 794, *856*

Ikusima, M., 567, 571, 572, 578, *608*, *609*

Ilgenfritz, G., 740, 765, 766, 767, 768, *770*, *771*

Imai, H., 627, *727*
Inman, D., 893, 894, 895, 896, 898, 899, *900*, *901*
Inskeep, W. H., 751, *771*
Interrante, L. V., 828, *856*
Isgaryshev, N. A., 598, 599, *607*
Ishibashi, K., 627, *730*
Ishii, T., 626, 627, *727*
Ivanov, J., 828, *855*
Ivcher, T. S., 627, *728*
Ives, D. J. G., 610, 712, *726*

J

Jablokowa, I. E., 599, *607*
Jaffe, S. S., 827, *857*
Jaget, C. W., 750, 751, *770*
James, D. W., 873, *901*
Janz, G. J., 865, 875, 876, 880, 881, 882, 883, 884, 885, 887, 894, 895, *900*, *901*
Jensen, R. P., 740, 748, 750, 751, *769*, *770*, *771*
Jofa, J. A., 581, 582, *609*
Jofa, S., *569*
John, H. F., 696, *725*
Johnson, G. K., 604, 606, *608*, *609*
Johnson, G. R., 666, 668, *730*
Johnson, K. E., 893, *901*
Johnson, P. R., 605, *609*
Jones, D. L., 751, *771*
Jones, H., 706, *727*
Jovanović, D., 626, 627, 629, 632, 633, 714, *725*, *727*
Justi, E., 800, *856*
Juza, V. A., 619, 627, 631, *729*

K

Kabanov, B., 639, *727*
Kaishev, R., 666, 671, 678, 680, *724*, *727*, *730*
Kambara, T., 626, 627, *727*
Kammermaier, H., 581, *607*
Kardos, O., 658, 659, *726*, *727*
Kasper, C., 653, 705, *727*
Katchalsky, A., 907, *980*
Katz, B., 905, 912, 920, 923, 924, 949, 953, 957, 971, *980*
Katz, J. L., 863, 864, 870, *901*
Kaupp, E., 670, 673, *726*
Kavanau, J. L., 764, *771*

Kay, R. L., 736, *771*
Keck, J. C., 546, *609*
Kedem, O., 961, *978*
Keenan, A. G., 877, *901*
Keene, J. P., 741, *770*
Keii, T., 558, 559, 566, 572, 573, 574, 575, 586, *608*, *609*
Kelm, H., 759, *770*
Ketelaar, J. A. A., 877, *900*
Keynes, R. D., 912, 913, 918, 922, 924, 957, 959, 973, *980*
Kimball, G. E., 712, 714, *727*
Kirkaldy, J. S., 691, *727*
Kirkwood, J. G., 861, *902*
Kishimoto, U., 950, *978*
Kita, H., 564, 565, 574, 575, 582, 583, 584, *608*, *609*, *610*, 626, 629, 630, 633, 650, 651, 652, 653, 667, *724*, *727*
Kleitman, D., 758, *771*
Klemm, A., 878, 879, 887, 888, 889, 890, *901*
Kleppa, O. J., 866, 869, 870, 871, 885, 886, *900*, *901*
Klimasenko, N. L., 705, *730*
Knorr, C. A., 579, 580, 581, *607*, *610*, 826, *855*
Kober, F. P., 841, *856*
Kochergin, S. M., 670, *727*
Kodera, T., 558, 572, 573, 574, 599, *609*
Koechlin, B. A., 913, 914, *981*
Köhling, A., *607*
Kohlschütter, V., 662, 666, *727*
Kolotyrkin, Ya. M., *569*, 639, *725*
Kolthoff, I. M., 827, *856*
Komobuchi, Y., *569*, *608*
Kopaczyk, K., 938, 939, *979*, *981*
Kor, S. K., 752, *771*
Kordesch, K., 831, *856*
Korman, E. F., 938, 939, *979*
Korn, E. D., 929, 937, *981*
Koryta, J., 627, 629, *727*
Kossel, W., 612, 674, *727*
Kotzeva, A., 678, *724*
Kozlovskii, M. I., 680, *727*
Krasilishchikov (Krasilischichikov), A. I., 599, *609*, 827, 828, 829, 831, 832, *856*
Krause, D., 706, *729*
Krause, M., 627, *726*
Krishtalik, L., 1058, *1072*
Kromhout, R. A., 740, *770*
Kromrey, G. G., 556, *608*
Kronenberg, M. L., 874, *901*

Kruglikov, S. S., 660, 661, *727*
Krupp, H., *607*
Kruse, W., 740, 751, 765, 766, 767, 768, *770*
Kryukova, T. A., 705, *728*
Ksenzhek, O. S., 805, 811, 812, *856*
Kubshenko, I., 721, *729*
Kudryavtsev, N. T., 660, 661, *727*
Kuhn, A. T., 604, 605, *608, 609*
Kumar, D. M., 669, *727*
Kummer, J. T., 877, *902*
Kundu, K. K., 748, *771*
Kustin, K., 732, 740, 745, 753, 758, 759, *769, 770, 771*
Kutschker, A., 605, *609*
Kuznetsov, A., 1006, 1008, 1018, 1039, 1060, 1061, 1071, 1072, *1072, 1073*

L

Labanon, F., 937, *981*
Laidler, K. J., 544, 545, 546, *608*, 732, *770*, 791, *855*
Lainer, V. I., 705, 706, *727*
Laitinen, H. A., 598, *609*, 627, *727*, 827, *856*, 893, 894, 896, 898, 899, *901*
Laity, R. W., 876, 879, 890, 892, *900, 901*
Landau, L., 1063, *1073*
Landsberg, R., 841, *856*
Lantratov, M. F., 876, *899*
Law, J. T., 616, *727*
Layton, D. N., 670, *727*
Lecar, H., 950, 959, *977, 981*
Lechkova, N. V., 748, *770*
Lee, J. M., 845, *856*
Legare, R. J., 732, *770*
Lemmlein, G. G., 680, *727*
Lenard, J., 929, *981*
Leonardi, J., 881, *901*
Lettvin, J. Y., 959, *981*
Levich, B., 828, *855*
Levich, V. G., 794, *856*, 998, 1008, 1018, 1024, 1039, 1060, 1063, 1071, 1072, *1072, 1073, 1074*
Levy, H. A., 861, *901*
Lewis, E. S., *769*
Lewis, G. N., 882, *901*
Lewis, P. F., 912, *980*
Libby, W., 1071, *1073*
Lifshitz, E., 1063, *1073*
Liler, M., 685, *725*

Lindley, B. D., 961, 974, *981*
Lindstrom, H. V., 666, *730*
Ling, G. N., 912, 959, *981*
Lipton, S., 938, 939, *979*
Liu, C. H., 893, *901*
Llopis, J., 594, *609*
Loewe, L., 767, *769*
Logan, H. L., 717, *727*
Logan, S. R., 732, *771*
Lopushanskaya, A. I., 627, *728*
Lorenz, W., 627, 631, 638, 642, 645, *727, 728*, 792, 793, *856*
Losev, V. V., 629, *728*
Loshkarev, M. A., 705, *728*
Lossew, W. W., 627, 630, *728*
Lovreček, B., 629, *728*
Lozier, G. S., 848, *857*
Lucy, J. A., 937, 938, *981*
Luks, K. D., 863, 864, 871, 872, *901*
Lumry, R., 732, *770*, 968, 970, *978*
Lustman, B., 707, *728*
Lutz, K., 767, *771*

M

Maass, G., 740, 751, 760, 765, *770*
McBeth, R. L., 873, *900*
McConnel, D. G., 938, 939, *979*
McCulloch, W. S., 959, *981*
Macey, R. I., 957, *981*
MacLennan, D. H., 938, 939, *979*
Maisel, G. W., 918, *980*
Makrides, A. C., 578, *609*
Malinovski, J., 666, 680, *727*
Malyszko, J., 627, *724*
Marcus, R. A., 623, *728*, 1031, 1060, 1071, 1072, *1073*
Marcus, R. J., 1071, *1074*
Margerum, D. W., 759, *771*
Marinčić, N., 629, *728*
Markov, B. F., 875, 880, 883, 887, 893, 894, *900*
Marmont, G., 949, *981*
Maron, S. H., 765, *772*
Martell, A. E., 762, *771*
Martinola, F., 831, *856*
Martirosyan, A. P., 705, *728*
Matsuda, A., 567, 574, 578, 582, *608, 609*
Matsuda, H., 627, *728*

Matsunaga, M., 670, *728*
Matsushima, T., 579, *607*
Matthew, H. I., 666, *728*
Matthews, D. B., 621, 622, *724*, 1063, 1069, *1072*
Mattock, G., 959, 961, *977*
Mattson, E., 630, *728*
Mawston, I., 759, *769*
Mayer, J. E., 860, *901*
Mayer, M. G., 860, *901*
Mayer, S. W., 862, 863, 864, *901*
Mehl, W., 612, 632, 638, 641, 642, 643, 644, 645, 667, *728*
Mellors, G. W., 877, 881, 886, 896, *901*
Meves, H., 950, 951, 959, 960, 961, 977, *981*
Miceli, J., 761, 762, *771*
Michailova, W., 670, 673, *728*
Michelsen, K., 639, *724*
Miles, M. H., 750, 751, *770, 771*
Milner, P. C., 847, *856*
Minc, S., 583, *609*
Mituya, A., 567, *609*
Moelwyn-Hughes, E. A., 622, *728*
Molodov, A. I., 629, *728*
Monse, E. U., *901*
Moore, J. W., 949, 950, 957, 958, 961, 973, 974, *978, 981, 982*
Moore, L. E., 957, *979*
Moore, W. J., 750, *771*
Morachevskii, A. G., 876, *899*
Morel, 638, *725*
Morgenstern, W., 666, *726*
Morinaga, K., 627, *728*
Morrell, M. L., 753, 754, *770*
Morris, J. B., 696, *728*
Mott, N. F., 706, *728*
Mueller, K., 574, *608*, 615, *724*
Mueller, W. A., 627, *728*
Mullins, L. J., 913, 959, 974, 975, *977, 981*
Murgulescu, I. G., 881, *901*
Mussini, T., 599, *607*
Mutucumarana, T. de S., 666, *728*
Myamlin, V., 998, 1039, *1073, 1074*

N

Nakamura, T., 544, 545, 546, 548, 574, 575, *608*
Napolitano, L., 937, *981*
Narahashi, T., 958, 960, 961, 973, 974, *981*

Navrotsky, A., 886, *900*
Neil, D. E., 877, *901*
Nekrasov, L. N., 597, 599, *607*, 828, *855*
Newman, J. S., 813, *856*
Newton, R. F., 881, *901*
Niedrach, L. W., 606, *609*, 833, 834, *856*
Nielsen, G. B., 639, *724*
Nikolaeva-Fedorovich, N., 1030, 1072, *1072*
Nilsson, S. E. G., 935, *981*
Nomura, O., 565, *609*
Nonnenmacher, W., 800, 812, 813, *855*
Nord, H., 639, *724*
Notz, K., 877, *901*
Nowinski, W., 909, *978*
Noyes, R. M., 733, 738, 764, *771*
Nürnberg, H., 1057, *1073*

O

O'Brien, J. S., 928, 929, *982*
Odell, A. L., 759, *769*
Ogburn, F., 662, 696, *726, 728*
Ohmori, T., 582, *609*
Oikawa, T., 950, *982*
Okamoto, G., 544, 558, 563, 569, 574, 575, *608, 609*
Okaniwa, H., 594, *609*
Oliver, R. M., 957, *981*
Onsager, L., 735, 749, *771*
Oomen, J. J. C., 627, *729*
Oppenheimer, J. R., 573, *607*
Osteryoung, R. A., 893, 894, 896, 898, *901*
Ostlund, R. E., 750, *771*
Ostvold, T., 877, 883, *900, 901*
Ostwald, W., 778, *856*
Ovenston, T. S. J., 666, *728*
Owen, J. D., 763, 764, *769*
Owens, B., 879, *900*
Oxley, J. E., 604, 606, *609*

P

Palm, U. W., *569*
Pamfilov, A. V., 627, *728*
Pangarov, N. A., 670, 671, 673, *728*
Panikar, S. K., 707, *729*
Panish, M. B., 881, *901*
Paretzkin, B., 696, *728*
Parker, C. A., 666, *728*
Parker, R. C., 760, 768, *771*
Parsonce, R., 1031, 1072, *1073*

Parsons, R., 615, *728*
Pashley, D. W., 669, *728*
Passeron, E., 1031, 1072, *1073*
Past, V. E., *569*, 581, 582, *609*
Pasternack, R. F., 753, 759, *769*, *771*
Patton, H. D., 906, 910, 916, 927, 941, 945, 951, *982*
Pauling, L., 764, *771*
Paunović, M., 662, 666, 668, 677, 680, 681, 684, *724*, *725*
Pearson, R. G., 732, *769*, *770*
Pease, D. C., 935, *982*
Peiser, H. S., 696, *728*
Pekar, S., 1008, 1009, 1017, *1073*
Pentland, N., *569*
Perdue, J. F., 938, 939, *979*, *981*
Persiantseva, V. P., 705, *728*
Petrov, P. S., *569*
Petry, O. A., 603, 605, *609*
Petzii, O., 1030, 1072, *1073*
Philbert, G., 626, *728*
Phillips, J. N., 767, *769*
Pick, H. J., 662, 666, 668, 680, *723*, *729*
Pickard, W. F., 959, *981*
Piersma, B. J., 606, *607*
Pilkhun, M., 800, *856*
Pitts, W., 959, *981*
Plambeck, J. A., 897, *899*
Plane, R. A., 759, 761, *771*
Platzmann, R., 1071, *1074*
Pleskov, In., 998, 1039, *1073*
Pliskin, W., 603, *607*
Podlovchenko, B. I., 603, 605, *609*
Polanyi, M., 555, 557, 573, 585, *608*, 623, *726*, 1048, *1073*
Poli, G., 669, 670, *729*
Polukarov, Yu. M., 671, 698, 699, 702, 705, 707, 708, 709, *726*, *729*
Poluyan, E. S., 619, 627, 631, *729*
Popoff, P., 627, *726*
Popova, O. S., 671, *726*
Porter, G., 741, *771*
Porter, K. R., 915, 918, *977*
Porter, R. F., 860, *900*
Posner, A. M., 785, *856*
Post, B., *569*
Potter, E. C., *569*
Power, P. D., 892, 893, *900*
Price, P. B., 666, 696, *729*
Prigogine, I., 691, *729*
Pshenichnikov, A. G., 606, *609*

Q

Quarrell, A. G., 670, *725*

R

Raicheff, R. G., 716, 717, 718, 720, *725*, *729*
Rajčeva, S., *569*
Rama-Char, T. L., 701, 703, 707, *729*
Rampton, D. T., 740, 748, *769*, *771*
Randall, M., 882, *901*
Randles, J. E. B., 619, 626, 627, *729*
Rashkov, St., 670, 671, 673, *728*
Raub, E., 706, *729*
Raynor, G. V., 706, *726*, *729*
Razumney, G. A., 675, 677, 685, *724*
Reddy, A. K. N., 670, 673, 674, 710, 711, *724*, *729*
Reddy, T. B., 692, 696, *729*
Ree, T., 862, *900*
Reiss, H., 863, 864, 870, *901*
Reynolds, W. L., 760, *772*
Rice, S. A., 871, *900*
Rich, L. D., 761, 763, 764, *769*, *771*
Richardson, S. H., 928, *982*
Richter, K., *607*
Rickert, H., 716, 717, *726*
Rideal, E. K., 920, *978*
Rieske, J. S., 938, 939, *979*
Ritterman, P., 848, *857*
Rius, A., 627, *729*
Roberts, J. L., Jr., 603, *609*
Robertson, J. D., 915, 918, 937, 938, *977*, *982*
Robertson, W. D., 717, *723*
Robinson, A. E., 666, *728*
Robinson, G. C., 760, *772*
Robinson, R. A., 736, *771*, 913, *982*
Roe, D. K., 627, *727*, 898, 899, *901*
Roe, R. M., 598, *609*
Roger, G. L., 845, *856*
Rogers, G. T., 660, 661, *729*
Roiter, V. A., 619, 627, 631, *729*
Rojas, E., 960, *982*
Roper, J. H., 929, *979*
Rosen, H. M., 759, *771*
Rosental, K. J., 598, *609*, 627, *725*
Roughton, F. J. W., 741, *771*
Rubin, B., 715, 723, *724*
Ruch, T. C., 906, 910, 916, 927, 941, 945, 951, *982*

Rüetschi, P., 584, *609*
Rushton, W. A. H., 913, 946, 947, *980*
Russ, R., 554, 557, *608*
Ruth, J. M., 767, *769*
Ryser, H. J.-P., 915, *982*

S

Sacristan, A., 627, *729*
Saegusa, F., 895, *901*
Saez, F., 909, *978*
Saito, T., 572, 574, *609*
Salkind, A. J., 851, *856*
Salomon, M., 1068, *1074*
Salstrom, E. J., 867, 881, *900, 901*
Sammon, D. C., 748, *769*
Sandblom, J. P., 961, *982*
Sandstede, G., *607*
Sato, R., 662, 670, 674, 680, *729*
Sawyer, D. T., 599, 603, *609*, 828, 829, *856*
Scaletti, J., 937, *981*
Scheibe, W., 800, *856*
Scheraga, H. A., 768, *770*
Schiele, J., 746, *772*
Schmid, A., 786, *856*
Schmitt, F. O., 912, *977*
Schottky, W., 778, *856*
Schuldiner, S., 598, *609*
Schuldiner, W., 603, *610*
Schumilowa, N. A., 597, 599, *607*
Schwan, H. P., 934, 936, *982*
Schwarz, K., 878, *901*
Schwarzenbach, G., 766, 767, *771*
Scorcheletti, C., 721, *729*
Scott, A. B., 886, *900*
Seewald, D., 763, *771*
Seidensticker, R. G., 696, *726*
Seiter, H., 662, 666, 667, 680, *729*
Seitz, F., 622, *729*
Senderoff, S., 873, 877, 881, 886, 896, *901*
Seo, E. T., 829, *856*
Šepa, D., 678, 679, *725*
Serjeant, E. P., 750, *769*
Setty, T. H. V., 668, 670, 681, *725, 729*
Shams, El Din, 894, *901*
Shanes, A. M., 914, *982*
Shaw, T. I., 950, 958, 959, 960, 973, *977, 981*
Shear, D., 758, *771*
Shearer, R. E., 844, *857*
Shimamura, M., 960, *982*
Shimotake, H., 845, *856*

Shugar, D., 767, *770*
Silfvast, W. T., 751, *771*
Sillen, L. G., 762, *771*
Silverman, H. P., 845, *855*
Silzars, A., 740, *769*
Singer, I., 950, 959, *982*
Singer, S. J., 929, *981*
Sityagina, A. A., 671, *726*
Sjöstrand, F. S., 935, *982*
Slutsky, L. J., 760, 768, *771*
Sluyters, J. H., 627, *729*
Slygin, A. I., 602, *608*
Smid, J., 759, 760, *771*
Smith, B. E., 759, *769*
Smith, D. L., 627, *724*
Smith, E. L., 906, *982*
Smith, G. P., 873, *901*
Sobkowski, J., 583, *609*
Sobolyev, R. P., 660, *727*
Sokolova, L. A., 639, *725*
Somerton, K. W., 626, 627, *729*
Spyropoulos, C. S., 950, *982*
Sree, V., 707, *729*
Srinivasan, S., 584, *607*
Sroka, R., 662, *729*
Stabrovsky, A. N., 705, *729*
Steigerwald, R. F., *729*
Stein, W. D., 915, 918, 923, *982*
Steinfeld, J. I., 753, 756, 759, *770*
Stender, V. V., 811, 812, *856*
Stepanov, A., 721, *729*
Stern, K., 881, *901*
Stern, M., *729*
Sternberg, S., 877, 881, *901*
Stetten, D., Jr., 906, *982*
Stevenson, D. P., 584, 586, *609*
Stillinger, F. H., 861, 862, *901, 902*
Stockdale, G., 848, *857*
Stoeckenius, W., 937, 939, *979, 982*
Stojnov, Z., 678, *724*
Stokes, R. H., 736, *771*, 913, *982*
Storey, G. G., 662, 666, 668, *723, 729*
Stover, B. J., 907, 968, 974, *978*
Stranski, I. N., 612, 671, 674, *730*
Strehlow, H., 741, 762, *772*
Streinberg, M. A., 666, 680, *730*
Stuehr, J., 741, 755, 761, 762, 766, *771, 772*
Sun, C. H., 670, *725*
Sundheim, B. R., 887, 888, 889, 890, *902*
Sutin, N., 763, *771*
Sutyagina, A. A., 670, *726*

Swallow, A. J., 741, *770*
Swan, R. C., 922, 924, *980*
Swift, T. J., 753, *772*
Swinehart, J. H., 759, 763, *772*
Swinkels, D. M., 668, *724*

T

Tafel, J., 551, 554, 557, *609*, 825, 826, *856*
Takahashi, N., 670, *730*
Takenaka, T., 950, *982*
Tamamushi, R., 626, 627, *730*
Tamm, K., 753, 760, *770*
Tanaka, N., 626, 627, *730*
Tasaki, I., 950, 959, 960, *982*
Taube, H., 759, *771*
Taylor, K. J., 660, 661, *729*
Taylor, R. E., 946, 949, 950, *981*, *982*
Tedoradse, G., 592, 593, 594, *608*
Temkin, N., 866, *902*
Teorell, T., 950, *982*
Thirsk, H. R., 649, 650, 667, 668, 685, 686, 688, 710, *723*, *725*, *730*, 793, *855*
Thomas, S., 760, *772*
Thomas, U. B., 847, *856*
Tieffenberg, M., 923, *982*
Tischer, R., 627, 727, 898, 899, *901*
Tisdale, H. D., 928, 929, *978*
Titov, P. S., 705, *728*
Tobias, C. W., 791, 794, 800, 802, 803, 813, 815, 817, 819, 835, *855*, *856*, *857*
Tobias, J. M., 959, 961, *982*
Tolmatchev, V., *1072*
Toman, J. E. P., 906, *982*
Topley, B., 558, 574, *609*
Topol, L. E., 887, *900*
Tordesillas, I. M., 627, *729*
Torrichelli, A., 662, *727*
Toshima, S., 594, *609*
Tosteson, D. C., 923, *982*
Toya, T., 563, 586, *609*
Trachtenberg, I., 627, 652, *725*
Trivich, D., 662, 666, 668, *725*
Trumpler, G., 791, *856*
Tucker, G. F., 767, *769*
Turnbull, D., 765, *772*
Turner, D. R., 666, 668, *730*
Tyurin, V. S., 606, *609*
Tzagoloff, A., 928, 929, 938, 939, *979*

U

Übersax, F., 662, 666, *727*
Uhler, E. F., 848, *857*
Ukshe, E. A., 892, *902*
Ulbrich, W., 961, *981*
Usachev, D. N., 705, *730*

V

Vagramyan, A. T., 662, 703, 705, 706, *730*
Vaid, J., 701, 703, 707, *729*
VanArtsdalen, E. R., 872, 887, 888, 889, *900*, *901*
Van Cakenberghe, J., 626, *730*
Van der Merve, J. H., 668, *726*
Van Der Meulen, P. A., 666, *730*
Vasilyev, Yu. B., 601, 602, 606, *607*
Vaughan, T. B., 662, 666, 669, *729*, *730*
Vazquez, M., 594, *609*
Vdovin, Yn., 998, 1039, *1074*
Velinov, V., 670, 671, *728*
Verma, A. R., 680, *730*
Vermilyea, D. A., 638, 666, 667, 669, 676, 690, 696, *724*, *729*, *730*
Veselovsky, V. J., 598, *609*
Vetter, K. J., 589, 590, 594, 597, 599, *609*, 630, 632, *730*, 789, 793, 803, 851, 853, *857*
Vielstich, W., 579, 603, 605, *607*, *609*, 627, *730*, 826, *857*
Vitanov, T., 678, *724*
Vitkova, S. D., 670, 671, 673, *728*
Völkl, W., 579, 580, *607*
Vogel, U., 603, 605, *609*
Volkl, W., 826, *855*
Volmer, M., 556, 557, 559, 561, 563, *608*, 612, 662, 674, 688, *725*, *730*, 791, 826, *855*
Volta, A., 774, *857*
von Smoluchowski, M., 732, *772*
Vorobyeva, G. F., 660, 661, *727*

W

Wagner, C., 653, *730*, 878, *902*
Walton, H. F., 558, 559, *609*, *610*
Warner, T. B., 603, *610*
Washi, V., 670, *730*
Watanabe, A., 959, *982*
Watanabe, N., 597, *610*

Watson, S. A., 660, *730*
Watts, A. M., 739, *772*
Webb, M. B., 666, 696, *729*
Weertman, J., 696, *724*
Weiss, J., 1071, *1074*
Weiss, R. S., 827, *857*
Weissberger, A., *769*
Wendt, H., 764, *772*
Werner, R. C., 844, *857*
West, H., 841, *856*
West, J. M., 716, *726*
White, A., 906, *982*
White, S. H., 938, *982*
Whittaker, M. P., 763, *772*
Wick, H., 591, 592, 594, *607*
Wien, M., 746, *772*
Wigner, E. P., 546, *610*
Wilburn, N. T., 851, *857*
Wilcock, A., 662, *730*
Wilke, C. R., 791, 794, *857*
Wilkins, R. G., 753, 760, 763, *769, 770*
Will, F. G., 580, 581, *610*, 800, 804, *857*
Williams, A. G., 670, *725*
Williams, J. A., 927, *983*
Wilman, H., 666, 669, 670, *725, 727, 728, 729, 730*
Wilmshurst, J. K., 873, *902*
Winkelmann, D., 596, 599, *610*, 828, *857*
Winsel, A., 800, 811, *856, 857*
Winstein, S., 760, *772*
Wiswall, R. H., 877, *901*
Witt, H. T., 742, *772*

Wojtowicz, P. J., 861, *902*
Wolfenden, J. H., 558, 559, *609, 610*
Wolfgram, F., 929, *983*
Wood, W. A., 671, *730*
Woodbury, J. W., 905, 917, 924, 947, 953, 968, 970, 978, 983
Wranglén, G., 662, 666, 680, 696, *730*
Wroblowa, H., 606, *607*
Wullhorst, B., 706, *729*
Wynne-Jones, W. F. K., 688, *725*

Y

Yablokova, I. E., 828, *854*
Yaffe, I. S., 872, 888, 889, *902*
Yamamoto, T., 935, *983*
Yamazaki, T., 565, *609, 610*
Yang, L., 669, 670, 696, *725, 730*
Yao, Y. Y., 877, *902*
Yarlikov, M. M., 660, 661, 727
Yeager, E., 599, *607*, 741, 755, *771, 772*, 828, 855
Yokoyama, T., 587, 579, *607*
Yosim, S. T., 871, 887, *900*
Young, J. Z., 912, *977*
Yu, T.-J., 705, *727*

Z

Zhukova, A. I., 666, 696, *726*
Zwolinski, B., 1071, *1074*

Subject Index

A

Alloy deposition, 697–708
  kinetics of, 701–706
  phase formation, 706–708
  thermodynamics of codeposition in, 698–701
Amalgam electrodes, 784
Axoplasm, conductivity of, 913–914

B

Batteries, electrochemical energy conversion of, 846–853
Bioelectrochemistry, 903–983
  definition of, 904
  of electrolytes in living organisms, 908–914
  of excitability, 939–955
  of ion channel characteristics, 955–961
  of membranes, 914–927
  of properties, 961–976
  of structure studies, 927–939
  units used in, 907–908
Blood plasma, electrolytes in, 909
Bromine evolution reaction, 594–595

C

Cable equation, 946–947
Cable properties of cylindrical cells, 944–947
Carbon dioxide, evolution in oxidation of simple organics, 601–607
Cells
  electrolytes in, 909–914
  membrane role in function of, 914–917
Charge transport, kinetics of reactions with, 986–1072
Chlorine evolution reaction, 591–594
Cylindrical cells, cable properties of, 944–947
Cytoplasm, electrolytes in, 911–914

E

Electrical transport processes of fused salts, 887–893
Electrocardiogram, definition of, 906
Electrochemical energy conversion, 773–857
  of batteries, 846–853
  efficiency considerations, 795–797
  electrode kinetic aspects of, 787–797
    charge-transfer polarization, 790–792
    concentration polarization, 793–794
    crystallization polarization, 792–793
    EMF of closed circuit cells, 787–790
    gas side concentration polarization, 794–795
  in fuel cells and related dynamic systems, 824–842
  of porous electrodes, 797–824
  symbols used in study of, 853–854
  thermodynamics of galvanic cells, 775–787
Electrodes
  in galvanic systems, 781–787
  porous, theory of, 797–824
  reactions and electrode current in, 998–1004
  theoretical aspects, 1063–1070
Electroencephalogram, definition of, 906
Electrolytes, in living organisms, 908–914
Electromyogram, definition of, 906
EMF measurements of fused salts, 874–887
Excitability of nerves, bioelectrochemistry of, 939–955
  voltage clamping, 949–950
Eyring rate kinetics, membrane ion penetration and, 967–976

F

Fast ionic reactions, 731–772
Fuel cells
  energy conversion in, 824–842

Fuel cells—*cont.*
  energy conversion in—*cont.*
    fuel-cell systems, 839–840
    hydrocarbon anodic electrodes, 833–835
    hydrogen electrode, 825–826
    metal-gas hybrid cells, 840–842
    oxygen electrode, 827–833
  regenerative types, 842–846
Fused-salt electrochemistry, 859–902
  electrical transport processes in, 887–893
    electrical conductivity and transference numbers, 887–892
    electrical double layer, 892–893
  electrochemical kinetics, 893–899
    impedence and pulse techniques, 897–899
    steady-state and intermediate-time methods, 894–896
  EMF measurements, 874–887
    on cells with liquid junction, 878–880
    on formation cells, 880–883
    on reversible electrodes, 876–877
    of standard states, 875–876
  nature of fused salts, 859–873
    pure type, 860–865
    in solution, 866–873

G

Galvanic cells
  electrodes occurring in, 781–787
  thermodynamics of, 775–787
    free energy and enthalpy, 775–778
    pressure effect, 780–781
    reversible electrode potential, 779–780
    temperature effect, 780
Gas diffusion electrodes, 786–787
  pore structure, 806–811
  theoretical aspects, 799–811
Gas evolution reactions, 543–610
  of halogens, 587–595
  of hydrogen, 550–587
  in oxidation of simple organics with CO_2 evolution, 601–607
  of oxygen, 595–601
  statistical mechanical methods for, 544–549
Generalized theory of absolute reaction rate, 544

Giant axons, voltage clamping studies, 948–950
Goldman equation, 919

H

Halogen evolution reaction, 587–595
  characteristics of, 588–591
  of chlorine, 591–594
  of iodine and bromine, 594–595
Hodgkin-Huxley equations, 953–955
Hydrogen electrode, mechanism of, 825–826
Hydrogen evolution reaction, 550–587
  dependence of i_0 on electrode materials, 582–586
  double layer and hydrogen intermediates, 559–560
  dual mechanism, 558–559
  historical survey, 551–559
  separation factor, 571–577
  stoichiometric number of rate-determining step, 577–579
  τ_+-values of, 552–553
  transient phenomena of, 579–582

I

Interstitial fluid, electrolytes in, 909, 911
Intracellular fluid, electrolytes in, 911–912
Iodine evolution reaction, 594–595
Ion channel characteristics, bioelectro-chemistry of, 955–961
Ionic reactions (fast), 731–772
  diffusion controlled type, 732–749
    measurement, 745–749
    relaxation experiments, 741–749
    specific rates, 742–745
  slower types, 749–768
    complex ion formation, 752–761
    hydrogen bond rupture, 750–752
    metal ion hydrolysis and polymeriza-tion, 761–764
    pseudo or carbon acid dissociations, 765–768
Iron, deposition mechanism of, 634–640

K

Kinetics of reactions with charge transport, 985–1072
  adiabatic approximation and quantum transition probability, 1010–1014
  charge transfer reaction with chemical bond breaking, 1014-1046
  current density, 1049–1060
  of electrode reactions, 998–1004, 1060–1062
    theoretical aspects, 1063–1070
  flow of current through metal-solution interface, 1024–1031
  flow of current through semiconductor-solution interface, 1031–1039
  model of a polar solution, 1004–1010
  physical pattern of reaction with electron transfer in polar solvents, 989–998
  accompanied by breaking of chemical bonds, 1039–1041
  probability of electron transfer, 1015–1019
  transition probability, 1046–1049
    in reactions without bond breaking, 1019–1023

L

Liquid diffusion electrodes, in fuel cells, 835–839
Liquid fuel electrodes, 787

M

Membrane(s)
  architectural models of, 936–939
  ion transport through neutral, homogeneous types, 971–923
  ionic current of, 950–952
  properties of, theoretical aspects, 961–976
  steady-state ion transport in, 914–927
  structure studies on, 927–939
    chemical composition, 928–929
    electron microscopy, 934–935
    physical and morphological characteristics, 929–936
    thickness, 935–936
Metal deposition and dissolution, 611–730
  of alloys, 697–708
    kinetics of dissolution, 720–722
  basic steps in, 613–660
    discharge process, 613–638

Metal deposition and dissolution—*cont.*
  basic steps in—*cont.*
    surface diffusion and incorporation, 638–652
    transport in solution, 653–660
  crystallographic aspects of, 660–697
    in growth of a single crystal, 674–697
    morphology and texture, 660–674
  kinetics of, 708–723
    acceleration caused by strain, 716–720
    dissolution and deposition, 708–712
    dissolution into solutions free of corresponding ions, 712–716
Metal-gas hybrid cells, 840–842

N

Nerve impulse, initiation and propagation of, 940–944
Nerves, bioelectrochemistry of excitability of, 939–955

O

Oxygen electrode, mechanism of, 827–833
Oxygen evolution reaction, 595–601
  kinetics of, 598–600
  oxygen deposit, 597–598
  at reversible oxygen electrode, 595–597

P

Porous electrodes
  flooded battery type, 811
  gas diffusion type, 799–811
  theory of, 797–824

R

Redox electrodes, 784–785

T

Tafel's law, 551–554
Tissues, electrolytes in, 908–909
Transference numbers of fused salts, 887–892
Transmembrane potential, determination of, 923–927
Transport equation, 920

V

Voltage clamping, 948–950